Student Solutions Manual

for

College Mathematics for the Managerial, Life, and Social Sciences

Sixth Edition

S. T. Tan
Stonehill College

Australia • Canada • Mexico • Singapore • Spain • United Kingdom • United States

Printed in the United States of America
1 2 3 4 5 6 7 08 07 06 05 04

Printer: Malloy Incorporated

ISBN: 0-534-49240-1

For more information about our products,
contact us at:
**Thomson Learning Academic Resource Center
1-800-423-0563**

For permission to use material from this text or
product, submit a request online at
http://www.thomsonrights.com.
Any additional questions about permissions can be
submitted by email to **thomsonrights@thomson.com.**

**Thomson Brooks/Cole
10 Davis Drive
Belmont, CA 94002-3098
USA**

Asia
Thomson Learning
5 Shenton Way #01-01
UIC Building
Singapore 068808

Australia/New Zealand
Thomson Learning
102 Dodds Street
Southbank, Victoria 3006
Australia

Canada
Nelson
1120 Birchmount Road
Toronto, Ontario M1K 5G4
Canada

Europe/Middle East/South Africa
Thomson Learning
High Holborn House
50/51 Bedford Row
London WC1R 4LR
United Kingdom

Latin America
Thomson Learning
Seneca, 53
Colonia Polanco
11560 Mexico D.F.
Mexico

Spain/Portugal
Paraninfo
Calle/Magallanes, 25
28015 Madrid, Spain

COLLEGE MATHEMATICS

CONTENTS

CHAPTER 11 DIFFERENTIATION

CHAPTER 12 APPLICATIONS OF THE DERIVATIVE

CHAPTER 13 EXPONENTIAL AND LOGARITHMIC FUNCTIONS

CHAPTER 14 INTEGRATION

CHAPTER 15 ADDITIONAL TOPICS IN INTEGRATION

CHAPTER 16 CALCULUS OF SEVERAL VARIABLES

CHAPTER 1

EXERCISES 1.1, page 8

1. The coordinates of A are (3,3) and it is located in Quadrant I.

3. The coordinates of C are (2,-2) and it is located in Quadrant IV.

5. The coordinates of E are (-4,-6) and it is located in Quadrant III.

7. A 9. E, F, and G 11. F

For Exercises 13-19, refer to the following figure.

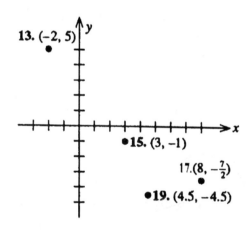

21. Using the distance formula, we find that $\sqrt{(4-1)^2 + (7-3)^2} = \sqrt{3^2 + 4^2} = \sqrt{25} = 5$.

23. Using the distance formula, we find that
$$\sqrt{(4-(-1))^2 + (9-3)^2} = \sqrt{5^2 + 6^2} = \sqrt{25+36} = \sqrt{61}.$$

25. The coordinates of the points have the form $(x,-6)$. Since the points are 10 units away from the origin, we have
$$(x-0)^2 + (-6-0)^2 = 10^2$$

$$x^2 = 64,$$
or $x = \pm 8$. Therefore, the required points are $(-8,-6)$ and $(8,-6)$.

27. The points are shown in the diagram that follows.

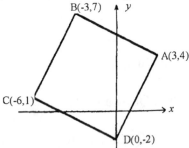

To show that the four sides are equal, we compute the following:
$$d(A,B) = \sqrt{(-3-3)^2 + (7-4)^2} = \sqrt{(-6)^2 + 3^2} = \sqrt{45}$$
$$d(B,C) = \sqrt{[(-6-(-3)]^2 + (1-7)^2} = \sqrt{(-3)^2 + (-6)^2} = \sqrt{45}$$
$$d(C,D) = \sqrt{[0-(-6)]^2 + [(-2)-1]^2} = \sqrt{(6)^2 + (-3)^2} = \sqrt{45}$$

$$d(A,D) = \sqrt{(0-3)^2 + (-2-4)^2} = \sqrt{(3)^2 + (-6)^2} = \sqrt{45}.$$

Next, to show that $\triangle ABC$ is a right triangle, we show that it satisfies the Pythagorean Theorem. Thus,
$$d(A,C) = \sqrt{(-6-3)^2 + (1-4)^2} = \sqrt{(-9)^2 + (-3)^2} = \sqrt{90} = 3\sqrt{10}$$
and $[d(A,B)]^2 + [d(B,C)]^2 = 90 = [d(A,C)]^2$. Similarly, $d(B,D) = \sqrt{90} = 3\sqrt{10}$, so $\triangle BAD$ is a right triangle as well. It follows that $\angle B$ and $\angle D$ are right angles, and we conclude that $ADCB$ is a square

29. The equation of the circle with radius 5 and center $(2,-3)$ is given by
$$(x-2)^2 + [y-(-3)]^2 = 5^2$$
or $$(x-2)^2 + (y+3)^2 = 25.$$

31. The equation of the circle with radius 5 and center $(0, 0)$ is given by
$$(x-0)^2 + (y-0)^2 = 5^2$$
or $$x^2 + y^2 = 25$$

33. The distance between the points (5,2) and (2,-3) is given by
$$d = \sqrt{(5-2)^2 + (2-(-3))^2} = \sqrt{3^2 + 5^2} = \sqrt{34}.$$
Therefore $r = \sqrt{34}$ and the equation of the circle passing through (5,2) and centered at (2,-3) is
$$(x-2)^2 + [y-(-3)]^2 = 34$$
or $\qquad (x-2)^2 + (y+3)^2 = 34.$

35. Referring to the diagram on page 9 of the text, we see that the distance from A to B is given by $d(A,B) = \sqrt{400^2 + 300^2} = \sqrt{250,000} = 500$. The distance from B to C is given by
$$d(B,C) = \sqrt{(-800-400)^2 + (800-300)^2} = \sqrt{(-1200)^2 + (500)^2}$$
$$= \sqrt{1,690,000} = 1300.$$
The distance from C to D is given by
$$d(C,D) = \sqrt{[-800-(-800)]^2 + (800-0)^2} = \sqrt{0+800^2} = 800 .$$
The distance from D to A is given by
$$d(D,A) = \sqrt{[0-(-800)]^2 + (0-0)} = \sqrt{640000} = 800.$$

Therefore, the total distance covered on the tour, is
$$d(A,B) + d(B,C) + d(C,D) + d(D,A) = 500 + 1300 + 800 + 800$$
$$= 3400, \quad \text{or } 3400 \text{ miles.}$$

37. Referring to the following diagram,

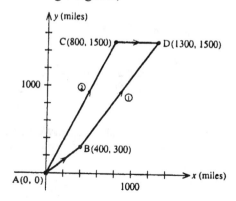

we see that the distance he would cover if he took Route (1) is given by

$$d(A,B)+d(B,D)=\sqrt{400^2+300^2}+\sqrt{(1300-400)^2+(1500-300)^2}$$
$$=\sqrt{250,000}+\sqrt{2,250,000}=500+1500=2000,$$

or 2000 miles. On the other hand, the distance he would cover if he took Route (2) is given by

$$d(A,C)+d(C,D)=\sqrt{800^2+1500^2}+\sqrt{(1300-800)^2}$$
$$=\sqrt{2,890,000}+\sqrt{250,000}=1700+500=2200,$$

or 2200 miles. Comparing these results, we see that he should take Route (1).

39. Calculations to determine VHF requirements:
$$d=\sqrt{25^2+35^2}=\sqrt{625+1225}=\sqrt{1850}\approx43.01.$$
Models B through D satisfy this requirement.
Calculations to determine UHF requirements:
$$d=\sqrt{20^2+32^2}=\sqrt{400+1024}=\sqrt{1424}\approx37.74$$
Models C through D satisfy this requirement. Therefore, Model C will allow him to receive both channels at the least cost.

41. a. Let the position of ship A and ship B after t hours be $A(0, y)$ and $B(x, 0)$, respectively. Then $x=30t$ and $y=20t$. Therefore, the distance between the two ships is

$$D=\sqrt{(30t)^2+(20t)^2}=\sqrt{900t^2+400t^2}=10\sqrt{13}t.$$

b. The required distance is obtained by letting $t=2$ giving $D=10\sqrt{13}(2)$ or approximately 72.11 miles.

43. False. The distance between $P_1(a,b)$ and $P_3(kc,kd)$ is

$$d=\sqrt{(kc-a)^2+(kd-b)^2}$$
$$\neq|k|D=|k|\sqrt{(c-a)^2+(d-b)^2}=\sqrt{k^2(c-a)^2+k^2(d-b)^2}=\sqrt{[k(c-a)]^2+[k(d-b)]^2}$$

45. Referring to the figure in the text, we see that the distance between the two points is given by the length of the hypotenuse of the right triangle. That is,
$$d=\sqrt{(x_2-x_1)^2+(y_2-y_1)^2}$$

EXERCISES 1.2, page 21

1. Referring to the figure shown in the text, we see that $m = \dfrac{2-0}{0-(-4)} = \dfrac{1}{2}$.

3. This is a vertical line, and hence its slope is undefined.

5. $m = \dfrac{y_2 - y_1}{x_2 - x_1} = \dfrac{8-3}{5-4} = 5$.

7. $m = \dfrac{y_2 - y_1}{x_2 - x_1} = \dfrac{8-3}{4-(-2)} = \dfrac{5}{6}$.

9. $m = \dfrac{y_2 - y_1}{x_2 - x_1} = \dfrac{d-b}{c-a}$.

11. Since the equation is in the slope-intercept form, we read off the slope $m = 4$.
    a. If x increases by 1 unit, then y increases by 4 units.
    b. If x decreases by 2 units, y decreases by $4(-2) = -8$ units.

13. The slope of the line through A and B is $\dfrac{-10-(-2)}{-3-1} = \dfrac{-8}{-4} = 2$.

    The slope of the line through C and D is $\dfrac{1-5}{-1-1} = \dfrac{-4}{-2} = 2$.

    Since the slopes of these two lines are equal, the lines are parallel.

15. The slope of the line through A and B is $\dfrac{2-5}{4-(-2)} = -\dfrac{3}{6} = -\dfrac{1}{2}$.

    The slope of the line through C and D is $\dfrac{6-(-2)}{3-(-1)} = \dfrac{8}{4} = 2$.

    Since the slopes of these two lines are the negative reciprocals of each other, the lines are perpendicular.

17. The slope of the line through the point $(1, a)$ and $(4, -2)$ is $m_1 = \dfrac{-2-a}{4-1}$ and the

    slope of the line through $(2, 8)$ and $(-7, a+4)$ is $m_2 = \dfrac{a+4-8}{-7-2}$. Since these two lines are parallel, m_1 is equal to m_2. Therefore,

$$\frac{-2-a}{3} = \frac{a-4}{-9}$$
$$-9(-2-a) = 3(a-4)$$
$$18+9a = 3a-12$$
$$6a = -30 \qquad \text{and} \quad a = -5$$

19. An equation of a horizontal line is of the form $y = b$. In this case $b = -3$, so $y = -3$ is an equation of the line.

21. e 23. a 25. f

27. We use the point-slope form of an equation of a line with the point $(3,-4)$ and slope $m = 2$. Thus
$$y - y_1 = m(x - x_1),$$
and
$$y - (-4) = 2(x - 3)$$
$$y + 4 = 2x - 6$$
$$y = 2x - 10.$$

29. Since the slope $m = 0$, we know that the line is a horizontal line of the form $y = b$. Since the line passes through $(-3,2)$, we see that $b = 2$, and an equation of the line is $y = 2$.

31. We first compute the slope of the line joining the points $(2,4)$ and $(3,7)$. Thus,
$$m = \frac{7-4}{3-2} = 3.$$
Using the point-slope form of an equation of a line with the point $(2,4)$ and slope $m = 3$, we find
$$y - 4 = 3(x - 2)$$
$$y = 3x - 2.$$

33. We first compute the slope of the line joining the points $(1,2)$ and $(-3,-2)$. Thus,
$$m = \frac{-2-2}{-3-1} = \frac{-4}{-4} = 1.$$
Using the point-slope form of an equation of a line with the point $(1,2)$ and slope $m = 1$, we find
$$y - 2 = x - 1$$
$$y = x + 1.$$

35. We use the slope-intercept form of an equation of a line: $y = mx + b$. Since $m = 3$, and $b = 4$, the equation is $y = 3x + 4$.

37. We use the slope-intercept form of an equation of a line: $y = mx + b$. Since $m = 0$, and $b = 5$, the equation is $y = 5$.

39. We first write the given equation in the slope-intercept form:
$$x - 2y = 0$$
$$-2y = -x$$
$$y = \tfrac{1}{2}x \ .$$
From this equation, we see that $m = 1/2$ and $b = 0$.

41. We write the equation in slope-intercept form:
$$2x - 3y - 9 = 0$$
$$-3y = -2x + 9$$
$$y = \tfrac{2}{3}x - 3.$$
From this equation, we see that $m = 2/3$ and $b = -3$.

43. We write the equation in slope-intercept form:
$$2x + 4y = 14$$
$$4y = -2x + 14$$
$$y = -\tfrac{2}{4}x + \tfrac{14}{4}$$
$$= -\tfrac{1}{2}x + \tfrac{7}{2}.$$
From this equation, we see that $m = -1/2$ and $b = 7/2$.

45. We first write the equation $2x - 4y - 8 = 0$ in slope-intercept form:
$$2x - 4y - 8 = 0$$
$$4y = 2x - 8$$
$$y = \tfrac{1}{2}x - 2$$
Now the required line is parallel to this line, and hence has the same slope. Using the point-slope equation of a line with $m = 1/2$ and the point $(-2, 2)$, we have
$$y - 2 = \tfrac{1}{2}[x - (-2)] = \tfrac{1}{2}x + 1$$
$$y = \tfrac{1}{2}x + 3.$$

47. A line parallel to the x-axis has slope 0 and is of the form $y = b$. Since the line is 6 units below the axis, it passes through $(0,-6)$ and its equation is $y = -6$.

49. We use the point-slope form of an equation of a line to obtain
$$y - b = 0(x - a) \quad \text{or} \quad y = b.$$

51. Since the required line is parallel to the line joining $(-3,2)$ and $(6,8)$, it has slope
$$m = \frac{8-2}{6-(-3)} = \frac{6}{9} = \frac{2}{3}.$$
We also know that the required line passes through $(-5,-4)$. Using the point-slope form of an equation of a line, we find
$$y - (-4) = \tfrac{2}{3}(x - (-5)); \quad y = \tfrac{2}{3}x + \tfrac{10}{3} - 4, \text{ or } y = \tfrac{2}{3}x - \tfrac{2}{3} \quad .$$

53. Since the point $(-3,5)$ lies on the line $kx + 3y + 9 = 0$, it satisfies the equation. Substituting $x = -3$ and $y = 5$ into the equation gives
$$-3k + 15 + 9 = 0 \quad \text{or} \quad k = 8.$$

55. $3x - 2y + 6 = 0$

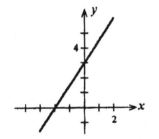

57. $x + 2y - 4 = 0$

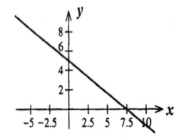

59. $y + 5 = 0$

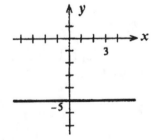

61. Since the line passes through the points $(a, 0)$ and $(0, b)$, its slope is $m = \dfrac{b-0}{0-a} = -\dfrac{b}{a}$.

Then, using the point-slope form of an equation of a line with the point $(a, 0)$ we have

$$y - 0 = -\tfrac{b}{a}(x - a)$$
$$y = -\tfrac{b}{a}x + b$$

which may be written in the form $\tfrac{b}{a}x + y = b$.

Multiplying this last equation by $1/b$, we have $\dfrac{x}{a} + \dfrac{y}{b} = 1$.

63. Using the equation $\dfrac{x}{a} + \dfrac{y}{b} = 1$ with $a = -2$ and $b = -4$, we have $-\dfrac{x}{2} - \dfrac{y}{4} = 1$.

Then

$$-4x - 2y = 8$$
$$2y = -8 - 4x$$
$$y = -2x - 4.$$

65. Using the equation $\dfrac{x}{a} + \dfrac{y}{b} = 1$ with $a = 4$ and $b = -1/2$, we have

$$\dfrac{x}{4} + \dfrac{y}{-\frac{1}{2}} = 1$$
$$-\tfrac{1}{4}x + 2y = -1$$
$$2y = \tfrac{1}{4}x - 1, \quad y = \tfrac{1}{8}x - \tfrac{1}{2}.$$

67. The slope of the line passing through A and B is $m = \dfrac{7-1}{1-(-2)} = \dfrac{6}{3} = 2$,

and the slope of the line passing through B and C is $m = \dfrac{13-7}{4-1} = \dfrac{6}{3} = 2$.

Since the slopes are equal, the points lie on the same line.

69. a.

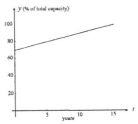

b. The slope is 1.9467 and the y-intercept is 70.082.

c. The output is increasing at the rate of 1.9467%/yr; the output at the beginning of 1990 was 70.082%.

d. We solve the equation $1.9467t + 70.082 = 100$ giving $t = 15.37$. We conclude that the plants will be generating at maximum capacity shortly after 2005.

71. a. $y = 0.55x$

b. Solving the equation $1100 = 0.55x$ for x, we have $x = \dfrac{1100}{0.55} = 2000$.

73. Using the points $(0, 0.68)$ and $(10, 0.80)$, we see that the slope of the required line is
$$m = \frac{0.80 - 0.68}{10 - 0} = \frac{0.12}{10} = .012.$$
Next, using the point-slope form of the equation of a line, we have
$$y - 0.68 = 0.012(t - 0)$$
or
$$y = 0.012t + 0.68.$$
Therefore, when $t = 14$, we have
$$y = 0.012(14) + 0.68$$
$$= .848$$
or 84.8%. That is, in 2004 women's wages are expected to be 84.8% of men's wages.

75. a. – b.

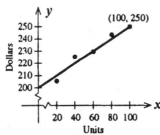

c. Using the points $(0,200)$ and $(100,250)$, we see that the slope of the required line is $m = \dfrac{250-200}{100} = \dfrac{1}{2}$. Therefore, the required equation is

$$y - 200 = \tfrac{1}{2}x \quad \text{or} \quad y = \tfrac{1}{2}x + 200.$$

d. The approximate cost for producing 54 units of the commodity is
$\tfrac{1}{2}(54) + 200,$ or \$227.

77. a. – b.

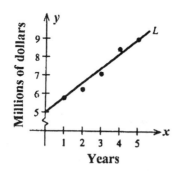

c. The slope of L is $m = \dfrac{9.0-5.8}{5-1} = \dfrac{3.2}{4} = 0.8$ .Using the point-slope form of an

equation of a line, we have $y - 5.8 = 0.8(x-1) = 0.8x - 0.8$, or $\quad y = 0.8x + 5$.

d. Using the equation of part (c) with $x = 9$, we have
$$y = 0.8(9) + 5 = 12.2, \quad \text{or } \$12.2 \text{ million.}$$

79. a. We obtain a family of parallel lines each having slope m.
b. We obtain a family of straight lines all of which pass through the point $(0,b)$.

81. False. Substituting $x = -1$ and $y = 1$ into the equation gives $3(-1) + 7(1) = 4$, and this is not equal to the right-hand side of the equation. Therefore, the equation is not satisfied and so the given point does not lie on the line.

83. True. The slope of the line $Ax + By + C = 0$ is $-A/B$. (Write it in the slope-intercept form.) Similarly, the slope of the line $ax + by + c = 0$ is $-a/b$. They are parallel if and only if $-\dfrac{A}{B} = -\dfrac{a}{b},$ $Ab = aB$, or $Ab - aB = 0$.

85. True. The slope of the line $ax + by + c_1 = 0$ is $m_1 = -a/b$. The slope of the line $bx - ay + c_2 = 0$ is $m_2 = b/a$. Since $m_1 m_2 = -1$, the straight lines are indeed

perpendicular.

87. Writing each equation in the slope-intercept form, we have

$$y = -\frac{a_1}{b_1}x - \frac{c_1}{b_1} \quad (b_1 \neq 0) \quad \text{and} \quad y = -\frac{a_2}{b_2}x - \frac{c_2}{b_2} \quad (b_2 \neq 0)$$

Since two lines are parallel if and only if their slopes are equal, we see that the lines

are parallel if and only if $-\frac{a_1}{b_1} = -\frac{a_2}{b_2}$, or $a_1b_2 - b_1a_2 = 0$.

USING TECHNOLOGY EXERCISES 1.2, page 27

1.

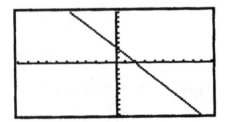

3.

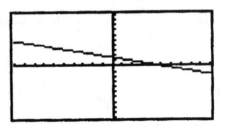

5.

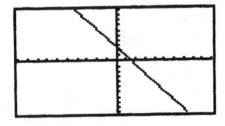

7. a.

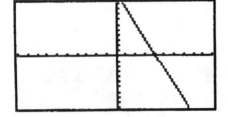

b.

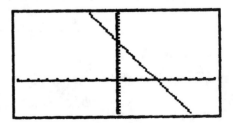

9. a.

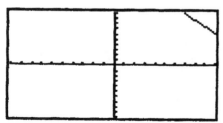

b.

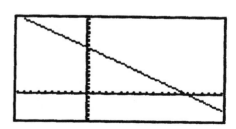

11.

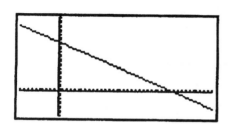

13.

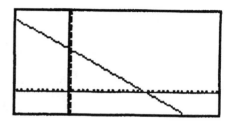

15.

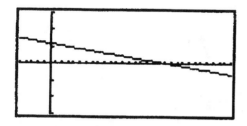

17.

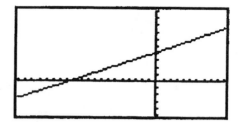

EXCEL

1.

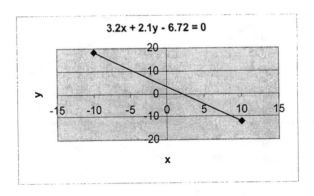

3.

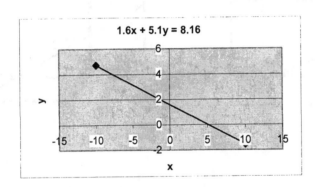

5.

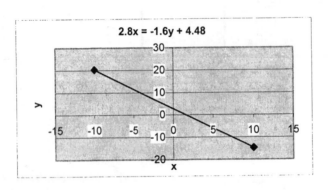

7.

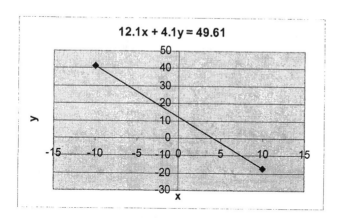

$12.1x + 4.1y = 49.61$

9.

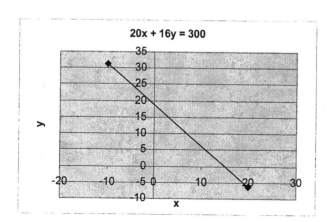

$20x + 16y = 300$

11.

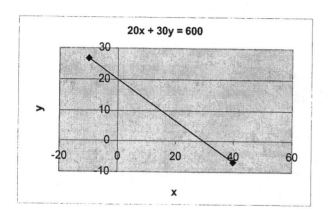

$20x + 30y = 600$

13.

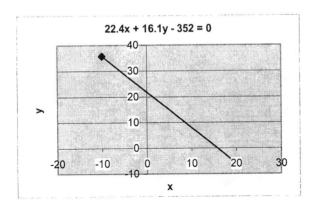

15.

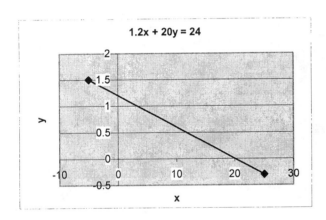

17.

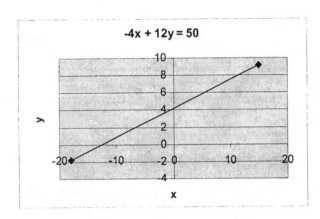

EXERCISES 1.3, page 37

1. Yes. Solving for y in terms of x, we find $3y = -2x + 6$, or $y = -\frac{2}{3}x + 2$.

3. Yes. Solving for y in terms of x, we find $2y = x + 4$, or $y = \frac{1}{2}x + 2$.

5. Yes. Solving for y in terms of x, we have $4y = 2x + 9$, or $y = \frac{1}{2}x + \frac{9}{4}$.

7. y is not a linear function of x because of the quadratic term $2x^2$.

9. y is not a linear function of x because of the nonlinear term $-3y^2$.

11. a. $C(x) = 8x + 40,000$, where x is the number of units produced.
    b. $R(x) = 12x$, where x is the number of units sold.
    c. $P(x) = R(x) - C(x) = 12x - (8x + 40,000) = 4x - 40,000$.
    d. $P(8,000) = 4(8,000) - 40,000 = -8,000$, or a loss of $8,000.
       $P(12,000) = 4(12,000) - 40,000 = 8,000$ or a profit of $8,000.

13. $f(0) = 2$ gives $m(0) + b = 2$, or $b = 2$. So, $f(x) = mx + 2$. Next, $f(3) = -1$ gives $m(3) + 2 = -1$, or $m = -1$.

15. Let V be the book value of the office building after 2003. Since $V = 1,000,000$ when $t = 0$, the line passes through $(0, 1,000,000)$. Similarly, when $t = 50$, $V = 0$, so the line passes through $(50, 0)$. Then the slope of the line is given by
$$m = \frac{0 - 1,000,000}{50 - 0} = -20,000.$$
Using the point-slope form of the equation of a line with the point $(0, 1,000,000)$, we have $V - 1,000,000 = -20,000(t - 0)$,
    or $\qquad\qquad V = -20,000t + 1,000,000.$

    In 2008, $t = 5$ and $V = -20,000(5) + 1,000,000 = 900,000$, or $900,000.
    In 2013, $t = 10$ and $V = -20,000(10) + 1,000,000 = 800,000$, or $800,000.

17. The consumption function is given by $C(x) = 0.75x + 6$. When $x = 0$, we have $C(0) = 0.75(0) + 6 = 6$, or $6 billion dollars.
    When $x = 50$, $\qquad\qquad C(50) = 0.75(50) + 6 = 43.5$, or $43.5 billion dollars.

When $x = 100$, $C(100) = 0.75(100) + 6 = 81$,   or $81 billion dollars.

19. a. $y = 1.053x$, where x is the monthly benefit before adjustment, and y is the adjusted monthly benefit.
    b. His adjusted monthly benefit will be $(1.053)(620) = 652.86$, or $652.86.

21. Let the number of tapes produced and sold be x. Then
$$C(x) = 12,100 + 0.60x; \qquad R(x) = 1.15x$$
and $\quad P(x) = R(x) - C(x) = 1.15x - (12,100 + 0.60x)$
$$= 0.55x - 12,100.$$

23. Let the value of the server after t years be V. When $t = 0$, $V = 60,000$ and when $t = 4$, $V = 12,000$.

    a. Since $\quad m = \dfrac{12,000 - 60,000}{4} = -\dfrac{48,000}{4} = -12,000$

    the rate of depreciation $(-m)$ is $12,000.
    b. Using the point-slope form of the equation of a line with the point $(4, 12,000)$, we
    have $\qquad V - 12,000 = -12,000(t - 4)$
    or $\qquad\qquad V = -12,000t + 60,000.$

    c.

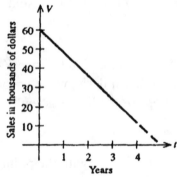

    d. When $t = 3$,
    $$V = -12,000(3) + 60,000 = 24,000, \text{ or } \$24,000.$$

25. The formula given in Exercise 24 is $V = C - \dfrac{C - S}{N} t$.

    Then, when $C = 1,000,000$, $N = 50$, and $S = 0$, we have
    $$V = 1,000,000 - \frac{1,000,000 - 0}{50} t \text{ or } \qquad V = 1,000,000 - 20,000t.$$
    In 2008, $t = 5$ and $\quad V = 1,000,000 - 20,000(5) = 900,000,$ or $900,000.

In 2013, $t = 10$ and $V = 1,000,000 - 20,000(10) = 800,000$ or \$800,000.

27. a. $D(S) = \dfrac{Sa}{1.7}$. If we think of D as having the form $D(S) = mS + b$, then

$m = \dfrac{a}{1.7}$, $b = 0$, and D is a linear function of S.

b. $D(0.4) = \dfrac{500(0.4)}{1.7} = 117.647$, or approximately 117.65 mg.

29. a. Since the relationship is linear, we can write $F = mC + b$, where m and b are constants. Using the condition $C = 0$ when $F = 32$, we have $32 = b$, and so $F = mC + 32$. Next, using the condition $C = 100$ when $F = 212$, we have
$$212 = 100m + 32 \quad \text{or} \quad m = \tfrac{9}{5}.$$
Therefore, $F = \tfrac{9}{5}C + 32$.

b. From (a), we see $F = \tfrac{9}{5}C + 32$. Next, when $C = 20$, $F = \tfrac{9}{5}(20) + 32 = 68$ and so the temperature equivalent to 20°C is 68°F.

c. Solving for C in terms of F, we find $\tfrac{9}{5}C = F - 32$, or $C = \tfrac{5}{9}F - \tfrac{160}{9}$.
When $F = 70$, $C = \tfrac{5}{9}(70) - \tfrac{160}{9} = \tfrac{190}{9}$, or approximately 21.1°C.

31. The slope of L_2 is greater than that of L_1. This tells us that if the manufacturer lowers the unit price for each model clock radio by the same amount, the additional quantity demanded of model B radios will be greater than that of the model A radios.

33. a. Setting $x = 0$, gives $3p = 18$, or $p = 6$. Next, setting $p = 0$, gives $2x = 18$, or $x = 9$.

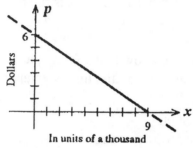

b. When $p = 4$, $\qquad 2x + 3(4) - 18 = 0$

$$2x = 18 - 12 = 6$$

and $x = 3$. Therefore, the quantity demanded when $p = 4$ is 3000. (Remember x is given in units of a thousand.)

35. a. When $x = 0$, $p = 60$ and when $p = 0$, $-3x = -60$, or $x = 20$.

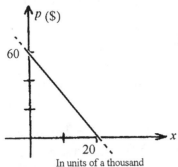

P ($)

60

20

In units of a thousand

x

b. When $p = 30$, $30 = -3x + 60$

                           $3x = 30$ and $x = 10$.

Therefore, the quantity demanded when $p = 30$ is 10,000 units.

37. When $x = 1000$, $p = 55$, and when $x = 600$, $p = 85$. Therefore, the graph of the linear demand equation is the straight line passing through the points (1000, 55) and (600, 85). The slope of the line is

$$\frac{85-55}{600-1000} = -\frac{3}{40}.$$

Using the point (1000, 55) and the slope just found, we find that the required equation is $p - 55 = -\frac{3}{40}(x - 1000)$

$$p = -\frac{3}{40}x + 130 \quad .$$

When $x = 0$, $p = 130$ which means that there will be no demand above $130.
When $p = 0$, $x = 1733.33$, which means that 1733 units is the maximum quantity demanded.

39. Since the demand equation is linear, we know that the line passes through the points (1000,9) and (6000,4). Therefore, the slope of the line is given by

$$m = \frac{4-9}{6000-1000} = -\frac{5}{5000} = -0.001 .$$

Since the equation of the line has the form $p = ax + b$,

$$9 = -0.001(1000) + b, \quad \text{or} \quad b = 10.$$

Therefore, the equation of the line is $p = -0.001x + 10$.
If $p = 7.50$, $7.50 = -0.001x + 10$
$$0.001x = 2.50, \text{ or } x = 2500.$$
So, the quantity demanded when the unit price is $7.50 is 2500 units.

41. a. Setting $x = 0$, we obtain $3(0) - 4p + 24 = 0$
$$-4p = -24$$
or
$$p = 6.$$
Setting $p = 0$, we obtain $3x - 4(0) + 24 = 0$
$$3x = -24$$
or
$$x = -8.$$

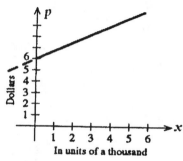

In units of a thousand

b. When $p = 8$, $3x - 4(8) + 24 = 0$
$$3x = 32 - 24 = 8$$
$$x = 8/3.$$
Therefore, 2667 units of the commodity would be supplied at a unit price of $8. (Here again x is measured in units of thousands.)

43. a. When $x = 0$, $p = 10$, and when $p = 0$, $x = -5$.

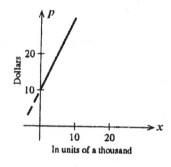

In units of a thousand

b. $p = 2x + 10$, $14 = 2x + 10$, $2x = 4$, and $x = 2$. Therefore, when $p = 14$, the supplier will make 2000 units of the commodity available.

45. When $x = 10,000$, $p = 45$ and when $x = 20,000$, $p = 50$. Therefore, the slope of the line passing $(10,000, 45)$ and $(20,000, 50)$ is

$$m = \frac{50 - 45}{20,000 - 10,000} = \frac{5}{10,000} = 0.0005$$

Using the point- slope form of an equation of a line with the point $(10,000, 45)$, we have $\qquad p - 45 = 0.0005(x - 10,000)$
$$p = 0.0005x - 5 + 45$$
or $\qquad p = 0.0005x + 40.$
If $p = 70$, $\qquad 70 = 0.0005x + 40$

$$0.0005x = 30, \qquad \text{or} \quad x = \frac{30}{0.0005} = 60,000 \quad .$$

(If x is expressed in units of a thousand, then then the equation may be written in the form $p = \frac{1}{2}x + 40$.)

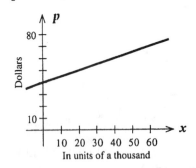

47. False. $P(x) = R(x) - C(x) = sx - (cx + F) = (s - c)x - F$. Therefore, the firm is making a profit if $P(x) = (s - c)x - F > 0$ or $x > \dfrac{F}{s - c}$.

USING TECHNOLOGY EXERCISES 1.3, page 43

1. 2.2875   3. 2.880952381   5. 7.2851648352   7. 2.4680851064

EXERCISES 1.4, page 51

1. We solve the system $y = 3x + 4$
$$y = -2x + 14.$$

Substituting the first equation into the second yields
$$3x + 4 = -2x + 14$$
$$5x = 10,$$
and $x = 2$. Substituting this value of x into the first equation yields
$$y = 3(2) + 4,$$
or $y = 10$. Thus, the point of intersection is (2,10).

3. We solve the system $\quad 2x - 3y = 6$
$$3x + 6y = 16.$$
Solving the first equation for y, we obtain
$$3y = 2x - 6$$
$$y = \tfrac{2}{3}x - 2 \quad.$$
Substituting this value of y into the second equation, we obtain
$$3x + 6(\tfrac{2}{3}x - 2) = 16$$
$$3x + 4x - 12 = 16$$
$$7x = 28$$
and $x = 4$. Then $\qquad y = \tfrac{2}{3}(4) - 2 = \tfrac{2}{3}.$
Therefore, the point of intersection is $(4, \tfrac{2}{3})$.

5. We solve the system $\begin{cases} y = \tfrac{1}{4}x - 5 \\ 2x - \tfrac{3}{2}y = 1 \end{cases}$. Substituting the value of y given in the first

equation into the second equation, we obtain
$$2x - \tfrac{3}{2}(\tfrac{1}{4}x - 5) = 1$$
$$2x - \tfrac{3}{8}x + \tfrac{15}{2} = 1$$
$$16x - 3x + 60 = 8$$
$$13x = -52,$$
or $x = -4$. Substituting this value of x in the first equation, we have
$$y = \tfrac{1}{4}(-4) - 5 = -1 - 5,$$
or $y = -6$. Therefore, the point of intersection is (–4,–6).

7. We solve the equation $R(x) = C(x)$, or $15x = 5x + 10,000$, obtaining $10x = 10,000$, or $x = 1000$. Substituting this value of x into the equation $R(x) = 15x$, we find $R(1000) = 15,000$. Therefore, the breakeven point is (1000, 15,000).

9. We solve the equation $R(x) = C(x)$, or $0.4x = 0.2x + 120$, obtaining $0.2x = 120$,

or $x = 600$. Substituting this value of x into the equation $R(x) = 0.4x$, we find $R(600) = 240$. Therefore, the breakeven point is $(600, 240)$.

11. a.

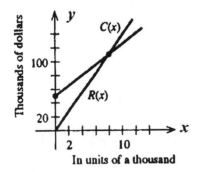

In units of a thousand

b. We solve the equation $R(x) = C(x)$ or $14x = 8x + 48,000$, obtaining $6x = 48,000$ or $x = 8000$. Substituting this value of x into the equation $R(x) = 14x$, we find $R(8000) = 14(8000) = 112,000$. Therefore, the breakeven point is $(8000, 112,000)$.

c.

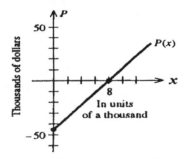

In units of a thousand

d. $P(x) = R(x) - C(x) = 14x - 8x - 48,000 = 6x - 48,000$.
The graph of the profit function crosses the x-axis when $P(x) = 0$, or $6x = 48,000$ and $x = 8000$. This means that the revenue is equal to the cost when 8000 units are produced and consequently the company breaks even at this point.

13. Let x denote the number of units sold. Then, the revenue function R is given by
$$R(x) = 9x.$$
Since the variable cost is 40 percent of the selling price and the monthly fixed costs are $50,000, the cost function C is given by
$$C(x) = 0.4(9x) + 50,000$$
$$= 3.6x + 50,000.$$
To find the breakeven point, we set $R(x) = C(x)$, obtaining

$$9x = 3.6x + 50,000$$
$$5.4x = 50,000$$
$$x \approx 9259, \text{ or } 9259 \text{ units.}$$
Substituting this value of x into the equation $R(x) = 9x$ gives
$$R(9259) = 9(9259) = 83,331.$$
Thus, for a breakeven operation, the firm should manufacture 9259 bicycle pumps resulting in a breakeven revenue of $83,331.

15. a. The cost function associated with using machine I is given by
$$C_1(x) = 18,000 + 15x.$$
The cost function associated with using machine II is given by
$$C_2(x) = 15,000 + 20x.$$
   b.

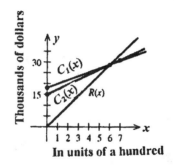

In units of a hundred

c. Comparing the cost of producing 450 units on each machine, we find
$$C_1(450) = 18,000 + 15(450)$$
$$= 24,750 \quad \text{or } \$24,750 \text{ on machine } I,$$
and $\qquad C_2(450) = 15,000 + 20(450)$
$$= 24,000 \text{ or } \$24,000 \text{ on machine } II.$$
Therefore, machine II should be used in this case.
Next, comparing the costs of producing 550 units on each machine, we find
$$C_1(550) = 18,000 + 15(550)$$
$$= 26,250 \text{ or } \$26,250 \text{ on machine } I,$$
and $\qquad C_2(550) = 15,000 + 20(550)$
$$= 26,000$$
on machine II. Therefore, machine II should be used in this instance. Once again, we compare the cost of producing 650 units on each machine and find that
$$C_1(650) = 18,000 + 15(650)$$
$$= 27,750, \text{ or } \$27,750 \text{ on machine } I \text{ and}$$
$$C_2(650) = 15,000 + 20(650)$$

$$= 28,000,$$

or \$28,000 on machine *II*. Therefore, machine *I* should be used in this case.

d. We use the equation $P(x) = R(x) - C(x)$ and find
$$P(450) = 50(450) - 24,000 = -1500,$$
or a loss of \$1500 when machine *II* is used to produce 450 units. Similarly,
$$P(550) = 50(550) - 26,000 = 1500,$$
or a profit of \$1500 when machine *II* is used to produce 550 units.
Finally, $P(650) = 50(650) - 27,750 = 4750,$
or a profit of \$4750 when machine *I* is used to produce 650 units.

17. We solve the system
$$4x + 3p = 59$$
$$5x - 6p = -14.$$
Solving the first equation for p, we find $p = -\frac{4}{3}x + \frac{59}{3}$.
Substituting this value of p into the second equation, we have
$$5x - 6(-\tfrac{4}{3}x + \tfrac{59}{3}) = -14$$
$$5x + 8x - 118 = -14$$
$$13x = 104$$
$$x = 8.$$
Substituting this value of x into the equation
$$p = -\tfrac{4}{3}x + \tfrac{59}{3}$$
we have
$$p = -\tfrac{4}{3}(8) + \tfrac{59}{3} = \tfrac{27}{3} = 9$$
Thus, the equilibrium quantity is 8000 units and the equilibrium price is \$9.

19. We solve the system $\quad p = -2x + 22$
$$p = 3x + 12 \quad .$$
Substituting the first equation into the second, we find
$$-2x + 22 = 3x + 12$$
$$5x = 10$$
and $\qquad\qquad x = 2.$
Substituting this value of x into the first equation, we obtain
$$p = -2(2) + 22 = 18.$$
Thus, the equilibrium quantity is 2000 units and the equilibrium price is \$18.

21. Let x denote the number of DVD's produced per week, and p denote the price of each DVD.

a. The slope of the demand curve is given by $\dfrac{\Delta p}{\Delta x} = -\dfrac{20}{250} = -\dfrac{2}{25}$.

Using the point-slope form of the equation of a line with the point (3000, 485), we have

$$p - 485 = -\tfrac{2}{25}(x - 3000)$$
$$p = -\tfrac{2}{25}x + 240 + 485$$

or
$$p = -0.08x + 725.$$

b. From the given information, we know that the graph of the supply equation passes through the points (0, 300) and (2500, 525). Therefore, the slope of the supply curve is
$$m = \dfrac{525 - 300}{2500 - 0} = \dfrac{225}{2500} = 0.09 .$$

Using the point-slope form of the equation of a line with the point (0, 300), we find that

$$p - 300 = 0.09x$$
$$p = 0.09x + 300.$$

c. Equating the supply and demand equations, we have
$$-0.08x + 725 = 0.09x + 300$$
$$0.17x = 425$$

or
$$x = 2500.$$

Then
$$p = -0.08(2500) + 725 = 525.$$

We conclude that the equilibrium quantity is 2500 and the equilibrium price is $525.

23. We solve the system $\quad 3x + \quad p = \quad 1500$
$$\qquad\qquad\qquad 2x - \ 3p = -1200.$$

Solving the first equation for p, we obtain
$$p = 1500 - 3x.$$

Substituting this value of p into the second equation, we obtain
$$2x - 3(1500 - 3x) = -1200$$
$$11x = 3300$$

or
$$x = 300.$$

Next,
$$p = 1500 - 3(300) = 600.$$

Thus, the equilibrium quantity is 300 and the equilibrium price is $600.

25. a. We solve the system of equations $p = cx + d$ and $p = ax + b$. Substituting the first into the second gives

$$cx + d = ax + b$$
$$(c - a)x = b - d$$

or
$$x = \frac{b-d}{c-a}.$$

Since $a < 0$ and $c > 0$, and $b > d > 0$, and $c - a \neq 0$, x is well-defined. Substituting this value of x into the second equation, we obtain

$$p = a\left(\frac{b-d}{c-a}\right) + b = \frac{ab - ad + bc - ab}{c-a} = \frac{bc - ad}{c-a} \qquad (1)$$

Therefore, the equilibrium quantity is $\dfrac{b-d}{c-a}$ and the equilibrium price is $\dfrac{bc-ad}{c-a}$.

b. If c is increased, the denominator in the expression for x increases and so x gets smaller. At the same time, the first term in the first equation (1) for p decreases (since a is a negative number) and so p gets larger. This analysis shows that if the unit price for producing the product is increased then the equilibrium quantity decreases while the equilibrium price increases.

c. If b is decreased, the numerator of the expression for x decreases while the denominator stays the same. Therefore x decreases. The expression for p also shows that p decreases. This analysis shows that if the (theoretical) upper bound for the unit price of a commodity is lowered, then both the equilibrium quantity and the equilibrium price drop.

27. True. $P(x) = R(x) - C(x) = sx - (cx + F) = (s-c)x - F$. Therefore, the firm is making a profit if $P(x) = (s-c)x - F > 0$, or $x > \frac{F}{s-c}$ $(s \neq c)$.

29. Solving the two equations simultaneously to find the point(s) of intersection of L_1 and L_2, we obtain $\qquad m_1 x + b_1 = m_2 x + b_2$

$$(m_1 - m_2)x = b_2 - b_1 \qquad (1)$$

a. If $m_1 = m_2$ and $b_2 \neq b_1$, then there is no solution for (1) and in this case L_1 and L_2 do not intersect.

b. If $m_1 \neq m_2$, then Equation (1) can be solved (uniquely) for x and this shows that L_1 and L_2 intersect at precisely one point.

c. If $m_1 = m_2$ and $b_1 = b_2$, then (1) is satisfied for all values of x and this shows that L_1 and L_2 intersect at infinitely many points.

USING TECHNOLOGY EXERCISES 1.4, page 53

1. $(0.6, 6.2)$ 3. $(3.8261, 0.1304)$ 5. $(386.9091, \ 145.3939)$

7.
   a.

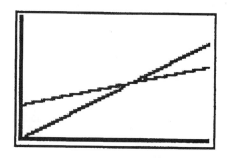

b. (3548, 27,997)

   c. x-intercept is approximately 3548

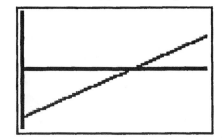

9. a. $C_1(x) = 34 + 0.18x$; $C_2(x) = 28 + 0.22x$
   b.

c. (150, 61)

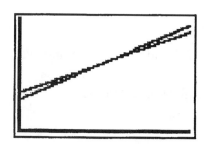

   d. If the distance driven is less than or equal to 150 mi, rent from Acme Truck Leasing; if the distance driven is more than 150 mi, rent from Ace Truck Leasing.

11. a. $p = -\dfrac{1}{10}x + 284$; $p = \dfrac{1}{60}x + 60$

b. (1920, 92)

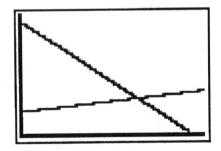

c. 1920/wk; $92/radio

EXERCISES 1.5, page 60

1. a. We first summarize the data:

x	y	x^2	xy
1	4	1	4
2	6	4	12
3	8	9	24
4	11	16	44
10	29	30	84

The normal equations are $4b + 10m = 29$
$10b + 30m = 84$.

Solving this system of equations, we obtain $m = 2.3$ and $b = 1.5$. So an equation is
$y = 2.3x + 1.5$.

b. The scatter diagram and the least squares line for this data follow:

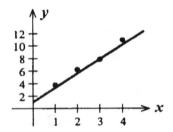

3. a. We first summarize the data:

x	y	x^2	xy
1	4.5	1	4.5
2	5	4	10
3	3	9	9
4	2	16	8
4	3.5	16	14
6	1	36	6
20	19	82	51.5

The normal equations are $6b + 20m = 19$

$$20b + 82m = 51.5.$$

The solutions are $m \approx -0.7717$ and $b \approx 5.7391$ and so a required equation is $y = -0.772x + 5.739$.

b. The scatter diagram and the least-squares line for these data follow.

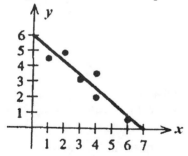

5. a. We first summarize the data:

x	y	x^2	xy
1	3	1	3
2	5	4	10
3	5	9	15
4	7	16	28
5	8	25	40
15	28	55	96

The normal equations are $55m + 15b = 96$

$$15m + 5b = 28.$$

Solving, we find $m = 1.2$ and $b = 2$, so that the required equation is $y = 1.2x + 2$.

b. The scatter diagram and the least-squares line for the given data follow.

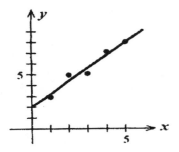

7. a. We first summarize the data:

x	y	x^2	xy
4	0.5	16	2
4.5	0.6	20.25	2.7
5	0.8	25	4
5.5	0.9	30.25	4.95
6	1.2	36	7.2
25	4	127.5	20.85

The normal equations are
$$5b + 25m = 4$$
$$25b + 127.5m = 20.85.$$

The solutions are $m = 0.34$ and $b = -0.9$, and so a required equation is $y = 0.34x - 0.9$.

b. The scatter diagram and the least-squares line for these data follow.

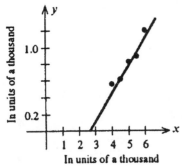

c. If $x = 6.4$, then $y = 0.34(6.4) - 0.9 = 1.276$ and so 1276 completed applications might be expected.

9. a. We first summarize the data:

x	y	x^2	xy
1	436	1	436
2	438	4	876
3	428	9	1284
4	430	16	1720
5	426	25	2130
15	2158	55	6446

The normal equations are
$$5b + 15m = 2158$$
$$15b + 55m = 6446.$$
Solving this system, we find $m = -2.8$ and $b = 440$.
Thus, the equation of the least-squares line is $y = -2.8x + 440$.
b. The scatter diagram and the least-squares line for this data are shown in the figure that follows.

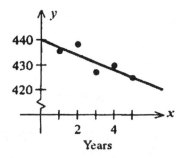

c. Two years from now, the average SAT verbal score in that area will be
$y = -2.8(7) + 440 = 420.4$.

11. a. We first summarize the data:

x	y	x^2	xy
0	168	0	0
10	213	100	2130
20	297	400	5940
30	374	900	11220
40	427	1600	17080
57	471	3249	26847
157	1950	6249	63217

The normal equations are

$$6b + 157m = 1950$$

$$157b + 6249m = 63217$$

The solutions are $m = 5.695 \approx 5.70$ and $b = 175.98 \approx 176$ and so a required equation is $y = 5.70x + 176$.

b. In 1985, $x = 45$, $y = 5.70 (45) + 176 = 432.5 \approx 433$ acres. Hence, the expected size of the average farm was 433 acres.

13. a. We first summarize the data:

x	y	x^2	xy
1	20	1	20
2	24	4	48
3	26	9	78
4	28	16	112
5	32	25	160
15	130	55	418

The normal equations are

$$5b + 15m = 130$$
$$15b + 55m = 418.$$

The solutions are $m = 2.8$ and $b = 17.6$, and so an equation of the line is

$$y = 2.8x + 17.6.$$

b. When $x = 8$, $y = 2.8(8) + 17.6 = 40$. Hence, the state subsidy is expected to be $40 million for the eighth year.

15. a. We first summarize the data:

x	y	x^2	xy
1	16.7	1	16.7
3	26	9	78
5	33.3	25	166.5
7	48.3	49	338.1
9	57	81	513
11	65.8	121	723.8
13	74.2	169	964.6
15	83.3	225	1249.5
64	404.6	680	4050.2

The normal equations are $\quad 8b + 64m = \quad 404.6$

$$64b + 680m = 4050.2.$$

The solutions are $m = 4.8417$ and $b = 11.8417$ and so a required equation is $y = 4.842x + 11.842$.

b. In 1993, $x = 19$, and so $y = 4.842(19) + 11.842 = 103.84$. Hence the estimated number of cans produced in 1993 is 103.8 billion.

17. a. We first summarize the data:

x	y	x^2	xy
0	21.7	0	0
1	32.1	1	32.1
2	45.0	4	90
3	58.3	9	174.9
4	69.6	16	278.4
10	226.7	30	575.4

The normal equations are $\quad \begin{aligned} 5b + 10m &= 226.7 \\ 10b + 30m &= 575.4 \end{aligned}$.

The solutions are $m = 12.2$ and $b = 20.9$ and so a required equation is $y = 12.2x + 20.9$.

b. In 2005, $x = 7$ so $y = 12.2(7) + 20.9 = 106.3$, or 106.3 million computers are expected to be connected to the internet in Europe in that year.

19. a. We first summarize the data.

x	y	x^2	xy
0	19.5	0	0
10	20	100	200
20	20.6	400	412
30	21.2	900	636
40	21.8	1600	872
50	22.4	2500	1120
150	125.5	5500	3240

The normal equations are

$$6b + 150m = 125.5$$

$$150b + 5500m = 3240$$

The solutions are $b = 19.45$ and $m = 0.0586$. Therefore, $y = 0.059x + 19.5$.

b. The life expectancy at 65 of a female in 2040 is

$$y = 0.059(40) + 19.5 = 21.86 \quad \text{or } 21.9 \text{ years.}$$

The datum gives a life expectancy of 21.8 years.

c. The life expectancy at 65 of a female in 2030 is

$$y = 0.059(30) + 19.5 = 21.27 \quad \text{or } 21.3 \text{ years.}$$

The datum gives a life expectancy of 21.2 years.

21. a. We first summarize the given data:

x	y	x^2	xy
0	90.4	0	0
1	100.0	1	100
2	110.4	4	220.8
3	120.4	9	361.2
4	130.8	16	523.2
5	140.4	25	702
6	150	36	900
21	842.4	91	2807.2

The normal equations are $\quad \begin{aligned} 7b + 21m &= 842.4 \\ 21b + 91m &= 2807.2 \end{aligned}$.

The solutions are $m = 10$ and $b = 90.34$. Therefore, the required equation is $y = 10x + 90.34$.

b. If $x = 6$, then $y = 10(6) + 90.34 = 150.34$, or 150,340,000. This compares well with the actual data for that year-- 150,000,000 subscribers.

23. False. See Example 1, page 57.

25. True. Since there exists one and only one line passing through two distinct points, the two lines must be the same.

USING TECHNOLOGY EXERCISES 1.5, page 67

1. $y = 2.3596x + 3.8639$ 3. $y = -1.1948x + 3.5525$

5. a. $y = 0.45x + 1.28$ b. \$3.08 billion

7. a. $y = 13.321x + 72.57$ b. 192 million tons

9. a. $1.95x + 12.19$; \$23.89 billion

11. a. $y = 3.76x + 87.46$ b. \$140,100

CHAPTER 1 REVIEW EXERCISES, page 69

1. The distance is $\quad d = \sqrt{(6-2)^2 + (4-1)^2} = \sqrt{4^2 + 3^2} = \sqrt{25} = 5$.

3. The distance is
$$d = \sqrt{[1-(-2)]^2 + [-7-(-3)]^2} = \sqrt{3^2 + (-4)^2} = \sqrt{9+16} = \sqrt{25} = 5.$$

5. An equation is $x = -2$.

7. The slope of L is $m = \dfrac{\frac{7}{2}-4}{3-(-2)} = \dfrac{\frac{7-8}{2}}{5} = -\dfrac{1}{10}$ and an equation of L is
$$y - 4 = -\tfrac{1}{10}[x - (-2)] = -\tfrac{1}{10}x - \tfrac{1}{5},$$
or $\qquad\qquad y = -\tfrac{1}{10}x + \tfrac{19}{5}$
The general form of this equation is $x + 10y - 38 = 0$.

9. Writing the given equation in the form $y = \tfrac{5}{2}x - 3$, we see that the slope of the given line is 5/2. So a required equation is $\quad y - 4 = \tfrac{5}{2}(x+2) \quad$ or $\quad y = \tfrac{5}{2}x + 9$
The general form of this equation is $5x - 2y + 18 = 0$.

11. Using the slope-intercept form of the equation of a line, we have $y = -\tfrac{1}{2}x - 3$.

13. Rewriting the given equation in the slope-intercept form, we have $4y = -3x + 8$
or $\qquad\qquad y = -\tfrac{3}{4}x + 2$
and conclude that the slope of the required line is $-3/4$. Using the point-slope form of the equation of a line with the point (2,3) and slope $-3/4$, we obtain

$$y - 3 = -\tfrac{3}{4}(x - 2)$$
$$y = -\tfrac{3}{4}x + \tfrac{6}{4} + 3$$
$$= -\tfrac{3}{4}x + \tfrac{9}{2}.$$

The general form of this equation is $3x + 4y - 18 = 0$.

15. Rewriting the given equation in the slope-intercept form $y = \tfrac{2}{3}x - 8$, we see that the slope of the line with this equation is 2/3. The slope of the required line is $-3/2$. Using the point-slope form of the equation of a line with the point $(-2, -4)$ and slope $-3/2$, we have

$$y - (-4) = -\tfrac{3}{2}[x - (-2)]$$

or $\qquad\qquad y = -\tfrac{3}{2}x - 7$.

The general form of this equation is $3x + 2y + 14 = 0$.

17. Setting $x = 0$, gives $5y = 15$, or $y = 3$. Setting $y = 0$, gives $-2x = 15$, or $x = -15/2$. The graph of the equation $-2x + 5y = 15$ follows.

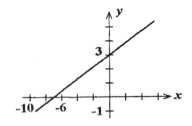

19. Let x denote the time in years. Since the function is linear, we know that it has the form $f(x) = mx + b$.

a. The slope of the line passing through $(0, 2.4)$ and $(5, 7.4)$ is $\; m = \dfrac{7.4 - 2.4}{5} = 1$.

Since the line passes through $(0, 2.4)$, we know that the y-intercept is 2.4. Therefore, the required function is $f(x) = x + 2.4$.

b. In 2002 $(x = 3)$, the sales were $f(3) = 3 + 2.4 = 5.4$, or \$5.4 million dollars.

21. a. $\quad D(w) = \dfrac{a}{150}w$. The given equation can be expressed in the form $y = mx + b$,

where $m = \dfrac{a}{150}$ and $b = 0$.

b. If $a = 500$ and $w = 35$, $D(35) = \frac{500}{150}(35) = 116\frac{2}{3}$, or approximately 117 mg.

23. Let V denote the value of the machine after t years.
   a. The rate of depreciation is
   $$-\frac{\Delta V}{\Delta t} = \frac{300{,}000 - 30{,}000}{12} = \frac{270{,}000}{12} = 22{,}500, \quad \text{or } \$22{,}500/\text{year.}$$

   b. Using the point-slope form of the equation of a line with the point $(0, 300{,}000)$ and $m = -22{,}500$, we have
   $$V - 300{,}000 = -22{,}500(t - 0)$$
   $$V = -22{,}500t + 300{,}000.$$

25. The slope of the demand curve is $\dfrac{\Delta p}{\Delta x} = -\dfrac{10}{200} = -0.05$.

   Using the point-slope form of the equation of a line with the point $(0, 200)$, we have
   $$p - 200 = -0.05(x), \quad \text{or} \qquad p = -0.05x + 200.$$
   The graph of the demand equation follows.

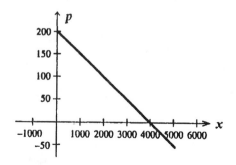

27. We solve the system
   $$3x + 4y = -6$$
   $$2x + 5y = -11.$$
   Solving the first equation for x, we have $3x = -4y - 6$ and
   $$x = -\tfrac{4}{3}y - 2.$$
   Substituting this value of x into the second equation yields
   $$2(-\tfrac{4}{3}y - 2) + 5y = -11$$
   $$-\tfrac{8}{3}y - 4 + 5y = -11$$
   $$\tfrac{7}{3}y = -7, \quad \text{or} \quad y = -3.$$
   Then
   $$x = -\tfrac{4}{3}(-3) - 2 = 4 - 2 = 2.$$

Therefore, the point of intersection is $(2, -3)$.

29. Setting $C(x) = R(x)$, we have $12x + 20{,}000 = 20x$
$$8x = 20{,}000$$
or $$x = 2500.$$
Next, $$R(2500) = 20(2500) = 50{,}000,$$
and we conclude that the breakeven point is $(2500, 50{,}000)$.

31. a. The slope of the line is $m = \dfrac{1 - 0.5}{4 - 2} = 0.25$.

Using the point-slope form of an equation of a line, we have
$$y - 1 = 0.25(x - 4)$$
$$y = 0.25x$$
b. $$y = 0.25(6.4) = 1.6, \quad \text{or } 1600 \text{ applications.}$$

CHAPTER 2

EXERCISES 2.1, page 79

1. Solving the first equation for x, we find $x = 3y - 1$. Substituting this value of x into the second equation yields

$$4(3y - 1) + 3y = 11$$
$$12y - 4 + 3y = 11$$

or
$$y = 1.$$

Substituting this value of y into the first equation gives $x = 3(1) - 1 = 2$.
Therefore, the unique solution of the system is $(2,1)$.

3. Solving the first equation for x, we have $x = 7 - 4y$. Substituting this value of x into the second equation, we have

$$\tfrac{1}{2}(7 - 4y) + 2y = 5$$
$$7 - 4y + 4y = 10$$
$$7 = 10.$$

Clearly, this is impossible and we conclude that the system of equations has no solution.

5. Solving the first equation for x, we obtain $x = 7 - 2y$.
Substituting this value of x into the second equation, we have

$$2(7 - 2y) - y = 4$$
$$14 - 4y - y = 4$$
$$-5y = -10$$

and
$$y = 2.$$
Then
$$x = 7 - 2(2) = 7 - 4 = 3.$$
We conclude that the solution to the system is $(3,2)$.

7. Solving the first equation for x, we have

$$2x = 5y + 10$$

and
$$x = \tfrac{5}{2}y + 5.$$

Substituting this value of x into the second equation, we have

$$6(\tfrac{5}{2}y + 5) - 15y = 30$$
$$15y + 30 - 15y = 30$$

or
$$0 = 0.$$

This result tells us that the second equation is equivalent to the first. Thus, any ordered pair of numbers (x, y) satisfying the equation
$$2x - 5y = 10 \qquad (\text{or } 6x - 15y = 30)$$

is a solution to the system. In particular, by assigning the value t to x, where t is any real number, we find that $y = -2 + \frac{2}{5}t$ so the ordered pair, $(t, \frac{2}{5}t - 2)$ is a solution to the system, and we conclude that the system has infinitely many solutions.

9. Solving the first equation for x, we obtain
$$4x - 5y = 14$$
$$4x = 14 + 5y$$
$$x = \frac{14}{4} + \frac{5}{4}y = \frac{7}{2} + \frac{5}{4}y.$$
Substituting this value of x into the second equation gives
$$2(\frac{7}{2} + \frac{5}{4}y) + 3y = -4$$
$$7 + \frac{5}{2}y + 3y = -4$$
$$\frac{11}{2}y = -11$$
or
$$y = -2.$$
Then,
$$x = \frac{7}{2} + \frac{5}{4}(-2) = 1.$$
We conclude that the ordered pair $(1, -2)$ satisfies the given system of equations.

11. Solving the first equation for x, we obtain
$$2x = 3y + 6$$
$$x = \frac{3}{2}y + 3$$
Substituting this value of x into the second equation gives
$$6(\frac{3}{2}y + 3) - 9y = 12$$
$$9y + 18 - 9y = 12$$
$$18 = 12.$$
which is impossible. We conclude that the system of equations has no solution.

13. Solving the first equation for y, we obtain $y = 2x - 3$. Substituting this value of y into the second equation yields
$$4x + k(2x - 3) = 4,$$
$$4x + 2xk - 3k = 4$$
$$2x(2 + k) = 4 + 3k$$
$$x = \frac{4 + 3k}{2(2 + k)}.$$
Since x is not defined when the denominator of this last expression is zero, we

conclude that the system has no solution when $k = -2$.

15. Let x and y denote the number of acres of corn and wheat planted, respectively. Then $x + y = 500$. Since the cost of cultivating corn is $42/acre and that of wheat $30/acre and Mr. Johnson has $18,600 available for cultivation, we have
$$42x + 30y = 18600.$$
Thus, the solution is found by solving the system of equations
$$x + \quad y = \quad 500$$
$$42x + 30y = 18,600 \quad ^\cdot$$

17. Let x denote the number of pounds of the $2.50/lb coffee and y denote the number of pounds of the $3/lb coffee. Then
$$x + y = 100.$$
Since the blended coffee sells for $2.80/lb, we know that the blended mixture is worth $(2.80)(100) = \$280$. Therefore,
$$2.50x + 3y = 280.$$
Thus, the solution is found by solving the system of equations
$$x + \quad y = 100$$
$$2.50x + 3y = 280$$

19. Let x denote the number of children who rode the bus during the morning shift and y denote the number of adults who rode the bus during the morning shift. Then $x + y = 1000$. Since the total fare collected was $650, we have $0.25x + 0.75y = 650$. Thus, the solution to the problem can be found by solving the system of equations
$$x + \quad \quad y = 1000$$
$$0.25x + 0.75y = \quad 650.$$

21. Let $\quad x =$ the amount of money invested at 6 percent in a savings account
$$y =$$ the amount of money invested at 8 percent in mutual funds
and $\quad z =$ the amount of money invested at 12 percent in bonds.
Since the total interest was $21,600, we have
$$0.6x + 0.8y + 0.12z = 21,600.$$
Also, since the amount of Mr. Sid's investment in bonds is twice the amount of the investment in the savings account, we have
$$z = 2x.$$
Finally, the interest earned from his investment in bonds was equal to the interest earned on his money mutual funds, so
$$0.08y = 0.12z.$$

Thus, the solution to the problem can be found by solving the system of equations

$$0.06x + 0.08y + 0.12z = 21{,}600$$

$$2x \qquad - \quad z = \quad 0$$

$$0.08y - 0.12z = \quad 0.$$

23. Let x, y, and z denote the number of 100-lb bags of grade-A, grade-B, and grade-C fertilizers to be produced. The amount of nitrogen required is $18x + 20y + 24z$, and this must be equal to 26,400. So, we have $18x + 20y + 24z = 26{,}400$. Similarly, the constraints on the use of phosphate and potassium lead to the equations $4x + 4y + 3z = 4900$, and $5x + 4y + 6z = 6200$, respectively. Thus we have the problem of finding the solution to the system

$$18x + 20y + 24z = 26{,}400 \qquad (Nitrogen)$$

$$4x + 4y + 3z = 4{,}900 \qquad (Phosphate).$$

$$5x + 4y + 6z = 6{,}200 \qquad (Potassium)$$

25. Let x, y, and z denote the number of compact, intermediate, and full-size cars, respectively, to be purchased. The cost incurred in buying the specified number of cars is $10000x + 15000y + 20000z$. Since the budget is $1.25 million, we have the system

$$10{,}000x + 15{,}000y + 20{,}000z = 1{,}250{,}000$$

$$x - \qquad 2y \qquad\qquad = \qquad 0$$

$$x + \qquad y + \qquad z = \qquad 100.$$

27. Let x = the number of ounces of Food I used in the meal

y = the number of ounces of Food II used in the meal

and z = the number of ounces of Food III used in the meal.

Since 100 percent of the daily requirement of proteins, carbohydrates, and iron is to be met by this meal, we have the following system of linear equations:

$$10x + 6y + 8z = 100$$

$$10x + 12y + 6z = 100$$

$$5x + 4y + 12z = 100.$$

29. True. If the three lines coincide, then the system has infinitely many solutions-- corresponding to all the points on the (common) line. If at least one line is distinct from the others, then the system has no solution.

EXERCISES 2.2, page 92

1. $\begin{bmatrix} 2 & -3 & | & 7 \\ 3 & 1 & | & 4 \end{bmatrix}$

3. $\begin{bmatrix} 0 & -1 & 2 & | & 6 \\ 2 & 2 & -8 & | & 7 \\ 0 & 3 & 4 & | & 0 \end{bmatrix}$

5. $3x + 2y = -4$
   $x - y = 5$

7. $x + 3y + 2z = 4$
   $2x \qquad\qquad = 5$
   $3x - 3y + 2z = 6$

9. Yes. Conditions 1-4 are satisfied (see page 85 of the text).

11. No. Condition 3 is violated. The first nonzero entry in the second row does not lie to the right of the first nonzero entry 1 in row 1.

13. Yes. Conditions 1-4 are satisfied.

15. No. Condition 2 and consequently condition 4 are not satisfied. The first nonzero entry in the last row is not a 1 and the column containing that entry does not have zeros elsewhere.

17. No. Condition 1 is violated. The first row consists entirely of zeros and it lies above row 2.

19. $\begin{bmatrix} \boxed{2} & 4 & | & 8 \\ 3 & 1 & | & 2 \end{bmatrix} \xrightarrow{\frac{1}{2}R_1} \begin{bmatrix} 1 & 2 & | & 4 \\ \boxed{3} & 1 & | & 2 \end{bmatrix} \xrightarrow{R_2 - 3R_1} \begin{bmatrix} 1 & 2 & | & 4 \\ 0 & -5 & | & -10 \end{bmatrix}$

21. $\begin{bmatrix} \boxed{-1} & 2 & | & 3 \\ 6 & 4 & | & 2 \end{bmatrix} \xrightarrow{-R_1} \begin{bmatrix} 1 & -2 & | & -3 \\ 6 & 4 & | & 2 \end{bmatrix} \xrightarrow{R_2 - 6R_1} \begin{bmatrix} 1 & -2 & | & -3 \\ 0 & 16 & | & 20 \end{bmatrix}$

23. $\begin{bmatrix} \boxed{2} & 4 & 6 & | & 12 \\ 2 & 3 & 1 & | & 5 \\ 3 & -1 & 2 & | & 4 \end{bmatrix} \xrightarrow{\frac{1}{2}R_1} \begin{bmatrix} 1 & 2 & 3 & | & 6 \\ 2 & 3 & 1 & | & 5 \\ 3 & -1 & 2 & | & 4 \end{bmatrix} \xrightarrow[R_3 - 3R_1]{R_2 - 2R_1} \begin{bmatrix} 1 & 2 & 3 & | & 6 \\ 0 & -1 & -5 & | & -7 \\ 0 & -7 & -7 & | & -14 \end{bmatrix}$

$$25. \quad \begin{bmatrix} 0 & 1 & 3 & | & 4 \\ 2 & 4 & \boxed{1} & | & 3 \\ 5 & 6 & 2 & | & -4 \end{bmatrix} \xrightarrow[R_3-2R_2]{R_1-3R_2} \begin{bmatrix} -6 & -11 & 0 & | & -5 \\ 2 & 4 & 1 & | & 3 \\ 1 & -2 & 0 & | & -10 \end{bmatrix}$$

$$27. \quad \begin{bmatrix} 3 & 9 & | & 6 \\ 2 & 1 & | & 4 \end{bmatrix} \xrightarrow{\frac{1}{3}R_1} \begin{bmatrix} 1 & 3 & | & 2 \\ 2 & 1 & | & 4 \end{bmatrix} \xrightarrow{R_2-2R_1} \begin{bmatrix} 1 & 3 & | & 2 \\ 0 & -5 & | & 0 \end{bmatrix} \xrightarrow{-\frac{1}{5}R_2}$$

$$\begin{bmatrix} 1 & 3 & | & 2 \\ 0 & 1 & | & 0 \end{bmatrix} \xrightarrow{R_1-3R_2} \begin{bmatrix} 1 & 0 & | & 2 \\ 0 & 1 & | & 0 \end{bmatrix}$$

$$29. \quad \begin{bmatrix} 1 & 3 & 1 & | & 3 \\ 3 & 8 & 3 & | & 7 \\ 2 & -3 & 1 & | & -10 \end{bmatrix} \xrightarrow[R_3-2R_1]{R_2-3R_1} \begin{bmatrix} 1 & 3 & 1 & | & 3 \\ 0 & -1 & 0 & | & -2 \\ 0 & -9 & -1 & | & -16 \end{bmatrix} \xrightarrow{-R_2} \begin{bmatrix} 1 & 3 & 1 & | & 3 \\ 0 & 1 & 0 & | & 2 \\ 0 & -9 & -1 & | & -16 \end{bmatrix}$$

$$\xrightarrow[R_3+9R_2]{R_1-3R_2} \begin{bmatrix} 1 & 0 & 1 & | & -3 \\ 0 & 1 & 0 & | & 2 \\ 0 & 0 & -1 & | & 2 \end{bmatrix} \xrightarrow[-R_3]{R_1+R_3} \begin{bmatrix} 1 & 0 & 0 & | & -1 \\ 0 & 1 & 0 & | & 2 \\ 0 & 0 & 1 & | & -2 \end{bmatrix}$$

31. The augmented matrix is equivalent to the system of linear equations
$$3x + 9y = 6$$
$$2x + y = 4$$
The ordered pair (2,0) is the solution to the system.

33. The augmented matrix is equivalent to the system of linear equations
$$x + 3y + z = 3$$
$$3x + 8y + 3z = 7$$
$$2x - 3y + z = -10$$
Reading off the solution from the last augmented matrix,
$$\begin{bmatrix} 1 & 0 & 0 & | & -1 \\ 0 & 1 & 0 & | & 2 \\ 0 & 0 & 1 & | & -2 \end{bmatrix},$$
which is in row-reduced form, we have $x = -1, y = 2$, and $z = -2$.

35. Using the Gauss-Jordan method, we have

$$\begin{bmatrix} 1 & -2 & | & 8 \\ 3 & 4 & | & 4 \end{bmatrix} \xrightarrow{R_2-3R_1} \begin{bmatrix} 1 & -2 & | & 8 \\ 0 & 10 & | & -20 \end{bmatrix} \xrightarrow{\frac{1}{10}R_2} \begin{bmatrix} 1 & -2 & | & 8 \\ 0 & 1 & | & -2 \end{bmatrix} \xrightarrow{R_1+2R_2} \begin{bmatrix} 1 & 0 & | & 4 \\ 0 & 1 & | & -2 \end{bmatrix}$$

The solution is $(4,-2)$.

37. Using the Gauss-Jordan method, we have

$$\begin{bmatrix} 2 & -3 & | & -8 \\ 4 & 1 & | & -2 \end{bmatrix} \xrightarrow{\frac{1}{2}R_1} \begin{bmatrix} 1 & -\frac{3}{2} & | & -4 \\ 4 & 1 & | & -2 \end{bmatrix} \xrightarrow{R_2-4R_1} \begin{bmatrix} 1 & -\frac{3}{2} & | & -4 \\ 0 & 7 & | & 14 \end{bmatrix} \xrightarrow{\frac{1}{7}R_2}$$

$$\begin{bmatrix} 1 & -\frac{3}{2} & | & -4 \\ 0 & 1 & | & 2 \end{bmatrix} \xrightarrow{R_1+\frac{3}{2}R_2} \begin{bmatrix} 1 & 0 & | & -1 \\ 0 & 1 & | & 2 \end{bmatrix}.$$

The solution is $(-1,2)$.

39. Using the Gauss-Jordan method, we have

$$\begin{bmatrix} 1 & 1 & 1 & | & 0 \\ 2 & -1 & 1 & | & 1 \\ 1 & 1 & -2 & | & 2 \end{bmatrix} \xrightarrow[R_3-R_1]{R_2-2R_1} \begin{bmatrix} 1 & 1 & 1 & | & 0 \\ 0 & -3 & -1 & | & 1 \\ 0 & 0 & -3 & | & 2 \end{bmatrix} \xrightarrow{-\frac{1}{3}R_2} \begin{bmatrix} 1 & 1 & 1 & | & 0 \\ 0 & 1 & \frac{1}{3} & | & -\frac{1}{3} \\ 0 & 0 & -3 & | & 2 \end{bmatrix} \xrightarrow{R_1-R_2}$$

$$\begin{bmatrix} 1 & 0 & \frac{2}{3} & | & \frac{1}{3} \\ 0 & 1 & \frac{1}{3} & | & -\frac{1}{3} \\ 0 & 0 & -3 & | & 2 \end{bmatrix} \xrightarrow{-\frac{1}{3}R_3} \begin{bmatrix} 1 & 0 & \frac{2}{3} & | & \frac{1}{3} \\ 0 & 1 & \frac{1}{3} & | & -\frac{1}{3} \\ 0 & 0 & 1 & | & -\frac{2}{3} \end{bmatrix} \xrightarrow[R_2-\frac{1}{3}R_3]{R_1-\frac{2}{3}R_3} \begin{bmatrix} 1 & 0 & 0 & | & \frac{7}{9} \\ 0 & 1 & 0 & | & -\frac{1}{9} \\ 0 & 0 & 1 & | & -\frac{2}{3} \end{bmatrix}.$$

The solution is $\left(\frac{7}{9},-\frac{1}{9},-\frac{2}{3}\right)$.

41.

$$\begin{bmatrix} 2 & 2 & 1 & | & 9 \\ 1 & 0 & 1 & | & 4 \\ 0 & 4 & -3 & | & 17 \end{bmatrix} \xrightarrow{R_1\leftrightarrow R_2} \begin{bmatrix} 1 & 0 & 1 & | & 4 \\ 2 & 2 & 1 & | & 9 \\ 0 & 4 & -3 & | & 17 \end{bmatrix} \xrightarrow{R_2-2R_1} \begin{bmatrix} 1 & 0 & 1 & | & 4 \\ 0 & 2 & -1 & | & 1 \\ 0 & 4 & -3 & | & 17 \end{bmatrix} \xrightarrow{\frac{1}{2}R_2}$$

$$\begin{bmatrix} 1 & 0 & 1 & | & 4 \\ 0 & 1 & -\frac{1}{2} & | & \frac{1}{2} \\ 0 & 4 & -3 & | & 17 \end{bmatrix} \xrightarrow{R_3-4R_2} \begin{bmatrix} 1 & 0 & 1 & | & 4 \\ 0 & 1 & -\frac{1}{2} & | & \frac{1}{2} \\ 0 & 0 & -1 & | & 15 \end{bmatrix} \xrightarrow{-R_3} \begin{bmatrix} 1 & 0 & 1 & | & 4 \\ 0 & 1 & -\frac{1}{2} & | & \frac{1}{2} \\ 0 & 0 & 1 & | & -15 \end{bmatrix} \xrightarrow[R_2+\frac{1}{2}R_3]{R_1-R_3}$$

$$\begin{bmatrix} 1 & 0 & 0 & | & 19 \\ 0 & 1 & 0 & | & -7 \\ 0 & 0 & 1 & | & -15 \end{bmatrix}. \qquad \text{The solution is } (19,-7,-15).$$

43. $\begin{bmatrix} 0 & -1 & 1 & | & 2 \\ 4 & -3 & 2 & | & 16 \\ 3 & 2 & 1 & | & 11 \end{bmatrix} \xrightarrow{R_1 \leftrightarrow R_2} \begin{bmatrix} 4 & -3 & 2 & | & 16 \\ 0 & -1 & 1 & | & 2 \\ 3 & 2 & 1 & | & 11 \end{bmatrix} \xrightarrow{R_1-R_3} \begin{bmatrix} 1 & -5 & 1 & | & 5 \\ 0 & -1 & 1 & | & 2 \\ 3 & 2 & 1 & | & 11 \end{bmatrix}$

$$\xrightarrow[R_3-3R_1]{-R_2} \begin{bmatrix} 1 & -5 & 1 & | & 5 \\ 0 & 1 & -1 & | & -2 \\ 0 & 17 & -2 & | & -4 \end{bmatrix} \xrightarrow[R_3-17R_2]{R_1+5R_2} \begin{bmatrix} 1 & 0 & -4 & | & -5 \\ 0 & 1 & -1 & | & -2 \\ 0 & 0 & 15 & | & 30 \end{bmatrix} \xrightarrow{\frac{1}{15}R_3}$$

$$\begin{bmatrix} 1 & 0 & -4 & | & -5 \\ 0 & 1 & -1 & | & -2 \\ 0 & 0 & 1 & | & 2 \end{bmatrix} \xrightarrow[R_2+R_3]{R_1+4R_3} \begin{bmatrix} 1 & 0 & 0 & | & 3 \\ 0 & 1 & 0 & | & 0 \\ 0 & 0 & 1 & | & 2 \end{bmatrix}.$$

The solution is $(3,0,2)$.

45. Using the Gauss-Jordan method, we have

$$\begin{bmatrix} 1 & -2 & 1 & | & 6 \\ 2 & 1 & -3 & | & -3 \\ 1 & -3 & 3 & | & 10 \end{bmatrix} \xrightarrow[R_3-R_1]{R_2-2R_1} \begin{bmatrix} 1 & -2 & 1 & | & 6 \\ 0 & 5 & -5 & | & -15 \\ 0 & -1 & 2 & | & 4 \end{bmatrix} \xrightarrow{\frac{1}{5}R_2} \begin{bmatrix} 1 & -2 & 1 & | & 6 \\ 0 & 1 & -1 & | & -3 \\ 0 & -1 & 2 & | & 4 \end{bmatrix}$$

$$\xrightarrow[R_3+R_2]{R_1+2R_2} \begin{bmatrix} 1 & 0 & -1 & | & 0 \\ 0 & 1 & -1 & | & -3 \\ 0 & 0 & 1 & | & 1 \end{bmatrix} \xrightarrow[R_2+R_3]{R_1+R_3} \begin{bmatrix} 1 & 0 & 0 & | & 1 \\ 0 & 1 & 0 & | & -2 \\ 0 & 0 & 1 & | & 1 \end{bmatrix}.$$

Therefore, the solution is $(1,-2,1)$.

47. Using the Gauss-Jordan method, we have

$$\begin{bmatrix} 2 & 0 & 3 & | & -1 \\ 3 & -2 & 1 & | & 9 \\ 1 & 1 & 4 & | & 4 \end{bmatrix} \xrightarrow{R_1 \leftrightarrow R_3} \begin{bmatrix} 1 & 1 & 4 & | & 4 \\ 3 & -2 & 1 & | & 9 \\ 2 & 0 & 3 & | & -1 \end{bmatrix} \xrightarrow[R_3-2R_1]{R_2-3R_1} \begin{bmatrix} 1 & 1 & 4 & | & 4 \\ 0 & -5 & -11 & | & -3 \\ 0 & -2 & -5 & | & -9 \end{bmatrix}$$

$$\xrightarrow{-\frac{1}{5}R_2} \begin{bmatrix} 1 & 1 & 4 & | & 4 \\ 0 & 1 & \frac{11}{5} & | & \frac{3}{5} \\ 0 & -2 & -5 & | & -9 \end{bmatrix} \xrightarrow[R_3+2R_2]{R_1-R_2} \begin{bmatrix} 1 & 0 & \frac{9}{5} & | & \frac{17}{5} \\ 0 & 1 & \frac{11}{5} & | & \frac{3}{5} \\ 0 & 0 & -\frac{3}{5} & | & -\frac{39}{5} \end{bmatrix} \xrightarrow{-\frac{5}{3}R_3}$$

$$\begin{bmatrix} 1 & 0 & \frac{9}{5} & | & \frac{17}{5} \\ 0 & 1 & \frac{11}{5} & | & \frac{3}{5} \\ 0 & 0 & 1 & | & 13 \end{bmatrix} \xrightarrow[R_2-\frac{11}{5}R_3]{R_1-\frac{9}{5}R_3} \begin{bmatrix} 1 & 0 & 0 & | & -20 \\ 0 & 1 & 0 & | & -28 \\ 0 & 0 & 1 & | & 13 \end{bmatrix}.$$

Therefore, the solution is (−20, −28, 13).

49. Using the Gauss-Jordan method, we have

$$\begin{bmatrix} 1 & -1 & 3 & | & 14 \\ 1 & 1 & 1 & | & 6 \\ -2 & -1 & 1 & | & -4 \end{bmatrix} \xrightarrow[R_3+2R_1]{R_2-R_1} \begin{bmatrix} 1 & -1 & 3 & | & 14 \\ 0 & 2 & -2 & | & -8 \\ 0 & -3 & 7 & | & 24 \end{bmatrix} \xrightarrow{\frac{1}{2}R_2} \begin{bmatrix} 1 & -1 & 3 & | & 14 \\ 0 & 1 & -1 & | & -4 \\ 0 & -3 & 7 & | & 24 \end{bmatrix}$$

$$\xrightarrow[R_3+3R_2]{R_1+R_2} \begin{bmatrix} 1 & 0 & 2 & | & 10 \\ 0 & 1 & -1 & | & -4 \\ 0 & 0 & 4 & | & 12 \end{bmatrix} \xrightarrow{\frac{1}{4}R_3} \begin{bmatrix} 1 & 0 & 2 & | & 10 \\ 0 & 1 & -1 & | & -4 \\ 0 & 0 & 1 & | & 3 \end{bmatrix} \xrightarrow[R_2+R_3]{R_1-2R_3} \begin{bmatrix} 1 & 0 & 0 & | & 4 \\ 0 & 1 & 0 & | & -1 \\ 0 & 0 & 1 & | & 3 \end{bmatrix}$$

Therefore, the solution is (4, −1, 3).

51. We wish to solve the system of equations

$$\begin{aligned} x + y &= 500 \\ 42x + 30y &= 18,600 \end{aligned} \qquad \begin{aligned} &(x = \text{the number of acres of corn planted}) \\ &(y = \text{the number of acres of wheat planted}) \end{aligned}$$

Using the Gauss-Jordan method, we find

$$\begin{bmatrix} 1 & 1 & | & 500 \\ 42 & 30 & | & 18600 \end{bmatrix} \xrightarrow{R_2 - 42R_1} \begin{bmatrix} 1 & 1 & | & 500 \\ 0 & -12 & | & -2400 \end{bmatrix} \xrightarrow{-\frac{1}{12}R_2} \begin{bmatrix} 1 & 1 & | & 500 \\ 0 & 1 & | & 200 \end{bmatrix}$$

$$\xrightarrow{R_1 - R_2} \begin{bmatrix} 1 & 0 & | & 300 \\ 0 & 1 & | & 200 \end{bmatrix}.$$

The solution to this system of equations is $x = 300$ and $y = 200$. We conclude that Jacob should plant 300 acres of corn and 200 acres of wheat.

53. Let x denote the number of pounds of the \$2.50/lb coffee and y denote the number of pounds of the \$3.00/lb coffee. Then we are required to solve the system

$$x + \quad y = 100$$
$$2.50x + 3.00y = 280$$

Using the Gauss-Jordan method of elimination, we have

$$\begin{bmatrix} 1 & 1 & | & 100 \\ 2.5 & 3 & | & 280 \end{bmatrix} \xrightarrow{R_2 - 2.5R_1} \begin{bmatrix} 1 & 1 & | & 100 \\ 0 & 0.5 & | & 30 \end{bmatrix} \xrightarrow{2R_2} \begin{bmatrix} 1 & 1 & | & 100 \\ 0 & 1 & | & 60 \end{bmatrix}$$

$$\xrightarrow{R_1 - R_2} \begin{bmatrix} 1 & 0 & | & 40 \\ 0 & 1 & | & 60 \end{bmatrix}.$$

Therefore, 40 pounds of the \$2.50/lb coffee and 60 pounds of the \$3.00/lb coffee should be used in the 100 lb mixture.

55. Let x and y denote the number of children and adults who rode the bus during the morning shift, respectively. Then the solution to the problem can be found by solving the system of equations

$$x + \quad y = 1000$$
$$0.25x + 0.75y = \quad 650$$

Using the Gauss-Jordan elimination method, we have

$$\begin{bmatrix} 1 & 1 & | & 1000 \\ 0.25 & 0.75 & | & 650 \end{bmatrix} \xrightarrow{R_2 - 0.25R_1} \begin{bmatrix} 1 & 1 & | & 1000 \\ 0 & 0.5 & | & 400 \end{bmatrix} \xrightarrow{2R_2} \begin{bmatrix} 1 & 1 & | & 1000 \\ 0 & 1 & | & 800 \end{bmatrix}$$

$$\xrightarrow{R_1 - R_2} \begin{bmatrix} 1 & 0 & | & 200 \\ 0 & 1 & | & 800 \end{bmatrix}.$$

We conclude that 800 adults and 200 children rode the bus during the morning shift.

57. Let x, y, and z, denote the amount of money he should invest in a savings account, in mutual funds, and in bonds, respectively. Then, we are required to solve the system

$$0.06x + 0.08y + 0.12z = 21{,}600$$
$$2x \qquad - \quad z = 0$$
$$0.08y - 0.12z = 0$$

Using the Gauss-Jordan method, we find

$$\begin{bmatrix} 0.06 & 0.08 & 0.12 & | & 21{,}600 \\ 2 & 0 & -1 & | & 0 \\ 0 & 0.08 & -0.12 & | & 0 \end{bmatrix} \xrightarrow[\frac{1}{0.08}R_3]{\frac{1}{0.06}R_1} \begin{bmatrix} 1 & \frac{4}{3} & 2 & | & 360{,}000 \\ 2 & 0 & -1 & | & 0 \\ 0 & 1 & -\frac{3}{2} & | & 0 \end{bmatrix} \xrightarrow{R_2 - 2R_1}$$

$$\begin{bmatrix} 1 & \frac{4}{3} & 2 & | & 360{,}000 \\ 0 & -\frac{8}{3} & -5 & | & -720{,}000 \\ 0 & 1 & -\frac{3}{2} & | & 0 \end{bmatrix} \xrightarrow{-\frac{3}{8}R_2} \begin{bmatrix} 1 & \frac{4}{3} & 2 & | & 360{,}000 \\ 0 & 1 & \frac{15}{8} & | & 270{,}000 \\ 0 & 1 & -\frac{3}{2} & | & 0 \end{bmatrix}$$

$$\xrightarrow[R_3 - R_2]{R_1 - \frac{4}{3}R_2} \begin{bmatrix} 1 & 0 & -\frac{1}{2} & | & 0 \\ 0 & 1 & \frac{15}{8} & | & 270{,}000 \\ 0 & 0 & -\frac{27}{8} & | & -270{,}000 \end{bmatrix} \xrightarrow{-\frac{8}{27}R_3} \begin{bmatrix} 1 & 0 & -\frac{1}{2} & | & 0 \\ 0 & 1 & \frac{15}{8} & | & 270{,}000 \\ 0 & 0 & 1 & | & 80{,}000 \end{bmatrix}$$

$$\xrightarrow[R_2 - \frac{15}{8}R_3]{R_1 + \frac{1}{2}R_3} \begin{bmatrix} 1 & 0 & 0 & | & 40{,}000 \\ 0 & 1 & 0 & | & 120{,}000 \\ 0 & 0 & 1 & | & 80{,}000 \end{bmatrix}$$

Therefore, Sid should invest \$40,000 in a savings account, \$120,000 in mutual funds, and \$80,000 in bonds.

59. Refer to Exercise 23, page 44. We obtain the following augmented matrices.

$$\begin{bmatrix} 18 & 20 & 24 & | & 26400 \\ 4 & 4 & 3 & | & 4900 \\ 5 & 4 & 6 & | & 6200 \end{bmatrix} \xrightarrow{R_1 \leftrightarrow R_3} \begin{bmatrix} 5 & 4 & 6 & | & 6200 \\ 4 & 4 & 3 & | & 4900 \\ 18 & 20 & 24 & | & 26400 \end{bmatrix} \xrightarrow{R_1 - R_2}$$

$$\begin{bmatrix} 1 & 0 & 3 & | & 1300 \\ 4 & 4 & 3 & | & 4900 \\ 18 & 20 & 24 & | & 26400 \end{bmatrix} \xrightarrow[R_3-18R_1]{R_2-4R_1} \begin{bmatrix} 1 & 0 & 3 & | & 1300 \\ 0 & 4 & -9 & | & -300 \\ 0 & 20 & -30 & | & 3000 \end{bmatrix} \xrightarrow{\frac{1}{4}R_2}$$

$$\begin{bmatrix} 1 & 0 & 3 & | & 1300 \\ 0 & 1 & -\frac{9}{4} & | & -75 \\ 0 & 20 & -30 & | & 3000 \end{bmatrix} \xrightarrow{R_3-20R_2} \begin{bmatrix} 1 & 0 & 3 & | & 1300 \\ 0 & 1 & -\frac{9}{4} & | & -75 \\ 0 & 0 & 15 & | & 4500 \end{bmatrix} \xrightarrow{\frac{1}{15}R_3}$$

$$\begin{bmatrix} 1 & 0 & 3 & | & 1300 \\ 0 & 1 & -\frac{9}{4} & | & -75 \\ 0 & 0 & 1 & | & 300 \end{bmatrix} \xrightarrow[R_2+\frac{9}{4}R_3]{R_1-3R_3} \begin{bmatrix} 1 & 0 & 0 & | & 400 \\ 0 & 1 & 0 & | & 600 \\ 0 & 0 & 1 & | & 300 \end{bmatrix}$$

We see that $x = 400$, $y = 600$, and $z = 300$. Therefore, Lawnco should produce 400, 600, and 300 100-lb bags of grade-A, grade-B, and grade-C fertilizer.

61. Let x, y, and z denote the number of compact, intermediate, and full-size cars, respectively, to be purchased. Then the problem can be solved by solving the system

$$10000x + 15,000y + 20,000z = 1,250,000$$
$$x - 2y \qquad = \qquad 0$$
$$x + y + z = \qquad 100$$

Using the Gauss-Jordan method, we have

$$\begin{bmatrix} 10,000 & 15,000 & 20,000 & | & 1,250,000 \\ 1 & -2 & 0 & | & 0 \\ 1 & 1 & 1 & | & 100 \end{bmatrix} \xrightarrow{R_1 \leftrightarrow R_3} \begin{bmatrix} 1 & 1 & 1 & | & 100 \\ 1 & -2 & 0 & | & 0 \\ 10000 & 15000 & 20000 & | & 1,250,000 \end{bmatrix}$$

$$\xrightarrow[R_3-10,000R_1]{R_2-R_1} \begin{bmatrix} 1 & 1 & 1 & | & 100 \\ 0 & -3 & -1 & | & -100 \\ 0 & 5000 & 10000 & | & 250,000 \end{bmatrix} \xrightarrow{-\frac{1}{3}R_2} \begin{bmatrix} 1 & 1 & 1 & | & 100 \\ 0 & 1 & \frac{1}{3} & | & \frac{100}{3} \\ 0 & 5000 & 10000 & | & 250,000 \end{bmatrix}$$

$$\xrightarrow[R_3-5000R_2]{R_1-R_2} \begin{bmatrix} 1 & 0 & \frac{2}{3} & | & \frac{200}{3} \\ 0 & 1 & \frac{1}{3} & | & \frac{100}{3} \\ 0 & 0 & \frac{25,000}{3} & | & \frac{250,000}{3} \end{bmatrix} \xrightarrow{\frac{3}{25,000}R_3} \begin{bmatrix} 1 & 0 & \frac{2}{3} & | & \frac{200}{3} \\ 0 & 1 & \frac{1}{3} & | & \frac{100}{3} \\ 0 & 0 & 1 & | & 10 \end{bmatrix} \xrightarrow[R_2-\frac{1}{3}R_3]{R_1-\frac{2}{3}R_3} \begin{bmatrix} 1 & 0 & 0 & | & 60 \\ 0 & 1 & 0 & | & 30 \\ 0 & 0 & 1 & | & 10 \end{bmatrix}.$$

We conclude that 60 compact cars, 30 intermediate-size cars, and 10 full-size cars will be purchased.

63. Let x, y, and z, represent the number of ounces of Food I, Food II, and Food III used in the meal, respectively. Then the problem reduces to solving the following system of linear equations:

$$10x + 6y + 8z = 100$$
$$10x + 12y + 6z = 100$$
$$5x + 4y + 12z = 100.$$

Using the Gauss-Jordan method, we obtain

$$\begin{bmatrix} 10 & 6 & 8 & | & 100 \\ 10 & 12 & 6 & | & 100 \\ 5 & 4 & 12 & | & 100 \end{bmatrix} \xrightarrow{\frac{1}{10}R_1} \begin{bmatrix} 1 & \frac{3}{5} & \frac{4}{5} & | & 10 \\ 10 & 12 & 6 & | & 100 \\ 5 & 4 & 12 & | & 100 \end{bmatrix} \xrightarrow[R_3-5R_1]{R_2-10R_1}$$

$$\begin{bmatrix} 1 & \frac{3}{5} & \frac{4}{5} & | & 10 \\ 0 & 6 & -2 & | & 0 \\ 0 & 1 & 8 & | & 50 \end{bmatrix} \xrightarrow{\frac{1}{6}R_2} \begin{bmatrix} 1 & \frac{3}{5} & \frac{4}{5} & | & 10 \\ 0 & 1 & -\frac{1}{3} & | & 0 \\ 0 & 1 & 8 & | & 50 \end{bmatrix} \xrightarrow[R_3-R_2]{R_1-\frac{3}{5}R_2}$$

$$\begin{bmatrix} 1 & 0 & 1 & | & 10 \\ 0 & 1 & -\frac{1}{3} & | & 0 \\ 0 & 0 & \frac{25}{3} & | & 50 \end{bmatrix} \xrightarrow{\frac{3}{25}R_3} \begin{bmatrix} 1 & 0 & 1 & | & 10 \\ 0 & 1 & -\frac{1}{3} & | & 0 \\ 0 & 0 & 1 & | & 6 \end{bmatrix} \xrightarrow[R_2+\frac{1}{3}R_3]{R_1-R_3} \begin{bmatrix} 1 & 0 & 0 & | & 4 \\ 0 & 1 & 0 & | & 2 \\ 0 & 0 & 1 & | & 6 \end{bmatrix}.$$

We conclude that 4 oz of Food I, 2 oz of Food II, and 6 oz of Food III should be used to prepare the meal.

65. Let $x =$ the number of front orchestra seats sold
            $y =$ the number of rear orchestra seats sold
   and $z =$ the number of front balcony seats sold for this performance.
Then, we are required to solve the system

$$x + y + z = 1{,}000$$
$$80x + 60y + 50z = 62{,}800$$
$$x + y - 2z = 400.$$

Using the Gauss-Jordan method, we find

$$\begin{bmatrix} 1 & 1 & 1 & \bigm| & 1{,}000 \\ 80 & 60 & 50 & \bigm| & 62{,}800 \\ 1 & 1 & -2 & \bigm| & 400 \end{bmatrix} \xrightarrow[R_3-R_1]{R_2-80R_1} \begin{bmatrix} 1 & 1 & 1 & \bigm| & 1{,}000 \\ 0 & -20 & -30 & \bigm| & -17{,}200 \\ 0 & 0 & -3 & \bigm| & -600 \end{bmatrix} \xrightarrow[-\frac{1}{3}R_3]{-\frac{1}{20}R_2}$$

$$\begin{bmatrix} 1 & 1 & 1 & \bigm| & 1{,}000 \\ 0 & 1 & \frac{3}{2} & \bigm| & 860 \\ 0 & 0 & 1 & \bigm| & 200 \end{bmatrix} \xrightarrow{R_1-R_2} \begin{bmatrix} 1 & 0 & -\frac{1}{2} & \bigm| & 140 \\ 0 & 1 & \frac{3}{2} & \bigm| & 860 \\ 0 & 0 & 1 & \bigm| & 200 \end{bmatrix} \xrightarrow[R_2-\frac{3}{2}R_3]{R_1+\frac{1}{2}R_3}$$

$$\begin{bmatrix} 1 & 0 & 0 & \bigm| & 240 \\ 0 & 1 & 0 & \bigm| & 560 \\ 0 & 0 & 1 & \bigm| & 200 \end{bmatrix}.$$

We conclude that tickets for 240 front orchestra seats, 560 rear orchestra seats, and 200 front balcony seats were sold.

67. Let x, y, and z denote the number of days he spent in London, Paris, and Rome, respectively. We have

$$180x + 230y + 160z = 2660$$
$$110x + 120y + 90z = 1520$$
$$x - y - z = 0 \qquad \text{(since } x = y + z\text{)}$$

Using the Gauss-Jordan method to solve the system, we have

$$\begin{bmatrix} 180 & 230 & 160 & \bigm| & 2660 \\ 110 & 120 & 90 & \bigm| & 1520 \\ 1 & -1 & -1 & \bigm| & 0 \end{bmatrix} \xrightarrow{R_1 \leftrightarrow R_3} \begin{bmatrix} 1 & -1 & -1 & \bigm| & 0 \\ 110 & 120 & 90 & \bigm| & 1520 \\ 180 & 230 & 160 & \bigm| & 2660 \end{bmatrix} \xrightarrow[R_3-180R_1]{R_2-110R_1}$$

$$\begin{bmatrix} 1 & -1 & -1 & \bigm| & 0 \\ 0 & 230 & 200 & \bigm| & 1520 \\ 0 & 410 & 340 & \bigm| & 2660 \end{bmatrix} \xrightarrow{\frac{1}{230}R_2} \begin{bmatrix} 1 & -1 & -1 & \bigm| & 0 \\ 0 & 1 & \frac{20}{23} & \bigm| & \frac{152}{23} \\ 0 & 410 & 340 & \bigm| & 2660 \end{bmatrix} \xrightarrow[R_3-410R_2]{R_1+R_2}$$

$$\begin{bmatrix} 1 & 0 & -\frac{3}{23} & \bigm| & \frac{152}{23} \\ 0 & 1 & \frac{20}{23} & \bigm| & \frac{152}{23} \\ 0 & 0 & -\frac{380}{23} & \bigm| & -\frac{1140}{23} \end{bmatrix} \xrightarrow{-\frac{23}{380}R_3} \begin{bmatrix} 1 & 0 & -\frac{3}{23} & \bigm| & \frac{152}{23} \\ 0 & 1 & \frac{20}{23} & \bigm| & \frac{152}{23} \\ 0 & 0 & 1 & \bigm| & 3 \end{bmatrix} \xrightarrow[R_2-\frac{20}{23}R_1]{R_1+\frac{3}{23}R_3}$$

$$\begin{bmatrix} 1 & 0 & 0 & | & 7 \\ 0 & 1 & 0 & | & 4 \\ 0 & 0 & 1 & | & 3 \end{bmatrix}$$

The solution is $x = 7$, $y = 4$, and $z = 3$. Therefore, he spent 7 days in London, 4 days in Paris, and 3 days in Rome.

69. False. The constant cannot be zero. The system

$$2x + y = 1$$
$$3x - y = 2$$

is not equivalent to

$$2x + y = 1$$
$$0(3x - y) = 0(2)$$

or

$$2x + y = 1$$
$$0 = 0$$.

USING TECHNOLOGY EXERCISES 2.2, page 96

1. $(3,1,-1,2)$ 3. $(5,4,-3,-4)$ 5. $(1,-1,2,0,3)$

EXERCISES 2.3, page 105

1. a. The system has one solution.   b. The solution is $(3, -1, 2)$.

3. a. The system has one solution.   b. The solution is $(2, 4)$.

5. a. The system has infinitely many solutions.
   b. Letting $x_3 = t$, we see that the solutions are given by $(4 - t, -2, t)$, where t is a parameter.

7. a. The system has no solution. The last row contains all zeros to the left of the vertical line and a nonzero number (1) to its right.

9. a. The system has infinitely many solutions.
   b. Letting $x_4 = t$, we see that the solutions are given by $(2, -1, 2 - t, t)$, where t is a parameter.

11. a. The system has infinitely many solutions.

b. Letting $x_3 = s$ and $x_4 = t$, the solutions are given by $(2 - 3s, 1 + s, s, t)$, where s and t are parameters.

13. Using the Gauss-Jordan method, we have

$$\begin{bmatrix} 2 & -1 & 3 \\ 1 & 2 & 4 \\ 2 & 3 & 7 \end{bmatrix} \xrightarrow{R_1 \leftrightarrow R_2} \begin{bmatrix} 1 & 2 & 4 \\ 2 & -1 & 3 \\ 2 & 3 & 7 \end{bmatrix} \xrightarrow[R_3 - 2R_1]{R_2 - 2R_1} \begin{bmatrix} 1 & 2 & 4 \\ 0 & -5 & -5 \\ 0 & -1 & -1 \end{bmatrix} \xrightarrow{-\frac{1}{5}R_2}$$

$$\begin{bmatrix} 1 & 2 & 4 \\ 0 & 1 & 1 \\ 0 & -1 & -1 \end{bmatrix} \xrightarrow[R_3 + R_2]{R_1 - 2R_2} \begin{bmatrix} 1 & 0 & 2 \\ 0 & 1 & 1 \\ 0 & 0 & 0 \end{bmatrix}. \qquad \text{The solution is } (2,1).$$

15. Using the Gauss-Jordan method, we have

$$\begin{bmatrix} 3 & -2 & -3 \\ 2 & 1 & 3 \\ 1 & -2 & -5 \end{bmatrix} \xrightarrow{R_1 \leftrightarrow R_3} \begin{bmatrix} 1 & -2 & -5 \\ 2 & 1 & 3 \\ 3 & -2 & -3 \end{bmatrix} \xrightarrow[R_3 - 3R_1]{R_2 - 2R_1} \begin{bmatrix} 1 & -2 & -5 \\ 0 & 5 & 13 \\ 0 & 4 & 12 \end{bmatrix} \xrightarrow{\frac{1}{5}R_2}$$

$$\begin{bmatrix} 1 & -2 & -5 \\ 0 & 1 & \frac{13}{5} \\ 0 & 4 & 12 \end{bmatrix} \xrightarrow[R_3 - 4R_2]{R_1 + 2R_2} \begin{bmatrix} 1 & 0 & \frac{1}{5} \\ 0 & 1 & \frac{13}{5} \\ 0 & 0 & \frac{8}{5} \end{bmatrix}.$$

Since the last row implies the $0 = 8/5$, we conclude that the system of equations is inconsistent and has no solution.

17. $$\begin{bmatrix} 3 & -2 & 5 \\ -1 & 3 & -4 \\ 2 & -4 & 6 \end{bmatrix} \xrightarrow{R_1 \leftrightarrow R_2} \begin{bmatrix} -1 & 3 & -4 \\ 3 & -2 & 5 \\ 2 & -4 & 6 \end{bmatrix} \xrightarrow{-R_1} \begin{bmatrix} 1 & -3 & 4 \\ 3 & -2 & 5 \\ 2 & -4 & 6 \end{bmatrix} \xrightarrow[R_3 - 2R_1]{R_2 - 3R_1}$$

$$\begin{bmatrix} 1 & -3 & 4 \\ 0 & 7 & -7 \\ 0 & 2 & -2 \end{bmatrix} \xrightarrow{\frac{1}{7}R_2} \begin{bmatrix} 1 & -3 & 4 \\ 0 & 1 & -1 \\ 0 & 2 & -2 \end{bmatrix} \xrightarrow[R_3 - 2R_2]{R_1 + 3R_2} \begin{bmatrix} 1 & 0 & 1 \\ 0 & 1 & -1 \\ 0 & 0 & 0 \end{bmatrix}.$$

We conclude that the solution is $(1,-1)$.

19. $\begin{bmatrix} 1 & -2 & | & 2 \\ 7 & -14 & | & 14 \\ 3 & -6 & | & 6 \end{bmatrix} \xrightarrow[R_3-3R_1]{R_2-7R_1} \begin{bmatrix} 1 & -2 & | & 2 \\ 0 & 0 & | & 0 \\ 0 & 0 & | & 0 \end{bmatrix}.$

We conclude that the infinitely many solutions are given by $(2t+2, t)$, where t is a parameter.

21. $\begin{bmatrix} 3 & 2 & | & 4 \\ -\frac{3}{2} & -1 & | & -2 \\ 6 & 4 & | & 8 \end{bmatrix} \xrightarrow{\frac{1}{3}R_1} \begin{bmatrix} 1 & \frac{2}{3} & | & \frac{4}{3} \\ -\frac{3}{2} & -1 & | & -2 \\ 6 & 4 & | & 8 \end{bmatrix} \xrightarrow[R_3-6R_1]{R_2+\frac{3}{2}R_1} \begin{bmatrix} 1 & \frac{2}{3} & | & \frac{4}{3} \\ 0 & 0 & | & 0 \\ 0 & 0 & | & 0 \end{bmatrix}.$

We conclude that the infinitely many solutions are given by $(\frac{4}{3}-\frac{2}{3}t, t)$, where t is a parameter.

23. $\begin{bmatrix} 2 & -1 & 1 & | & -4 \\ 3 & -\frac{3}{2} & \frac{3}{2} & | & -6 \\ -6 & 3 & -3 & | & 12 \end{bmatrix} \xrightarrow{\frac{1}{2}R_1} \begin{bmatrix} 1 & -\frac{1}{2} & \frac{1}{2} & | & -2 \\ 3 & -\frac{3}{2} & \frac{3}{2} & | & -6 \\ -6 & 3 & -3 & | & 12 \end{bmatrix} \xrightarrow[R_3+6R_1]{R_2-3R_1} \begin{bmatrix} 1 & -\frac{1}{2} & \frac{1}{2} & | & -2 \\ 0 & 0 & 0 & | & 0 \\ 0 & 0 & 0 & | & 0 \end{bmatrix}.$

We conclude that the infinitely many solutions are given by $(-2+\frac{1}{2}s-\frac{1}{2}t, s, t)$ where s and t are parameters.

25. $\begin{bmatrix} 1 & -2 & 3 & | & 4 \\ 2 & 3 & -1 & | & 2 \\ 1 & 2 & -3 & | & -6 \end{bmatrix} \xrightarrow[R_3-R_1]{R_2-2R_1} \begin{bmatrix} 1 & -2 & 3 & | & 4 \\ 0 & 7 & -7 & | & -6 \\ 0 & 4 & -6 & | & -10 \end{bmatrix} \xrightarrow{\frac{1}{7}R_2} \begin{bmatrix} 1 & -2 & 3 & | & 4 \\ 0 & 1 & -1 & | & -\frac{6}{7} \\ 0 & 4 & -6 & | & -10 \end{bmatrix}$

$\xrightarrow[R_3-4R_2]{R_1+2R_2} \begin{bmatrix} 1 & 0 & 1 & | & \frac{16}{7} \\ 0 & 1 & -1 & | & -\frac{6}{7} \\ 0 & 0 & -2 & | & -\frac{46}{7} \end{bmatrix} \xrightarrow{-\frac{1}{2}R_3} \begin{bmatrix} 1 & 0 & 1 & | & \frac{16}{7} \\ 0 & 1 & -1 & | & -\frac{6}{7} \\ 0 & 0 & 1 & | & \frac{23}{7} \end{bmatrix} \xrightarrow[R_2+R_3]{R_1-R_3}$

$\begin{bmatrix} 1 & 0 & 0 & | & -1 \\ 0 & 1 & 0 & | & \frac{17}{7} \\ 0 & 0 & 1 & | & \frac{23}{7} \end{bmatrix}.$

We conclude that the solution is $(-1, \frac{17}{7}, \frac{23}{7})$.

27. $\begin{bmatrix} 4 & 1 & -1 & | & 4 \\ 8 & 2 & -2 & | & 8 \end{bmatrix} \xrightarrow{\frac{1}{4}R_1} \begin{bmatrix} 1 & \frac{1}{4} & \frac{1}{4} & | & 1 \\ 8 & 2 & -2 & | & 8 \end{bmatrix} \xrightarrow{R_2-8R_1} \begin{bmatrix} 1 & \frac{1}{4} & -\frac{1}{4} & | & 1 \\ 0 & 0 & 0 & | & 0 \end{bmatrix}$

We conclude that the infinitely many solutions are given by $(1-\frac{1}{4}s+\frac{1}{4}t, s, t)$, where s and t are parameters.

29. $\begin{bmatrix} 2 & 1 & -3 & | & 1 \\ 1 & -1 & 2 & | & 1 \\ 5 & -2 & 3 & | & 6 \end{bmatrix} \xrightarrow{R_1 \leftrightarrow R_2} \begin{bmatrix} 1 & -1 & 2 & | & 1 \\ 2 & 1 & -3 & | & 1 \\ 5 & -2 & 3 & | & 6 \end{bmatrix} \xrightarrow[R_3-5R_1]{R_2-2R_1} \begin{bmatrix} 1 & -1 & 2 & | & 1 \\ 0 & 3 & -7 & | & -1 \\ 0 & 3 & -7 & | & 1 \end{bmatrix} \xrightarrow{\frac{1}{3}R_2}$

$\begin{bmatrix} 1 & -1 & 2 & | & 1 \\ 0 & 1 & -\frac{7}{3} & | & -\frac{1}{3} \\ 0 & 3 & -7 & | & 1 \end{bmatrix} \xrightarrow[R_3-3R_2]{R_1+R_2} \begin{bmatrix} 1 & 0 & -\frac{1}{3} & | & \frac{2}{3} \\ 0 & 1 & -\frac{7}{3} & | & -\frac{1}{3} \\ 0 & 0 & 0 & | & 2 \end{bmatrix}.$

This last row implies that $0 = 2$, which is impossible. We conclude that the system of equations is inconsistent and has no solution.

31. $\begin{bmatrix} 1 & 2 & -1 & | & -4 \\ 2 & 1 & 1 & | & 7 \\ 1 & 3 & 2 & | & 7 \\ 1 & -3 & 1 & | & 9 \end{bmatrix} \xrightarrow[\substack{R_3-R_1 \\ R_4-R_1}]{R_2-2R_1} \begin{bmatrix} 1 & 2 & -1 & | & -4 \\ 0 & -3 & 3 & | & 15 \\ 0 & 1 & 3 & | & 11 \\ 0 & -5 & 2 & | & 13 \end{bmatrix} \xrightarrow{-\frac{1}{3}R_2} \begin{bmatrix} 1 & 2 & -1 & | & -4 \\ 0 & 1 & -1 & | & -5 \\ 0 & 1 & 3 & | & 11 \\ 0 & -5 & 2 & | & 13 \end{bmatrix}$

$\xrightarrow[\substack{R_3-R_2 \\ R_4+5R_2}]{R_1-2R_2} \begin{bmatrix} 1 & 0 & 1 & | & 6 \\ 0 & 1 & -1 & | & -5 \\ 0 & 0 & 4 & | & 16 \\ 0 & 0 & -3 & | & -12 \end{bmatrix} \xrightarrow{\frac{1}{4}R_3} \begin{bmatrix} 1 & 0 & 1 & | & 6 \\ 0 & 1 & -1 & | & -5 \\ 0 & 0 & 1 & | & 4 \\ 0 & 0 & -3 & | & -12 \end{bmatrix} \xrightarrow[\substack{R_2+R_3 \\ R_4+3R_3}]{R_1-R_3} \begin{bmatrix} 1 & 0 & 0 & | & 2 \\ 0 & 1 & 0 & | & -1 \\ 0 & 0 & 1 & | & 4 \\ 0 & 0 & 0 & | & 0 \end{bmatrix}.$

We conclude that the solution of the system is $(2,-1,4)$.

33. Let x, y, and z represent the number of compact, mid-sized, and full-size cars, respectively, to be purchased. Then the problem can be solved by solving the system

$$x + \quad y + \quad z = \quad 60$$
$$10000x + 16000y + 22000z = 840000 .$$

Using the Gauss-Jordan method, we have

$$\begin{bmatrix} 1 & 1 & 1 & | & 60 \\ 10000 & 16000 & 22000 & | & 840000 \end{bmatrix} \xrightarrow{R_2-10,000R_1} \begin{bmatrix} 1 & 1 & 1 & | & 60 \\ 0 & 6000 & 12000 & | & 240000 \end{bmatrix}$$

$$\xrightarrow{\frac{1}{6000}R_2} \begin{bmatrix} 1 & 1 & 1 & | & 60 \\ 0 & 1 & 2 & | & 40 \end{bmatrix} \xrightarrow{R_1-R_2} \begin{bmatrix} 1 & 0 & -1 & | & 20 \\ 0 & 1 & 2 & | & 40 \end{bmatrix}$$

and we conclude that the solution is $(20 + z, 40 - 2z, z)$. Letting $z = 5$, we see that one possible solution is $(25,30,5)$; that is Hartman should buy 25 compact, 30 mid-sized cars, and 5 full-sized cars. Letting $z = 10$, we see that another possible solution is $(30,20,10)$; that is, 30 compact cars, 20 mid-sized cars, and 10 full-sized cars.

35. Let x, y, and z denote the number of ounces of Food I, Food II, and Food III, respectively, that the dietician includes in the meal. Then the problem can be solved by solving the system

$$400x + 1200y + 800z = 8800$$
$$110x + 570y + 340z = 2160$$
$$90x + 30y + 60z = 1020.$$

Using the Gauss-Jordan method, we have

$$\begin{bmatrix} 400 & 1200 & 800 & | & 8800 \\ 110 & 570 & 340 & | & 2160 \\ 90 & 30 & 60 & | & 1020 \end{bmatrix} \xrightarrow{\frac{1}{400}R_1} \begin{bmatrix} 1 & 3 & 2 & | & 22 \\ 110 & 570 & 340 & | & 2160 \\ 90 & 30 & 60 & | & 1020 \end{bmatrix} \xrightarrow[R_3-90R_1]{R_2-110R_1}$$

$$\begin{bmatrix} 1 & 3 & 2 & | & 22 \\ 0 & 240 & 120 & | & -260 \\ 0 & -240 & -120 & | & -960 \end{bmatrix} \xrightarrow{\frac{1}{240}R_2} \begin{bmatrix} 1 & 3 & 2 & | & 22 \\ 0 & 1 & \frac{1}{2} & | & -\frac{13}{12} \\ 0 & -240 & -120 & | & -960 \end{bmatrix} \xrightarrow[R_3+240R_2]{R_1-3R_2}$$

$$\begin{bmatrix} 1 & 0 & \frac{1}{2} & | & \frac{101}{4} \\ 0 & 1 & \frac{1}{2} & | & -\frac{13}{12} \\ 0 & 0 & 0 & | & -1220 \end{bmatrix}.$$

This last row implies that $0 = -1220$, which is impossible. We conclude that the system of equations is inconsistent and has no solution--that is, the dietician cannot prepare a meal from these foods and meet the given requirements.

37. a.
$$\begin{aligned}
x_1 - x_2 \qquad\qquad\qquad &= 200 \\
x_1 \qquad\qquad -x_5 \quad\;\; &= 100 \\
-x_2 + x_3 \qquad\quad +x_6 &= 600 \\
-x_3 + x_4 \qquad\qquad &= 200 \\
x_4 - x_5 + x_6 &= 700.
\end{aligned}$$

b.

$$\begin{bmatrix}
1 & -1 & 0 & 0 & 0 & 0 & 200 \\
1 & 0 & 0 & 0 & -1 & 0 & 100 \\
0 & -1 & 1 & 0 & 0 & 1 & 600 \\
0 & 0 & -1 & 1 & 0 & 0 & 200 \\
0 & 0 & 0 & 1 & -1 & 1 & 700
\end{bmatrix}
\xrightarrow{R_2 - R_1}
\begin{bmatrix}
1 & -1 & 0 & 0 & 0 & 0 & 200 \\
0 & 1 & 0 & 0 & -1 & 0 & -100 \\
0 & -1 & 1 & 0 & 0 & 1 & 600 \\
0 & 0 & -1 & 1 & 0 & 0 & 200 \\
0 & 0 & 0 & 1 & -1 & 1 & 700
\end{bmatrix}
\xrightarrow[R_1 + R_2]{R_3 + R_2}$$

$$\begin{bmatrix}
1 & 0 & 0 & 0 & -1 & 0 & 100 \\
0 & 1 & 0 & 0 & -1 & 0 & -100 \\
0 & 0 & 1 & 0 & -1 & 1 & 500 \\
0 & 0 & -1 & 1 & 0 & 0 & 200 \\
0 & 0 & 0 & 1 & -1 & 1 & 700
\end{bmatrix}
\xrightarrow{R_4 + R_3}
\begin{bmatrix}
1 & 0 & 0 & 0 & -1 & 0 & 100 \\
0 & 1 & 0 & 0 & -1 & 0 & -100 \\
0 & 0 & 1 & 0 & -1 & 1 & 500 \\
0 & 0 & 0 & 1 & -1 & 1 & 700 \\
0 & 0 & 0 & 1 & -1 & 1 & 700
\end{bmatrix}
\xrightarrow{R_5 - R_4}$$

$$\begin{bmatrix}
1 & 0 & 0 & 0 & -1 & 0 & 100 \\
0 & 1 & 0 & 0 & 1 & 0 & -100 \\
0 & 0 & 1 & 0 & -1 & 1 & 500 \\
0 & 0 & 0 & 1 & -1 & 1 & 700 \\
0 & 0 & 0 & 0 & 0 & 0 & 0
\end{bmatrix}.$$

We conclude that the solution is
$$(s+100,\; s-100,\; s-t+500,\; s-t+700,\; s,\; t).$$
Taking $s = 150$ and $t = 50$, we see that one possible traffic pattern is
$$(250, 50, 600, 800, 150, 50).$$
Similarly, taking $s = 200$, and $t = 100$, we see that another possible traffic pattern is
$$(300, 100, 600, 800, 200, 100).$$
c. Taking $t = 0$ and $s = 200$, we see that another possible traffic pattern is
$$(300, 100, 700, 900, 200, 0).$$

39. We solve the given system by using the Gauss-Jordan method. We have

$$\begin{bmatrix} 2 & 3 & | & 2 \\ 1 & 4 & | & 6 \\ 5 & k & | & 2 \end{bmatrix} \xrightarrow{R_1 \leftrightarrow R_2} \begin{bmatrix} 1 & 4 & | & 6 \\ 2 & 3 & | & 2 \\ 5 & k & | & 2 \end{bmatrix} \xrightarrow[R_3-5R_1]{R_2-2R_1} \begin{bmatrix} 1 & 4 & | & 6 \\ 0 & -5 & | & -10 \\ 0 & k-20 & | & -28 \end{bmatrix} \xrightarrow{-\frac{1}{5}R_2}$$

$$\begin{bmatrix} 1 & 4 & | & 6 \\ 0 & 1 & | & 2 \\ 0 & k-20 & | & -28 \end{bmatrix} \xrightarrow[R_3+aR_2]{R_1-4R_2} \begin{bmatrix} 1 & 0 & | & -2 \\ 0 & 1 & | & 2 \\ 0 & k+a-20 & | & -28+2a \end{bmatrix}$$

From the last matrix, we see that the system has a solution if and only if $x=-2$, $y=2$, and

$$-28+2a=0, \text{ or } a=14$$

and $\quad k+a-20=k-6=0, \text{ or } k=6.$

(All the entries in the last row of the matrix must be equal to zero.)

41. False. Such a system cannot have a unique solution.

USING TECHNOLOGY EXERCISES 2.3, page 109

1. $(1+t, 2+t, t)$; t, a parameter   3. $\left(-\frac{17}{7}+\frac{6}{7}t, 3-t, -\frac{18}{7}+\frac{1}{7}t, t\right)$ 5. No solution

EXERCISES 2.4, page 117

1. The size of A is 4×4; the size of B is 4×3; the size of C is 1×5, and the size of D is 4×1.

3. These are entries of the matrix B. The entry b_{13} refers to the entry in the first row and third column and is equal to 2. Similarly, $b_{31}=3$, and $b_{43}=8$.

5. The column matrix is the matrix D. The transpose of the matrix D is
$$D^T = [1 \ 3 \ -2 \ 0].$$

7. A is of size 3×2; B is of size 3×2; C and D are of size 3×3.

9.
$$A + B = \begin{bmatrix} -1 & 2 \\ 3 & -2 \\ 4 & 0 \end{bmatrix} + \begin{bmatrix} 2 & 4 \\ 3 & 1 \\ -2 & 2 \end{bmatrix} = \begin{bmatrix} 1 & 6 \\ 6 & -1 \\ 2 & 2 \end{bmatrix}.$$

11.
$$\begin{bmatrix} 3 & -1 & 0 \\ 2 & -2 & 3 \\ 4 & 6 & 2 \end{bmatrix} - \begin{bmatrix} 2 & -2 & 4 \\ 3 & 6 & 2 \\ -2 & 3 & 1 \end{bmatrix} = \begin{bmatrix} 1 & 1 & -4 \\ -1 & -8 & 1 \\ 6 & 3 & 1 \end{bmatrix}.$$

13.
$$\begin{bmatrix} 6 & 3 & 8 \\ 4 & 5 & 6 \end{bmatrix} - \begin{bmatrix} 3 & -2 & -1 \\ 0 & -5 & -7 \end{bmatrix} = \begin{bmatrix} 3 & 5 & 9 \\ 4 & 10 & 13 \end{bmatrix}.$$

15.
$$\begin{bmatrix} 1 & 4 & -5 \\ 3 & -8 & 6 \end{bmatrix} + \begin{bmatrix} 4 & 0 & -2 \\ 3 & 6 & 5 \end{bmatrix} - \begin{bmatrix} 2 & 8 & 9 \\ -11 & 2 & -5 \end{bmatrix} = \begin{bmatrix} 3 & -4 & -16 \\ 17 & -4 & 16 \end{bmatrix}.$$

17.
$$\begin{bmatrix} 1.2 & 4.5 & -4.2 \\ 8.2 & 6.3 & -3.2 \end{bmatrix} - \begin{bmatrix} 3.1 & 1.5 & -3.6 \\ 2.2 & -3.3 & -4.4 \end{bmatrix} = \begin{bmatrix} -1.9 & 3.0 & -0.6 \\ 6.0 & 9.6 & 1.2 \end{bmatrix}.$$

19.
$$\frac{1}{2}\begin{bmatrix} 1 & 0 & 0 & -4 \\ 3 & 0 & -1 & 6 \\ -2 & 1 & -4 & 2 \end{bmatrix} + \frac{4}{3}\begin{bmatrix} 3 & 0 & -1 & 4 \\ -2 & 1 & -6 & 2 \\ 8 & 2 & 0 & -2 \end{bmatrix} - \frac{1}{3}\begin{bmatrix} 3 & -9 & -1 & 0 \\ 6 & 2 & 0 & -6 \\ 0 & 1 & -3 & 1 \end{bmatrix}$$

$$= \begin{bmatrix} \frac{7}{2} & 3 & -1 & \frac{10}{3} \\ -\frac{19}{6} & \frac{2}{3} & -\frac{17}{2} & \frac{23}{3} \\ \frac{29}{3} & \frac{17}{6} & -1 & -2 \end{bmatrix}.$$

21.
$$\begin{bmatrix} 2x-2 & 3 & 2 \\ 2 & 4 & y-2 \\ 2z & -3 & 2 \end{bmatrix} = \begin{bmatrix} 3 & u & 2 \\ 2 & 4 & 5 \\ 4 & -3 & 2 \end{bmatrix}.$$

Now, by the definition of equality of matrices,

$$u = 3$$
$$2x - 2 = 3 \text{ and } 2x = 5, \text{ or } x = 5/2,$$
$$y - 2 = 5, \text{ and } y = 7,$$
$$2z = 4, \text{ and } z = 2.$$

23. $\begin{bmatrix} 1 & x \\ 2y & -3 \end{bmatrix} - 4\begin{bmatrix} 2 & -2 \\ 0 & 3 \end{bmatrix} = \begin{bmatrix} 3z & 10 \\ 4 & -u \end{bmatrix}; \begin{bmatrix} -7 & x+8 \\ 2y & -15 \end{bmatrix} = \begin{bmatrix} 3z & 10 \\ 4 & -u \end{bmatrix}.$

Now, by the definition of equality of matrices,
$$-u = -15, \text{ so } u = 15$$
$$x + 8 = 10, \text{ so } x = 2$$
$$2y = 4, \text{ so } y = 2$$
$$3z = -7, \text{ so } z = -7/3.$$

25. To verify the Commutative Law for matrix addition, let us show that $A + B = B + A$.

Now, $A + B = \begin{bmatrix} 2 & -4 & 3 \\ 4 & 2 & 1 \end{bmatrix} + \begin{bmatrix} 4 & -3 & 2 \\ 1 & 0 & 4 \end{bmatrix} = \begin{bmatrix} 6 & -7 & 5 \\ 5 & 2 & 5 \end{bmatrix}$

$= \begin{bmatrix} 4 & -3 & 2 \\ 1 & 0 & 4 \end{bmatrix} + \begin{bmatrix} 2 & -4 & 3 \\ 4 & 2 & 1 \end{bmatrix} = B + A.$

27. $(3+5)A = 8A = 8\begin{bmatrix} 3 & 1 \\ 2 & 4 \\ -4 & 0 \end{bmatrix} = \begin{bmatrix} 24 & 8 \\ 16 & 32 \\ -32 & 0 \end{bmatrix} = 3\begin{bmatrix} 3 & 1 \\ 2 & 4 \\ -4 & 0 \end{bmatrix} + 5\begin{bmatrix} 3 & 1 \\ 2 & 4 \\ -4 & 0 \end{bmatrix}$

$= 3A + 5A.$

29. $4(A + B) = 4\left(\begin{bmatrix} 3 & 1 \\ 2 & 4 \\ -4 & 0 \end{bmatrix} + \begin{bmatrix} 1 & 2 \\ -1 & 0 \\ 3 & 2 \end{bmatrix} \right) = 4\begin{bmatrix} 4 & 3 \\ 1 & 4 \\ -1 & 2 \end{bmatrix} = \begin{bmatrix} 16 & 12 \\ 4 & 16 \\ -4 & 8 \end{bmatrix}$

$4A + 4B = 4\begin{bmatrix} 3 & 1 \\ 2 & 4 \\ -4 & 0 \end{bmatrix} + 4\begin{bmatrix} 1 & 2 \\ -1 & 0 \\ 3 & 2 \end{bmatrix} = \begin{bmatrix} 16 & 12 \\ 4 & 16 \\ -4 & 8 \end{bmatrix}.$

31. $\begin{bmatrix} 3 & 2 & -1 & 5 \end{bmatrix}^T = \begin{bmatrix} 3 \\ 2 \\ -1 \\ 5 \end{bmatrix}$.

33. $\begin{bmatrix} 1 & -1 & 2 \\ 3 & 4 & 2 \\ 0 & 1 & 0 \end{bmatrix}^T = \begin{bmatrix} 1 & 3 & 0 \\ -1 & 4 & 1 \\ 2 & 2 & 0 \end{bmatrix}$.

35.
$$\begin{array}{c} \\ Mr.\ Cross \\ Mr.\ Jones \\ Mr.\ Smith \end{array} \begin{array}{cccc} 1 & 2 & 3 & 4 \\ \begin{bmatrix} 220 & 215 & 210 & 205 \\ 220 & 210 & 200 & 195 \\ 215 & 205 & 195 & 190 \end{bmatrix} \end{array}$$

37. a. $D = A + B - C$

$$= \begin{bmatrix} 2820 & 1470 & 1120 \\ 1030 & 520 & 480 \\ 1170 & 540 & 460 \end{bmatrix} + \begin{bmatrix} 260 & 120 & 110 \\ 140 & 60 & 50 \\ 120 & 70 & 50 \end{bmatrix} - \begin{bmatrix} 120 & 80 & 80 \\ 70 & 30 & 40 \\ 60 & 20 & 40 \end{bmatrix}$$

$$= \begin{bmatrix} 2960 & 1510 & 1150 \\ 1100 & 550 & 490 \\ 1230 & 590 & 470 \end{bmatrix}.$$

  b. $E = 1.1D = 1.1 \begin{bmatrix} 2960 & 1510 & 1150 \\ 1100 & 550 & 490 \\ 1230 & 590 & 470 \end{bmatrix} = \begin{bmatrix} 3256 & 1661 & 1265 \\ 1210 & 605 & 539 \\ 1353 & 649 & 517 \end{bmatrix}$.

39. True. Each element in $A + B$ is obtained by adding together the corresponding elements in A and B. Therefore, the matrix $c(A + B)$ is obtained by multiplying each element in $A + B$ by c. On the other hand, cA is obtained by multiplying each element in A by c

and cB is obtained by multiplying each element in B by c and $cA + cB$ is obtained by adding the corresponding elements in cA and cB. Thus $c(A + B) = cA + cB$.

41. False. Take $\begin{bmatrix} 1 & 2 \\ 3 & 4 \end{bmatrix}$ and $c = 2$. Then

$$cA = 2\begin{bmatrix} 1 & 2 \\ 3 & 4 \end{bmatrix} = \begin{bmatrix} 2 & 4 \\ 6 & 8 \end{bmatrix} \text{ and } (cA)^T = \begin{bmatrix} 2 & 6 \\ 4 & 8 \end{bmatrix}.$$

On the other hand, $\dfrac{1}{c} A^T = \dfrac{1}{2}\begin{bmatrix} 1 & 3 \\ 2 & 4 \end{bmatrix} = \begin{bmatrix} \frac{1}{2} & \frac{3}{2} \\ 1 & 2 \end{bmatrix} \neq (cA)^T$

USING TECHNOLOGY EXERCISES 2.4, page 121

1. $\begin{bmatrix} 15 & 38.75 & -67.5 & 33.75 \\ 51.25 & 40 & 52.5 & -38.75 \\ 21.25 & 35 & -65 & 105 \end{bmatrix}$

3. $\begin{bmatrix} -5 & 6.3 & -6.8 & 3.9 \\ 1 & 0.5 & 5.4 & -4.8 \\ 0.5 & 4.2 & -3.5 & 5.6 \end{bmatrix}$

5. $\begin{bmatrix} 16.44 & -3.65 & -3.66 & 0.63 \\ 12.77 & 10.64 & 2.58 & 0.05 \\ 5.09 & 0.28 & -10.84 & 17.64 \end{bmatrix}$

7. $\begin{bmatrix} 22.2 & -0.3 & -12 & 4.5 \\ 21.6 & 17.7 & 9 & -4.2 \\ 8.7 & 4.2 & -20.7 & 33.6 \end{bmatrix}$

EXERCISES 2.5, page 132

1. $(2 \times 3)(3 \times 5)$ so AB has order 2×5.
    ↑ ↑
    =

    $(3 \times 5)(2 \times 3)$ so BA is not defined.
    ↑ ↑
    ≠

3. $(1 \times 7)\ (7 \times 1)$ so AB has order 1×1.
    ↑ ↑
    =

$(7 \times 1)(1 \times 7)$ so BA has order 7×7.

$$\uparrow \; \uparrow$$
$$=$$

5. If AB and BA are defined then $n = s$ and $m = t$.

7. $\begin{bmatrix} 1 & 2 \\ 3 & 0 \end{bmatrix}\begin{bmatrix} 1 \\ -1 \end{bmatrix} = \begin{bmatrix} -1 \\ 3 \end{bmatrix}$

9. $\begin{bmatrix} 3 & 1 & 2 \\ -1 & 2 & 4 \end{bmatrix}\begin{bmatrix} 4 \\ 1 \\ -2 \end{bmatrix} = \begin{bmatrix} 9 \\ -10 \end{bmatrix}$

11. $\begin{bmatrix} -1 & 2 \\ 3 & 1 \end{bmatrix}\begin{bmatrix} 2 & 4 \\ 3 & 1 \end{bmatrix} = \begin{bmatrix} 4 & -2 \\ 9 & 13 \end{bmatrix}$

13. $\begin{bmatrix} 2 & 1 & 2 \\ 3 & 2 & 4 \end{bmatrix}\begin{bmatrix} -1 & 2 \\ 4 & 3 \\ 0 & 1 \end{bmatrix} = \begin{bmatrix} 2 & 9 \\ 5 & 16 \end{bmatrix}$

15. $\begin{bmatrix} 0.1 & 0.9 \\ 0.2 & 0.8 \end{bmatrix}\begin{bmatrix} 1.2 & 0.4 \\ 0.5 & 2.1 \end{bmatrix} = \begin{bmatrix} 0.1(1.2)+0.9(0.5) & 0.1(0.4)+0.9(2.1) \\ 0.2(1.2)+0.8(0.5) & 0.2(0.4)+0.8(2.1) \end{bmatrix} = \begin{bmatrix} 0.57 & 1.93 \\ 0.64 & 1.76 \end{bmatrix}$

17. $\begin{bmatrix} 6 & -3 & 0 \\ -2 & 1 & -8 \\ 4 & -4 & 9 \end{bmatrix}\begin{bmatrix} 1 & 0 & 0 \\ 0 & 1 & 0 \\ 0 & 0 & 1 \end{bmatrix} = \begin{bmatrix} 6 & -3 & 0 \\ -2 & 1 & -8 \\ 4 & -4 & 9 \end{bmatrix}.$

19. $\begin{bmatrix} 3 & 0 & -2 & 1 \\ 1 & 2 & 0 & -1 \end{bmatrix}\begin{bmatrix} 2 & 1 & -1 \\ -1 & 2 & 0 \\ 0 & 0 & 1 \\ -1 & -2 & 2 \end{bmatrix} = \begin{bmatrix} 5 & 1 & -3 \\ 1 & 7 & -3 \end{bmatrix}.$

21. $4\begin{bmatrix} 1 & -2 & 0 \\ 2 & -1 & 1 \\ 3 & 0 & -1 \end{bmatrix}\begin{bmatrix} 1 & 3 & 1 \\ 1 & 4 & 0 \\ 0 & 1 & -2 \end{bmatrix} = \begin{bmatrix} -4 & -20 & 4 \\ 4 & 12 & 0 \\ 12 & 32 & 20 \end{bmatrix}$

23. $$\begin{bmatrix} 1 & 0 \\ 0 & 1 \end{bmatrix}\begin{bmatrix} 4 & -3 & 2 \\ 7 & 1 & -5 \end{bmatrix}\begin{bmatrix} 1 & 0 & 0 \\ 0 & 1 & 0 \\ 0 & 0 & 1 \end{bmatrix} = \begin{bmatrix} 1 & 0 \\ 0 & 1 \end{bmatrix}\begin{bmatrix} 4 & -3 & 2 \\ 7 & 1 & -5 \end{bmatrix} = \begin{bmatrix} 4 & -3 & 2 \\ 7 & 1 & -5 \end{bmatrix}.$$

25. To verify the associative law for matrix multiplication, we will show that
$(AB)C = A(BC)$.

$$AB = \begin{bmatrix} 1 & 0 & -2 \\ 1 & -3 & 2 \\ -2 & 1 & 1 \end{bmatrix}\begin{bmatrix} 3 & 1 & 0 \\ 2 & 2 & 0 \\ 1 & -3 & -1 \end{bmatrix} = \begin{bmatrix} 1 & 7 & 2 \\ -1 & -11 & -2 \\ -3 & -3 & -1 \end{bmatrix}$$

$$(AB)C = \begin{bmatrix} 1 & 7 & 2 \\ -1 & -11 & -2 \\ -3 & -3 & -1 \end{bmatrix}\begin{bmatrix} 2 & -1 & 0 \\ 1 & -1 & 2 \\ 3 & -2 & 1 \end{bmatrix} = \begin{bmatrix} 15 & -12 & 16 \\ -19 & 16 & -24 \\ -12 & 8 & -7 \end{bmatrix}$$

$$BC = \begin{bmatrix} 3 & 1 & 0 \\ 2 & 2 & 0 \\ 1 & -3 & -1 \end{bmatrix}\begin{bmatrix} 2 & -1 & 0 \\ 1 & -1 & 2 \\ 3 & -2 & 1 \end{bmatrix} = \begin{bmatrix} 7 & -4 & 2 \\ 6 & -4 & 4 \\ -4 & 4 & -7 \end{bmatrix}$$

$$A(BC) = \begin{bmatrix} 1 & 0 & -2 \\ 1 & -3 & 2 \\ -2 & 1 & 1 \end{bmatrix}\begin{bmatrix} 7 & -4 & 2 \\ 6 & -4 & 4 \\ -4 & 4 & -7 \end{bmatrix} = \begin{bmatrix} 15 & -12 & 16 \\ -19 & 16 & -24 \\ -12 & 8 & -7 \end{bmatrix}.$$

27. $$AB = \begin{bmatrix} 1 & 2 \\ 3 & 4 \end{bmatrix}\begin{bmatrix} 2 & 1 \\ 4 & 3 \end{bmatrix} = \begin{bmatrix} 10 & 7 \\ 22 & 15 \end{bmatrix}$$

$$BA = \begin{bmatrix} 2 & 1 \\ 4 & 3 \end{bmatrix}\begin{bmatrix} 1 & 2 \\ 3 & 4 \end{bmatrix} = \begin{bmatrix} 5 & 8 \\ 13 & 20 \end{bmatrix}$$

Therefore, $AB \neq BA$ and matrix multiplication is not commutative.

29. $$AB = \begin{bmatrix} 3 & 0 \\ 8 & 0 \end{bmatrix}\begin{bmatrix} 0 & 0 \\ 4 & 5 \end{bmatrix} = \begin{bmatrix} 0 & 0 \\ 0 & 0 \end{bmatrix}$$

$AB = 0$, but neither A nor B is the zero matrix. Therefore, $AB = 0$, does not imply that A or B is the zero matrix.

31.
$$\begin{bmatrix} a & b \\ c & d \end{bmatrix}\begin{bmatrix} 1 & 0 \\ -1 & 3 \end{bmatrix} = \begin{bmatrix} a-b & 3b \\ c-d & 3d \end{bmatrix} = \begin{bmatrix} -1 & -3 \\ 3 & 6 \end{bmatrix}$$

Then $\quad 3b = -3, \quad$ and $b = -1$

$3d = 6, \quad$ and $d = 2$

$a - b = -1$, and $a = b - 1 = -2.$

$c - d = 3, \quad$ and $c = d + 3 = 5$

Therefore, $A = \begin{bmatrix} -2 & -1 \\ 5 & 2 \end{bmatrix}.$

33.   a.
$$A^T = \begin{bmatrix} 2 & 5 \\ 4 & -6 \end{bmatrix} \text{ and } (A^T)^T = \begin{bmatrix} 2 & 4 \\ 5 & -6 \end{bmatrix} = A$$

b. $(A + B)^T = \begin{bmatrix} 6 & 12 \\ -2 & -3 \end{bmatrix}^T = \begin{bmatrix} 6 & -2 \\ 12 & -3 \end{bmatrix}$

$$A^T + B^T = \begin{bmatrix} 2 & 5 \\ 4 & -6 \end{bmatrix} + \begin{bmatrix} 4 & -7 \\ 8 & 3 \end{bmatrix} = \begin{bmatrix} 6 & -2 \\ 12 & -3 \end{bmatrix}$$

c. $AB = \begin{bmatrix} 2 & 4 \\ 5 & -6 \end{bmatrix}\begin{bmatrix} 4 & 8 \\ -7 & 3 \end{bmatrix} = \begin{bmatrix} -20 & 28 \\ 62 & 22 \end{bmatrix}$, so $(AB)^T = \begin{bmatrix} -20 & 62 \\ 28 & 22 \end{bmatrix}.$

$$B^T A^T = \begin{bmatrix} 4 & -7 \\ 8 & 3 \end{bmatrix}\begin{bmatrix} 2 & 5 \\ 4 & -6 \end{bmatrix} = \begin{bmatrix} -20 & 62 \\ 28 & 22 \end{bmatrix} = (AB)^T$$

35.   The given system of linear equations can be represented by the matrix equation $AX = B$, where
$$A = \begin{bmatrix} 2 & -3 \\ 3 & -4 \end{bmatrix}, \quad X = \begin{bmatrix} x \\ y \end{bmatrix}, \text{ and } B = \begin{bmatrix} 7 \\ 8 \end{bmatrix}.$$

37. The given system of linear equations can be represented by the matrix equation $AX = B$, where

$$A = \begin{bmatrix} 2 & -3 & 4 \\ 0 & 2 & -3 \\ 1 & -1 & 2 \end{bmatrix}, \quad X = \begin{bmatrix} x \\ y \\ z \end{bmatrix}, \quad B = \begin{bmatrix} 6 \\ 7 \\ 4 \end{bmatrix}.$$

39. The given system of linear equations can be represented by the matrix equation $AX = B$, where

$$A = \begin{bmatrix} -1 & 1 & 1 \\ 2 & -1 & -1 \\ -3 & 2 & 4 \end{bmatrix}, \quad X = \begin{bmatrix} x_1 \\ x_2 \\ x_3 \end{bmatrix}, \quad B = \begin{bmatrix} 0 \\ 2 \\ 4 \end{bmatrix}.$$

41. a. $$AB = \begin{bmatrix} 200 & 300 & 100 & 200 \\ 100 & 200 & 400 & 0 \end{bmatrix} \begin{bmatrix} 54 \\ 48 \\ 98 \\ 82 \end{bmatrix} = \begin{bmatrix} 51,400 \\ 54,200 \end{bmatrix}$$

b. The first entry shows that William's total stock holdings are $51,400, while the second entry shows that Michael's stockholdings are $54,200.

43. The column vector that represents the profit for each type of house is

$$B = \begin{bmatrix} 20,000 \\ 22,000 \\ 25,000 \\ 30,000 \end{bmatrix}.$$

The column vector that gives the total profit for Bond Brothers is

$$AB = \begin{bmatrix} 60 & 80 & 120 & 40 \\ 20 & 30 & 60 & 10 \\ 10 & 15 & 30 & 5 \end{bmatrix} \begin{bmatrix} 20,000 \\ 22,000 \\ 25,000 \\ 30,000 \end{bmatrix}$$

$$= \begin{bmatrix} 7,160,000 \\ 2,860,000 \\ 1,430,000 \end{bmatrix}.$$

Therefore, Bond Brothers expects to make $7,160,000 in New York, $2,860,000 in Connecticut, and $1,430,000 in Massachusetts, and the total profit is $11,450,000.

45.
$$BA = \begin{bmatrix} 30,000 & 40,000 & 20,000 \end{bmatrix} \begin{array}{ccc} D & R & I \\ \begin{bmatrix} 0.50 & 0.30 & 0.20 \\ 0.45 & 0.40 & 0.15 \\ 0.40 & 0.50 & 0.10 \end{bmatrix} \end{array}$$

$$= \begin{array}{ccc} D & R & I \\ \begin{bmatrix} 41,000 & 35,000 & 14,000 \end{bmatrix} \end{array}$$

47.
$$AB = \begin{bmatrix} 2700 & 3000 \\ 800 & 700 \\ 500 & 300 \end{bmatrix} \begin{bmatrix} 0.25 & 0.20 & 0.30 & 0.25 \\ 0.30 & 0.35 & 0.25 & 0.10 \end{bmatrix} = \begin{bmatrix} 1575 & 1590 & 1560 & 975 \\ 410 & 405 & 415 & 270 \\ 215 & 205 & 225 & 155 \end{bmatrix}$$

49. $AC = \begin{bmatrix} 80 & 60 & 40 \end{bmatrix} \begin{bmatrix} 0.34 \\ 0.42 \\ 0.48 \end{bmatrix} = 71.6 \qquad BD = \begin{bmatrix} 300 & 150 & 250 \end{bmatrix} \begin{bmatrix} 0.24 \\ 0.31 \\ 0.35 \end{bmatrix} = 206$

$AC + BD = \begin{bmatrix} 277.60 \end{bmatrix}$, or $277.60. It represents Cindy's long distance bill for phone calls to those 3 cities.

51. a. $\quad MA^T = \begin{bmatrix} 400 & 1200 & 800 \\ 110 & 570 & 340 \\ 90 & 30 & 60 \end{bmatrix} \begin{bmatrix} 7 \\ 1 \\ 6 \end{bmatrix} = \begin{bmatrix} 8800 \\ 3380 \\ 1020 \end{bmatrix}.$

The amounts of vitamin A, vitamin C, and calcium taken by a girl in the first meal are 8800, 3380, and 1020 units respectively.

b.
$$MB^T = \begin{bmatrix} 400 & 1200 & 800 \\ 110 & 570 & 340 \\ 90 & 30 & 60 \end{bmatrix} \begin{bmatrix} 9 \\ 3 \\ 2 \end{bmatrix} = \begin{bmatrix} 8800 \\ 3380 \\ 1020 \end{bmatrix}$$

The amounts of vitamin A, vitamin C, and calcium taken by a girl in the second meal are 8,800, 3380, and 1020 units, respectively.

c. $$M(A+B)^T = \begin{bmatrix} 400 & 1200 & 800 \\ 110 & 570 & 340 \\ 90 & 30 & 60 \end{bmatrix} \begin{bmatrix} 16 \\ 4 \\ 8 \end{bmatrix} = \begin{bmatrix} 17,600 \\ 6,760 \\ 2,040 \end{bmatrix}.$$

The amounts of vitamin A, vitamin C, and calcium taken by a girl in the two meals are 17,600, 6,760, and 2,040 units respectively.

53. False. Let A be a matrix of order 2×3 and let B be a matrix of order 3×2. Then AB and BA are both defined. But, evidently, neither A nor B is a square matrix.

USING TECHNOLOGY EXERCISES 2.5, page 136

1. $$\begin{bmatrix} 18.66 & 15.2 & -12 \\ 24.48 & 41.88 & 89.82 \\ 15.39 & 7.16 & -1.25 \end{bmatrix}$$

3. $$\begin{bmatrix} 20.09 & 20.61 & -1.3 \\ 44.42 & 71.6 & 64.89 \\ 20.97 & 7.17 & -60.65 \end{bmatrix}$$

5. $$\begin{bmatrix} 32.89 & 13.63 & -57.17 \\ -12.85 & -8.37 & 256.92 \\ 13.48 & 14.29 & 181.64 \end{bmatrix}$$

7. $$\begin{bmatrix} 128.59 & 123.08 & -32.50 \\ 246.73 & 403.12 & 481.52 \\ 125.06 & 47.01 & -264.81 \end{bmatrix}$$

9. $$\begin{bmatrix} 87 & 68 & 110 & 82 \\ 119 & 176 & 221 & 143 \\ 51 & 128 & 142 & 94 \\ 28 & 174 & 174 & 112 \end{bmatrix}$$

$$\begin{bmatrix} 113 & 117 & 72 & 101 & 90 \\ 72 & 85 & 36 & 72 & 76 \\ 81 & 69 & 76 & 87 & 30 \\ 133 & 157 & 56 & 121 & 146 \\ 154 & 157 & 94 & 127 & 122 \end{bmatrix}$$

11. $$\begin{bmatrix} 170 & 18.1 & 133.1 & -106.3 & 341.3 \\ 349 & 226.5 & 324.1 & 164 & 506.4 \\ 245.2 & 157.7 & 231.5 & 125.5 & 312.9 \\ 310 & 245.2 & 291 & 274.3 & 354.2 \end{bmatrix}$$

EXERCISES 2.6, page 149

1. $\begin{bmatrix} 1 & -3 \\ 1 & -2 \end{bmatrix}\begin{bmatrix} -2 & 3 \\ -1 & 1 \end{bmatrix} = \begin{bmatrix} 1 & 0 \\ 0 & 1 \end{bmatrix};\qquad \begin{bmatrix} -2 & 3 \\ -1 & 1 \end{bmatrix}\begin{bmatrix} 1 & -3 \\ 1 & -2 \end{bmatrix} = \begin{bmatrix} 1 & 0 \\ 0 & 1 \end{bmatrix}$

3. $\begin{bmatrix} 3 & 2 & 3 \\ 2 & 2 & 1 \\ 2 & 1 & 1 \end{bmatrix}\begin{bmatrix} -\frac{1}{3} & -\frac{1}{3} & \frac{4}{3} \\ 0 & 1 & -1 \\ \frac{2}{3} & -\frac{1}{3} & -\frac{2}{3} \end{bmatrix} = \begin{bmatrix} 1 & 0 & 0 \\ 0 & 1 & 0 \\ 0 & 0 & 1 \end{bmatrix}$ and

$\begin{bmatrix} -\frac{1}{3} & -\frac{1}{3} & \frac{4}{3} \\ 0 & 1 & -1 \\ \frac{2}{3} & -\frac{1}{3} & -\frac{2}{3} \end{bmatrix}\begin{bmatrix} 3 & 2 & 3 \\ 2 & 2 & 1 \\ 2 & 1 & 1 \end{bmatrix} = \begin{bmatrix} 1 & 0 & 0 \\ 0 & 1 & 0 \\ 0 & 0 & 1 \end{bmatrix}$

5. Using Formula (13), we find $A^{-1} = \dfrac{1}{(2)(3)-(1)(5)}\begin{bmatrix} 3 & -5 \\ -1 & 2 \end{bmatrix} = \begin{bmatrix} 3 & -5 \\ -1 & 2 \end{bmatrix}$.

7. Since $ad - bc = (3)(2) - (-2)(-3) = 6 - 6 = 0$, the inverse does not exist.

9. $\left[\begin{array}{ccc|ccc} 2 & -3 & -4 & 1 & 0 & 0 \\ 0 & 0 & -1 & 0 & 1 & 0 \\ 1 & -2 & 1 & 0 & 0 & 1 \end{array}\right] \xrightarrow{R_1 \leftrightarrow R_3} \left[\begin{array}{ccc|ccc} 1 & -2 & 1 & 0 & 0 & 1 \\ 0 & 0 & -1 & 0 & 1 & 0 \\ 2 & -3 & -4 & 1 & 0 & 0 \end{array}\right] \xrightarrow{R_3 - 2R_1}$

$\left[\begin{array}{ccc|ccc} 1 & -2 & 1 & 0 & 0 & 1 \\ 0 & 0 & -1 & 0 & 1 & 0 \\ 0 & 1 & -6 & 1 & 0 & -2 \end{array}\right] \xrightarrow{R_2 \leftrightarrow R_3} \left[\begin{array}{ccc|ccc} 1 & -2 & 1 & 0 & 0 & 1 \\ 0 & 1 & -6 & 1 & 0 & -2 \\ 0 & 0 & -1 & 0 & 1 & 0 \end{array}\right] \xrightarrow[-R_3]{R_1 + 2R_2}$

$\left[\begin{array}{ccc|ccc} 1 & 0 & -11 & 2 & 0 & -3 \\ 0 & 1 & -6 & 1 & 0 & -2 \\ 0 & 0 & 1 & 0 & -1 & 0 \end{array}\right] \xrightarrow[R_2 + 6R_3]{R_1 + 11R_3} \left[\begin{array}{ccc|ccc} 1 & 0 & 0 & 2 & -11 & -3 \\ 0 & 1 & 0 & 1 & -6 & -2 \\ 0 & 0 & 1 & 0 & -1 & 0 \end{array}\right]$.

Therefore, the required inverse is $\begin{bmatrix} 2 & -11 & -3 \\ 1 & -6 & -2 \\ 0 & -1 & 0 \end{bmatrix}$.

11.

$$\begin{bmatrix} 4 & 2 & 2 & | & 1 & 0 & 0 \\ -1 & -3 & 4 & | & 0 & 1 & 0 \\ 3 & -1 & 6 & | & 0 & 0 & 1 \end{bmatrix} \xrightarrow{R_1 - R_3} \begin{bmatrix} 1 & 3 & -4 & | & 1 & 0 & -1 \\ -1 & -3 & 4 & | & 0 & 1 & 0 \\ 3 & -1 & 6 & | & 0 & 0 & 1 \end{bmatrix}$$

$$\xrightarrow{R_2 + R_1} \begin{bmatrix} 1 & 3 & -4 & | & 1 & 0 & -1 \\ 0 & 0 & 0 & | & 1 & 1 & -1 \\ 3 & -1 & 6 & | & 0 & 0 & 1 \end{bmatrix}$$

Because there is a row of zeros to the left of the vertical line, we see that the inverse does not exist.

13.

$$\begin{bmatrix} 1 & 4 & -1 & | & 1 & 0 & 0 \\ 2 & 3 & -2 & | & 0 & 1 & 0 \\ -1 & 2 & 3 & | & 0 & 0 & 1 \end{bmatrix} \begin{array}{c} R_2 - 2R_1 \\ \xrightarrow{R_3 + R_1} \end{array} \begin{bmatrix} 1 & 4 & -1 & | & 1 & 0 & 0 \\ 0 & -5 & 0 & | & -2 & 1 & 0 \\ 0 & 6 & 2 & | & 1 & 0 & 1 \end{bmatrix} \xrightarrow{R_2 + R_3}$$

$$\begin{bmatrix} 1 & 4 & -1 & | & 1 & 0 & 0 \\ 0 & 1 & 2 & | & -1 & 1 & 1 \\ 0 & 6 & 2 & | & 1 & 0 & 1 \end{bmatrix} \begin{array}{c} R_1 - 4R_2 \\ \xrightarrow{R_3 - 6R_2} \end{array} \begin{bmatrix} 1 & 0 & -9 & | & 5 & -4 & -4 \\ 0 & 1 & 2 & | & -1 & 1 & 1 \\ 0 & 0 & -10 & | & 7 & -6 & -5 \end{bmatrix} \xrightarrow{-\frac{1}{10}R_3}$$

$$\begin{bmatrix} 1 & 0 & -9 & | & 5 & -4 & -4 \\ 0 & 1 & 2 & | & -1 & 1 & 1 \\ 0 & 0 & 1 & | & -\frac{7}{10} & \frac{3}{5} & \frac{1}{2} \end{bmatrix} \begin{array}{c} R_1 + 9R_3 \\ \xrightarrow{R_2 - 2R_3} \end{array} \begin{bmatrix} 1 & 0 & 0 & | & -\frac{13}{10} & \frac{7}{5} & \frac{1}{2} \\ 0 & 1 & 0 & | & \frac{2}{5} & -\frac{1}{5} & 0 \\ 0 & 0 & 1 & | & -\frac{7}{10} & \frac{3}{5} & \frac{1}{2} \end{bmatrix}$$

So $A^{-1} = \begin{bmatrix} -\frac{13}{10} & \frac{7}{5} & \frac{1}{2} \\ \frac{2}{5} & -\frac{1}{5} & 0 \\ -\frac{7}{10} & \frac{3}{5} & \frac{1}{2} \end{bmatrix}$.

15.

$$\begin{bmatrix} 1 & 1 & -1 & 1 & | & 1 & 0 & 0 & 0 \\ 2 & 1 & 1 & 0 & | & 0 & 1 & 0 & 0 \\ 2 & 1 & 0 & 1 & | & 0 & 0 & 1 & 0 \\ 2 & -1 & -1 & 3 & | & 0 & 0 & 0 & 1 \end{bmatrix} \begin{array}{c} R_2 - 2R_1 \\ R_3 - 2R_1 \\ \xrightarrow{R_4 - 2R_1} \end{array} \begin{bmatrix} 1 & 1 & -1 & 1 & | & 1 & 0 & 0 & 0 \\ 0 & -1 & 3 & -2 & | & -2 & 1 & 0 & 0 \\ 0 & -1 & 2 & -1 & | & -2 & 0 & 1 & 0 \\ 0 & -3 & 1 & 1 & | & -2 & 0 & 0 & 1 \end{bmatrix} \xrightarrow{-R_2}$$

$$\begin{bmatrix} 1 & 1 & -1 & 1 & | & 1 & 0 & 0 & 0 \\ 0 & 1 & -3 & 2 & | & 2 & -1 & 0 & 0 \\ 0 & -1 & 2 & -1 & | & -2 & 0 & 1 & 0 \\ 0 & -3 & 1 & 1 & | & -2 & 0 & 0 & 1 \end{bmatrix} \xrightarrow[\substack{R_1-R_2 \\ R_3+R_2 \\ R_4+3R_2}]{} \begin{bmatrix} 1 & 0 & 2 & -1 & | & -1 & 1 & 0 & 0 \\ 0 & 1 & -3 & 2 & | & 2 & -1 & 0 & 0 \\ 0 & 0 & -1 & 1 & | & 0 & -1 & 1 & 0 \\ 0 & 0 & -8 & 7 & | & 4 & -3 & 0 & 1 \end{bmatrix} \xrightarrow{-R_3}$$

$$\begin{bmatrix} 1 & 0 & 2 & -1 & | & -1 & 1 & 0 & 0 \\ 0 & 1 & -3 & 2 & | & 2 & -1 & 0 & 0 \\ 0 & 0 & 1 & -1 & | & 0 & 1 & -1 & 0 \\ 0 & 0 & -8 & 7 & | & 4 & -3 & 0 & 1 \end{bmatrix} \xrightarrow[\substack{R_1-2R_3 \\ R_2+3R_3 \\ R_4+8R_3}]{} \begin{bmatrix} 1 & 0 & 0 & 1 & | & -1 & -1 & 2 & 0 \\ 0 & 1 & 0 & -1 & | & 2 & 2 & -3 & 0 \\ 0 & 0 & 1 & -1 & | & 0 & 1 & -1 & 0 \\ 0 & 0 & 0 & -1 & | & 4 & 5 & -8 & 1 \end{bmatrix}$$

$$\xrightarrow[\substack{R_1+R_4 \\ R_2-R_4 \\ R_3-R_4 \\ -R_4}]{} \begin{bmatrix} 1 & 0 & 0 & 0 & | & 3 & 4 & -6 & 1 \\ 0 & 1 & 0 & 0 & | & -2 & -3 & 5 & -1 \\ 0 & 0 & 1 & 0 & | & -4 & -4 & 7 & -1 \\ 0 & 0 & 0 & 1 & | & -4 & -5 & 8 & -1 \end{bmatrix}.$$

So the required inverse is

$$A^{-1} = \begin{bmatrix} 3 & 4 & -6 & 1 \\ -2 & -3 & 5 & -1 \\ -4 & -4 & 7 & -1 \\ -4 & -5 & 8 & -1 \end{bmatrix}.$$

We can verify our result by showing that $A^{-1}A = A$. Thus,

$$\begin{bmatrix} 3 & 4 & -6 & 1 \\ -2 & -3 & 5 & -1 \\ -4 & -4 & 7 & -1 \\ -4 & -5 & 8 & -1 \end{bmatrix} \begin{bmatrix} 1 & 1 & -1 & 1 \\ 2 & 1 & 1 & 0 \\ 2 & 1 & 0 & 1 \\ 2 & -1 & -1 & 3 \end{bmatrix} = \begin{bmatrix} 1 & 0 & 0 & 0 \\ 0 & 1 & 0 & 0 \\ 0 & 0 & 1 & 0 \\ 0 & 0 & 0 & 1 \end{bmatrix}.$$

17.   a. $A = \begin{bmatrix} 2 & 5 \\ 1 & 3 \end{bmatrix}$, $X = \begin{bmatrix} x \\ y \end{bmatrix}$, $B = \begin{bmatrix} 3 \\ 2 \end{bmatrix}$;

b. $X = A^{-1}B = \begin{bmatrix} 3 & -5 \\ -1 & 2 \end{bmatrix} \begin{bmatrix} 3 \\ 2 \end{bmatrix} = \begin{bmatrix} -1 \\ 1 \end{bmatrix}$;

19. a. $A = \begin{bmatrix} 2 & -3 & -4 \\ 0 & 0 & -1 \\ 1 & -2 & 1 \end{bmatrix}$, $X = \begin{bmatrix} x \\ y \\ z \end{bmatrix}$, $B = \begin{bmatrix} 4 \\ 3 \\ -8 \end{bmatrix}$

$$X = A^{-1}B = \begin{bmatrix} 2 & -11 & -3 \\ 1 & -6 & -2 \\ 0 & -1 & 0 \end{bmatrix} \begin{bmatrix} 4 \\ 3 \\ -8 \end{bmatrix} = \begin{bmatrix} -1 \\ 2 \\ -3 \end{bmatrix}$$

21. a. $A = \begin{bmatrix} 1 & 4 & -1 \\ 2 & 3 & -2 \\ -1 & 2 & 3 \end{bmatrix}$, $X = \begin{bmatrix} x \\ y \\ z \end{bmatrix}$, $B = \begin{bmatrix} 3 \\ 1 \\ 7 \end{bmatrix}$; b. $X = A^{-1}B = \begin{bmatrix} -\frac{13}{10} & \frac{7}{5} & \frac{1}{2} \\ \frac{2}{5} & -\frac{1}{5} & 0 \\ -\frac{7}{10} & \frac{3}{5} & \frac{1}{2} \end{bmatrix} \begin{bmatrix} 3 \\ 1 \\ 7 \end{bmatrix} = \begin{bmatrix} 1 \\ 1 \\ 2 \end{bmatrix}$.

23. a. $A = \begin{bmatrix} 1 & 1 & -1 & 1 \\ 2 & 1 & 1 & 0 \\ 2 & 1 & 0 & 1 \\ 2 & -1 & -1 & 3 \end{bmatrix}$, $X = \begin{bmatrix} x_1 \\ x_2 \\ x_3 \\ x_4 \end{bmatrix}$, $B = \begin{bmatrix} 6 \\ 4 \\ 7 \\ 9 \end{bmatrix}$.

b. $X = A^{-1}B = \begin{bmatrix} 3 & 4 & -6 & 1 \\ -2 & -3 & 5 & -1 \\ -4 & -4 & 7 & -1 \\ -4 & -5 & 8 & -1 \end{bmatrix} \begin{bmatrix} 6 \\ 4 \\ 7 \\ 9 \end{bmatrix} = \begin{bmatrix} 1 \\ 2 \\ 0 \\ 3 \end{bmatrix}$.

25. a. $A = \begin{bmatrix} 1 & 2 \\ 2 & -1 \end{bmatrix}$, $X = \begin{bmatrix} x \\ y \end{bmatrix}$, $B = \begin{bmatrix} b_1 \\ b_2 \end{bmatrix}$;

b (i). $X = A^{-1}B = \begin{bmatrix} 0.2 & 0.4 \\ 0.4 & -0.2 \end{bmatrix} \begin{bmatrix} 14 \\ 5 \end{bmatrix} = \begin{bmatrix} 4.8 \\ 4.6 \end{bmatrix}$ and we conclude that $x = 4.8$ and $y = 4.6$.

(ii). $X = A^{-1}B = \begin{bmatrix} 0.2 & 0.4 \\ 0.4 & -0.2 \end{bmatrix} \begin{bmatrix} 4 \\ -1 \end{bmatrix} = \begin{bmatrix} 0.4 \\ 1.8 \end{bmatrix}$ and we conclude that $x = 0.4$ and $y = 1.8$.

27. a. First we find A^{-1}.

$$\begin{bmatrix} 1 & 2 & 1 & | & 1 & 0 & 0 \\ 1 & 1 & 1 & | & 0 & 1 & 0 \\ 3 & 1 & 1 & | & 0 & 0 & 1 \end{bmatrix} \xrightarrow[\begin{subarray}{c} R_2 - R_1 \\ R_3 - 3R_1 \end{subarray}]{} \begin{bmatrix} 1 & 2 & 1 & | & 1 & 0 & 0 \\ 0 & -1 & 0 & | & -1 & 1 & 0 \\ 0 & -5 & -2 & | & -3 & 0 & 1 \end{bmatrix} \xrightarrow{-R_2}$$

$$\begin{bmatrix} 1 & 2 & 1 & | & 1 & 0 & 0 \\ 0 & 1 & 0 & | & 1 & -1 & 0 \\ 0 & -5 & -2 & | & -3 & 0 & 1 \end{bmatrix} \xrightarrow[\begin{subarray}{c} R_1 - 2R_2 \\ R_3 + 5R_2 \end{subarray}]{} \begin{bmatrix} 1 & 0 & 1 & | & -1 & 2 & 0 \\ 0 & 1 & 0 & | & 1 & -1 & 0 \\ 0 & 0 & -2 & | & 2 & -5 & 1 \end{bmatrix} \xrightarrow{-\frac{1}{2}R_3}$$

$$\begin{bmatrix} 1 & 0 & 1 & | & -1 & 2 & 0 \\ 0 & 1 & 0 & | & 1 & -1 & 0 \\ 0 & 0 & 1 & | & -1 & \frac{5}{2} & -\frac{1}{2} \end{bmatrix} \xrightarrow{R_1 - R_3} \begin{bmatrix} 1 & 0 & 0 & | & 0 & -\frac{1}{2} & \frac{1}{2} \\ 0 & 1 & 0 & | & 1 & -1 & 0 \\ 0 & 0 & 1 & | & -1 & \frac{5}{2} & -\frac{1}{2} \end{bmatrix}$$

$$\begin{bmatrix} 1 & 2 & 1 \\ 1 & 1 & 1 \\ 3 & 1 & 1 \end{bmatrix} \begin{bmatrix} x \\ y \\ z \end{bmatrix} = \begin{bmatrix} b_1 \\ b_2 \\ b_3 \end{bmatrix}$$

b. (i). $\begin{bmatrix} x \\ y \\ z \end{bmatrix} = \begin{bmatrix} 0 & -\frac{1}{2} & \frac{1}{2} \\ 1 & -1 & 0 \\ -1 & \frac{5}{2} & -\frac{1}{2} \end{bmatrix} \begin{bmatrix} 7 \\ 4 \\ 2 \end{bmatrix} = \begin{bmatrix} -1 \\ 3 \\ 2 \end{bmatrix}$

and we conclude that $x = -1$, $y = 3$, and $z = 2$

(ii). $\begin{bmatrix} x \\ y \\ z \end{bmatrix} = \begin{bmatrix} 0 & -\frac{1}{2} & \frac{1}{2} \\ 1 & -1 & 0 \\ -1 & \frac{5}{2} & -\frac{1}{2} \end{bmatrix} \begin{bmatrix} 5 \\ -3 \\ -1 \end{bmatrix} = \begin{bmatrix} 1 \\ 8 \\ -12 \end{bmatrix}$

and we conclude that $x = 1$, $y = 8$, and $z = -12$.

29. a. $\begin{bmatrix} 3 & 2 & -1 & | & 1 & 0 & 0 \\ 2 & -3 & 1 & | & 0 & 1 & 0 \\ 1 & -1 & -1 & | & 0 & 0 & 1 \end{bmatrix} \xrightarrow{R_1 \leftrightarrow R_3} \begin{bmatrix} 1 & -1 & -1 & | & 0 & 0 & 1 \\ 2 & -3 & 1 & | & 0 & 1 & 0 \\ 3 & 2 & -1 & | & 1 & 0 & 0 \end{bmatrix} \xrightarrow[\begin{subarray}{c} R_2 - 2R_1 \\ R_3 - 3R_1 \end{subarray}]{}$

$$\begin{bmatrix} 1 & -1 & -1 & 0 & 0 & 1 \\ 0 & -1 & 3 & 0 & 1 & -2 \\ 0 & 5 & 2 & 1 & 0 & -3 \end{bmatrix} \xrightarrow{-R_2} \begin{bmatrix} 1 & -1 & -1 & 0 & 0 & 1 \\ 0 & 1 & -3 & 0 & -1 & 2 \\ 0 & 5 & 2 & 1 & 0 & -3 \end{bmatrix} \xrightarrow[R_3-5R_2]{R_1+R_2}$$

$$\begin{bmatrix} 1 & 0 & -4 & 0 & -1 & 3 \\ 0 & 1 & -3 & 0 & -1 & 2 \\ 0 & 0 & 17 & 1 & 5 & -13 \end{bmatrix} \xrightarrow{\frac{1}{17}R_3} \begin{bmatrix} 1 & 0 & -4 & 0 & -1 & 3 \\ 0 & 1 & -3 & 0 & -1 & 2 \\ 0 & 0 & 1 & \frac{1}{17} & \frac{5}{17} & -\frac{13}{17} \end{bmatrix}$$

$$\xrightarrow[R_2+3R_3]{R_1+4R_3} \begin{bmatrix} 1 & 0 & 0 & \frac{4}{17} & \frac{3}{17} & -\frac{1}{17} \\ 0 & 1 & 0 & \frac{3}{17} & -\frac{2}{17} & -\frac{5}{17} \\ 0 & 0 & 1 & \frac{1}{17} & \frac{5}{17} & -\frac{13}{17} \end{bmatrix}. \quad \text{Therefore } A^{-1} = \begin{bmatrix} \frac{4}{17} & \frac{3}{17} & -\frac{1}{17} \\ \frac{3}{17} & -\frac{2}{17} & -\frac{5}{17} \\ \frac{1}{17} & \frac{5}{17} & -\frac{13}{17} \end{bmatrix}.$$

Next, $\begin{bmatrix} 3 & 2 & -1 \\ 2 & -3 & 1 \\ 1 & -1 & -1 \end{bmatrix} \begin{bmatrix} x \\ y \\ z \end{bmatrix} = \begin{bmatrix} b_1 \\ b_2 \\ b_3 \end{bmatrix}$

b. (i) $\begin{bmatrix} x \\ y \\ z \end{bmatrix} = \begin{bmatrix} \frac{4}{17} & \frac{3}{17} & -\frac{1}{17} \\ \frac{3}{17} & -\frac{2}{17} & -\frac{5}{17} \\ \frac{1}{17} & \frac{5}{17} & -\frac{13}{17} \end{bmatrix} \begin{bmatrix} 2 \\ -2 \\ 4 \end{bmatrix} = \begin{bmatrix} -\frac{2}{17} \\ -\frac{10}{17} \\ -\frac{60}{17} \end{bmatrix}$

We conclude that $x = -2/17$, $y = -10/17$, and $z = -60/17$.

(ii) $\begin{bmatrix} x \\ y \\ z \end{bmatrix} = \begin{bmatrix} \frac{4}{17} & \frac{3}{17} & -\frac{1}{17} \\ \frac{3}{17} & -\frac{2}{17} & -\frac{5}{17} \\ \frac{1}{17} & \frac{5}{17} & -\frac{13}{17} \end{bmatrix} \begin{bmatrix} 8 \\ -3 \\ 6 \end{bmatrix} = \begin{bmatrix} 1 \\ 0 \\ -5 \end{bmatrix}$. We conclude that $x = 1$, $y = 0$, and $z = -5$.

31. a. $AX = B_1$ and $AX = B_2$, where

$$A = \begin{bmatrix} 1 & 1 & 1 & 1 \\ 1 & -1 & -1 & 1 \\ 0 & 1 & 2 & 2 \\ 1 & 2 & 1 & -2 \end{bmatrix}, \; X = \begin{bmatrix} x_1 \\ x_2 \\ x_3 \\ x_4 \end{bmatrix}, \; B_1 = \begin{bmatrix} 1 \\ -1 \\ 4 \\ 0 \end{bmatrix} \text{ and } B_2 = \begin{bmatrix} 2 \\ 8 \\ 4 \\ -1 \end{bmatrix}.$$

We first find A^{-1}.

$$\begin{bmatrix} 1 & 1 & 1 & 1 & | & 1 & 0 & 0 & 0 \\ 1 & -1 & -1 & 1 & | & 0 & 1 & 0 & 0 \\ 0 & 1 & 2 & 2 & | & 0 & 0 & 1 & 0 \\ 1 & 2 & 1 & -2 & | & 0 & 0 & 0 & 1 \end{bmatrix} \xrightarrow[R_4-R_1]{R_2-R_1} \begin{bmatrix} 1 & 1 & 1 & 1 & | & 1 & 0 & 0 & 0 \\ 0 & -2 & -2 & 0 & | & -1 & 1 & 0 & 0 \\ 0 & 1 & 2 & 2 & | & 0 & 0 & 1 & 0 \\ 0 & 1 & 0 & -3 & | & -1 & 0 & 0 & 1 \end{bmatrix}$$

$$\xrightarrow{R_2 \leftrightarrow R_3} \begin{bmatrix} 1 & 1 & 1 & 1 & | & 1 & 0 & 0 & 0 \\ 0 & 1 & 2 & 2 & | & 0 & 0 & 1 & 0 \\ 0 & -2 & -2 & 0 & | & -1 & 1 & 0 & 0 \\ 0 & 1 & 0 & -3 & | & -1 & 0 & 0 & 1 \end{bmatrix} \xrightarrow[\substack{R_3+2R_2 \\ R_4-R_2}]{R_1-R_2}$$

$$\begin{bmatrix} 1 & 0 & -1 & -1 & | & 1 & 0 & -1 & 0 \\ 0 & 1 & 2 & 2 & | & 0 & 0 & 1 & 0 \\ 0 & 0 & 2 & 4 & | & -1 & 1 & 2 & 0 \\ 0 & 0 & -2 & -5 & | & -1 & 0 & -1 & 1 \end{bmatrix} \xrightarrow{\frac{1}{2}R_3} \begin{bmatrix} 1 & 0 & -1 & -1 & | & 1 & 0 & -1 & 0 \\ 0 & 1 & 2 & 2 & | & 0 & 0 & 1 & 0 \\ 0 & 0 & 1 & 2 & | & -\frac{1}{2} & \frac{1}{2} & 1 & 0 \\ 0 & 0 & -2 & -5 & | & -1 & 0 & -1 & 1 \end{bmatrix}$$

$$\xrightarrow[\substack{R_2-2R_3 \\ R_4+2R_3}]{R_1+R_3} \begin{bmatrix} 1 & 0 & 0 & 1 & | & \frac{1}{2} & \frac{1}{2} & 0 & 0 \\ 0 & 1 & 0 & -2 & | & 1 & -1 & -1 & 0 \\ 0 & 0 & 1 & 2 & | & -\frac{1}{2} & \frac{1}{2} & 1 & 0 \\ 0 & 0 & 0 & -1 & | & -2 & 1 & 1 & 1 \end{bmatrix} \xrightarrow[\substack{R_2-2R_4 \\ R_3+2R_4 \\ -R_4}]{R_1+R_4}$$

$$\begin{bmatrix} 1 & 0 & 0 & 0 & | & -\frac{3}{2} & \frac{3}{2} & 1 & 1 \\ 0 & 1 & 0 & 0 & | & 5 & -3 & -3 & -2 \\ 0 & 0 & 1 & 0 & | & -\frac{9}{2} & \frac{5}{2} & 3 & 2 \\ 0 & 0 & 0 & 1 & | & 2 & -1 & -1 & -1 \end{bmatrix}. \quad \text{So} \quad A^{-1} = \begin{bmatrix} -\frac{3}{2} & \frac{3}{2} & 1 & 1 \\ 5 & -3 & -3 & -2 \\ -\frac{9}{2} & \frac{5}{2} & 3 & 2 \\ 2 & -1 & -1 & -1 \end{bmatrix}.$$

b. (i).
$$\begin{bmatrix} x_1 \\ x_2 \\ x_3 \\ x_4 \end{bmatrix} = \begin{bmatrix} -\frac{3}{2} & \frac{3}{2} & 1 & 1 \\ 5 & -3 & -3 & -2 \\ -\frac{9}{2} & \frac{5}{2} & 3 & 2 \\ 2 & -1 & -1 & -1 \end{bmatrix} \begin{bmatrix} 1 \\ -1 \\ 4 \\ 0 \end{bmatrix} = \begin{bmatrix} 1 \\ -4 \\ 5 \\ -1 \end{bmatrix}$$

and we conclude that $x_1 = 1$, $x_2 = -4$, $x_3 = 5$, and $x_4 = -1$.

$$(ii). \quad \begin{bmatrix} x_1 \\ x_2 \\ x_3 \\ x_4 \end{bmatrix} = \begin{bmatrix} -\frac{3}{2} & \frac{3}{2} & 1 & 1 \\ 5 & -3 & -3 & -2 \\ -\frac{9}{2} & \frac{5}{2} & 3 & 2 \\ 2 & -1 & -1 & -1 \end{bmatrix} \begin{bmatrix} 2 \\ 8 \\ 4 \\ -1 \end{bmatrix} = \begin{bmatrix} 12 \\ -24 \\ 21 \\ -7 \end{bmatrix}$$

and we conclude that $x_1 = 12$, $x_2 = -24$, $x_3 = 21$, and $x_4 = -7$.

33.   a. Using Formula (13), we find $A^{-1} = \dfrac{1}{(2)(-5)-(-4)(3)} \begin{bmatrix} -5 & -3 \\ 4 & 2 \end{bmatrix} = \begin{bmatrix} -\frac{5}{2} & -\frac{3}{2} \\ 2 & 1 \end{bmatrix}$.

b. Using Formula (13) once again, we find

$$\left(A^{-1}\right)^{-1} = \frac{1}{(-\frac{5}{2})(1)-2(-\frac{3}{2})} \begin{bmatrix} 1 & \frac{3}{2} \\ -2 & -\frac{5}{2} \end{bmatrix} = \begin{bmatrix} 2 & 3 \\ -4 & -5 \end{bmatrix} = A.$$

35.   a. $ABC = \begin{bmatrix} 2 & -5 \\ 1 & -3 \end{bmatrix} \begin{bmatrix} 4 & 3 \\ 1 & 1 \end{bmatrix} \begin{bmatrix} 2 & 3 \\ -2 & 1 \end{bmatrix} = \begin{bmatrix} 2 & -5 \\ 1 & -3 \end{bmatrix} \begin{bmatrix} 2 & 15 \\ 0 & 4 \end{bmatrix} = \begin{bmatrix} 4 & 10 \\ 2 & 3 \end{bmatrix}.$

Using the formula for finding the inverse of a 2×2 matrix, we find

$$A^{-1} = \begin{bmatrix} 3 & -5 \\ 1 & -2 \end{bmatrix}, \quad B^{-1} = \begin{bmatrix} 1 & -3 \\ -1 & 4 \end{bmatrix}, \quad C^{-1} = \begin{bmatrix} \frac{1}{8} & -\frac{3}{8} \\ \frac{1}{4} & \frac{1}{4} \end{bmatrix}.$$

b. Using the formula for finding the inverse of a 2×2 matrix, we find

$$(ABC)^{-1} = \begin{bmatrix} -\frac{3}{8} & \frac{5}{4} \\ \frac{1}{4} & -\frac{1}{2} \end{bmatrix}$$

$$C^{-1}B^{-1}A^{-1} = \begin{bmatrix} \frac{1}{8} & -\frac{3}{8} \\ \frac{1}{4} & \frac{1}{4} \end{bmatrix} \begin{bmatrix} 1 & -3 \\ -1 & 4 \end{bmatrix} \begin{bmatrix} 3 & -5 \\ 1 & -2 \end{bmatrix}$$

$$= \begin{bmatrix} \frac{1}{8} & -\frac{3}{8} \\ \frac{1}{4} & \frac{1}{4} \end{bmatrix} \begin{bmatrix} 0 & 1 \\ 1 & -3 \end{bmatrix} = \begin{bmatrix} -\frac{3}{8} & \frac{5}{4} \\ \frac{1}{4} & -\frac{1}{2} \end{bmatrix}.$$

Therefore, $(ABC)^{-1} = C^{-1}B^{-1}A^{-1}$.

37. Let x denote the number of copies of the deluxe edition and y the number of copies of the standard edition demanded per month when the unit prices are p and q dollars, respectively. Then the three systems of linear equations

$$5x + y = 20000 \quad 5x + y = 25000 \quad 5x + y = 25000$$
$$x + 3y = 15000 \quad x + 3y = 15000 \quad x + 3y = 20000$$

give the quantity demanded of each edition at the stated price. These systems may be written in the form $AX = B_1$, $AX = B_2$, and $AX = B_3$, where

$$A = \begin{bmatrix} 5 & 1 \\ 1 & 3 \end{bmatrix}, \quad B_1 = \begin{bmatrix} 20000 \\ 15000 \end{bmatrix}, \quad B_2 = \begin{bmatrix} 25000 \\ 15000 \end{bmatrix}, \quad \text{and} \quad B_3 = \begin{bmatrix} 25000 \\ 20000 \end{bmatrix}$$

Using the formula for finding the inverse of a 2×2 matrix, with $a = 5$, $b = 1$, $c = 1$, $d = 3$, and $D = ad - bc = (5)(3) - (1)(1) = 14$, we find that $A^{-1} = \begin{bmatrix} \frac{3}{14} & -\frac{1}{14} \\ -\frac{1}{14} & \frac{5}{14} \end{bmatrix}$.

a. $\begin{bmatrix} x \\ y \end{bmatrix} = \begin{bmatrix} \frac{3}{14} & -\frac{1}{14} \\ -\frac{1}{14} & \frac{5}{14} \end{bmatrix} \begin{bmatrix} 20,000 \\ 15,000 \end{bmatrix} = \begin{bmatrix} 3,214 \\ 3,929 \end{bmatrix}$ b. $\begin{bmatrix} x \\ y \end{bmatrix} = \begin{bmatrix} \frac{3}{14} & -\frac{1}{14} \\ -\frac{1}{14} & \frac{5}{14} \end{bmatrix} \begin{bmatrix} 25,000 \\ 15,000 \end{bmatrix} = \begin{bmatrix} 4,286 \\ 3,571 \end{bmatrix}$

c. $\begin{bmatrix} x \\ y \end{bmatrix} = \begin{bmatrix} \frac{3}{14} & -\frac{1}{14} \\ -\frac{1}{14} & \frac{5}{14} \end{bmatrix} \begin{bmatrix} 25,000 \\ 20,000 \end{bmatrix} = \begin{bmatrix} 3,929 \\ 5,357 \end{bmatrix}$.

39. Let x, y, and z denote the number of acres of soybeans, corn, and wheat to be cultivated, respectively. Furthermore, let a, b, and c denote the amount of land available; the amount of labor available, and the amount of money available for seeds, respectively. Then we have the system

$$x + y + z = a \quad \text{(land)}$$
$$2x + 6y + 6z = b \quad \text{(labor)}$$
$$12x + 20y + 8z = c \quad \text{(seeds)}$$

The system can be written in the form $AX = B$, where

$$A = \begin{bmatrix} 1 & 1 & 1 \\ 2 & 6 & 6 \\ 12 & 20 & 8 \end{bmatrix}, \quad X = \begin{bmatrix} x \\ y \\ z \end{bmatrix}, \quad \text{and} \quad B = \begin{bmatrix} a \\ b \\ c \end{bmatrix}$$

Using the technique for finding A^{-1} developed in this section, we find

$$A^{-1} = \begin{bmatrix} \frac{3}{2} & -\frac{1}{4} & 0 \\ -\frac{7}{6} & \frac{1}{12} & \frac{1}{12} \\ \frac{2}{3} & \frac{1}{6} & -\frac{1}{12} \end{bmatrix}.$$

a. Here $a = 1000$, $b = 4400$, and $c = 13,200$. Therefore

$$X = A^{-1}B = \begin{bmatrix} \frac{3}{2} & -\frac{1}{4} & 0 \\ -\frac{7}{6} & \frac{1}{12} & \frac{1}{12} \\ \frac{2}{3} & \frac{1}{6} & -\frac{1}{12} \end{bmatrix} \begin{bmatrix} 1,000 \\ 4,400 \\ 13,200 \end{bmatrix} = \begin{bmatrix} 400 \\ 300 \\ 300 \end{bmatrix}$$

So, Jackson Farms should cultivate 400, 300, and 300 acres of soybeans, corn, and wheat, respectively.

b. Here $a = 1200$, $b = 5200$, and $c = 16,400$. Therefore,

$$X = A^{-1}B = \begin{bmatrix} \frac{3}{2} & -\frac{1}{4} & 0 \\ -\frac{7}{6} & \frac{1}{12} & \frac{1}{12} \\ \frac{2}{3} & \frac{1}{6} & -\frac{1}{12} \end{bmatrix} \begin{bmatrix} 1,200 \\ 5,200 \\ 16,400 \end{bmatrix} = \begin{bmatrix} 500 \\ 400 \\ 300 \end{bmatrix}$$

So, Jackson Farms should cultivate 500, 400, and 300 acres of soybeans, corn, and wheat, respectively.

41. Let x, y, and z denote the amount to be invested in high-risk, medium-risk, and low-risk stocks, respectively. Next, let a denote the amount to be invested and let c denote the return on the investments. Then, we have the system

$$
\begin{aligned}
x + \quad y + \quad z &= a \\
x + \quad y - \quad z &= 0 \qquad \text{(since } z = x + y\text{)} \\
0.15x + 0.1y + 0.06z &= c
\end{aligned}
$$

The system is equivalent to the matrix equation $AX = B$, where

$$A = \begin{bmatrix} 1 & 1 & 1 \\ 1 & 1 & -1 \\ .15 & .10 & .06 \end{bmatrix}, \quad X = \begin{bmatrix} x \\ y \\ z \end{bmatrix}, \quad \text{and} \quad B = \begin{bmatrix} a \\ 0 \\ c \end{bmatrix}.$$

We find $A^{-1} = \begin{bmatrix} -1.6 & -0.4 & 20 \\ 2.1 & 0.9 & -20 \\ 0.5 & -0.5 & 0 \end{bmatrix}.$

a. Here $a = 200,000$ and $c = 20,000$. Therefore,

$$X = A^{-1}B = \begin{bmatrix} -1.6 & -0.4 & 20 \\ 2.1 & 0.9 & -20 \\ 0.5 & -0.5 & 0 \end{bmatrix} \begin{bmatrix} 200,000 \\ 0 \\ 20,000 \end{bmatrix} = \begin{bmatrix} 80,000 \\ 20,000 \\ 100,000 \end{bmatrix}.$$

So, the club should invest \$80,000 in high-risk, \$20,000 in medium risk, and \$100,000 in low risk stocks.

b. Here $a = 220,000$ and $c = 22,000$. The solution is $x = 88,000$, $y = 22,000$, and $z = 110,000$; that is, the club should invest \$88,000 in high-risk, \$22,000 in medium-risk, and \$110,000 in low-risk stocks.

c. Here $a = 240,000$ and $c = 22,000$. The result is \$56,000 in high-risk stocks, \$64,000 in medium-risk stocks, and \$120,000 in low-risk stocks.

43. True. Multiplying both sides of the equation by cA yields

$$I = (cA)(cA)^{-1} = (cA)\left[\frac{1}{c}(A^{-1})\right] = c\left(\frac{1}{c}\right)AA^{-1} = I.$$

45. True. $AX = B$ can have a unique solution only if A^{-1} exists, in which case the solution is found as follows:

$$A^{-1}(AX) = A^{-1}B$$
$$(A^{-1}A)X = A^{-1}B$$
$$IX = A^{-1}B$$
$$X = A^{-1}B$$

USING TECHNOLOGY EXERCISES 2.6, page 154

1. $\begin{bmatrix} 0.36 & 0.04 & -0.36 \\ 0.06 & 0.05 & 0.20 \\ -0.19 & 0.10 & 0.09 \end{bmatrix}$

3. $\begin{bmatrix} 0.01 & -0.09 & 0.31 & -0.11 \\ -0.25 & 0.58 & -0.15 & -0.02 \\ 0.86 & -0.42 & 0.07 & -0.37 \\ -0.27 & 0.01 & -0.05 & 0.31 \end{bmatrix}$

5. $\begin{bmatrix} 0.30 & 0.85 & -0.10 & -0.77 & -0.11 \\ -0.21 & 0.10 & 0.01 & -0.26 & 0.21 \\ 0.03 & -0.16 & 0.12 & -0.01 & 0.03 \\ -0.14 & -0.46 & 0.13 & 0.71 & -0.05 \\ 0.10 & -0.05 & -0.10 & -0.03 & 0.11 \end{bmatrix}$

7. $x = 1.2$, $y = 3.6$, and $z = 2.7$.

9. $x_1 = 2.50$, $x_2 = -0.88$, $x_3 = 0.70$, and $x_4 = 0.51$.

EXERCISES 2.7, page 161

1. a. The amount of agricultural products consumed in the production of $100 million worth of manufactured goods is given by $(100)(0.10)$, or $10 million.
    b. The amount of manufactured goods required to produce $200 million of all goods in the economy is given by $200(0.1 + 0.4 + 0.3) = 160$, or $160 million.
    c. From the input-output matrix, we see that the agricultural sector consumes the greatest amount of agricultural products, namely, 0.4 units, in the production of each unit of goods in that sector. The manufacturing and transportation sectors consume the least, 0.1 units each.

3. Multiplying both sides of the given equation on the left by $(I - A)^{-1}$, we see that
    $$X = (I - A)^{-1}D.$$

    Now, $(I - A) = \begin{bmatrix} 1 & 0 \\ 0 & 1 \end{bmatrix} - \begin{bmatrix} 0.4 & 0.2 \\ 0.3 & 0.1 \end{bmatrix} = \begin{bmatrix} 0.6 & -0.2 \\ -0.3 & 0.9 \end{bmatrix}.$

    Using the formula for finding the inverse of a 2×2 matrix, we find
    $$(I - A)^{-1} = \begin{bmatrix} 1.875 & 0.417 \\ 0.625 & 1.25 \end{bmatrix}.$$

    Then, $(I - A)^{-1}X = \begin{bmatrix} 1.875 & 0.417 \\ 0.625 & 1.25 \end{bmatrix}\begin{bmatrix} 10 \\ 12 \end{bmatrix} = \begin{bmatrix} 23.754 \\ 21.25 \end{bmatrix}.$

5. We first compute $(I - A) = \begin{bmatrix} 1 & 0 \\ 0 & 1 \end{bmatrix} - \begin{bmatrix} 0.5 & 0.2 \\ 0.2 & 0.5 \end{bmatrix} = \begin{bmatrix} 0.5 & -0.2 \\ -0.2 & 0.5 \end{bmatrix}$

    Using the formula for finding the inverse of a 2×2 matrix, we find
    $$(I - A)^{-1} = \begin{bmatrix} 2.381 & 0.952 \\ 0.952 & 2.381 \end{bmatrix}.$$

Then $\begin{bmatrix} x \\ y \end{bmatrix} = \begin{bmatrix} 2.381 & 0.952 \\ 0.952 & 2.381 \end{bmatrix} \begin{bmatrix} 10 \\ 20 \end{bmatrix} = \begin{bmatrix} 42.85 \\ 57.14 \end{bmatrix}$

7. We verify

$$(I - A)(I - A)^{-1} = \begin{bmatrix} 0.92 & -0.60 & -0.30 \\ -0.04 & 0.98 & -0.01 \\ -0.02 & 0 & 0.94 \end{bmatrix} \begin{bmatrix} 1.13 & 0.69 & 0.37 \\ 0.05 & 1.05 & 0.03 \\ 0.02 & 0.02 & 1.07 \end{bmatrix} = \begin{bmatrix} 1 & 0 & 0 \\ 0 & 1 & 0 \\ 0 & 0 & 1 \end{bmatrix}$$

9. a. $A = \begin{bmatrix} 0.2 & 0.4 \\ 0.3 & 0.3 \end{bmatrix}$ and

$$(I - A) = \begin{bmatrix} 1 & 0 \\ 0 & 1 \end{bmatrix} - \begin{bmatrix} 0.2 & 0.4 \\ 0.3 & 0.3 \end{bmatrix} = \begin{bmatrix} 0.8 & -0.4 \\ -0.3 & 0.7 \end{bmatrix}.$$

Using the formula for finding the inverse of a 2×2 matrix, we find

$$(I - A)^{-1} = \begin{bmatrix} 1.591 & 0.909 \\ 0.682 & 1.818 \end{bmatrix}.$$

Then $\begin{bmatrix} x \\ y \end{bmatrix} = \begin{bmatrix} 1.591 & 0.909 \\ 0.682 & 1.818 \end{bmatrix} \begin{bmatrix} 120 \\ 140 \end{bmatrix} = \begin{bmatrix} 318.18 \\ 336.36 \end{bmatrix}$

To fullfill consumer demand, $318.2 million worth of agricultural goods and $336.4 million worth of manufactured goods should be produced.

b. The net value of goods consumed in the internal process of production is

$$AX = X - D = \begin{bmatrix} 318.18 \\ 336.36 \end{bmatrix} - \begin{bmatrix} 120 \\ 140 \end{bmatrix} = \begin{bmatrix} 198.18 \\ 196.36 \end{bmatrix}.$$

or $198.2 million of agricultural goods and $196.4 million worth of manufactured goods.

11. a.

$$(I - A) = \begin{bmatrix} 1 & 0 & 0 \\ 0 & 1 & 0 \\ 0 & 0 & 1 \end{bmatrix} - \begin{bmatrix} 0.4 & 0.1 & 0.1 \\ 0.1 & 0.4 & 0.3 \\ 0.2 & 0.2 & 0.2 \end{bmatrix} = \begin{bmatrix} 0.6 & -0.1 & -0.1 \\ -0.1 & 0.6 & -0.3 \\ -0.2 & -0.2 & 0.8 \end{bmatrix}$$

Using the methods of Section 2.6 we next compute the inverse of $(1 - A)^{-1}$ and use this value to find

$$X = (1-A)^{-1}D = \begin{bmatrix} 1.875 & 0.446 & 0.402 \\ 0.625 & 2.054 & 0.848 \\ 0.625 & 0.625 & 1.563 \end{bmatrix} \begin{bmatrix} 200 \\ 100 \\ 60 \end{bmatrix} = \begin{bmatrix} 443.7 \\ 381.3 \\ 281.3 \end{bmatrix}.$$

Therefore, to fulfull demand, $443.7 million worth of agricultural products, $381.3 million worth of manufactured products, and $281.3 million worth of transportation services should be produced.

b. To meet the gross output, the value of goods and transportation consumed in the internal process of production is

$$AX = X - D = \begin{bmatrix} 443.7 \\ 381.3 \\ 281.3 \end{bmatrix} - \begin{bmatrix} 200 \\ 100 \\ 60 \end{bmatrix} = \begin{bmatrix} 243.7 \\ 281.3 \\ 221.3 \end{bmatrix},$$

or $243.7 million worth of agricultural products, $281.3 million worth of manufactured services, and $221.3 million worth of transportation services.

13. We want to solve the equation $(I - A)X = D$ for X, the total output matrix. First, we compute

$$(I - A) = \begin{bmatrix} 1 & 0 \\ 0 & 1 \end{bmatrix} - \begin{bmatrix} 0.4 & 0.2 \\ 0.3 & 0.5 \end{bmatrix} = \begin{bmatrix} 0.6 & -0.2 \\ -0.3 & 0.5 \end{bmatrix}.$$

Using the formula for finding the inverse of a 2×2 matrix, we find

$$(I - A)^{-1} = \begin{bmatrix} 2.08 & 0.833 \\ 1.25 & 2.5 \end{bmatrix}$$

Therefore,

$$X = (I - A)^{-1}D = \begin{bmatrix} 2.08 & 0.833 \\ 1.25 & 2.5 \end{bmatrix} \begin{bmatrix} 12 \\ 24 \end{bmatrix} = \begin{bmatrix} 45 \\ 75 \end{bmatrix} .$$

We conclude that $45 million worth of goods of one industry and $75 million worth of goods of the other industry must be produced.

15. First, we compute

$$I - A = \begin{bmatrix} 1 & 0 & 0 \\ 0 & 1 & 0 \\ 0 & 0 & 1 \end{bmatrix} - \begin{bmatrix} 0.2 & 0.4 & 0.2 \\ 0.5 & 0 & 0.5 \\ 0 & 0.2 & 0 \end{bmatrix} = \begin{bmatrix} 0.8 & -0.4 & -0.2 \\ -0.5 & 1 & -0.5 \\ 0 & -0.2 & 1 \end{bmatrix}.$$

Next, using the Gauss-Jordan method, we find

$$(I - A)^{-1} = \begin{bmatrix} 1.8 & 0.88 & 0.80 \\ 1 & 1.6 & 1 \\ 0.2 & 0.32 & 1.20 \end{bmatrix}$$

Then $\begin{bmatrix} x \\ y \\ z \end{bmatrix} = \begin{bmatrix} 1.8 & 0.88 & 0.80 \\ 1 & 1.6 & 1 \\ 0.2 & 0.32 & 1.20 \end{bmatrix} \begin{bmatrix} 10 \\ 5 \\ 15 \end{bmatrix} = \begin{bmatrix} 34.4 \\ 33 \\ 21.6 \end{bmatrix}.$

We conclude that $34.4 million worth of goods of one industry, $33 million worth of a second industry, and $21.6 million worth of a third industry should be produced.

USING TECHNOLOGY EXERCISES 2.7, page 166

1. The final outputs of the first, second, third, and fourth industries are 602.62, 502.30, 572.57, and 523.46 units, respectively.

3. The final outputs of the first, second, third, and fourth industries are 143.06, 132.98, 188.59, and 125.53 units, respectively.

CHAPTER 2, REVIEW EXERCISES, page 168

1. $\begin{bmatrix} 1 & 2 \\ -1 & 3 \\ 2 & 1 \end{bmatrix} + \begin{bmatrix} 1 & 0 \\ 0 & 1 \\ 1 & 2 \end{bmatrix} = \begin{bmatrix} 2 & 2 \\ -1 & 4 \\ 3 & 3 \end{bmatrix}.$

3. $\begin{bmatrix} -3 & 2 & 1 \end{bmatrix} \begin{bmatrix} 2 & 1 \\ -1 & 0 \\ 2 & 1 \end{bmatrix} = \begin{bmatrix} -6 & -2 \end{bmatrix}.$

5. By the equality of matrices, $x = 2$, $z = 1$, $y = 3$ and $w = 3$.

7. By the equality of matrices,
$a + 3 = 6$, or $a = 3$.
$-1 = e + 2$, or $e = -3$; $b = 4$
$c + 1 = -1$, or $c = -2$; $d = 2$
$e + 2 = -1$, and $e = -3$.

9. $2A + 3B = 2\begin{bmatrix} 1 & 3 & 1 \\ -2 & 1 & 3 \\ 4 & 0 & 2 \end{bmatrix} + 3\begin{bmatrix} 2 & 1 & 3 \\ -2 & -1 & -1 \\ 1 & 4 & 2 \end{bmatrix} = \begin{bmatrix} 2 & 6 & 2 \\ -4 & 2 & 6 \\ 8 & 0 & 4 \end{bmatrix} + \begin{bmatrix} 6 & 3 & 9 \\ -6 & -3 & -3 \\ 3 & 12 & 6 \end{bmatrix}$

$$= \begin{bmatrix} 8 & 9 & 11 \\ -10 & -1 & 3 \\ 11 & 12 & 10 \end{bmatrix}.$$

11.
$$3A = 3 \begin{bmatrix} 1 & 3 & 1 \\ -2 & 1 & 3 \\ 4 & 0 & 2 \end{bmatrix} = \begin{bmatrix} 3 & 9 & 3 \\ -6 & 3 & 9 \\ 12 & 0 & 6 \end{bmatrix}$$

and $\quad 2(3A) = 2 \begin{bmatrix} 3 & 9 & 3 \\ -6 & 3 & 9 \\ 12 & 0 & 6 \end{bmatrix} = \begin{bmatrix} 6 & 18 & 6 \\ -12 & 6 & 18 \\ 24 & 0 & 12 \end{bmatrix}.$

13.
$$B - C = \begin{bmatrix} 2 & 1 & 3 \\ -2 & -1 & -1 \\ 1 & 4 & 2 \end{bmatrix} - \begin{bmatrix} 3 & -1 & 2 \\ 1 & 6 & 4 \\ 2 & 1 & 3 \end{bmatrix} = \begin{bmatrix} -1 & 2 & 1 \\ -3 & -7 & -5 \\ -1 & 3 & -1 \end{bmatrix}$$

and so $\quad A(B - C) = \begin{bmatrix} 1 & 3 & 1 \\ -2 & 1 & 3 \\ 4 & 0 & 2 \end{bmatrix} \begin{bmatrix} -1 & 2 & 1 \\ -3 & -7 & -5 \\ -1 & 3 & -1 \end{bmatrix} = \begin{bmatrix} -11 & -16 & -15 \\ -4 & -2 & -10 \\ -6 & 14 & 2 \end{bmatrix}.$

15.
$$BC = \begin{bmatrix} 2 & 1 & 3 \\ -2 & -1 & -1 \\ 1 & 4 & 2 \end{bmatrix} \begin{bmatrix} 3 & -1 & 2 \\ 1 & 6 & 4 \\ 2 & 1 & 3 \end{bmatrix} = \begin{bmatrix} 13 & 7 & 17 \\ -9 & -5 & -11 \\ 11 & 25 & 24 \end{bmatrix}$$

$$ABC = \begin{bmatrix} 1 & 3 & 1 \\ -2 & 1 & 3 \\ 4 & 0 & 2 \end{bmatrix} \begin{bmatrix} 13 & 7 & 17 \\ -9 & -5 & -11 \\ 11 & 25 & 24 \end{bmatrix} = \begin{bmatrix} -3 & 17 & 8 \\ -2 & 56 & 27 \\ 74 & 78 & 116 \end{bmatrix}.$$

17. Using the Gauss-Jordan elimination method, we find

$$\begin{bmatrix} 2 & -3 & 5 \\ 3 & 4 & -1 \end{bmatrix} \xrightarrow{\frac{1}{2}R_1} \begin{bmatrix} 1 & -\frac{3}{2} & \frac{5}{2} \\ 3 & 4 & -1 \end{bmatrix} \xrightarrow{R_2 - 3R_1} \begin{bmatrix} 1 & -\frac{3}{2} & \frac{5}{2} \\ 0 & \frac{17}{2} & -\frac{17}{2} \end{bmatrix}$$

$$\xrightarrow{\frac{2}{17}R_2} \begin{bmatrix} 1 & -\frac{3}{2} & \frac{5}{2} \\ 0 & 1 & -1 \end{bmatrix} \xrightarrow{R_1+\frac{3}{2}R_2} \begin{bmatrix} 1 & 0 & 1 \\ 0 & 1 & -1 \end{bmatrix}.$$ We conclude that $x = 1$ and $y = -1$.

19. $\begin{bmatrix} 1 & -1 & 2 & 5 \\ 3 & 2 & 1 & 10 \\ 2 & -3 & -2 & -10 \end{bmatrix} \xrightarrow[R_3-2R_1]{R_2-3R_1} \begin{bmatrix} 1 & -1 & 2 & 5 \\ 0 & 5 & -5 & -5 \\ 0 & -1 & -6 & -20 \end{bmatrix} \xrightarrow{\frac{1}{5}R_2} \begin{bmatrix} 1 & -1 & 2 & 5 \\ 0 & 1 & -1 & -1 \\ 0 & -1 & -6 & -20 \end{bmatrix}$

$\xrightarrow[R_3+R_2]{R_1+R_2} \begin{bmatrix} 1 & 0 & 1 & 4 \\ 0 & 1 & -1 & -1 \\ 0 & 0 & -7 & -21 \end{bmatrix} \xrightarrow{-\frac{1}{7}R_3} \begin{bmatrix} 1 & 0 & 1 & 4 \\ 0 & 1 & -1 & -1 \\ 0 & 0 & 1 & 3 \end{bmatrix} \xrightarrow[R_2+R_3]{R_1-R_3}$

$= \begin{bmatrix} 1 & 0 & 0 & 1 \\ 0 & 1 & 0 & 2 \\ 0 & 0 & 1 & 3 \end{bmatrix}.$ Therefore, $x = 1$, $y = 2$, and $z = 3$.

21. $\begin{bmatrix} 3 & -2 & 4 & 11 \\ 2 & -4 & 5 & 4 \\ 1 & 2 & -1 & 10 \end{bmatrix} \xrightarrow{R_1-R_2} \begin{bmatrix} 1 & 2 & -1 & 7 \\ 2 & -4 & 5 & 4 \\ 1 & 2 & -1 & 10 \end{bmatrix} \xrightarrow[R_3-R_1]{R_2-2R_1} \begin{bmatrix} 1 & 2 & -1 & 7 \\ 0 & -8 & 7 & -10 \\ 0 & 0 & 0 & 3 \end{bmatrix}.$

Since this last row implies that $0 = 3!$, we conclude that the system has no solution.

23. $\begin{bmatrix} 3 & -2 & 1 & 4 \\ 1 & 3 & -4 & -3 \\ 2 & -3 & 5 & 7 \\ 1 & -8 & 9 & 10 \end{bmatrix} \xrightarrow{R_1-R_3} \begin{bmatrix} 1 & 1 & -4 & -3 \\ 1 & 3 & -4 & -3 \\ 2 & -3 & 5 & 7 \\ 1 & -8 & 9 & 10 \end{bmatrix} \xrightarrow[R_4-R_1]{\substack{R_2-R_1 \\ R_3-2R_1}} \begin{bmatrix} 1 & 1 & -4 & -3 \\ 0 & 2 & 0 & 0 \\ 0 & -5 & 13 & 13 \\ 0 & -9 & 13 & 13 \end{bmatrix}$

$\xrightarrow{\frac{1}{2}R_2} \begin{bmatrix} 1 & 1 & -4 & -3 \\ 0 & 1 & 0 & 0 \\ 0 & -5 & 13 & 13 \\ 0 & -9 & 13 & 13 \end{bmatrix} \xrightarrow[R_4+9R_2]{\substack{R_1-R_2 \\ R_3+5R_2}} \begin{bmatrix} 1 & 0 & -4 & -3 \\ 0 & 1 & 0 & 0 \\ 0 & 0 & 13 & 13 \\ 0 & 0 & 13 & 13 \end{bmatrix} \xrightarrow{\frac{1}{13}R_3} \begin{bmatrix} 1 & 0 & -4 & -3 \\ 0 & 1 & 0 & 0 \\ 0 & 0 & 1 & 1 \\ 0 & 0 & 13 & 13 \end{bmatrix}$

$\xrightarrow[R_4-13R_3]{R_1+4R_3} \begin{bmatrix} 1 & 0 & 0 & 1 \\ 0 & 1 & 0 & 0 \\ 0 & 0 & 1 & 1 \\ 0 & 0 & 0 & 0 \end{bmatrix}.$ Therefore, $x = 1$, $y = 0$, and $z = 1$.

25. $A^{-1} = \dfrac{1}{(3)(2)-(1)(1)}\begin{bmatrix} 2 & -1 \\ -1 & 3 \end{bmatrix} = \begin{bmatrix} \frac{2}{5} & -\frac{1}{5} \\ -\frac{1}{5} & \frac{3}{5} \end{bmatrix}.$

27. $A^{-1} = \dfrac{1}{(3)(2)-(2)(4)}\begin{bmatrix} 2 & -4 \\ -2 & 3 \end{bmatrix} = \begin{bmatrix} -1 & 2 \\ 1 & -\frac{3}{2} \end{bmatrix}.$

29. $\left[\begin{array}{ccc|ccc} 2 & 3 & 1 & 1 & 0 & 0 \\ 1 & -1 & 2 & 0 & 1 & 0 \\ 1 & 2 & 1 & 0 & 0 & 1 \end{array}\right] \xrightarrow{R_1 - R_2} \left[\begin{array}{ccc|ccc} 1 & 4 & -1 & 1 & -1 & 0 \\ 1 & -1 & 2 & 0 & 1 & 0 \\ 1 & 2 & 1 & 0 & 0 & 1 \end{array}\right] \xrightarrow[R_3 - R_1]{R_2 - R_1}$

$\left[\begin{array}{ccc|ccc} 1 & 4 & -1 & 1 & -1 & 0 \\ 0 & -5 & 3 & -1 & 2 & 0 \\ 0 & -2 & 2 & -1 & 1 & 1 \end{array}\right] \xrightarrow{R_2 - 3R_3} \left[\begin{array}{ccc|ccc} 1 & 4 & -1 & 1 & -1 & 0 \\ 0 & 1 & -3 & 2 & -1 & -3 \\ 0 & -2 & 2 & -1 & 1 & 1 \end{array}\right] \xrightarrow[R_3 + 2R_2]{R_1 - 4R_2}$

$\left[\begin{array}{ccc|ccc} 1 & 0 & 11 & -7 & 3 & 12 \\ 0 & 1 & -3 & 2 & -1 & -3 \\ 0 & 0 & -4 & 3 & -1 & -5 \end{array}\right] \xrightarrow{-\frac{1}{4}R_3} \left[\begin{array}{ccc|ccc} 1 & 0 & 11 & -7 & 3 & 12 \\ 0 & 1 & -3 & 2 & -1 & -3 \\ 0 & 0 & 1 & -\frac{3}{4} & \frac{1}{4} & \frac{5}{4} \end{array}\right] \xrightarrow[R_2 + 3R_3]{R_1 - 11R_3}$

$\left[\begin{array}{ccc|ccc} 1 & 0 & 0 & \frac{5}{4} & \frac{1}{4} & -\frac{7}{4} \\ 0 & 1 & 0 & -\frac{1}{4} & -\frac{1}{4} & \frac{3}{4} \\ 0 & 0 & 1 & -\frac{3}{4} & \frac{1}{4} & \frac{5}{4} \end{array}\right].$ So $A^{-1} = \begin{bmatrix} \frac{5}{4} & \frac{1}{4} & -\frac{7}{4} \\ -\frac{1}{4} & -\frac{1}{4} & \frac{3}{4} \\ -\frac{3}{4} & \frac{1}{4} & \frac{5}{4} \end{bmatrix}.$

31. $\left[\begin{array}{ccc|ccc} 1 & 2 & 4 & 1 & 0 & 0 \\ 3 & 1 & 2 & 0 & 1 & 0 \\ 1 & 0 & -6 & 0 & 0 & 1 \end{array}\right] \xrightarrow[R_3 - R_1]{R_2 - 3R_1} \left[\begin{array}{ccc|ccc} 1 & 2 & 4 & 1 & 0 & 0 \\ 0 & -5 & -10 & -3 & 1 & 0 \\ 0 & -2 & -10 & -1 & 0 & 1 \end{array}\right] \xrightarrow{R_2 - 3R_3}$

$\left[\begin{array}{ccc|ccc} 1 & 2 & 4 & 1 & 0 & 0 \\ 0 & 1 & 20 & 0 & 1 & -3 \\ 0 & -2 & -10 & -1 & 0 & 1 \end{array}\right] \xrightarrow[R_3 + 2R_2]{R_1 - 2R_2} \left[\begin{array}{ccc|ccc} 1 & 0 & -36 & 1 & -2 & 6 \\ 0 & 1 & 20 & 0 & 1 & -3 \\ 0 & 0 & 30 & -1 & 2 & -5 \end{array}\right] \xrightarrow{\frac{1}{30}R_3}$

$$\left[\begin{array}{ccc|ccc} 1 & 0 & -36 & 1 & -2 & 6 \\ 0 & 1 & 20 & 0 & 1 & -3 \\ 0 & 0 & 1 & -\frac{1}{30} & \frac{1}{15} & -\frac{1}{6} \end{array}\right] \xrightarrow[R_2-20R_3]{R_1+36R_3} \left[\begin{array}{ccc|ccc} 1 & 0 & 0 & -\frac{1}{5} & \frac{2}{5} & 0 \\ 0 & 1 & 0 & \frac{2}{3} & -\frac{1}{3} & \frac{1}{3} \\ 0 & 0 & 1 & -\frac{1}{30} & \frac{1}{15} & -\frac{1}{6} \end{array}\right]$$

So $\quad A^{-1} = \begin{bmatrix} -\frac{1}{5} & \frac{2}{5} & 0 \\ \frac{2}{3} & -\frac{1}{3} & \frac{1}{3} \\ -\frac{1}{30} & \frac{1}{15} & -\frac{1}{6} \end{bmatrix}.$

33. $(A^{-1}B)^{-1} = B^{-1}(A^{-1})^{-1} = B^{-1}A$. Now

$$B^{-1} = \frac{1}{(3)(2)-4(1)}\begin{bmatrix} 2 & -1 \\ -4 & 3 \end{bmatrix} = \begin{bmatrix} 1 & -\frac{1}{2} \\ -2 & \frac{3}{2} \end{bmatrix}. \quad B^{-1}A = \begin{bmatrix} 1 & -\frac{1}{2} \\ -2 & \frac{3}{2} \end{bmatrix}\begin{bmatrix} 1 & 2 \\ -1 & 2 \end{bmatrix} = \begin{bmatrix} \frac{3}{2} & 1 \\ -\frac{7}{2} & -1 \end{bmatrix}.$$

35. $2A-C = \begin{bmatrix} 2 & 4 \\ -2 & 4 \end{bmatrix} - \begin{bmatrix} 1 & 1 \\ -1 & 2 \end{bmatrix} = \begin{bmatrix} 1 & 3 \\ -1 & 2 \end{bmatrix}.$

$$(2A-C)^{-1} = \frac{1}{(1)(2)-(-1)(3)}\begin{bmatrix} 2 & -3 \\ 1 & 1 \end{bmatrix} = \begin{bmatrix} \frac{2}{5} & -\frac{3}{5} \\ \frac{1}{5} & \frac{1}{5} \end{bmatrix}.$$

37. $A = \begin{bmatrix} 2 & 3 \\ 1 & -2 \end{bmatrix}, \quad X = \begin{bmatrix} x \\ y \end{bmatrix}, \quad C = \begin{bmatrix} -8 \\ 3 \end{bmatrix}; \quad A^{-1} = \frac{1}{(-2)(2)-(1)(3)}\begin{bmatrix} -2 & -3 \\ -1 & 2 \end{bmatrix} = \begin{bmatrix} \frac{2}{7} & \frac{3}{7} \\ \frac{1}{7} & -\frac{2}{7} \end{bmatrix}$

$$\begin{bmatrix} x \\ y \end{bmatrix} = A^{-1}B = \begin{bmatrix} \frac{2}{7} & \frac{3}{7} \\ \frac{1}{7} & -\frac{2}{7} \end{bmatrix}\begin{bmatrix} -8 \\ 3 \end{bmatrix} = \begin{bmatrix} -1 \\ -2 \end{bmatrix}.$$

39. Put
$$X = \begin{bmatrix} x \\ y \\ z \end{bmatrix}, \quad A = \begin{bmatrix} 1 & -2 & 4 \\ 2 & 3 & -2 \\ 1 & 4 & -6 \end{bmatrix}, \quad C = \begin{bmatrix} 13 \\ 0 \\ -15 \end{bmatrix}.$$

Then $AX = C$ and $X = A^{-1}C$. To find A^{-1},

$$\begin{bmatrix} 1 & -2 & 4 & | & 1 & 0 & 0 \\ 2 & 3 & -2 & | & 0 & 1 & 0 \\ 1 & 4 & -6 & | & 0 & 0 & 1 \end{bmatrix} \xrightarrow[R_3-R_1]{R_2-2R_1} \begin{bmatrix} 1 & -2 & 4 & | & 1 & 0 & 0 \\ 0 & 7 & -10 & | & -2 & 1 & 0 \\ 0 & 6 & -10 & | & -1 & 0 & 1 \end{bmatrix} \xrightarrow{R_2-R_3}$$

$$\begin{bmatrix} 1 & -2 & 4 & | & 1 & 0 & 0 \\ 0 & 1 & 0 & | & -1 & 1 & -1 \\ 0 & 6 & -10 & | & -1 & 0 & 1 \end{bmatrix} \xrightarrow[R_3-6R_2]{R_1+2R_2} \begin{bmatrix} 1 & 0 & 4 & | & -1 & 2 & -2 \\ 0 & 1 & 0 & | & -1 & 1 & -1 \\ 0 & 0 & -10 & | & 5 & -6 & 7 \end{bmatrix} \xrightarrow{-\frac{1}{10}R_3}$$

$$\begin{bmatrix} 1 & 0 & 4 & | & -1 & 2 & -2 \\ 0 & 1 & 0 & | & -1 & 1 & -1 \\ 0 & 0 & 1 & | & -\frac{1}{2} & \frac{3}{5} & -\frac{7}{10} \end{bmatrix} \xrightarrow{R_1-4R_3} \begin{bmatrix} 1 & 0 & 0 & | & 1 & -\frac{2}{5} & \frac{4}{5} \\ 0 & 1 & 0 & | & -1 & 1 & -1 \\ 0 & 0 & 1 & | & -\frac{1}{2} & \frac{3}{5} & -\frac{7}{10} \end{bmatrix}.$$

So $A^{-1} = \begin{bmatrix} 1 & -\frac{2}{5} & \frac{4}{5} \\ -1 & 1 & -1 \\ -\frac{1}{2} & \frac{3}{5} & -\frac{7}{10} \end{bmatrix}$. Therefore, $X = A^{-1}C = \begin{bmatrix} 1 & -\frac{2}{5} & \frac{4}{5} \\ -1 & 1 & -1 \\ -\frac{1}{2} & \frac{3}{5} & -\frac{7}{10} \end{bmatrix} \begin{bmatrix} 13 \\ 0 \\ -15 \end{bmatrix} = \begin{bmatrix} 1 \\ 2 \\ 4 \end{bmatrix}$

that is, $x = 1$, $y = 2$, and $z = 4$.

41. Let

$$S = \begin{array}{cccc} & \text{Prem} & \text{Sup} & \text{Reg} & \text{Dies} \\ & \begin{bmatrix} 600 & 800 & 1000 & 700 \\ 700 & 600 & 1200 & 400 \\ 900 & 700 & 1400 & 800 \end{bmatrix} \end{array} \quad \text{and} \quad T = \begin{array}{c} \text{Prem} \\ \text{Sup} \\ \text{Reg} \\ \text{Dies} \end{array} \begin{bmatrix} 1.60 \\ 1.40 \\ 1.20 \\ 1.50 \end{bmatrix}$$

be the matrices representing the sales in the three gasoline stations and the unit prices for the various fuels, respectively. Then the total revenue at each station is found by computing

$$ST = \begin{bmatrix} 600 & 800 & 1000 & 700 \\ 700 & 600 & 1200 & 400 \\ 900 & 700 & 1400 & 800 \end{bmatrix} \begin{bmatrix} 1.60 \\ 1.40 \\ 1.20 \\ 1.50 \end{bmatrix} = \begin{bmatrix} 4330 \\ 4000 \\ 5300 \end{bmatrix}.$$

We conclude that the total revenue of station A is $4330, that of station B is $4000, and that of station C is $5300.

43. We wish to solve the system of equations

$$2x + 2y + 3z = 210$$
$$2x + 3y + 4z = 270$$
$$3x + 4y + 3z = 300.$$

Using the Gauss–Jordan method of elimination, we find

$$\begin{bmatrix} 2 & 2 & 3 & | & 210 \\ 2 & 3 & 4 & | & 270 \\ 3 & 4 & 3 & | & 300 \end{bmatrix} \xrightarrow{\frac{1}{2}R_1} \begin{bmatrix} 1 & 1 & \frac{3}{2} & | & 105 \\ 2 & 3 & 4 & | & 270 \\ 3 & 4 & 3 & | & 300 \end{bmatrix} \xrightarrow[R_3 - 3R_1]{R_2 - 2R_1} \begin{bmatrix} 1 & 1 & \frac{3}{2} & | & 105 \\ 0 & 1 & 1 & | & 60 \\ 0 & 1 & -\frac{3}{2} & | & -15 \end{bmatrix}$$

$$\xrightarrow[R_3 - R_2]{R_1 - R_2} \begin{bmatrix} 1 & 0 & \frac{1}{2} & | & 45 \\ 0 & 1 & 1 & | & 60 \\ 0 & 0 & -\frac{5}{2} & | & -75 \end{bmatrix} \xrightarrow{-\frac{2}{5}R_3} \begin{bmatrix} 1 & 0 & \frac{1}{2} & | & 45 \\ 0 & 1 & 1 & | & 60 \\ 0 & 0 & 1 & | & 30 \end{bmatrix} \xrightarrow[R_2 - R_3]{R_1 - \frac{1}{2}R_3} \begin{bmatrix} 1 & 0 & 0 & | & 30 \\ 0 & 1 & 0 & | & 30 \\ 0 & 0 & 1 & | & 30 \end{bmatrix}$$

So $x = y = z = 30$. Therefore, Desmond should produce 30 of each type of pendant.

CHAPTER 3

EXERCISES 3.1, page 176

1. $4x - 8 < 0$ implies $x < 2$. The graph of the inequality follows.

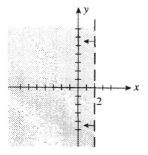

3. $x - y \leq 0$ implies $x \leq y$. The graph of the inequality follows.

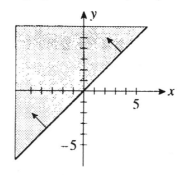

5. The graph of the inequality $x \leq -3$ follows.

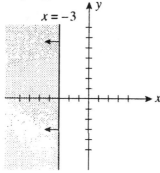

7. We first sketch the straight line with equation $2x + y = 4$. Next, picking the test point $(0,0)$, we have $2(0) + (0) = 0 \le 4$. We conclude that the half-plane containing the origin is the required half-plane.

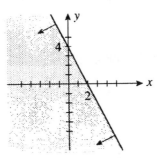

9. We first sketch the graph of the straight line $4x - 3y = -24$. Next, picking the test point $(0,0)$, we see that $4(0) - 3(0) = 0 \not< -24$. We conclude that the half-plane not containing the origin is the required half-plane. The graph of this inequality follows.

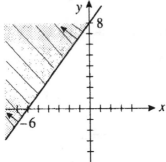

11. The system of linear inequalities that describes the shaded region is
$$x \ge 1, x \le 5, y \ge 2, \text{ and } y \le 4.$$
We may also combine the first and second inequalities and the third and fourth inequalities and write
$$1 \le x \le 5 \quad \text{and} \quad 2 \le y \le 4.$$

13. The system of linear inequalities that describes the shaded region is
$$2x - y \ge 2, 5x + 7y \ge 35, \text{ and } x \le 4.$$

15. The system of linear inequalities that describes the shaded region is
$$7x + 4y \le 140, x + 3y \ge 30, \text{ and } x - y \ge -10.$$

17. The system of linear inequalities that describes the shaded region is
$$x + y \ge 7, x \ge 2, y \ge 3, \text{ and } y \le 7.$$

19. The required solution set is shown below.

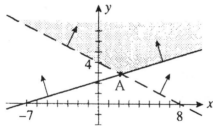

To find the coordinates of A, we solve the system
$$2x + 4y = 16$$
$$-x + 3y = 7,$$

giving $A = (2,3)$. Observe that a dotted line is used to show that no point on the line constitutes a solution to the given problem. Observe that this is an unbounded solution set.

21. The solution set is shown in the figure below. Observe that the set is unbounded.

To find the coordinates of A, we solve the system $\begin{cases} x - y = 0 \\ 2x + 3y = 10 \end{cases}$ giving $A = (2,2)$.

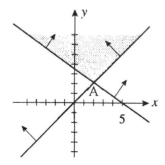

23. The half-planes defined by the two inequalities are shown in the figure at the right. Since the two half-planes have no points in common, we conclude that the given system of inequalities has no solution. (The empty set is a bounded set.)

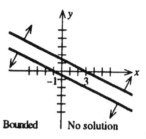

Bounded   No solution

25. The half-planes defined by the three inequalities are shown below. The point A is found by solving the system $\begin{cases} x + y = 6 \\ x = 3 \end{cases}$ giving $A = (3,3)$. Observe that this is a

bounded solution set.

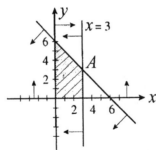

27. The half-planes defined by the given inequalities are shown below. Observe that the two lines described by the equations $3x - 6y = 12$ and $-x + 2y = 4$ do not intersect because they are parallel. The solution set is unbounded.

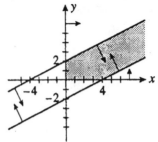

29. The required solution set is shown in the figure below. The coordinates of A are found by solving the system $\begin{cases} 3x - 7y = -24 \\ x + 3y = 8 \end{cases}$ giving $(-1,3)$. The solution set is unbounded.

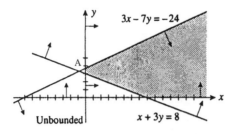

31. The required solution set is shown in the figure that follows. The solution set is bounded.

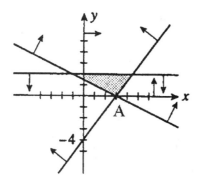

33. The required solution set is shown in the figure below. The solution set has vertices at (0,6), (5,0), (4,0), and (1,3). The solution set is bounded.

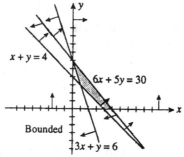

35. The required solution set is shown in the figure below. The unbounded solution set has vertices at (2,8), (0,6), (0,3),and (2,2).

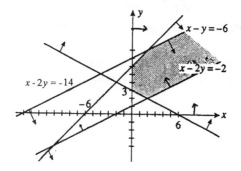

37. False. It is always a half-plane. A straight line is the graph of a linear equation and vice-versa.

39. True. Since a circle can always be enclosed by a rectangle, the solution set of such a

system is bounded if it can be enclosed by a rectangle.

EXERCISES 3.2, page 185

1. We tabulate the given information:

	Product A	Product B	Time Available
Machine I	6	9	300
Machine II	5	4	180
Profit per unit ($)	3	4	

Let x and y denote the number of units of Product A and Product B to be produced. Then the required linear programming problem is:
Maximize $P = 3x + 4y$ subject to the constraints
$$6x + 9y \le 300$$
$$5x + 4y \le 180$$
$$x \ge 0, y \ge 0$$

3. Let x denote the number of model A grates to be produced and y denote the number of model B grates to be produced. Since only 1000 pounds of cast iron are available, we must have
$$3x + 4y \le 1000.$$
The restriction that only 20 hours of labor are available per day implies that
$$6x + 3y \le 1200. \quad \text{(time in minutes)}$$
Then the profit on the production of these grates is given by
$$P = 2x + 1.5y.$$
The additional restriction that at least 150 model A grates be produced each day implies that $x \ge 150$. Summarizing, we have the following linear programming problem:
Maximize $P = 2x + 1.5y$ subject to
$$3x + 4y \le 1000$$
$$6x + 3y \le 1200$$
$$x \ge 150, y \ge 0$$

5. Let x and y denote the amount (in thousands of dollars) to be invested in project A

and project B, respectively. Since the amount available for investment is up to $500,000, we have $x + y \le 500$. Next, the condition on the allocation of the funds implies that
$$y \le 0.4(x+y), \quad -0.4x + 0.6y \le 0, \quad \text{or} \quad -2x + 3y \le 0$$
The linear programming problem at hand is
Maximize $P = 0.1x + 0.15y$ subject to
$$x + y \le 500$$
$$-2x + 3y \le 0$$
$$x \ge 0, y \ge 0$$

7. Let x denote the number of fully assembled units to be produced daily and let y denote the number of kits to be produced. Then the fraction of the day the fabrication department works on the fully assembled cabinets is $\frac{1}{200}x$. Similarly the fraction of the day the fabrication department works on kits is $\frac{1}{200}y$. Since the fraction of the day during which the fabrication department is busy cannot exceed one, we must have
$$\frac{1}{200}x + \frac{1}{200}y \le 1.$$
Similarly, the restrictions place on the assembly department leads to the inequality
$$\frac{1}{100}x + \frac{1}{300}y \le 1.$$
The profit (objective) function is $P = 50x + 40y$. Summarizing, the required linear programming problem is
Maximize $P = 50x + 40y$ subject to
$$\frac{1}{200}x + \frac{1}{200}y \le 1$$
$$\frac{1}{100}x + \frac{1}{300}y \le 1$$
$$x \ge 0, \ y \ge 0.$$

9. Let x and y denote the number of days the Saddle Mine and the Horseshoe Mine are operated, respectively. Then the operating cost is $C = 14{,}000x + 16{,}000y$. The amount of gold produced in the two mines is $(50x + 75y)$ oz, and this amount must be at least 650 oz. So we have $50x + 75y \ge 650$. Similarly, the requirement for silver production leads to the inequality $3000x + 1000y \ge 18{,}000$. So the problem is
Minimize $C = 14{,}000x + 16{,}000y$ subject to

$$50x + 75y \geq 650$$
$$3000x + 1000y \geq 18,000$$
$$x \geq 0, \; y \geq 0$$

11. Let x, y, and z denote the amount of money she invests in project A, project B, and project C, respectively. Since she plans to invest up to \$2 million, we must have $x + y + z \leq 2,000,000$. Because she decides to put not more than 20% of her total investment in project C, we have
$$z \leq 0.2(x + y + z) \quad \text{or} \quad -0.2x - 0.2y + 0.8z \leq 0$$
Since her investments in project B and C should not exceed 60% of her total investment, we have
$$y + z \leq 0.6(x + y + z) \quad \text{or} \quad -0.6x + 0.4y + 0.4z \leq 0$$
Also, since her investment in project A should be at least 60% of her investments in projects B and C, we have
$$x \geq 0.6(y + z) \quad \text{or} \quad -x + 0.6y + 0.6z \leq 0$$
Finally, the returns on her investments are given by $P = 0.1x + 0.15y + 0.2z$.
To summarize, the problem is
Maximize $P = 0.1x + 0.15y + 0.2z$ subject to
$$x + y + z \leq 2,000,000$$
$$-2x - 2y + 8z \leq 0$$
$$-6x + 4y + 4z \leq 0$$
$$-10x + 6y + 6z \leq 0$$
$$x \geq 0, \; y \geq 0, \; z \geq 0$$

13. Let x and y denote the amount of money given in aid to country A and country B, respectively. Then the requirement that country A receives between \$1 and \$1.5 million, inclusive, in aid, implies $1 \leq x \leq 1.5$. Similarly, the requirement that country B receive at least \$0.75 million in aid implies $y \geq 0.75$. The condition that between \$2 and \$2.5 million dollars in aid has been earmarked for these two countries implies that
$$x + y \leq 2.5$$
$$x + y \geq 2$$
Therefore, we have the following linear programming problem:
Maximize $P = 0.6x + 0.8y$ subject to

$$x + y \leq 2.5$$

$$x + y \geq 2$$

$$1 \leq x \leq 1.5$$

$$y \geq 0.75$$

15. Let x denote the number of picture tubes shipped from location I to city A and let y denote the number of picture tubes shipped from location I to city B. Since the number of picture tubes required by the two factories in city A and city B are 3000 and 4000, respectively, the number of picture tubes shipped from location II to city A and city B, are $(3000 - x)$ and $(4000 - y)$, respectively. These numbers are shown in the following schematic,

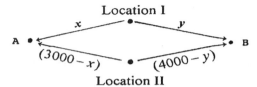

Location II

Referring to the schematic and the shipping schedule, we find that the total shipping costs incurred by the company are given by
$$C = 3x + 2y + 4(3000 - x) + 5(4000 - y)$$
$$= 32000 - x - 3y$$

The production constraints on Location I and II lead to the inequalities
$$x + \qquad y \leq 6000$$
$$(3000 - x) + (4000 - y) \leq 5000$$

This last inequality simplifies to $x + y \geq 2000$.

The requirements of the two factories lead to the inequalities
$$x \geq 0, y \geq 0, 3000 - x \geq 0, \text{ and } 4000 - y \geq 0.$$

These last two inequalities may be written as $x \leq 3000$ and $y \leq 4000$.

Summarizing, we have the following linear programming problem:

Minimize $C = 32,000 - x - 3y$ subject to
$$x + y \leq 6000$$
$$x + y \geq 2000$$
$$x \leq 3000$$
$$y \leq 4000$$
$$x \geq 0, \ y \geq 0$$

17. Let x, y, and z denote the number of units produced of products A, B, and C,

respectively. From the given information, we formulate the following linear programming problem:

Maximize $P = 18x + 12y + 15z$ subject to
$$2x + y + 2z \leq 900$$
$$3x + y + 2z \leq 1080$$
$$2x + 2y + z \leq 840$$
$$x \geq 0, y \geq 0, z \geq 0$$

19. We first tabulate the given information:

Department	Model A	Model B	Model C	Time Available
Fabrication	$\frac{5}{4}$	$\frac{3}{2}$	$\frac{3}{2}$	310
Assembly	1	1	$\frac{3}{4}$	205
Finishing	1	1	$\frac{1}{2}$	190

Let x, y, and z denote the number of units of model A, model B, and model C to be produced, respectively. Then the required linear programming problem is

Maximize $P = 26x + 28y + 24z$ subject to
$$\tfrac{5}{4}x + \tfrac{3}{2}y + \tfrac{3}{2}z \leq 310$$
$$x + y + \tfrac{3}{4}z \leq 205$$
$$x + y + \tfrac{1}{2}z \leq 190$$
$$x \geq 0, y \geq 0, z \geq 0$$

21. The shipping costs are tabulated in the following table.

	Warehouse A	Warehouse B	Warehouse C
Plant I	60	60	80
Plant II	80	70	50

Letting x_1 denote the number of pianos shipped from plant I to warehouse A, x_2 the number of pianos shipped from plant I to warehouse B, and so we have

	Warehouse A	Warehouse B	Warehouse C	Maximum Production
Plant I	x_1	x_2	x_3	300
Plant II	x_4	x_5	x_6	250
Minimum Requirement	200	150	200	

From the two tables we see that the total monthly shipping cost is given by
$$C = 60x_1 + 60x_2 + 80x_3 + 80x_4 + 70x_5 + 50x_6.$$
Next, the production constraints on plants I and II lead to the inequalities
$$x_1 + x_4 \geq 200$$
$$x_2 + x_5 \geq 150$$
$$x_3 + x_6 \geq 200$$
Summarizing we have the following linear programming problem:
$$\text{Minimize } C = 60x_1 + 60x_2 + 80x_3 + 80x_4 + 70x_5 + 50x_6 \text{ subject to}$$
$$x_1 + x_2 + x_3 \leq 300$$
$$x_4 + x_5 + x_6 \leq 250$$
$$x_1 + x_4 \geq 200$$
$$x_2 + x_5 \geq 150$$
$$x_3 + x_6 \geq 200$$
$$x_1 \geq 0, \ x_2 \geq 0, \ \ldots, \ x_6 \geq 0$$

23. The given data can be summarized as follows:

	Concentrates			
	Pineapple	Orange	Banana	Profit ($)
Pineapple-orange	8	8	0	1
Orange-banana	0	12	4	0.80
Pineapple-orange-banana	4	8	4	0.90
Maximum available (oz)	16,000	24,000	5,000	

Suppose x, y, and z cartons of pineapple-orange, orange-banana, and pineapple-orange-banana juice are to be produced, respectively. The linear programming problem is
$$\text{Maximize } P = x + 0.8y + 0.9z \text{ subject to}$$

$$8x + \quad 4z \le 16,000$$
$$8x + 12y + 8z \le 24,000$$
$$4y + 4z \le 5,000$$
$$z \le 800$$
$$x \ge 0, y \ge 0, z \le 0$$

25. True. It satisfies the definition of a linear programming problem.

EXERCISES 3.3, page 196

1. Evaluating the objective function at each of the corner points we obtain the following table.

Vertex	Z = 2x + 3y
(1,1)	5
(8,5)	31
(4,9)	35
(2,8)	28

From the table, we conclude that the maximum value of Z is 35 and it occurs at the vertex (4,9). The minimum value of Z is 5 and it occurs at the vertex (1,1).

3. Evaluating the objective function at each of the corner points we obtain the following table.

Vertex	Z = 3x + 4y
(0,20)	80
(3,10)	49
(4,6)	36
(9,0)	27

From the graph, we conclude that there is no maximum value since Z is unbounded. The minimum value of Z is 27 and it occurs at the vertex (9,0).

5. Evaluating the objective function at each of the corner points we obtain the following table.

Vertex	$Z = x + 4y$
(0,6)	24
(4,10)	$\boxed{44}$
(12,8)	$\boxed{44}$
(15,0)	15

From the table, we conclude that the maximum value of Z is 44 and it occurs at every point on the line segment joining the points (4,10) and (12,8). The minimum value of Z is 15 and it occurs at the vertex (15,0).

7. The problem is to maximize $P = 2x + 3y$ subject to
$$x + y \le 6$$
$$x \le 3$$
$$x \ge 0, y \ge 0$$
The feasible set S for the problem is shown in the following figure, and the values of the function P at the vertices of S are summarized in the accompanying table.

Vertex	$P = 2x + 3y$
A(0,0)	0
B(3,0)	6
C(3,3)	15
D(0,6)	$\boxed{18}$

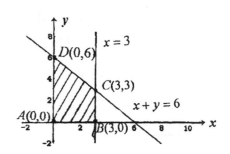

We conclude that P attains a maximum value of 18 when $x = 0$ and $y = 6$.

9. The problem is to maximize $P = 2x + y$ subject to
$$x + y \le 4$$
$$2x + y \le 5$$
$$x \ge 0, y \ge 0$$

Referring to the following figure and table

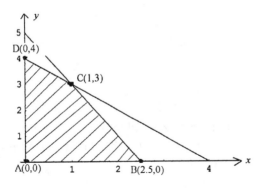

V	$P = 2x + y$
$(0,0)$	0
$(2.5,0)$	$\boxed{5}$
$(1,3)$	$\boxed{5}$
$(0,4)$	4

we conclude that P attains a maximum value of 5 at any point (x, y) lying on the line segment joining $(1, 3)$ to $(2.5,0)$.

11. The problem is

$$\text{Maximize } P = x + 8y \text{ subject to}$$
$$x + y \le 8$$
$$2x + y \le 10$$
$$x \ge 0, y \ge 0$$

From the following figure and table

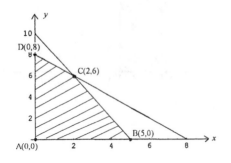

V	$P = x + 8y$
$A(0,0)$	0
$B(5,0)$	5
$C(2,6)$	50
$D(0,8)$	$\boxed{64}$

We conclude that P attains a maximum value of 64 when $x = 0$ and $y = 8$.

13. The linear programming problem is

$$\text{Maximize } P = x + 3y \text{ subject to}$$
$$2x + y \le 6$$
$$x + y \le 4$$
$$x \le 1$$
$$x \ge 0, y \ge 0$$

From the following figure and table,

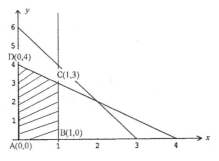

V	$P = x + 3y$
$A(0,0)$	0
$B(1,0)$	1
$C(1,3)$	10
$D(0,4)$	12

We conclude that P attains a maximum value of 12 when $x = 0$ and $y = 4$.

15. The linear programming problem is
$\quad\quad$ Minimize $C = 3x + 4y$ subject to
$\quad\quad x + \ y \geq 3$
$\quad\quad x + 2y \geq 4$
$\quad\quad x \geq 0, y \geq 0$

From the following figure and table,

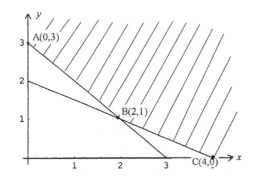

V	$C = 3x + 4y$
$A(0,3)$	12
$B(2,1)$	$\boxed{10}$
$C(4,0)$	12

We conclude that C attains a minimum value of 10 when $x = 2$ and $y = 1$.

17. The linear programming problem is
$\quad\quad$ Minimize $C = 3x + 6y$ subject to
$\quad\quad\quad x + 2y \geq 40$
$\quad\quad\quad x + \ y \geq 30$
$\quad\quad\quad x \geq 0, y \geq 0$

From the following figure and table,

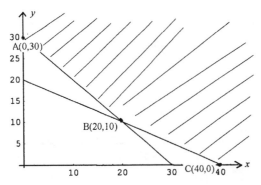

we conclude that C attains a minimum value of 120 at any point on the line segment joining (20,10) to (40,0).

V	$C = 3x + 6y$
A(0,30)	180
B(20,10)	120
C(40,0)	120

19. The problem is

Minimize $C = 2x + 10y$ subject to

$$5x + 2y \geq 40$$

$$x + 2y \geq 20$$

$$y \geq 3, \; x \geq 0$$

The feasible set S for the problem is shown in the following figure and the values of the function C at the vertices of S are summarized in the accompanying table.

Vertex	$C = 2x + 10y$
A(0,20)	200
B(5, $\frac{15}{2}$)	85
C(14,3)	$\boxed{58}$

We conclude that C attains a minimum value of 58 when $x = 14$ and $y = 3$.

21. The problem is to minimize $C = 10x + 15y$ subject to

$$x + y \leq 10$$

$$3x + y \geq 12$$

$$-2x + 3y \geq 3$$

$$x \geq 0, \; y \geq 0$$

The feasible set is shown in the following figure, and the values of C at each of the vertices of S are shown in the accompanying table.

Vertex	$C = 10x + 15y$
$A(3,3)$	$\boxed{75}$
$B(\frac{27}{5}, \frac{23}{5})$	123
$C(1,9)$	145

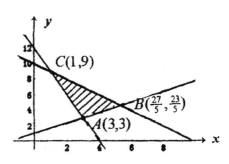

We conclude that C attains a minimum value of 75 when $x = 3$ and $y = 3$.

23. The problem is to maximize $P = 3x + 4y$ subject to
$$x + 2y \le 50$$
$$5x + 4y \le 145$$
$$2x + y \ge 25$$
$$y \ge 5, x \ge 0$$

The feasible set S is shown in the figure that follows, and the values of P at each of the vertices of S are shown in the accompanying table.

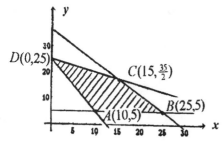

Vertex	$P = 3x + 4y$
$A(10,5)$	50
$B(25,5)$	95
$C(15, \frac{35}{2})$	$\boxed{115}$
$D(0,25)$	100

We conclude that P attains a maximum value of 115 when $x = 15$ and $y = 35/2$.

25. The problem is to maximize $P = 2x + 3y$ subject to
$$x + y \le 48$$
$$x + 3y \ge 60$$
$$9x + 5y \le 320$$
$$x \ge 10, y \ge 0$$

The feasible set S is shown in the figure that follows, and the values of P at each of the vertices of S are shown in the accompanying table.

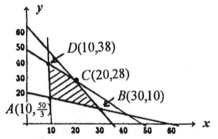

Vertex	$P = 2x + 3y$
$A(10, \frac{50}{3})$	70
$B(30,10)$	90
$C(20,28)$	124
$D(10,38)$	$\boxed{134}$

We conclude that P attains a maximum value of 134 when $x = 10$ and $y = 38$.

27. The problem is to find the maximum and minimum value of $P = 10x + 12y$ subject to

$$5x + 2y \geq 63$$
$$x + y \geq 18$$
$$3x + 2y \leq 51$$
$$x \geq 0, y \geq 0$$

The feasible set is shown at the right and the value of P at each of the vertices of S are shown in the accompanying table.

Vertex	$P = 10x + 12y$
$A(9,9)$	198
$B(15,3)$	186
$C(6, \frac{33}{2})$	258

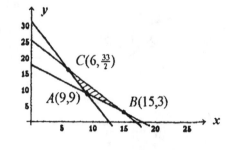

P attains a maximum value of 258 when $x = 6$ and $y = 33/2$. The minimum value of P is 186. It is attained when $x = 15$ and $y = 3$.

29. The problem is to find the maximum and minimum value of $P = 2x + 4y$ subject to

$$x + y \le 20$$
$$-x + y \le 10$$
$$x \le 10$$
$$x + y \ge 5$$
$$y \ge 5, \quad x \ge 0$$

The feasible set is shown in the figure that follows, and the value of P at each of the vertices of S are shown in the accompanying table.

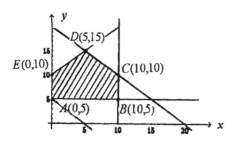

Vertex	$P = 2x + 4y$
A(0,5)	20
B(10,5)	40
C(10,10)	60
D(5,15)	70
E(0,10)	40

P attains a maximum value of 70 when $x = 5$ and $y = 15$. The minimum value of P is 20. It is attained when $x = 0$ and $y = 5$.

31. Let x and y denote the number of model A and model B fax machines produced in each shift. Then the restriction on manufacturing costs implies
$$100x + 150y \le 600{,}000,$$
and the limitation on the number produced implies
$$x + y \le 2{,}500.$$
The total profit is $P = 30x + 40y$. Summarizing, we have the following linear programming problem.

Maximize $P = 30x + 40y$ subject to
$$100x + 150y \le 600{,}000$$
$$x + \quad y \le \quad 2{,}500$$
$$x \ge 0, \ y \ge 0$$

The graph of the feasible set S and the associated table of values of P follow.

Vertex	$C = 30x + 40y$
A(0,0)	0
B(2500,0)	75,000
D(0,2500)	100,000

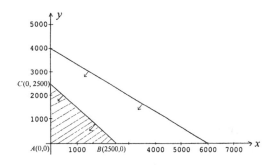

P attains a maximum value of 100,000 when $x = 0$ and $y = 2500$. Thus, by producing 2500 model B fax machines in each shift, the company will realize an optimal profit of $100,000.

33. Refer to the solution of Exercise 2, Section 3.2, The problem is

Maximize $P = 2x + 1.5y$ subject to
$$3x + 4y \leq 1000$$
$$6x + 3y \leq 1200$$
$$x \geq 0, \ y \geq 0$$

The graph of the feasible set S and the associated table of values of P follow.

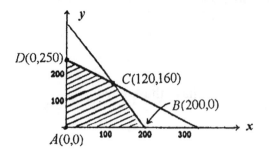

Vertex	$P = 2x + 1.5y$
A(0,0)	0
B(200,0)	400
C(120,160)	480
D(0,250)	375

P attains a maximum value of 480 when $x = 120$ and $y = 160$. Thus, by producing 120 model A grates and 160 model B grates in each shift, the company will realize an optimal profit of $480.

35. Refer to the solution of Exercise 4, Section 3.2. The linear programming problem is

Maximize $P = 0.1x + 0.12y$ subject to

$$x + y \le 20$$
$$x - 4y \ge 0$$
$$x \ge 0, \ y \ge 0$$

The feasible set S for the problem is shown in the figure at the right, and the value of P at each of the vertices of S is shown in the accompanying table.

Vertex	$C = 0.1x + 0.12y$
A(0,0)	0
B(16,4)	2.08
C(20,0)	2.00

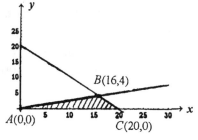

The maximum value of P is attained when $x = 16$ and $y = 4$. Thus, by extending \$16 million in housing loans and \$4 million in automobile loans, the company will realize a return of \$2.08 million on its loans.

37. Refer to Exercise 8, Section 3.2. The problem is
$$\text{Maximize } P = 150x + 200y \text{ subject to}$$
$$40x + 60y \le 7400$$
$$20x + 25y \le 3300$$
$$x \ge 0, \ y \ge 0$$

The graph of the feasible set S and the associated table of values of P follow.

Vertex	$P = 150x + 200y$
A(0,0)	0
B(165,0)	24,750
C(65,80)	25,750
D(0,123)	24,600

P attains a maximum value of 25,750 when $x = 65$ and $y = 80$. Thus, by producing 65 acres of crop A and 80 acres of crop B, the farmer will realize a maximum profit of \$25,750.

39. Let x and y denote the number of days the Saddle Mine and the Horseshoe Mine are operated, respectively. Then the operating cost is $C = 14{,}000x + 16{,}000y$. The amount of gold produced in the two mines is $(50x + 75y)$ oz, and this amount must

be at least 650 oz. So we have $50x + 75y \geq 650$. Similarly, the requirement for silver production leads to the inequality $3000x + 1000y \geq 18{,}000$. So the problem is

$$\text{Minimize } C = 14{,}000x + 16{,}000y \text{ subject to}$$

$$50x + 75y \geq 650$$

$$3000x + 1000y \geq 18{,}000$$

$$x \geq 0, \; y \geq 0$$

The feasible set is shown in the accompanying figure. From the table, we see that the minimum value of $C = 152{,}000$ is attained at $x = 4$ and $y = 6$. So, the Saddle Mine should be operated for 4 days and the Horseshoe Mine should be operated for 6 days at a minimum cost of $\$152{,}000$/day.

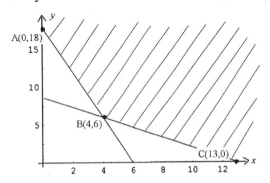

Vertex	$C = 14{,}000x + 16{,}000y$
A(0, 18)	288,000
B(4, 6)	152,000
C(13, 0)	182,000

41. Refer to Exercise 12, Section 3.2. The problem is

$$\text{Minimize } C = 2x + 5y \text{ subject to}$$

$$30x + 25y \geq 400$$

$$x + 0.5y \geq 10$$

$$2x + 5y \geq 40$$

$$x \geq 0, \; y \geq 0$$

The graph of the feasible set S and the associated table of values of C follow.

Vertex	$C = 2x + 5y$
A(0,20)	100
B(5,10)	60
C(10,4)	40
D(20,0)	40

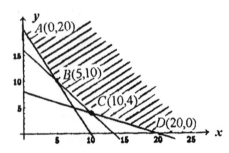

C attains a minimum value of 40 when $x = 10$ and $y = 4$ and $x = 20$, and $y = 0$. This means that any point lying on the line joining the points (10,4) and (20,0) will satisfy these constraints. For example, we could use 10 ounces of food A and 4 ounces of food B, or we could use 20 ounces of food A and zero ounces of food B.

43. Refer to the solution of Exercise 13, Section 3.2. The problem is
$\qquad$ Maximize $P = 0.6x + 0.8y$ subject to

$$x + y \le 2.5$$

$$x + y \ge 2$$

$$1 \le x \le 1.5$$

$$y \ge 0.75$$

The feasible set S for the problem is shown in the figure at the right, and the value of P at each of the vertices of S is given in the table that follows.

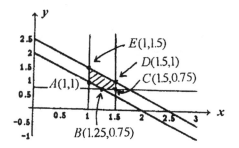

Vertex	$P = 0.6x + 0.8y$
A(1,1)	1.4
B(1.25,0.75)	1.35
C(1.5,0.75)	1.5
D(1.5,1)	1.7
E(1,1.5)	1.8

We conclude that the maximum value of P occurs when $x = 1$ and $y = 1.5$. Therefore, AntiFam should give $1 million in aid to country A and $1.5 million in aid to country B.

45. Refer to the solution of Exercise 15, Section 3.2.
$\qquad$ Minimize $C = 32{,}000 - x - 3y$ subject to

$$x + y \le 6000$$

$$x + y \ge 2000$$

$$x \le 3000$$

$$y \le 4000$$

$$x \ge 0, \; y \ge 0$$

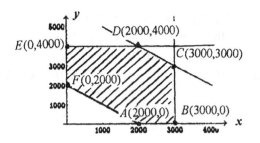

Vertex	$C = 32{,}000 - x - 3y$
A(2000,0)	30,000
B(3000,0)	29,000
C(3000,3000)	20,000
D(2000,4000)	18,000
E(0,4000)	20,000
F(0,2000)	26,000

Since x denotes the number of picture tubes shipped from location I to city A and y denotes the number of picture tubes shipped from location I to city B, we see that the company should ship 2000 tubes from location I to city A and 4000 tubes from location I to city B. Since the number of picture tubes required by the two factories in city A and city B are 3000 and 4000, respectively, the number of picture tubes shipped from location II to city A and city B, are

$(3000 - x) = 3000 - 2000 = 1000$ and $(4000 - y) = 4000 - 4000 = 0$

respectively. The minimum shipping cost will then be $18,000.

47. Let x denote Patricia's investment in growth stocks and y denote the value of her investment in speculative stocks, where both x and y are measured in thousands of dollars. Then the return on her investments is given by $P = 0.15x + 0.25y$. Since her investment may not exceed $30,000, we have the constraint $x + y \leq 30$. The condition that her investment in growth stocks be at least 3 times as much as her investment in speculative stocks translates into the inequality $x \geq 3y$. Thus, we have the following linear programming problem:

$$\text{Maximize } P = 0.15x + 0.25y \text{ subject to}$$
$$x + y \leq 30$$
$$x - 3y \geq 0$$
$$x \geq 0, \ y \geq 0$$

The graph of the feasible set S is shown in the figure and the value of P at each of the vertices of S is shown in the accompanying table. The maximum

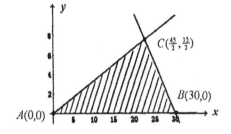

Vertex	$C = 0.15x + 0.25y$
A(0,0)	0
B(30,0)	4.5
$C(\frac{45}{2}, \frac{15}{2})$	5.25

value of P occurs when $x = 22.5$ and $y = 7.5$. Thus, by investing $22,500 in growth stocks and $7,500 in speculative stocks. Patricia will realize a return of $5250 on her investments.

49. True. The optimal solution of a linear programming problem is the largest of all the feasible solutions.

51. False. The feasible set could be empty, and therefore, bounded.

53. Since the point $Q(x_1,y_1)$ lies in the interior of the feasible set S, it is possible to find another point $P(x_1, y_2)$ lying to the right and above the point Q and contained in S. (See figure.) Clearly, $x_2 > x_1$ and $y_2 > y_1$. Therefore, $ax_2 + by_2 > ax_1 + by_1$, since $a > 0$ and $b > 0$ and this shows that the objective function $P = ax + by$ takes on a larger value at P than it does at Q. Therefore, the optimal solution cannot occur at Q.

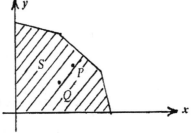

55. a.

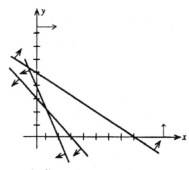

b. There is no point that satisfies all the given inequalities. Therefore, there is no solution.

CHAPTER 3 REVIEW EXERCISES, page 202

1. Evaluating Z at each of the corners of the feasible set S, we obtain the following table.

Vertex	$Z = 2x + 3y$
(0,0)	0
(5,0)	10
(3,4)	18
(0,6)	18

    We conclude that Z attains a minimum value of 0 when $x = 0$ and $y = 0$, and a maximum value of 18 when x and y lie on the line segment joining (3,4) and (0,6).

3. The graph of the feasible set S is shown at the right.

Vertex	$Z = 3x + 5y$
A(0,0)	0
B(5,0)	15
C(3,2)	19
D(0,4)	20

    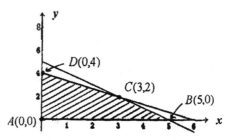

    We conclude that the maximum value of P is 20 when $x = 0$ and $y = 4$.

5. The values of the objective function $C = 2x + 5y$ at the corners of the feasible set are given in the following table. The graph of the feasible set S follows.

    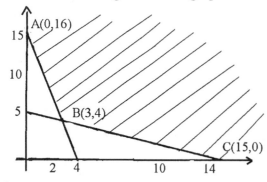

Vertex	$C = 2x + 5y$
A(0,16)	80
B(3,4))	26
C(15,0)	30

    We conclude that the minimum value of C is 26 when $x = 3$ and $y = 4$.

7. The values of the objective function $P = 3x + 2y$ at the vertices of the feasible set are given in the following table. The graph of the feasible set is shown at the right.

Vertex	$P = 3x + 2y$
$A(0, \frac{28}{5})$	$\frac{56}{5}$
$B(7,0)$	21
$C(8,0)$	24
$D(3,10)$	$\boxed{29}$
$E(0,12)$	24

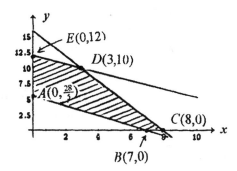

We conclude that P attains a maximum value of 29 when $x = 3$ and $y = 10$.

9. The graph of the feasible set S is shown at the right. The values of the objective function $C = 2x + 7y$ at each of the corner points of the feasible set S are shown in the table that follows.

Vertex	$C = 2x + 7y$
$A(20,0)$	$\boxed{40}$
$B(10,3)$	41
$C(0,9)$	63

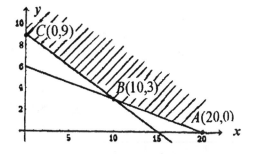

We conclude that C attains a minimum value of 40 when $x = 20$ and $y = 0$.

11. The graph of the feasible set S is shown in the following figure. We conclude that Q attains a maximum value of 22 when $x = 22$ and $y = 0$, and a minimum value of $5\frac{1}{2}$ when $x = 3$ and $y = \frac{5}{2}$.

Vertex	$Q = x + y$
$A(2,5)$	7
$B(3, \frac{5}{2})$	$5\frac{1}{2}$
$C(8,0)$	8
$D(22,0)$	22

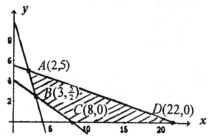

13. Suppose the investor puts x and y thousand dollars into the stocks of company A and company B, respectively. Then the mathematical formulation leads to the linear programming problem:

$$\text{Maximize } P = 0.14x + 0.20y \text{ subject to}$$
$$x + \quad y \le 80$$
$$0.01x + 0.04y \le 2$$
$$x \ge 0, \ y \ge 0$$

The feasible set S for this problem is shown in figure at the right and the values at each corner point are given in the accompanying table.

Vertex	$P = 0.14x + 0.20y$
A(0,0)	0
B(80,0)	11.2
C(40,40)	$\boxed{13.6}$
D(0,50)	10

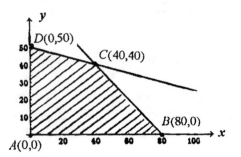

P attains a maximum value of 13.6 when $x = 40$ and $y = 40$. Thus, by investing $40,000 in the stocks of each company, the investor will achieve a maximum return of $13,600.

15. Let x denote the number of model A grates and y the number of model B grates to be produced. Then the constraint on the amount of cast iron available leads to

$$3x + 4y \le 1000$$

and the number of minutes of labor used each day leads to

$$6x + 3y \le 1200.$$

One additional constraint specifies that $y \ge 180$. The daily profit is $P = 2x + 1.5y$. Therefore, we have the following linear programming problem:

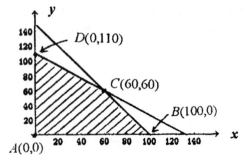

$$\text{Maximize } P = 2x + 1.5y \text{ subject to}$$
$$3x + 4y \le 1000$$
$$6x + 3y \le 1200$$
$$x \ge 0, \ y \ge 180$$

The graph of the feasible set S is shown in the figure that follows.

Vertex	$P = 2x + 1.5y$
A(0,180)	270
B(0,250)	375
C(93 $\frac{1}{3}$,180)	$\boxed{456\frac{2}{3}}$

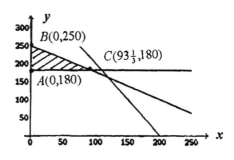

Thus, the optimal profit of $456 is realized when 93 units of model A grates and 180 units of model B grates are produced.

CHAPTER 4

EXERCISES 4.1, page 223

1. All entries in the last row of the simplex tableau are nonnegative and an optimal solution has been reached. We find
    $$x = 30/7, \ y = 20/7, \ u = 0, \ v = 0, \ \text{and} \ P = 220/7.$$

3. The simplex tableau is not in final form because there is an entry in the last row that is negative. The entry in the first row, second column, is the next pivot element and has a value of 1/2.

5. The simplex tableau is in final form. We find
    $$x = 1/3, \ y = 0, \ z = 13/3, \ u = 0, \ v = 6, \ w = 0 \ \text{and} \ P = 17.$$

7. The simplex tableau is not in final form because there are two entries in the last row that are negative. The entry in the third row, second column, is the pivot element and has a value of 1.

9. The simplex tableau is in final form. The solutions are
    $$x = 30, \ y = 0, \ z = 0, \ u = 10, \ v = 0, \ \text{and} \ P = 60,$$

    and $\quad x = 0, \ y = 30, \ z = 0, \ u = 10, \ v = 0, \ \text{and} \ P = 60.$
    (There are infinitely many answers).

11. We obtain the following sequence of tableaus:

	x	y	u	v	P	Const
p.r. →	1	$\boxed{1}$	1	0	0	4
	2	1	0	1	0	5
	−3	−4	0	0	1	0

Ratio
4
5

$\xrightarrow[\substack{R_2-R_1 \\ R_3+4R_1}]{}$

x	y	u	v	P	Const.
1	1	1	0	0	4
1	0	−1	1	0	1
1	0	4	0	1	16

↑
p.c.

The last tableau is in final form and we conclude that $x = 0, \ y = 4, \ u = 0, \ v = 1,$ and $P = 16.$

13.

	x	y	u	v	P	Const	Ratio
p.r. →	1	[2]	1	0	0	12	12/2 = 6
	3	2	0	1	0	24	24/2 = 12
	−10	−12	0	0	1	0	

p.c. (under y)

$\dfrac{1}{2}R_1 \longrightarrow$

x	y	u	v	P	Const
$\frac{1}{2}$	1	$\frac{1}{2}$	0	0	6
3	2	0	1	0	24
−10	−12	0	0	1	0

$\begin{array}{c} R_2-2R_1 \\ R_3+12R_1 \end{array} \longrightarrow$

	x	y	u	v	P	Const	Ratio
	$\frac{1}{2}$	1	$\frac{1}{2}$	0	0	6	6/(1/2) = 12
p.r. →	[2]	0	−1	1	0	12	12/2 = 6
	−4	0	6	0	1	72	

p.c. (under x)

$\dfrac{1}{2}R_2 \longrightarrow$

x	y	u	v	P	Const
$\frac{1}{2}$	1	$\frac{1}{2}$	0	0	6
1	0	$-\frac{1}{2}$	$\frac{1}{2}$	0	6
−4	0	6	0	1	72

$\begin{array}{c} R_1-\frac{1}{2}R_2 \\ R_3+4R_2 \end{array} \longrightarrow$

x	y	u	v	P	Const
0	1	$\frac{3}{4}$	$-\frac{1}{4}$	0	3
1	0	$-\frac{1}{2}$	$\frac{1}{2}$	0	6
0	0	4	2	1	96

The last tableau is in final form. We find that $x = 6$, $y = 3$, $u = 0$, $v = 0$, and $P = 96$.

15. We obtain the following sequence of tableaus:

	x	y	u	v	w	P	Const.		Ratio
	3	1	1	0	0	0	24		24
	2	1	0	1	0	0	18		18
p.r. →	1	[3]	0	0	1	0	24		8
	−4	−6	0	0	0	1	0		

$$\uparrow$$
$$p.c.$$

$\xrightarrow{\frac{1}{3}R_3}$

	x	y	u	v	w	P	Const.
	3	1	1	0	0	0	24
	2	1	0	1	0	0	18
	$\frac{1}{3}$	1	0	0	$\frac{1}{3}$	0	8
	−4	−6	0	0	0	1	0

$\begin{array}{l} R_1-R_3 \\ R_2-R_3 \\ R_4+6R_3 \end{array} \longrightarrow$

	x	y	u	v	w	P	Const.		Ratio
p.r. →	[$\frac{8}{3}$]	0	1	0	$-\frac{1}{3}$	0	16		6
	$\frac{5}{3}$	0	0	1	$-\frac{1}{3}$	0	10		6
	$\frac{1}{3}$	1	0	0	$\frac{1}{3}$	0	8		24
	−2	0	0	0	2	1	48		

$$\uparrow$$
$$p.c.$$

(Observe that we have a choice here.)

$\xrightarrow{\frac{3}{8}R_1}$

	x	y	u	v	w	P	Const.
	1	0	$\frac{3}{8}$	0	$-\frac{1}{8}$	0	6
	$\frac{5}{3}$	0	0	1	$-\frac{1}{3}$	0	10
	$\frac{1}{3}$	1	0	0	$\frac{1}{3}$	0	8
	−2	0	0	0	2	1	48

$\begin{array}{l} R_2-\frac{5}{3}R_1 \\ R_3-\frac{1}{3}R_1 \\ R_4+2R_1 \end{array} \longrightarrow$

x	y	u	v	w	P	Const.
1	0	$\frac{3}{8}$	0	$-\frac{1}{8}$	0	6
0	0	$-\frac{5}{8}$	1	$-\frac{1}{8}$	0	0
0	1	$-\frac{1}{8}$	0	$\frac{3}{8}$	0	6
0	0	$\frac{3}{4}$	0	$\frac{7}{4}$	1	60

We deduce that $x = 6$, $y = 6$, $u = 0$, $v = 0$, $w = 0$, and $P = 60$.

17. We obtain the following sequence of tableaus:

	x	y	z	u	v	P	Const.
	1	1	1	1	0	0	8
p.r. →	3	2	$\boxed{4}$	0	1	0	24
	−3	−4	−5	0	0	1	0

Ratio
$8/1 = 8$
$24/4 = 6$

$\xrightarrow{\frac{1}{4}R_2}$

↑
p.c.

x	y	z	u	v	P	Const.
1	1	1	1	0	0	8
$\frac{3}{4}$	$\frac{1}{2}$	1	0	$\frac{1}{4}$	0	6
−3	−4	−5	0	0	1	0

$\xrightarrow[R_3 + 5R_2]{R_1 - R_2}$

	x	y	z	u	v	P	Const.
p.r. →	$\frac{1}{4}$	$\boxed{\frac{1}{2}}$	0	1	$-\frac{1}{4}$	0	2
	$\frac{3}{4}$	$\frac{1}{2}$	1	0	$\frac{1}{4}$	0	6
	$\frac{3}{4}$	$-\frac{3}{2}$	0	0	$\frac{5}{4}$	1	30

Ratio
$2/(1/2) = 4$
$6/(1/2) = 12$

$\xrightarrow{2R_1}$

↑
p.c.

x	y	z	u	v	P	Const.
$\frac{1}{2}$	1	0	2	$-\frac{1}{2}$	0	4
$\frac{3}{4}$	$\frac{1}{2}$	1	0	$\frac{1}{4}$	0	6
$\frac{3}{4}$	$-\frac{3}{2}$	0	0	$\frac{5}{4}$	1	30

$\xrightarrow[R_3 + \frac{3}{2}R_1]{R_2 - \frac{1}{2}R_1}$

x	y	z	u	v	P	Const.
$\frac{1}{2}$	1	0	2	$-\frac{1}{2}$	0	4
$\frac{1}{2}$	0	1	-1	$\frac{1}{2}$	0	4
$\frac{3}{2}$	0	0	3	$\frac{1}{2}$	1	36

This last tableau is in final form. We find that $x = 0$, $y = 4$, $z = 4$, $u = 0$, $v = 0$, and $P = 36$.

19.

x	y	z	u	v	w	P	Const.
3	10	5	1	0	0	0	120
5	$\boxed{2}$	8	0	1	0	0	6
8	10	3	0	0	1	0	105
-3	-4	-1	0	0	0	1	0

p.r. → (row 2), p.c. ↑ (column y)

Ratio
$120/10 = 12$
$6/2 = 3$
$105/10 = 21/2$

$\frac{1}{2}R_2$ →

x	y	z	u	v	w	P	Const.
3	10	5	1	0	0	0	120
$\frac{5}{2}$	1	4	0	$\frac{1}{2}$	0	0	3
8	10	3	0	0	1	0	105
-3	-4	-1	0	0	0	1	0

$R_1 - 10R_2$
$R_3 - 10R_2$
$R_4 + 4R_2$ →

x	y	z	u	v	w	P	Const.
-22	0	-35	1	-5	0	0	90
$\frac{5}{2}$	1	4	0	$\frac{1}{2}$	0	0	3
-17	0	-37	0	-5	1	0	75
7	0	15	0	2	0	1	12

The last tableau is in final form. We find that $x = 0$, $y = 3$, $z = 0$, $u = 90$, $v = 0$, $w = 75$, and $P = 12$.

21. We obtain the following sequence of tableaus:

	x	y	z	u	v	w	P	Const.		Ratio
	1	1	1	1	0	0	0	20		20
p.r. →	2	[4]	3	0	1	0	0	42		$10\frac{1}{2}$
	2	0	3	0	0	1	0	30		---
	−4	−6	−5	0	0	0	1	0		

p.c. (under y) $\xrightarrow{\frac{1}{4}R_2}$

x	y	z	u	v	w	P	Const.
1	1	1	1	0	0	0	20
$\frac{1}{2}$	1	$\frac{3}{4}$	0	$\frac{1}{4}$	0	0	$\frac{21}{2}$
2	0	3	0	0	1	0	30
−4	−6	−5	0	0	0	1	0

$\xrightarrow[R_4+6R_2]{R_1-R_2}$

	x	y	z	u	v	w	P	Const.		Ratio
	$\frac{1}{2}$	0	$\frac{1}{4}$	1	$-\frac{1}{4}$	0	0	$\frac{19}{2}$		19
	$\frac{1}{2}$	1	$\frac{3}{4}$	0	$\frac{1}{4}$	0	0	$\frac{21}{2}$		21
p.r. →	[2]	0	3	0	0	1	0	30		15
	−1	0	$-\frac{1}{2}$	0	$\frac{3}{2}$	0	1	63		

p.c. (under x) $\xrightarrow{\frac{1}{2}R_3}$

x	y	z	u	v	w	P	Const.
$\frac{1}{2}$	0	$\frac{1}{4}$	1	$-\frac{1}{4}$	0	0	$\frac{19}{2}$
$\frac{1}{2}$	1	$\frac{3}{4}$	0	$\frac{1}{4}$	0	0	$\frac{21}{2}$
1	0	$\frac{3}{2}$	0	0	$\frac{1}{2}$	0	15
−1	0	$-\frac{1}{2}$	0	$\frac{3}{2}$	0	1	63

$\xrightarrow[R_4+R_3]{\substack{R_1-\frac{1}{2}R_3\\R_2-\frac{1}{2}R_3}}$

x	y	z	u	v	w	P	Const.
0	0	$-\frac{1}{2}$	1	$-\frac{1}{4}$	$-\frac{1}{4}$	0	2
0	1	0	0	$\frac{1}{4}$	$-\frac{1}{4}$	0	3
1	0	$\frac{3}{2}$	0	0	$\frac{1}{2}$	0	15
0	0	1	0	$\frac{3}{2}$	$\frac{1}{2}$	1	78

So the solution is $x = 15$, $y = 3$, $z = 0$, $u = 2$, $v = 0$, $w = 0$, and $P = 78$.

23. We obtain the following sequence of tableaus:

	x	y	z	u	v	w	P	Const.		Ratio
p.r. →	$\boxed{2}$	1	1	1	0	0	0	10		$10/2 = 5$
	3	5	1	0	1	0	0	45		$45/3 = 15$
	2	5	1	0	0	1	0	40		$40/2 = 20$
	−12	−10	−5	0	0	0	1	0		

↑
p.c.

$\xrightarrow{\frac{1}{2}R_1}$

x	y	z	u	v	w	P	Const.
1	$\frac{1}{2}$	$\frac{1}{2}$	$\frac{1}{2}$	0	0	0	5
3	5	1	0	1	0	0	45
2	5	1	0	0	1	0	40
−12	−10	−5	0	0	0	1	0

$\xrightarrow[\begin{array}{c}R_2-3R_1\\R_3-2R_1\\R_4+12R_1\end{array}]{}$

	x	y	z	u	v	w	P	Const.		Ratio
	1	$\frac{1}{2}$	$\frac{1}{2}$	$\frac{1}{2}$	0	0	0	5		$5/(1/2) = 10$
	0	$\frac{7}{2}$	$-\frac{1}{2}$	$-\frac{3}{2}$	1	0	0	30		$30/(7/2) = 60/7$
p.r. → 0		$\boxed{4}$	0	−1	0	1	0	30		$30/4 = 15/2$
	0	−4	1	6	0	0	1	60		

↑
.p.c

$\xrightarrow{\frac{1}{4}R_3}$

x	y	z	u	v	w	P	Const.
1	$\frac{1}{2}$	$\frac{1}{2}$	$\frac{1}{2}$	0	0	0	5
0	$\frac{7}{2}$	$-\frac{1}{2}$	$-\frac{3}{2}$	1	0	0	30
0	1	0	$-\frac{1}{4}$	0	$\frac{1}{4}$	0	$\frac{15}{2}$
0	−4	1	6	0	0	1	60

$\xrightarrow[\begin{array}{c}R_1-\frac{1}{2}R_3\\R_2-\frac{7}{2}R_3\\R_4+4R_3\end{array}]{}$

x	y	z	u	v	w	P	Const.
1	0	$\frac{1}{2}$	$\frac{5}{8}$	0	$-\frac{1}{8}$	0	$\frac{5}{4}$
0	0	$-\frac{1}{2}$	$-\frac{5}{8}$	1	$-\frac{7}{8}$	0	$\frac{15}{4}$
0	1	0	$-\frac{1}{4}$	0	$\frac{1}{4}$	0	$\frac{15}{2}$
0	0	1	5	0	1	1	90

This last tableau is in final form, and we conclude that $x = 5/4$, $y = 15/2$, $z = 0$, $u = 0$, $v = 15/4$, $w = 0$, and $P = 90$.

25. We obtain the following sequence of tableaus, where u, v and w are slack variables.

	x	y	z	u	v	w	P	Const.		Ratio	
p.r. →	[2]	1	2	1	0	0	0	7		$\frac{7}{2}$	
	2	3	1	0	1	0	0	8		4	$\xrightarrow{\frac{1}{2}R_1}$
	1	2	3	0	0	1	0	7		7	
	−24	−16	−23	0	0	0	1	0			

p.c. (under the x column)

x	y	z	u	v	w	P	Const.	
[1]	$\frac{1}{2}$	1	$\frac{1}{2}$	0	0	0	$\frac{7}{2}$	$R_2 - 2R_1$
2	3	1	0	1	0	0	8	$R_3 - R_1$
1	2	3	0	0	1	0	7	$R_4 + 24R_1$
−24	−16	−23	0	0	0	1	0	$\xrightarrow{}$

	x	y	z	u	v	w	P	Const.		Ratio	
	1	$\frac{1}{2}$	1	$\frac{1}{2}$	0	0	0	$\frac{7}{2}$		7	
p.r. →	0	[2]	−1	−1	1	0	0	1		$\frac{1}{2}$	$\xrightarrow{\frac{1}{2}R_2}$
	0	$\frac{3}{2}$	2	$-\frac{1}{2}$	0	1	0	$\frac{7}{2}$		$\frac{7}{3}$	
	0	−4	1	12	0	0	1	84			

p.c. (under the y column)

x	y	z	u	v	w	P	Const.
1	$\frac{1}{2}$	1	$\frac{1}{2}$	0	0	0	$\frac{7}{2}$
0	1	$-\frac{1}{2}$	$-\frac{1}{2}$	$\frac{1}{2}$	0	0	$\frac{1}{2}$
0	$\frac{3}{2}$	2	$-\frac{1}{2}$	0	1	0	$\frac{7}{2}$
0	-4	1	12	0	0	1	84

$$\begin{aligned} R_1 - \tfrac{1}{2}R_2 \\ R_3 - \tfrac{3}{2}R_2 \\ R_4 + 4R_2 \end{aligned} \longrightarrow$$

x	y	z	u	v	w	P	Const.		Ratio
1	0	$\frac{5}{4}$	$\frac{3}{4}$	$\frac{1}{4}$	0	0	$\frac{13}{4}$		$\frac{13}{5}$
0	1	$-\frac{1}{2}$	$-\frac{1}{2}$	$\frac{1}{2}$	0	0	$\frac{1}{2}$		--
p.r. $\to$ 0	0	$\boxed{\frac{11}{4}}$	$\frac{1}{4}$	$-\frac{3}{4}$	1	0	$\frac{11}{4}$		1
0	0	-1	10	2	0	1	86		

$$\uparrow \\ p.c.$$

$$\frac{4}{11}R_3 \longrightarrow$$

x	y	z	u	v	w	P	Const.
1	0	$\frac{5}{4}$	$\frac{3}{4}$	$\frac{1}{4}$	0	0	$\frac{13}{4}$
0	1	$-\frac{1}{2}$	$-\frac{1}{2}$	$\frac{1}{2}$	0	0	$\frac{1}{2}$
0	0	1	$\frac{1}{11}$	$-\frac{3}{11}$	$\frac{4}{11}$	0	1
0	0	-1	10	2	0	1	86

$$\begin{aligned} R_1 - \tfrac{5}{4}R_3 \\ R_2 + \tfrac{1}{2}R_3 \\ R_4 + R_3 \end{aligned} \longrightarrow$$

x	y	z	u	v	w	P	Const.
1	0	0	$\frac{7}{11}$	$\frac{13}{22}$	$-\frac{5}{11}$	0	2
0	1	0	$-\frac{5}{11}$	$\frac{4}{11}$	$\frac{2}{11}$	0	1
0	0	1	$\frac{1}{11}$	$-\frac{3}{11}$	$\frac{4}{11}$	0	1
0	0	0	$\frac{111}{11}$	$\frac{19}{11}$	$\frac{4}{11}$	1	87

This last tableau is in final form and we conclude that P attains a maximum value of 87 when $x = 2$, $y = 1$, and $z = 1$.

27. Pivoting about the x-column in the initial simplex tableau, we have

x	y	z	u	v	P	Const.
3	3	−2	1	0	0	100
5	5	3	0	1	0	150
−2	−2	4	0	0	1	0

p.r. → points to row with $\boxed{5}$

p.c. arrow points up to x column

Ratio: $100/3$, $150/5$, $\frac{1}{5}R_2$ →

x	y	z	u	v	P	Const.
3	3	−2	1	0	0	100
1	1	$\frac{3}{5}$	0	$\frac{1}{5}$	0	30
−2	−2	4	0	0	1	0

$\begin{array}{l}R_1-3R_2\\R_3+2R_2\end{array}$ →

x	y	z	u	v	P	Const.
0	0	$-\frac{19}{5}$	1	$-\frac{3}{5}$	0	10
1	1	$\frac{3}{5}$	0	$\frac{1}{5}$	0	30
0	0	$\frac{26}{5}$	0	$\frac{2}{5}$	1	60

and we see that one optimal solution occurs when $x = 30$, $y = 0$, $z = 0$, and $P = 60$. Similarly, pivoting about the y-column, we obtain another optimal solution: $x = 0$, $y = 30$, $z = 0$, and $P = 60$.

29. Let the number of model A and model B fax machines made each month be x and y, respectively. Then we have the following linear programming problem:

$$\text{Maximize } P = 30x + 40y \text{ subject to}$$
$$100x + 150y \le 600,000$$
$$x + y \le 2,500$$
$$x \ge 0,\ y \ge 0$$

Using the simplex method, we obtain the following sequence of tableaus:

x	y	u	v	P	Const
100	150	1	0	0	600,000
1	$\boxed{1}$	0	1	0	2,500
−30	−40	0	0	1	0

Ratio: 4000 , 2500 , $\begin{array}{l}R_1-150R_2\\R_3+40R_2\end{array}$ →

x	y	u	v	P	Const
-50	0	1	-150	0	225000
1	1	0	1	0	2500
10	0	0	40	1	100000

We conclude that the maximum monthly profit is $100,000, and this occurs when 0 model A and 2500 model B fax machines are produced.

31. Suppose the farmer plants x acres of Crop A and y acres of crop B. Then the problem is

$$\text{Maximize } P = 150x + 200y \text{ subject to}$$
$$x + y \leq 150$$
$$40x + 60y \leq 7400$$
$$20x + 25y \leq 3300$$
$$x \geq 0, \ y \geq 0$$

Using the simplex method, we obtain the following sequence of tableaus:

x	y	u	v	w	P	Const
1	1	1	0	0	0	150
40	60	0	1	0	0	7400
20	25	0	0	1	0	3300
-150	-200	0	0	0	1	0

$$\begin{array}{c} R_2 - 40R_1 \\ R_3 - 20R_1 \\ R_4 + 150R_1 \end{array} \longrightarrow$$

	x	y	u	v	w	P	Const	Ratio
	1	1	1	0	0	0	150	150
	0	20	-40	1	0	0	1400	700
p.r. →	0	5	-20	0	1	0	300	60
	0	-50	150	0	0	1	22500	

$$\frac{1}{5}R_3 \longrightarrow$$

↑
p.c.

x	y	u	v	w	P	Const	
1	1	1	0	0	0	150	$R_1 - R_3$
0	20	-40	1	0	0	1400	$R_2 - 20R_3$
0	1	-4	0	$\frac{1}{5}$	0	60	$R_4 + 50R_3$ $\longrightarrow$
0	-50	150	0	0	1	22500	

	x	y	u	v	w	P	Const	Ratio	
	1	0	5	0	$-\frac{1}{5}$	0	90	18	
p.r. $\rightarrow$	0	0	$\boxed{40}$	1	-4	0	200	5	$\frac{1}{40}R_2$ $\longrightarrow$
	0	1	-4	0	$\frac{1}{5}$	0	60	--	
	0	0	-50	0	10	1	25500		

$$\uparrow$$
$$p.c.$$

x	y	u	v	w	P	Const	
1	0	5	0	$-\frac{1}{5}$	0	90	$R_1 - 5R_2$
0	0	1	$\frac{1}{40}$	$-\frac{1}{10}$	0	5	$R_3 + 4R_2$
0	1	-4	0	$\frac{1}{5}$	0	60	$R_4 + 50R_2$ $\longrightarrow$
0	0	-50	0	10	1	25500	

x	y	u	v	w	P	Const
1	0	0	$-\frac{1}{8}$	$\frac{3}{10}$	0	65
0	0	1	$\frac{1}{40}$	$-\frac{1}{10}$	0	5
0	1	0	$\frac{1}{10}$	$-\frac{1}{5}$	0	80
0	0	0	$\frac{5}{4}$	5	1	25750

The last tableau is in final form. We find $x = 65$, $y = 80$, and $P = 25{,}750$. So the maximum profit of $25,750 is realized by planting 65 acres of Crop A and 80 acres of Crop B. Since $u = 5$, we see that there are 5 acres of land left unused.

33. Suppose Ashley invests x, y, and z dollars in the money market fund, the international equity fund, and the growth-and-income fund, respectively. Then the

objective function is $P = 0.06x + 0.1y + 0.15z$. The constraints are

$$x + y + z \leq 250,000; \quad z \leq 0.25(x + y + z); \text{ and } \quad y \leq 0.5(x + y + z).$$

The last two inequalities simplify to

$$-\tfrac{1}{4}x - \tfrac{1}{4}y + \tfrac{3}{4}z \leq 0 \quad \text{ or } \quad -x - y + 3z \leq 0$$

and

$$-\tfrac{1}{2}x + \tfrac{1}{2}y - \tfrac{1}{2}z \leq 0, \quad \text{ or } \quad -x + y - z \leq 0.$$

So the required linear programming problem is

$$\text{Maximize } P = 0.06x + 0.1y + 0.15z = \tfrac{3}{50}x + \tfrac{1}{10}y + \tfrac{3}{20}z \text{ subject to}$$

$$x + y + z \leq 250000$$

$$-x - y + 3z \leq 0$$

$$-x + y - z \leq 0$$

$$x \geq 0, y \geq 0, z \geq 0$$

Let u, v, and w be slack variables. We obtain the following tableaus:

	x	y	z	u	v	w	P	Const.	Ratio	
p.r. →	1	1	1	1	0	0	0	250000	250000	
	−1	−1	3	0	1	0	0	0	0	$\tfrac{1}{3}R_2$ →
	−1	1	−1	0	0	1	0	0	−−	
	$-\tfrac{3}{50}$	$-\tfrac{1}{10}$	$-\tfrac{3}{20}$	0	0	0	1	0		

$$\uparrow$$
$$\text{p.c}$$

x	y	z	u	v	w	P	Const.	
1	1	1	1	0	0	0	250000	
$-\tfrac{1}{3}$	$-\tfrac{1}{3}$	1	0	$\tfrac{1}{3}$	0	0	0	$\dfrac{R_1 - R_2}{R_3 + R_2}$ →
−1	1	−1	0	0	1	0	0	$R_4 + \tfrac{3}{20}R_2$
$-\tfrac{3}{50}$	$-\tfrac{1}{10}$	$-\tfrac{3}{20}$	0	0	0	1	0	

x	y	z	u	v	w	P	Const.		Ratio
$\frac{4}{3}$	$\frac{4}{3}$	0	1	$-\frac{1}{3}$	0	0	250000		$\frac{250000}{4/3}=187500$
$-\frac{1}{3}$	$-\frac{1}{3}$	1	0	$\frac{1}{3}$	0	0	0		$--$
$-\frac{4}{3}$	$\boxed{\frac{2}{3}}$	0	0	$\frac{1}{3}$	1	0	0		0
$-\frac{11}{100}$	$-\frac{3}{20}$	0	0	$\frac{1}{20}$	0	1	0		

$\xrightarrow{\frac{3}{2}R_3}$

$\uparrow$
p.c.

x	y	z	u	v	w	P	Const.
$\frac{4}{3}$	$\frac{4}{3}$	0	1	$-\frac{1}{3}$	0	0	250000
$-\frac{1}{3}$	$-\frac{1}{3}$	1	0	$\frac{1}{3}$	0	0	0
-2	$\boxed{1}$	0	0	$\frac{1}{2}$	$\frac{3}{2}$	0	0
$-\frac{11}{100}$	$-\frac{3}{20}$	0	0	$\frac{1}{20}$	0	1	0

$\xrightarrow[\substack{R_2+\frac{1}{3}R_3}]{\substack{R_1-\frac{4}{3}R_3}}$
$R_4+\frac{3}{20}R_2$

x	y	z	u	v	w	P	Const.	Ratio
$\boxed{4}$	0	0	1	-1	-2	0	250000	62500
-1	0	1	0	$\frac{1}{2}$	$\frac{1}{2}$	0	0	$--$
-2	1	0	0	$\frac{1}{2}$	$\frac{3}{2}$	0	0	0
$-\frac{41}{100}$	0	0	0	$\frac{3}{20}$	$\frac{9}{40}$	1	0	

$\xrightarrow{\frac{1}{4}R_1}$

$\uparrow$
p.c.

x	y	z	u	v	w	P	Const.
$\boxed{1}$	0	0	$\frac{1}{4}$	$-\frac{1}{4}$	$-\frac{1}{2}$	0	62500
-1	0	1	0	$\frac{1}{2}$	$\frac{1}{2}$	0	0
-2	1	0	0	$\frac{1}{2}$	$\frac{3}{2}$	0	0
$-\frac{41}{100}$	0	0	0	$\frac{3}{20}$	$\frac{9}{40}$	1	0

$\xrightarrow[\substack{R_3+2R_2}]{\substack{R_2+R_1}}$
$R_4+\frac{41}{100}R_1$

x	y	z	u	v	w	P	Const.
1	0	0	$\frac{1}{4}$	$-\frac{1}{4}$	$-\frac{1}{2}$	0	62500
0	0	1	$\frac{1}{4}$	$\frac{1}{4}$	0	0	62500
0	1	0	$\frac{1}{2}$	0	$\frac{1}{2}$	0	125000
0	0	0	$\frac{41}{400}$	$\frac{19}{400}$	$\frac{1}{50}$	1	25,625

The last tableau is in final form, and we see that $x = 62,500$, $y = 125,000$, $z = 62,500$, and $P = 25,625$. So, Ashley should invest $62,500 in the money market fund, $125,000 in the international equity fund, and $62,500 in the growth-and-income fund. Her maximum return will be $25,625.

35. Suppose the Excelsior Company buys x, y, and z minutes of morning, afternoon, and evening commercials, respectively. Then we wish to maximize
$$P = 200,000x + 100,000y + 600,000z \text{ subject to}$$
$$3000x + 1000y + 12,000z \leq 102,000$$
$$z \leq 6$$
$$x + y + z \leq 25$$
$$x \geq 0, y \geq 0, z \geq 0$$

Using the simplex method, we obtain the following sequence of tableaus.

x	y	z	u	v	w	P	Const.	Ratio	
3000	1000	12,000	1	0	0	0	102,000	17/2	$R_1 - 12,000R_2$
0	0	1	0	1	0	0	6	6	$R_3 - R_2$
1	1	1	0	0	1	0	25	25	$R_4 + 600,000R_2 \longrightarrow$
-200,000	-100,000	-600,000	0	0	0	1	0		

x	y	z	u	v	w	P	Const.	Ratio	
3000	1000	0	1	-12,000	0	0	30,000	10	$\frac{1}{3000}R_1$
0	0	1	0	1	0	0	6	--	$\longrightarrow$
1	1	0	0	-1	1	0	19	19	
-200,000	-100,000	0	0	600,000	0	1	3,600,000		

x	y	z	u	v	w	P	Const.	
1	$\frac{1}{3}$	0	$\frac{1}{3000}$	-4	0	0	10	$R_3 - R_1$
0	0	1	0	1	0	0	6	$R_4 + 200,000R_1 \longrightarrow$
1	1	0	0	-1	1	0	19	
-200,000	-100,000	0	0	600,000	0	1	3,600,000	

x	y	z	u	v	w	P	Const.
1	$\frac{1}{3}$	0	$\frac{1}{3000}$	-4	0	0	10
0	0	1	0	1	0	0	6
0	$\frac{2}{3}$	0	$-\frac{1}{3000}$	$\boxed{3}$	1	0	9
0	$-\frac{100,000}{3}$	0	$\frac{200}{3}$	$-200,000$	0	1	5,600,000

$$\xrightarrow{\;\frac{1}{3}R_3\;}$$

x	y	z	u	v	w	P	Const.
1	$\frac{1}{3}$	0	$\frac{1}{3000}$	-4	0	0	10
0	0	1	0	1	0	0	6
0	$\frac{2}{9}$	0	$-\frac{1}{9000}$	1	$\frac{1}{3}$	0	3
0	$-\frac{100,000}{3}$	0	$\frac{200}{3}$	$-200,000$	0	1	5,600,000

$$\xrightarrow[\;R_4+200,000R_3\;]{\substack{R_1+4R_3\\ R_2-R_3}}$$

x	y	z	u	v	w	P	Const.
1	$\frac{11}{9}$	0	$-\frac{1}{9000}$	0	$\frac{4}{3}$	0	22
0	$-\frac{2}{9}$	1	$\frac{1}{9000}$	0	$-\frac{1}{3}$	0	3
0	$\frac{2}{9}$	0	$-\frac{1}{9000}$	1	$\frac{1}{3}$	0	3
0	$\frac{100,000}{9}$	0	$\frac{400}{9}$	0	$\frac{200,000}{3}$	1	6,200,000

We conclude that $x = 22$, $y = 0$, $z = 3$, $u = 0$, $v = 3$, and $P = 6,200,000$. Therefore, the company should buy 22 minutes of morning and 3 minutes of evening advertising time, thereby maximizing their exposure to 6,200,000 viewers.

37. We first tabulate the given information:

Department	Model A	Model B	Model C	Time available
Fabrication	$\frac{5}{4}$	$\frac{3}{2}$	$\frac{3}{2}$	310
Assembly	1	1	$\frac{3}{4}$	205
Finishing	1	1	$\frac{1}{2}$	190
Profit	26	28	24	

Let x, y, and z denote the number of units of model A, model B, and model C to be produced, respectively. Then the required linear programming problem is

Maximize $P = 26x + 28y + 24z$ subject to

$$\tfrac{5}{4}x + \tfrac{3}{2}y + \tfrac{3}{2}z \leq 310$$
$$x + y + \tfrac{3}{4}z \leq 205$$
$$x + y + \tfrac{1}{2}z \leq 190$$
$$x \geq 0, y \geq 0, z \geq 0$$

Using the simplex method, we obtain the following tableaus:

	x	y	z	u	v	w	P	Const.		Ratio	
	$\frac{5}{4}$	$\frac{3}{2}$	$\frac{3}{2}$	1	0	0	0	310		$206\frac{2}{3}$	$R_1 - \frac{3}{2}R_3$
	1	1	$\frac{3}{4}$	0	1	0	0	205		205	$R_2 - R_3$
p.r. →	1	$\boxed{1}$	$\frac{1}{2}$	0	0	1	0	190		190	$R_4 + 28R_3$
	-26	-28	-24	0	0	0	1	0			

p.c.

	x	y	z	u	v	w	P	Const.		Ratio	
p.r. →	$-\frac{1}{4}$	0	$\boxed{\frac{3}{4}}$	1	0	$-\frac{3}{2}$	0	25		$33\frac{1}{3}$	
	0	0	$\frac{1}{4}$	0	1	-1	0	15		60	$-\frac{4}{3}R_1$
	1	1	$\frac{1}{2}$	0	0	1	0	190		380	
	2	0	-10	0	0	28	1	5320			

p.c.

	x	y	z	u	v	w	P	Const.	
	$-\frac{1}{3}$	0	1	$\frac{4}{3}$	0	-2	0	$\frac{100}{3}$	$R_2 - \frac{1}{4}R_1$
	0	0	$\frac{1}{4}$	0	1	-1	0	15	$R_3 - \frac{1}{2}R_1$
	1	1	$\frac{1}{2}$	0	0	1	0	190	$R_4 + 10R_1$
	2	0	-10	0	0	28	1	5320	

x	y	z	u	v	w	P	Const.		Ratio	
$-\frac{1}{3}$	0	1	$\frac{4}{3}$	0	-2	0	$\frac{100}{3}$		$--$	
$\boxed{\frac{1}{12}}$	0	0	$-\frac{1}{3}$	1	$-\frac{1}{2}$	0	$\frac{20}{3}$		80	$\xrightarrow{12R_2}$
$\frac{7}{6}$	1	0	$-\frac{2}{3}$	0	2	0	$\frac{520}{3}$		$148\frac{4}{7}$	
$-\frac{4}{3}$	0	0	$\frac{40}{3}$	0	8	1	$\frac{16960}{3}$			

p.r. → (row 2)   p.c. ↑ (column x)

x	y	z	u	v	w	P	Const.
$-\frac{1}{3}$	0	1	$\frac{4}{3}$	0	-2	0	$\frac{100}{3}$
1	0	0	-4	12	-6	0	80
$\frac{7}{6}$	1	0	$-\frac{2}{3}$	0	2	0	$\frac{520}{3}$
$-\frac{4}{3}$	0	0	$\frac{40}{3}$	0	8	1	$\frac{16960}{3}$

$R_1 + \frac{1}{3}R_2$
$R_3 - \frac{7}{6}R_2$
$R_4 + \frac{4}{3}R_2$

x	y	z	u	v	w	P	Const.
0	0	1	0	4	-4	0	60
1	0	0	-4	12	-6	0	80
0	1	0	4	-14	9	0	80
0	0	0	8	16	0	1	5760

The last tableau is in final form. We see that $x = 80$, $y = 80$, $z = 60$, $u = 0$, $v = 0$, $w = 0$, and $P = 5760$. So, by producing 80 units each of Models A and B, and 60 units of Model C, the company stands to make a profit of $5760. Since $u = v = w = 0$, there are no resources left over.

39. Let x, y, and z denote the number (in thousands) of bottles of formula I, formula II, and formula III, respectively, produced. The resulting linear programming problem is

$$\text{Maximize } P = 180x + 200y + 300z \text{ subject to}$$
$$\tfrac{5}{2}x + 3y + 4z \le 70$$
$$x \le 9$$
$$y \le 12$$
$$z \le 6$$
$$x \ge 0, \; y \ge 0, \; z \ge 0$$

Using the simplex method, we have

	x	y	z	s	t	u	v	P	Const.		Ratio
	$\frac{5}{2}$	3	4	1	0	0	0	0	70		$17\frac{1}{2}$
	1	0	0	0	1	0	0	0	9		$--$
	0	1	0	0	0	1	0	0	12		$--$
p.r. →	0	0	$\boxed{1}$	0	0	0	1	0	6		6
	−180	−200	−300	0	0	0	0	1	0		

p.c. (under z)

$$R_1 - 4R_4$$
$$R_5 + 300R_4$$

	x	y	z	s	t	u	v	P	Const.		Ratio
	$\frac{5}{2}$	3	0	1	0	0	−4	0	46		$15\frac{1}{3}$
	1	0	0	0	1	0	0	0	9		$--$
p.r. →	0	$\boxed{1}$	0	0	0	1	0	0	12		12
	0	0	1	0	0	0	1	0	6		$--$
	−180	−200	0	0	0	0	300	1	1800		

p.c. (under y)

$$R_1 - 3R_3$$
$$R_5 + 200R_3$$

	x	y	z	s	t	u	v	P	Const.	Ratio
	$\frac{5}{2}$	0	0	1	0	−3	−4	0	10	4
	1	0	0	0	1	0	0	0	9	9
	0	1	0	0	0	1	0	0	12	$--$
	0	0	1	0	0	0	1	0	6	$--$
	−180	0	0	0	0	200	300	1	4200	

$$\frac{2}{5}R_1$$

x	y	z	s	t	u	v	P	Const.
1	0	0	$\frac{2}{5}$	0	$-\frac{6}{5}$	$-\frac{8}{5}$	0	4
1	0	0	0	1	0	0	0	9
0	1	0	0	0	1	0	0	12
0	0	1	0	0	0	1	0	6
−180	0	0	0	0	200	300	1	4200

$$\xrightarrow{\begin{array}{c} R_2 - R_1 \\ R_5 + 180R_1 \end{array}}$$

x	y	z	s	t	u	v	P	Const.		Ratio
1	0	0	$\frac{2}{5}$	0	$-\frac{6}{5}$	$-\frac{8}{5}$	0	4		--
1	0	0	$-\frac{2}{5}$	1	$\boxed{\frac{6}{5}}$	$\frac{8}{5}$	0	5		$\frac{25}{6}$
0	1	0	0	0	1	0	0	12		12
0	0	1	0	0	0	1	0	6		--
0	0	0	72	0	−16	12	1	4920		

$$\xrightarrow{\frac{5}{6}R_2}$$

x	y	z	s	t	u	v	P	Const.
1	0	0	$\frac{2}{5}$	0	$-\frac{6}{5}$	$-\frac{8}{5}$	0	4
0	0	0	$-\frac{1}{3}$	$\frac{5}{6}$	1	$\frac{4}{3}$	0	$\frac{25}{6}$
0	1	0	0	0	1	0	0	12
0	0	1	0	0	0	1	0	6
0	0	0	72	0	−16	12	1	4920

$$\xrightarrow{\begin{array}{c} R_1 + \frac{6}{5}R_2 \\ R_3 - R_2 \\ R_5 + 16R_2 \end{array}}$$

x	y	z	s	t	u	v	P	Const.
1	0	0	0	1	0	0	0	9
0	0	0	$-\frac{1}{3}$	$\frac{5}{6}$	1	$\frac{4}{3}$	0	$\frac{25}{6}$
0	1	0	$\frac{1}{3}$	$-\frac{5}{6}$	0	$-\frac{4}{3}$	0	$\frac{47}{6}$
0	0	1	0	0	0	1	0	6
0	0	0	$\frac{200}{3}$	$\frac{40}{3}$	0	$\frac{100}{3}$	1	$4986\frac{2}{3}$

Therefore, $x = 9$, $y = 47/6$, $z = 6$, $s = 0$, $t = 0$, $u = \frac{25}{6}$ and $P \approx 4986.67$; that is, the company should manufacture 9000 bottles of formula *I*, 7833 bottles of formula *II*, and 6000 bottles of formula *III* for a maximum profit of $4986.67. Yes, ingredients for 4167 bottles of formula *II*.

41. Refer to Exercise 11, page 100 of this manual. Let u, v, w, and s be slack variables. We obtain the following tableaus.

	x	y	z	u	v	w	s	P	Const.	Ratio	
	1	1	1	1	0	0	0	0	2000000	2000000	
p.r.→	-2	-2	$\boxed{8}$	0	1	0	0	0	0	0	$\frac{1}{8}R_2$ →
	-6	4	4	0	0	1	0	0	0	0	
	-10	6	6	0	0	0	1	0	0	0	
	$-\frac{1}{10}$	$-\frac{3}{20}$	$-\frac{1}{5}$	0	0	0	0	1	0		

$\uparrow$
p.c.

x	y	z	u	v	w	s	P	Const.	
1	1	1	1	0	0	0	0	2000000	
$-\frac{1}{4}$	$-\frac{1}{4}$	$\boxed{1}$	0	$\frac{1}{8}$	0	0	0	0	R_1-R_2
-6	4	4	0	0	1	0	0	0	R_3-4R_2
-10	6	6	0	0	0	1	0	0	R_4-6R_2
$-\frac{1}{10}$	$-\frac{3}{20}$	$-\frac{1}{5}$	0	0	0	0	1	0	$R_5+\frac{1}{5}R_2$

x	y	z	u	v	w	s	P	Const.	Ratio	
$\frac{5}{4}$	$\frac{5}{4}$	0	1	$-\frac{1}{8}$	0	0	0	2000000	1600000	
$-\frac{1}{4}$	$-\frac{1}{4}$	1	0	$\frac{1}{8}$	0	0	0	0	--	$\frac{1}{5}R_3$ →
-5	$\boxed{5}$	0	0	$-\frac{1}{2}$	1	0	0	0	0	
$-\frac{17}{2}$	$\frac{15}{2}$	0	0	$-\frac{3}{4}$	0	1	0	0	0	
$-\frac{3}{20}$	$-\frac{1}{5}$	0	0	$\frac{1}{40}$	0	0	1	0		

x	y	z	u	v	w	s	P	Const.	
$\frac{5}{4}$	$\frac{5}{4}$	0	1	$-\frac{1}{8}$	0	0	0	2000000	
$-\frac{1}{4}$	$-\frac{1}{4}$	1	0	$\frac{1}{8}$	0	0	0	0	
-1	$\boxed{1}$	0	0	$-\frac{1}{10}$	$\frac{1}{5}$	0	0	0	$R_1-\frac{5}{4}R_3$
$-\frac{17}{2}$	$\frac{15}{2}$	0	0	$-\frac{3}{4}$	0	1	0	0	$R_2+\frac{1}{4}R_3$
$-\frac{3}{20}$	$-\frac{1}{5}$	0	0	$\frac{1}{40}$	0	0	1	0	$R_4-\frac{15}{2}R_3$
									$R_5+\frac{1}{5}R_3$

x	y	z	u	v	w	s	P	Const.	Ratio
$\frac{5}{2}$	0	0	1	$-\frac{1}{4}$	0	0	0	2000000	800000
$-\frac{1}{2}$	0	1	0	$\frac{1}{10}$	$\frac{1}{20}$	0	0	0	--
-1	1	0	0	$-\frac{1}{10}$	$\frac{1}{5}$	0	0	0	0
-1	0	0	0	0	$-\frac{3}{2}$	1	0	0	0
$-\frac{7}{20}$	0	0	0	$\frac{1}{200}$	$\frac{1}{25}$	0	1	0	

$$\xrightarrow{\frac{2}{5}R_1}$$

x	y	z	u	v	w	s	P	Const.
1	0	0	$\frac{2}{5}$	0	$-\frac{1}{10}$	0	0	800000
$-\frac{1}{2}$	0	1	0	$\frac{1}{10}$	$\frac{1}{20}$	0	0	0
-1	1	0	0	$-\frac{1}{10}$	$\frac{1}{5}$	0	0	0
-1	0	0	0	0	$-\frac{3}{2}$	1	0	0
$-\frac{7}{20}$	0	0	0	$\frac{1}{200}$	$\frac{1}{25}$	0	1	0

$$\xrightarrow[\substack{R_2+\frac{1}{2}R_1\\ R_3+R_1\\ R_4+R_1\\ R_5+\frac{7}{20}R_1}]{}$$

x	y	z	u	v	w	s	P	Const.
1	0	0	$\frac{2}{5}$	0	$-\frac{1}{10}$	0	0	800000
0	0	1	$\frac{1}{5}$	$\frac{1}{10}$	0	0	0	400000
0	1	0	$\frac{2}{5}$	$-\frac{1}{10}$	$\frac{1}{10}$	0	0	800000
0	0	0	$\frac{2}{5}$	0	$-\frac{8}{5}$	1	0	800000
0	0	0	$\frac{7}{50}$	$\frac{1}{200}$	$\frac{1}{200}$	0	1	280000

The last tableau is in final form, and we see that $x = 800{,}000$, $y = 800{,}000$, $Z = 400{,}000$, and $P = 280{,}000$. Thus, the financier should invest \$800,000 each in Projects A and B, and \$400,000 in project C. The maximum returns are \$280,000.

43. False. consider the linear programming problem

Maximize $P = 2x + 3y$ subject to
$$-x + y \le 0$$
$$x \ge 0, \; y \ge 0$$

45. True. Consider the objective function $P = c_1x + c_2x_2 + \cdots c_nx_n$, which may be written in the form

$$-c_1x_1 - c_2x_2 - \cdots - c_nx_n + P = 0$$

Observe that the most negative of the numbers $-c_1, -c_2, \cdots, -c_n$ (which are the numbers comprising the last row of the simplex tableau) is just the largest coefficient of x_i in the expression for P. Thus, moving in the direction of the variable with this coefficient ensures that P increases most.

USING TECHNOLOGY EXERCISES 4.1, page 232

1. $x = 1.2$, $y = 0$, $z = 1.6$, $w = 0$, and $P = 8.8$

3. $x = 1.6$, $y = 0$, $z = 0$, $w = 3.6$, and $P = 12.4$

EXERCISES 4.2, page 243

1. We solve the associated regular problem:

Maximize $P = -C = 2x - y$ subject to

$$x + 2y \le 6$$
$$3x + 2y \le 12$$
$$x \ge 0, \ y \ge 0$$

Using the simplex method where u and v are slack variables, we have

	x	y	u	v	P	Const.	Ratio	
	1	2	1	0	0	6	6	$\frac{1}{3}R_2$
p.r. →	3	2	0	1	0	12	4	
	−2	1	0	0	1	0		

↑

p.c.

x	y	u	v	P	Const.			x	y	u	v	P	Const.
1	2	1	0	0	6	$R_1 - R_2$		0	$\frac{4}{3}$	1	$-\frac{1}{3}$	0	2
1	$\frac{2}{3}$	0	$\frac{1}{3}$	0	4	$R_3 + 2R_2$ →		1	$\frac{2}{3}$	0	$\frac{1}{3}$	0	4
−2	1	0	0	1	0			0	$\frac{7}{3}$	0	$\frac{2}{3}$	1	8

Therefore, $x = 4$, $y = 0$, and $C = -P = -8$.

3. We maximize $P = -C = 3x + 2y$. Using the simplex method, we obtain

	x	y	u	v	P	Const.	Ratio	
	3	4	1	0	0	24	8	$\frac{1}{7}R_2$
p.r. →	$\boxed{7}$	−4	0	1	0	16	$\frac{16}{7}$	
	−3	−2	0	0	1	0		

p.c. ↑

x	y	u	v	P	Const.	
3	4	1	0	0	24	$R_1 - 3R_2$
1	$-\frac{4}{7}$	0	$\frac{1}{7}$	0	$\frac{16}{7}$	$R_3 + 3R_2$
−3	−2	0	0	1	0	

	x	y	u	v	P	Const.	Ratio	
p.r. →	0	$\boxed{\frac{40}{7}}$	1	$-\frac{3}{7}$	0	$\frac{120}{7}$	3	$\frac{7}{40}R_1$
	1	$-\frac{4}{7}$	0	$\frac{1}{7}$	0	$\frac{16}{7}$	−−	
	0	$-\frac{26}{7}$	0	$\frac{3}{7}$	1	$\frac{48}{7}$		

p.c. ↑

x	y	u	v	P	Const.	
0	1	$\frac{7}{40}$	$-\frac{3}{40}$	0	3	$R_2 + \frac{4}{7}R_1$
1	$-\frac{4}{7}$	0	$\frac{1}{7}$	0	$\frac{16}{7}$	$R_3 + \frac{26}{7}R_1$
0	$-\frac{26}{7}$	0	$\frac{3}{7}$	1	$\frac{48}{7}$	

x	y	u	v	P	Const.
0	1	$\frac{7}{40}$	$-\frac{3}{40}$	0	3
1	0	$\frac{1}{10}$	$\frac{1}{10}$	0	4
0	0	$\frac{13}{20}$	$\frac{3}{20}$	1	18

The last tableau is in final form. We find $x = 4$, $y = 3$, and $C = -P = -18$.

5. We maximize $P = -C = -2x + 3y + 4z$ subject to the given constraints. Using the simplex method we obtain

x	y	z	u	v	w	P	Const.	Ratio	
−1	2	−1	1	0	0	0	8	−−	
p.r. → 1	−2	$\boxed{2}$	0	1	0	0	10	5	$\xrightarrow{\frac{1}{2}R_2}$
2	4	−3	0	0	1	0	12	−−	
2	−3	−4	0	0	0	1	0		

$$\uparrow$$
$$p.c.$$

x	y	z	u	v	w	P	Const.	
−1	2	−1	1	0	0	0	8	R_1+R_2
$\frac{1}{2}$	−1	1	0	$\frac{1}{2}$	0	0	5	R_3+3R_2 $\xrightarrow{R_4+4R_2}$
2	4	−3	0	0	1	0	12	
2	−3	−4	0	0	0	1	0	

x	y	z	u	v	w	P	Const.	Ratio	
p.r. → −$\frac{1}{2}$	$\boxed{1}$	0	1	$\frac{1}{2}$	0	0	13	13	R_2+R_1
$\frac{1}{2}$	−1	1	0	$\frac{1}{2}$	0	0	5	−−	R_3-R_1 $\xrightarrow{R_4+7R_1}$
$\frac{7}{2}$	1	0	0	$\frac{3}{2}$	1	0	27	27	
4	−7	0	0	2	0	1	20		

$$\uparrow$$
$$p.c.$$

x	y	z	u	v	w	P	Const.
−$\frac{1}{2}$	1	0	1	$\frac{1}{2}$	0	0	13
0	0	1	1	1	0	0	18
4	0	0	−1	1	1	0	14
$\frac{1}{2}$	0	0	7	$\frac{11}{2}$	0	1	111

The last tableau is in final form. We see that $x = 0$, $y = 13$, $z = 18$, $w = 14$, and $C = -P = -111$.

7. $x = 5/4$, $y = 1/4$, $u = 2$, $v = 3$, and $C = P = 13$.

9. $x = 5, y = 10, z = 0, u = 1, v = 2,$ and $C = P = 80.$

11. We first write the tableau

x	y	Const.
1	2	4
3	2	6
2	5	

Then obtain the following by interchanging rows and columns:

u	v	Const.
1	3	2
2	2	5
4	6	

From this table we construct the dual problem:
Maximize the objective function
$$P = 4u + 6v \text{ subject to}$$
$$u + 3v \leq 2$$
$$2u + 2v \leq 5$$
$$u \geq 0, \ v \geq 0$$

Solving the dual problem using the simplex method with x and y as the slack variables, we obtain

	u	v	x	y	P	Const.	Ratio	
p.r. $\rightarrow$	1	③	1	0	0	2	$\frac{2}{3}$	$\frac{1}{3}R_1$ $\longrightarrow$
	2	2	0	1	0	5	$\frac{5}{2}$	
	−4	−6	0	0	1	0		

$$\uparrow$$
$$p.c.$$

u	v	x	y	P	Const.	
$\frac{1}{3}$	1	$\frac{1}{3}$	0	0	$\frac{2}{3}$	$\begin{array}{c} R_2-2R_1 \\ R_3+6R_1 \end{array}$ →
2	2	0	1	0	5	
-4	-6	0	0	1	0	

	u	v	x	y	P	Const.	Ratio	
p.r. →	$\boxed{\frac{1}{3}}$	1	$\frac{1}{3}$	0	0	$\frac{2}{3}$	2	$3R_1$ →
	$\frac{4}{3}$	0	$-\frac{2}{3}$	1	0	$\frac{11}{3}$	$2\frac{3}{4}$	
	-2	0	2	0	1	4		

$\uparrow$
p.c.

u	v	x	y	P	Const.	
1	3	1	0	0	2	$\begin{array}{c} R_2-\frac{4}{3}R_1 \\ R_3+2R_1 \end{array}$ →
$\frac{4}{3}$	0	$-\frac{2}{3}$	1	0	$\frac{11}{3}$	
-2	0	2	0	1	4	

u	v	x	y	P	Const.
1	3	1	0	0	2
0	-4	-2	1	0	1
0	6	4	0	1	8

Interpreting the final tableau, we see that $x = 4$, $y = 0$, and $P = C = 8$.

13.   We first write the tableau

x	y	Const.
6	1	60
2	1	40
1	1	30
6	4	

Then obtain the following by interchanging rows and columns:

u	v	w	Const.
6	2	1	6
1	1	1	4
60	40	30	

From this table we construct the dual problem:

Maximize $P = 60u + 40v + 30w$ subject to

$$6u + 2v + w \leq 6$$

$$u + v + w \leq 4$$

$$u \geq 0, \ v \geq 0, \ w \geq 0$$

We solve the problem as follows.

	u	v	w	x	y	P	Const.	Ratio	
$p.r. \rightarrow$	$\boxed{6}$	2	1	1	0	0	6	1	$\frac{1}{6}R_1$
	1	1	1	0	1	0	4	4	$\longrightarrow$
	-60	-40	-30	0	0	1	0	--	

$\uparrow$
$p.\varsigma$

u	v	w	x	y	P	Const.	
1	$\frac{1}{3}$	$\frac{1}{6}$	$\frac{1}{6}$	0	0	1	$R_2 - R_1$
1	1	1	0	1	0	4	$R_3 + 60R_1$
-60	-40	-30	0	0	1	0	$\longrightarrow$

	u	v	w	x	y	P	Const.	Ratio	
	1	$\frac{1}{3}$	$\frac{1}{6}$	$\frac{1}{6}$	0	0	1	6	$\frac{6}{5}R_2$
$p.r. \rightarrow$	0	$\frac{2}{3}$	$\boxed{\frac{5}{6}}$	$-\frac{1}{6}$	1	0	3	18/5	$\longrightarrow$
	0	-20	-20	10	0	1	60	--	

$\uparrow$
$p.c.$

u	v	w	x	y	P	Const.	
1	$\frac{1}{3}$	$\frac{1}{6}$	$\frac{1}{6}$	0	0	1	$R_1 - \frac{1}{6}R_2$
0	$\frac{4}{5}$	1	$-\frac{1}{5}$	$\frac{6}{5}$	0	$\frac{18}{5}$	$R_3 + 20R_2$
0	−20	−20	10	0	1	60	

	u	v	w	x	y	P	Const.	Ratio	
p.r. →	1	$\boxed{\frac{1}{5}}$	0	$\frac{1}{5}$	$-\frac{1}{5}$	0	$\frac{2}{5}$	2	$5R_1$
	0	$\frac{4}{5}$	1	$-\frac{1}{5}$	$\frac{6}{5}$	0	$\frac{18}{5}$	9/2	
	0	−4	0	6	24	1	132	--	

$$\uparrow \\ p.c.$$

u	v	w	x	y	P	Const.	
5	1	0	1	−1	0	2	$R_2 - \frac{4}{5}R_1$
0	$\frac{4}{5}$	1	$-\frac{1}{5}$	$\frac{6}{5}$	0	$\frac{18}{5}$	$R_3 + 4R_1$
0	−4	0	6	24	1	132	

u	v	w	x	y	P	Const.
5	1	0	1	−1	0	2
−4	0	1	−1	2	0	2
20	0	0	10	20	1	140

The last tableau is in final form. We find that $x = 10$, $y = 20$, and $C = 140$.

15. We first write the tableau

x	y	z	Const.
20	10	1	10
1	1	2	20
200	150	120	

Then obtain the following by interchanging rows and columns:

u	v	Const.
20	1	200
10	1	150
1	2	120
10	20	

From this table we construct the dual problem:

Maximize $P = 10u + 20v$ subject to

$$20u + v \le 200$$
$$10u + v \le 150$$
$$u + 2v \le 120$$
$$u \ge 0, v \ge 0$$

Solving this problem, we obtain the following tableaus:

	u	v	x	y	z	P	Const.	Ratio	
	20	1	1	0	0	0	200	200	
	10	1	0	1	0	0	150	150	$\frac{1}{2}R_3$
p.r. →	1	2	0	0	1	0	120	60	
	−10	−20	0	0	0	1	0		

$$\underset{p.c}{\uparrow}$$

	u	v	x	y	z	P	Const.	
	20	1	1	0	0	0	200	$R_1 - R_3$
	10	1	0	1	0	0	150	$R_2 - R_3$ $R_4 + 20R_3$
	$\frac{1}{2}$	1	0	0	$\frac{1}{2}$	0	60	
	−10	−20	0	0	0	1	0	

u	v	x	y	z	P	Const.
$\frac{39}{2}$	0	1	0	$-\frac{1}{2}$	0	140
$\frac{19}{2}$	0	0	1	$-\frac{1}{2}$	0	90
$\frac{1}{2}$	1	0	0	$\frac{1}{2}$	0	60
0	0	0	0	10	1	1200

This last tableau is in final form. We find that $x = 0$, $y = 0$, $z = 10$, and $C = 1200$.

17. We first write the tableau

x	y	z	Const.
1	2	2	10
2	1	1	24
1	1	1	16
6	8	4	

Then obtain the following by interchanging rows and columns:

u	v	w	Const.
1	2	1	6
2	1	1	8
2	1	1	4
10	24	16	

From this table we construct the dual problem:

Maximize the objective function

$P = 10u + 24v + 16w$ subject to

$$u + 2v + w \le 6$$
$$2u + v + w \le 8$$
$$2u + v + w \le 4$$
$$u \ge 0, v \ge 0, w \ge 0$$

Solving the dual problem using the simplex method with x, y, and z as slack variables, we obtain

	u	v	w	x	y	z	P	Const.		Ratio	
$p.r. \rightarrow$	1	$\boxed{2}$	1	1	0	0	0	6		3	
	2	1	1	0	1	0	0	8		8	$\frac{1}{2}R_1 \longrightarrow$
	2	1	1	0	0	1	0	4		4	
	−10	−24	−16	0	0	0	1	0			

$$\uparrow$$
$$p.c.$$

u	v	w	x	y	z	P	Const.	
$\frac{1}{2}$	1	$\frac{1}{2}$	$\frac{1}{2}$	0	0	0	3	$R_2 - R_1$
2	1	1	0	1	0	0	8	$R_3 - R_1$ $R_4 + 24R_1$ $\longrightarrow$
2	1	1	0	0	1	0	4	
-10	-24	-16	0	0	0	1	0	

u	v	w	x	y	z	P	Const.	Ratio	
$\frac{1}{2}$	1	$\frac{1}{2}$	$\frac{1}{2}$	0	0	0	3	6	
$\frac{3}{2}$	0	$\frac{1}{2}$	$-\frac{1}{2}$	1	0	0	5	10	$2R_3$ $\longrightarrow$
p.r. → $\frac{3}{2}$	0	$\boxed{\frac{1}{2}}$	$-\frac{1}{2}$	0	1	0	1	2	
2	0	-4	12	0	0	1	72		

$$\uparrow$$
$$p.c.$$

u	v	w	x	y	z	P	Const.	
$\frac{1}{2}$	1	$\frac{1}{2}$	$\frac{1}{2}$	0	0	0	3	$R_1 - \frac{1}{2}R_3$
$\frac{3}{2}$	0	$\frac{1}{2}$	$-\frac{1}{2}$	1	0	0	5	$R_2 - \frac{1}{2}R_3$
3	0	1	-1	0	2	0	2	$R_4 + 4R_3$ $\longrightarrow$
2	0	-4	12	0	0	1	72	

u	v	w	x	y	z	P	Const.
-1	1	0	1	0	-1	0	2
0	0	0	0	1	-1	0	4
3	0	1	-1	0	2	0	2
14	0	0	8	0	8	1	80

The solution to the primal problem is $x = 8$, $y = 0$, $z = 8$, and $C = 80$.

19.   We first write

x	y	z	Const.
2	4	3	6
6	0	1	2
0	6	2	4
30	12	20	

Then obtain the following by interchanging rows and columns:

u	v	w	Const.
2	6	0	30
4	0	6	12
3	1	2	20
6	2	4	

From this table we construct the dual problem:

Maximize $P = 6u + 2v + 4w$ subject to

$$2u + 6v \qquad \le 30$$
$$4u + \qquad 6w \le 12$$
$$3u + v + 2w \le 20$$
$$u \ge 0, \ v \ge 0, \ w \ge 0$$

Using the simplex method, we obtain

	u	v	w	x	y	z	P	Const.	Ratio
	2	6	0	1	0	0	0	30	15
p.r. $\to$	$\boxed{4}$	0	6	0	1	0	0	12	3
	3	1	2	0	0	1	0	20	$\frac{20}{3}$
	−6	−2	−4	0	0	0	1	0	

$$\uparrow$$
$$p.c.$$

$\frac{1}{4}R_2 \longrightarrow$

u	v	w	x	y	z	P	Const.
2	6	0	1	0	0	0	30
1	0	$\frac{3}{2}$	0	$\frac{1}{4}$	0	0	3
3	1	2	0	0	1	0	20
−6	−2	−4	0	0	0	1	0

$$\begin{array}{c} R_1 - 2R_2 \\ R_3 - 3R_2 \\ R_4 + 6R_2 \end{array} \longrightarrow$$

u	v	w	x	y	z	P	Const.		Ratio
p.r. → 0	$\boxed{6}$	-3	1	$-\frac{1}{2}$	0	0	24		4
1	0	$\frac{3}{2}$	0	$\frac{1}{4}$	0	0	3		---
0	1	$-\frac{5}{2}$	0	$-\frac{3}{4}$	1	0	11		11
0	-2	5	0	$\frac{3}{2}$	0	1	18		

$$\xrightarrow{\ \frac{1}{6}R_1\ }$$

p.c. (under v column)

u	v	w	x	y	z	P	Const.	
0	1	$-\frac{1}{2}$	$\frac{1}{6}$	$-\frac{1}{12}$	0	0	4	$R_3 - R_1$
1	0	$\frac{3}{2}$	0	$\frac{1}{4}$	0	0	3	$R_4 + 2R_1$
0	1	$-\frac{5}{2}$	0	$-\frac{3}{4}$	1	0	11	
0	-2	5	0	$\frac{3}{2}$	0	1	18	

$$\xrightarrow{\ R_3 - R_1,\ R_4 + 2R_1\ }$$

u	v	w	x	y	z	P	Const.
0	1	$-\frac{1}{2}$	$\frac{1}{6}$	$-\frac{1}{12}$	0	0	4
1	0	$\frac{3}{2}$	0	$\frac{1}{4}$	0	0	3
0	0	-2	$-\frac{1}{6}$	$-\frac{2}{3}$	1	0	7
0	0	4	$\frac{1}{3}$	$\frac{4}{3}$	0	1	26

The last tableau is in final form. We find $x = 1/3$, $y = 4/3$, $z = 0$, and $C = 26$.

21. Let x denote the number of type-A vessels and y the number of type-B vessels to be operated. Then the problem is

Maximize $C = 44,000x + 54,000y$ subject to

$$60x + 80y \geq 360$$
$$160x + 120y \geq 680$$
$$x \geq 0, y \geq 0$$

We first write down the following tableau for the primal problem.

x	y	Constant
60	80	360
160	120	680
44,000	54,000	

Next, we interchange the columns and rows of the frequency tableaus obtaining

u	v	Constant
60	160	44,000
80	120	54,000
360	680	

Proceeding, we are led to the dual problem

Maximize $P = 360u + 680v$ subject to

$$60u + 160v \le 44000$$

$$80u + 120v \le 54000$$

$$u \ge 0, v \ge 0$$

Let x and y be slack variables. We obtain the following tableaus:

	u	v	x	y	P	Const.	Ratio	
p.r. →	60	$\boxed{160}$	1	0	0	44,000	275	$\frac{1}{160}R_1 \rightarrow$
	80	120	0	1	0	54,000	450	
	−360	−680	0	0	1	0		

$\uparrow$ p.c.

u	v	x	y	P	Const.	
$\frac{3}{8}$	$\boxed{1}$	$\frac{1}{160}$	0	0	275	$R_2 - 120R_1 \rightarrow$
80	120	0	1	0	54,000	$R_3 + 680R_1$
−360	−680	0	0	1	0	

	u	v	x	y	P	Const.	Ratio	
	$\frac{3}{8}$	1	$\frac{1}{160}$	0	0	275	$733\frac{1}{3}$	$\frac{1}{35}R_1 \rightarrow$
p.r. →	$\boxed{35}$	0	$-\frac{3}{4}$	1	0	21000	600	
	−105	0	$\frac{17}{4}$	0	1	187000		

$\uparrow$ p.c.

u	v	x	y	P	Const.
$\frac{3}{8}$	1	$\frac{1}{160}$	0	0	275
$\boxed{1}$	0	$-\frac{3}{140}$	$\frac{1}{35}$	0	600
-105	0	$\frac{17}{4}$	0	1	187000

$$\xrightarrow[R_3+105R_2]{R_1-\frac{3}{8}R_2}$$

u	v	x	y	P	Const.
0	1	$\frac{1}{70}$	$-\frac{3}{280}$	0	50
1	0	$-\frac{3}{140}$	$\frac{1}{35}$	0	600
0	0	2	3	1	250000

The last tableau is in final form. The fundamental theorem of duality tells us that the solution to the primal problem is $x = 2$, $y = 3$ with a minimum value for C of 250,000. Thus, Deluxe River Cruises should use 2 type-A vessels and 3 type-B vessels. The minimum operating cost is \$250,000.

23. Let x and y denote, respectively, the number of advertisements to be placed in newspaper I and newspaper II. Then the problem is

Minimize $C = 1000x + 800y$ subject to

$$70000x + 10000y \geq 2000000$$
$$40000x + 20000y \geq 1400000$$
$$20000x + 40000y \geq 1000000$$
$$x \geq 0, y \geq 0$$

or upon simplification:

Minimize $C = 1000x + 800y$ subject to

$$7x + y \geq 200$$
$$2x + y \geq 70$$
$$x + 2y \geq 50$$
$$x \geq 0, y \geq 0$$

We first write down the following tableau for the primal problem

x	y	Const.
7	1	200
2	1	70
1	2	50
1000	800	

Next, we interchange the columns and rows of the foregoing tableau:

u	v	w	Const.
7	2	1	1000
1	1	2	800
200	70	50	

Proceeding, we obtain the following dual problem

Maximize $P = 200u + 70v + 50w$ subject to

$$7u + 2v + w \leq 1000$$

$$u + v + 2w \leq 800$$

$$u \geq 0, v \geq 0, w \geq 0$$

Let x and y denote the slack variables. We obtain the following tableaus:

	u	v	w	x	y	P	Const.	Ratio	
$p.r. \rightarrow$	$\boxed{7}$	2	1	1	0	0	1000	$142\frac{6}{7}$	$\frac{1}{7}R_1$
	1	1	2	0	1	0	800	800	
	-200	-70	-50	0	0	1	0		

$$\uparrow$$
$$p.c.$$

u	v	w	x	y	P	Const.	
$\boxed{1}$	$\frac{2}{7}$	$\frac{1}{7}$	$\frac{1}{7}$	0	0	$\frac{1000}{7}$	$R_2 - R_1$
1	1	2	0	1	0	800	$R_3 + 200R_1$
-200	-70	-50	0	0	1	0	

	u	v	w	x	y	P	Const.	Ratio	
	1	$\frac{2}{7}$	$\frac{1}{7}$	$\frac{1}{7}$	0	0	$\frac{1000}{7}$	1000	$\frac{7}{13}R_2$
$p.r. \rightarrow 0$		$\frac{5}{7}$	$\boxed{\frac{13}{7}}$	$-\frac{1}{7}$	1	0	$\frac{4600}{7}$	$657\frac{1}{7}$	
	0	$-\frac{90}{7}$	$-\frac{150}{7}$	$\frac{200}{7}$	0	1	$\frac{200000}{7}$		

$$\uparrow$$
$$p.c.$$

u	v	w	x	y	P	Const.	
1	$\frac{2}{7}$	$\frac{1}{7}$	$\frac{1}{7}$	0	0	$\frac{1000}{7}$	$R_1 - \frac{1}{7}R_2$
0	$\frac{5}{13}$	$\boxed{1}$	$-\frac{1}{13}$	$\frac{7}{13}$	0	$\frac{4600}{7}$	$R_1 + \frac{150}{7}R_2$ $\longrightarrow$
0	$-\frac{90}{7}$	$-\frac{150}{7}$	$\frac{200}{7}$	0	1	$\frac{200000}{7}$	

	u	v	w	x	y	P	Const.	Ratio	
$p.r. \rightarrow$	1	$\boxed{\frac{3}{13}}$	0	$\frac{2}{13}$	$-\frac{1}{13}$	0	$\frac{1200}{13}$	400	$\frac{13}{3}R_1$ $\longrightarrow$
	0	$\frac{5}{13}$	1	$-\frac{1}{13}$	$\frac{7}{13}$	0	$\frac{4600}{13}$	920	
	0	$-\frac{60}{13}$	0	$\frac{300}{13}$	$\frac{150}{13}$	1	$\frac{470000}{13}$		

$\uparrow$ $p.c.$

u	v	w	x	y	P	Const.	
$\frac{13}{3}$	$\boxed{1}$	0	$\frac{2}{3}$	$-\frac{1}{3}$	0	400	$R_2 - \frac{5}{13}R_1$
0	$\frac{5}{13}$	1	$-\frac{1}{13}$	$\frac{7}{13}$	0	$\frac{4600}{13}$	$R_3 + \frac{60}{13}R_1$ $\longrightarrow$
0	$-\frac{60}{13}$	0	$\frac{300}{13}$	$\frac{150}{13}$	1	$\frac{470000}{13}$	

u	v	w	x	y	P	Const.
$\frac{13}{3}$	1	0	$\frac{2}{3}$	$-\frac{1}{3}$	0	400
$-\frac{5}{2}$	0	1	$-\frac{1}{3}$	$\frac{2}{3}$	0	200
20	0	0	30	10	1	38000

The last tableau is in final form. The fundamental theorem of duality tells us that the solution to the primal problem is $x = 30$, $y = 10$, and $C = 38000$. Therefore, Everest Deluxe World Travel should place 30 advertisements in newspaper I and 10 advertisements in newspaper II at a minimum cost of $38,000.

25. The given data may be summarized as follows:

	Orange Juice	Grapefruit Juice
Vitamin A	60 I.U.	120 I.U.
Vitamin C	16 I.U.	12 I.U.
Calories	14	11

Suppose x ounces of orange juice and y ounces of pink-grapefruit juice are required for each glass of the blend. Then the problem is

$$\text{Minimize } C = 14x + 11y \text{ subject to}$$
$$60x + 120y \geq 1200$$
$$16x + 12y \geq 200$$
$$x \geq 0, y \geq 0$$

To construct the dual problem, we first write down the tableau

x	y	Const.
60	120	1200
16	12	200
14	11	

Then obtain the following by interchanging rows and columns:

u	v	Const.
60	16	14
120	12	11
1200	200	

From this table we construct the dual problem:

$$\text{Maximize } P = 1200u + 200v \text{ subject to}$$
$$60u + 16v \le 14$$
$$120u + 12v \le 11$$
$$u \ge 0, v \ge 0$$

The initial tableau is

u	v	x	y	P	Const.
60	16	1	0	0	14
120	12	0	1	0	11
−1200	−200	0	0	1	0

Using the following sequence of row operations,

1. $\frac{1}{120}R_2$ 2. $R_1 - 60R_2$, $R_3 + 1200R_2$ 3. $\frac{1}{10}R_1$ 4. $R_2 - \frac{1}{10}R_1$, $R_3 + 80R_1$

we obtain the final tableau

u	v	x	y	P	Const.
0	1	$\frac{1}{10}$	$-\frac{1}{20}$	0	$\frac{17}{20}$
0	0	$-\frac{1}{100}$	$\frac{1}{75}$	0	$\frac{1}{150}$
0	0	8	6	1	178

We conclude that the owner should use 8 ounces of orange juice and 6 ounces of grapefruit juice per glass of the blend for a minimal calorie count of 178.

27.   True. To maximize P, one maximizes -C. Since the minimization problem has a

unique solution, the negative of that solution is the solution of the maximization problem.

USING TECHNOLOGY EXERCISES 4.2, page 248

1. $x = \frac{4}{3}$, $y = \frac{10}{3}$, $z = 0$, and $C = \frac{14}{3}$

3. $x = 0.9524$, $y = 4.2857$, $z = 0$, and $C = 6.0952$.

CHAPTER 4 REVIEW EXERCISES, page 249

1.   This is a regular linear programming problem. Using the simplex method with u and v as slack variables, we obtain the following sequence of tableaus:

	x	y	u	v	P	Const
p.r. →	1	3	1	0	0	15
	4	1	0	1	0	16
	−3	−4	0	0	1	0

p.c. (under y column)

Ratio
5
16

$\xrightarrow{\frac{1}{3}R_1}$

	x	y	u	v	P	Const.
	$\frac{1}{3}$	1	$\frac{1}{3}$	0	0	5
	4	1	0	1	0	16
	−3	−4	0	0	1	0

$\xrightarrow[R_3+4R_1]{R_2-R_1}$

	x	y	u	v	P	Const.
	$\frac{1}{3}$	1	$\frac{1}{3}$	0	0	5
p.c. →	$\frac{11}{3}$	0	$-\frac{1}{3}$	1	0	11
	$-\frac{5}{3}$	0	$\frac{4}{3}$	0	1	20

p.c. (under x column)

Ratio
15
3

$\xrightarrow{\frac{3}{11}R_2}$

	x	y	u	v	P	Const.
	$\frac{1}{3}$	1	$\frac{1}{3}$	0	0	5
	1	0	$-\frac{1}{11}$	$\frac{3}{11}$	0	3
	$-\frac{5}{3}$	0	$\frac{4}{3}$	0	1	20

$\xrightarrow[R_3+\frac{5}{3}R_2]{R_1-\frac{1}{3}R_2}$

	x	y	u	v	P	Const.
	0	1	$\frac{4}{11}$	$-\frac{1}{11}$	0	4
	1	0	$-\frac{1}{11}$	$\frac{3}{11}$	0	3
	0	0	$\frac{13}{11}$	$\frac{5}{11}$	1	25

and conclude that $x = 3$, $y = 4$, $u = 0$, $v = 0$, and $P = 25$.

3. Using the simplex method to solve this regular linear programming problem we have

	x	y	z	u	v	P	Const.		Ratio	
p.r. →	1	2	$\boxed{3}$	1	0	0	12		4	$\frac{1}{3}R_1$
	1	-3	2	0	1	0	10		5	
	-2	-3	-5	0	0	1	0			

p.c. (arrow under z)

	x	y	z	u	v	P	Const.	
	$\frac{1}{3}$	$\frac{2}{3}$	1	$\frac{1}{3}$	0	0	4	R_2-2R_1
	1	-3	2	0	1	0	10	R_3+5R_1
	-2	-3	-5	0	0	1	0	

	x	y	z	u	v	P	Const.		Ratio	
	$\frac{1}{3}$	$\frac{2}{3}$	1	$\frac{1}{3}$	0	0	4		12	
p.r. →	$\boxed{\frac{1}{3}}$	$-\frac{13}{3}$	0	$-\frac{2}{3}$	1	0	2		6	$3R_2$
	$-\frac{1}{3}$	$\frac{1}{3}$	0	$\frac{5}{3}$	0	1	20			

p.c. (arrow under x)

	x	y	z	u	v	P	Const.	
	$\frac{1}{3}$	$\frac{2}{3}$	1	$\frac{1}{3}$	0	0	4	$R_1-\frac{1}{3}R_2$
	1	-13	0	-2	3	0	6	$R_3+\frac{1}{3}R_2$
	$-\frac{1}{3}$	$\frac{1}{3}$	0	$\frac{5}{3}$	0	1	20	

x	y	z	u	v	P	Const.	
0	1	$\frac{1}{5}$	$\frac{1}{5}$	$-\frac{1}{5}$	0	$\frac{2}{5}$	R_2+13R_1
0	-13	0	-2	3	0	6	R_3+4R_1
0	-4	0	1	1	1	22	

x	y	z	u	v	P	Const.
0	1	$\frac{1}{5}$	$\frac{1}{5}$	$-\frac{1}{5}$	0	$\frac{2}{5}$
1	0	$\frac{13}{5}$	$\frac{3}{5}$	$\frac{2}{5}$	0	$\frac{56}{5}$
0	0	$\frac{4}{5}$	$\frac{9}{5}$	$\frac{1}{5}$	1	$\frac{118}{5}$

We conclude that the P attains a maximum value of 23.6 when $x = 11.2$, $y = 0.4$, $z = 0$, $u = 0$, and $v = 0$.

5. We first write the tableau

x	y	Const.
2	3	6
2	1	4
3	2	

Then obtain the following by interchanging rows and columns:

u	v	Const.
2	2	3
3	1	2
6	4	

From this table we construct the dual problem:

Maximize the objective function $P = 6u + 4v$ subject to the constraints

$$2u + 2v \le 3$$

$$3u + v \le 2$$

$$u \ge 0, v \ge 0$$

Using the simplex method, we have

	u	v	x	y	P	Const
	2	2	1	0	0	3
p.r. →	[3]	1	0	1	0	2
	−6	−4	0	0	1	0

Ratio
3/2
2/3

$\dfrac{1}{3}R_2 \longrightarrow$

p.c.

u	v	x	y	P	Const
2	2	1	0	0	3
1	$\frac{1}{3}$	0	$\frac{1}{3}$	0	$\frac{2}{3}$
−6	−4	0	0	1	0

$\begin{array}{c} R_1 - 2R_2 \\ R_3 + 6R_2 \\ \longrightarrow \end{array}$

u	v	x	y	P	Const
0	$[\frac{4}{3}]$	1	$-\frac{2}{3}$	0	$\frac{5}{3}$
1	$\frac{1}{3}$	0	$\frac{1}{3}$	0	$\frac{2}{3}$
0	−2	0	2	1	4

Ratio
5/4
2

$\dfrac{3}{4}R_1 \longrightarrow$

u	v	x	y	P	Const
0	1	$\frac{3}{4}$	$-\frac{1}{2}$	0	$\frac{5}{4}$
1	$\frac{1}{3}$	0	$\frac{1}{3}$	0	$\frac{2}{3}$
0	−2	0	2	1	4

$\begin{array}{c} R_2 - \frac{1}{3}R_1 \\ R_3 + 2R_1 \\ \longrightarrow \end{array}$

u	v	x	y	P	Const
0	1	$\frac{3}{4}$	$-\frac{1}{2}$	0	$\frac{5}{4}$
1	0	$-\frac{1}{4}$	$\frac{1}{2}$	0	$\frac{1}{4}$
0	0	$\frac{3}{2}$	1	1	$\frac{13}{2}$

Therefore, C attains a minimum value of $13/2$ when $x = 3/2$, $y = 1$, $u = \frac{1}{4}$ and $v = \frac{5}{4}$.

7. We first write the tableau

x	y	z	Const.
3	2	1	4
1	1	3	6
24	18	24	

Then obtain the following by interchanging rows and columns:

u	v	Const.
3	1	24
2	1	18
1	3	24
4	6	

From this table we construct the dual problem:

Maximize the objective function $P = 4u + 6v$ subject to

$$3u + v \le 24$$
$$2u + v \le 18$$
$$u + 3v \le 24$$
$$u \ge 0, \, v \ge 0$$

The initial tableau is

u	v	x	y	z	P	Const.
3	1	1	0	0	0	24
2	1	0	1	0	0	18
1	3	0	0	1	0	24
−4	−6	0	0	0	1	0

Using the following sequence of row operations

1. $\frac{1}{3}R_3$ 2. $R_1 - R_3$, $R_2 - R_3$, $R_4 + 6R_3$ 3. $\frac{3}{8}R_1$ 4. $R_2 - \frac{5}{3}R_1$, $R_3 - \frac{1}{3}R_1$, $R_4 + 2R_1$

we obtain the final tableau

u	v	x	y	z	P	Const.
1	0	$\frac{3}{8}$	0	$-\frac{1}{8}$	0	6
0	0	$-\frac{5}{8}$	1	$-\frac{1}{8}$	0	0
0	1	$-\frac{1}{8}$	0	$\frac{3}{8}$	0	6
0	0	$\frac{3}{4}$	0	$\frac{7}{4}$	0	60

We conclude that C attains a minimum value of 60 when $x = 3/4$, $y = 0$, $z = 7/4$, $u = 6$, and $v = 6$.

9.   Refer to Exercise 9, page 99 of this manual. The problem simplifies to

Minimize $C = 14000x + 16000y$ subject to

$$2x + 3y \geq 26$$
$$3x + y \geq 18$$
$$x \geq 0, y \geq 0$$

We first write down the following tableaus for the primal problem:

x	y	Const.
2	3	26
3	1	18
14000	16000	

Next, we interchange the columns and rows of the foregoing tableau:

u	v	Const.
2	3	14000
3	1	16000
26	18	

This leads to the dual problem:

Maximize $P = 26u + 18v$ subject to

$$2u + 3v \leq 14000$$
$$3u + v \leq 16000$$
$$u \geq 0, v \geq 0$$

Let x and y be slack variables. We obtain the following tableaus:

The last tableau is in final form. The fundamental theorem of duality tells us that the solution to the primal problem is $x = 4$, $y = 6$, and $C = 152{,}000$. So, the Saddle Mine should be operated for 4 days and the Horseshoe Mine should be operated for 6 days at a minimum cost of \$152,000/day.

11. Let x, y, and z denote the number of units of products A, B, and C made, respectively. Then the problem is to maximize the profit
$$P = 4x + 6y + 8z \text{ subject to}$$
$$9x + 12y + 18z \le 360$$
$$6x + 6y + 10z \le 240$$
$$x \ge 0,\ y \ge 0,\ z \ge 0$$

The initial tableau is

x	y	z	u	v	P	Const.
9	12	18	1	0	0	360
6	6	10	0	1	0	240
-4	-6	-8	0	0	1	0

Using the sequence of row operations

1. $\frac{1}{18}R_1$ 2. $R_2 - 10R_1$ 3. $R_3 + 8R_1$ 4. $\frac{3}{2}R_1$ 5. $R_2 + \frac{2}{3}R_1$, $R_3 + \frac{2}{3}R_1$

we obtain the final tableau

x	y	z	u	v	P	Const.
$\frac{3}{4}$	1	$\frac{3}{2}$	$\frac{1}{12}$	0	0	30
$\frac{3}{2}$	0	1	$-\frac{1}{2}$	1	0	60
$\frac{1}{2}$	0	1	$\frac{1}{2}$	0	1	180

and conclude that the company should produce 0 units of product A, 30 units of product B, and 0 units of product C to realize a maximum profit of $180.

CHAPTER 5

EXERCISES 5.1, page 261

1. The interest is given by $I = (500)(2)(0.08) = 80$, or $80.
    The accumulated amount is $500 + 80$, or $580.

3. The interest is given by $I = (800)(0.06)(0.75) = 36$, or $36.
    The accumulated amount is $800 + 36$, or $836.

5. We are given that $A = 1160$, $t = 2$, and $r = 0.08$, and we are asked to find P. Since
    $$A = P(1 + rt)$$
    we see that $\quad P = \dfrac{A}{1+rt} = \dfrac{1160}{1+(0.08)(2)} = 1000$, or $1000.

7. We use the formula $I = Prt$ and solve for t when $I = 20$, $P = 1000$, and $r = 0.05$.
    Thus,
    $$20 = 1000(0.05)(\frac{t}{365}), \quad \text{and} \quad t = \frac{365(20)}{50} = 146, \quad \text{or 146 days.}$$

9. We use the formula $A = P(1 + rt)$ with $A = 1075$, $P = 1000$, $t = 0.75$, and solve for r.
    Thus,
    $$1075 = 1000(1 + 0.75r)$$
    $$75 = 750r$$
    or $\qquad\qquad\qquad r = 0.10$.
    Therefore, the interest rate is 10 percent per year.

11. $A = 1000(1 + 0.07)^8 \approx 1718.19$, or $1718.19.

13. $A = 2500\left(1+\dfrac{0.07}{2}\right)^{20} \approx 4974.47$, or $4974.47.

15. $A = 12000\left(1+\dfrac{0.08}{4}\right)^{42} \approx 27,566.93$, or $27,566.93.

17. $A = 150,000\left(1 + \dfrac{0.14}{12}\right)^{48} \approx 261,751.04$, or \$261,751.04.

19. $A = 150,000\left(1 + \dfrac{0.12}{365}\right)^{1095} = 214,986.69$, or \$214,986.69.

21. Using the formula
$$r_{eff} = \left(1 + \dfrac{r}{m}\right)^{m} - 1$$
with $r = 0.10$ and $m = 2$, we have
$$r_{eff} = \left(1 + \dfrac{0.10}{2}\right)^{2} - 1 = 0.1025, \quad \text{or } 10.25 \text{ percent}..$$

23. Using the formula $\quad r_{eff} = \left(1 + \dfrac{r}{m}\right)^{m} - 1$

with $r = 0.08$ and $m = 12$, we have
$$r_{eff} = \left(1 + \dfrac{0.08}{12}\right)^{12} - 1 \approx 0.08300, \quad \text{or } 8.3 \text{ percent per year}.$$

25. The present value is given by $P = 40,000\left(1 + \dfrac{0.08}{2}\right)^{-8} \approx 29,227.61$, or \$29,227.61.

27. The present value is given by
$$P = 40,000\left(1 + \dfrac{0.07}{12}\right)^{-48} \approx 30,255.95, \quad \text{or } \$30,255.95.$$

29. Think of \$300 as the principal and \$306 as the accumulated amount at the end of 30 days. If r denotes the simple interest rate per annum, then we have $P = 300$, $A = 306$, $t = 1/12$, and we are required to find r. Using (1b) we have
$$306 = 300\left(1 + \dfrac{r}{12}\right) = 300 + r\left(\dfrac{300}{12}\right)$$
and $\quad r = \left(\dfrac{12}{300}\right)6 = 0.24$, or 24 percent per year.

31. The Abdullahs will owe $A = P(1+rt) = 120,000[1+(0.12)(\frac{3}{12})] = 123,600$, or $123,600.

33. Here $P = 10,000$, $I = 3500$, and $t = 7$, and so from Formula (1a), we have
$$3500 = 10000\,(r)\,7 \quad \text{and so } r = \frac{3500}{70000} = 0.05$$
So the bond pays interest at the rate of 5% per year.

35. The rate that you would expect to pay is
$$A = 380(1 + 0.08)^5 \approx 705.28, \text{ or } \$705.28 \text{ per day.}$$
480

37. The amount that they can expect to pay is given by
$$A = 210,000(1 + 0.05)^4 \approx 255,256, \text{ or approximately } \$255,256.$$

39. The investment will be worth $A = 1.5\left(1 + \dfrac{0.055}{2}\right)^{20} = 2.58064$, or approximately

$2.58 million dollars.

41. We use Formula (3) with $P = 15,000$, $r = 0.098$, $m = 12$, and $t = 4$ giving the worth of Jodie's account as
$$A = 15,000\left(1 + \frac{0.098}{12}\right)^{(12)(4)} \approx 22,163.753, \text{ or approximately } \$22,163.75.$$

43. Using the formula $P = A\left(1 + \dfrac{r}{m}\right)^{-mt}$, we have
$$P = 40,000\left(1 + \frac{0.085}{4}\right)^{-20} \approx 26,267.49, \text{ or } \$26,267.49.$$

45. a. They should set aside
$$P = 100,000(1 + 0.085)^{-13} \approx 34,626.88, \text{ or } \$34,626.88.$$
   b. They should set aside
$$P = 100,000\left(1 + \frac{0.085}{2}\right)^{-26} \approx 33,886.16, \text{ or } \$33,886.16.$$

c. They should set aside

$$P = 100,000\left(1 + \frac{0.085}{4}\right)^{-52} \approx 33,506.76, \text{ or } \$33,506.76.$$

47. The effective rate of interest for the Bendix Mutual Fund is

$$r_{eff} = \left(1 + \frac{0.104}{4}\right)^4 - 1 \approx 0.1081 \quad \text{or} \quad 10.81\%/\text{yr}$$

whereas the effective rate of interest for the Acme Mutual fund is

$$r_{eff} = \left(1 + \frac{0.106}{2}\right)^4 - 1 \approx 0.1088 \quad \text{or} \quad 10.88\%/\text{yr}$$

We conclude that the Acme Mutual Fund has a better rate of return.

49. The present value of the $8000 loan due in 3 years is given by

$$P = 8000\left(1 + \frac{0.10}{2}\right)^{-6} - 1 = 5969.72, \text{ or } \$5969.72.$$

The present value of the $15,000 loan due in 6 years is given by

$$P = 15,000\left(1 + \frac{0.10}{2}\right)^{-12} - 1 = 8352.56, \text{ or } \$8352.56.$$

Therefore, the amount the proprietors of the inn will be required to pay at the end of 5 years is given by

$$A = 14,322.28\left(1 + \frac{0.10}{2}\right)^{10} = 23,329.48, \quad \text{or } \$23,329.48.$$

51. Using the compound interest formula with $A = 256,000$, $P = 200,000$ and $t = 6$, we have

$$256,000 = 200,000(1 + R)^6$$
$$(1 + R)^{1/6} = (1.28)^{1/6}$$
$$1 + R = 1.042,$$
$$R = 0.042, \quad \text{or } 4.2 \text{ percent.}$$

53. Let the effective rate of interest be R. Then R satisfies
$$A = P(1 + R)^t$$

or
$$10,000 = 6724.53\left(1 + \frac{R}{2}\right)^{14}$$

$$1 + \frac{R}{2} = (1.48709278)^{1/14} = 1.028750032$$

and $\quad R = 2(2.875) = 0.0575$, or 5.75 percent.

55. Let the effective rate of interest be R. Then R satisfies
$$A = P(1 + R)^t$$
or $\quad 5170.42 = 5000(1 + R)^{245/365}$
$$1 + R = (1.034084)^{365/245} = 1.051199629$$
and $\quad R = 0.5119962$, or approximately 5.12 percent.

57. Be definition, $A = P(1 + r_{eff})^t$. So
$$(1 + r_{eff})^t = \frac{A}{P}, \quad 1 + r_{eff} = \left(\frac{A}{P}\right)^{1/t} \quad \text{and} \quad r_{eff} = \left(\frac{A}{P}\right)^{1/t} - 1.$$

59. False. Under compound interest $A = P(1 + r)^t$ $(m = 1)$, whereas under simple interest, $A = P(1 + rt)$.

61. False. If Susan had gotten annual increases of 5 percent over 5 years, her salary would have been $A = 40,000(1 + 0.05)^5 = 51,051.26$, or approximately $51,051 after 5 years and not $50,000.

USING TECHNOLOGY EXERCISES 5.1, page 265

1. $5872.78   3. $475.49   5. 8.95%/yr   7. 10.20%/yr

9 . $29,743.30   11. $53,303.25

EXERCISES 5.2, page 275

1. $S = 1000\left[\dfrac{(1 + 0.1)^{10} - 1}{0.1}\right] = 15,937.42$, or $15,937.42.

3. $S = 1800 \left[\dfrac{\left(1 + \dfrac{0.08}{4}\right)^{24} - 1}{\dfrac{0.08}{4}} \right] \approx 54{,}759.35$, or \$54,759.35.

5. $S = 600 \left[\dfrac{\left(1 + \dfrac{0.12}{4}\right)^{36} - 1}{\dfrac{0.12}{4}} \right] \approx 37{,}965.57$, or \$37,965.57.

7. $S = 200 \left[\dfrac{\left(1 + \dfrac{0.09}{12}\right)^{243} - 1}{\dfrac{0.09}{12}} \right] \approx 137{,}209.97$, or \$137,209.97.

9. $P = 5000 \left[\dfrac{1 - (1 + 0.08)^{-8}}{0.08} \right] \approx 28{,}733.19$, or \$28,733.19.

11. $P = 4000 \left[\dfrac{1 - (1 + 0.09)^{-5}}{0.09} \right] \approx 15{,}558.61$, or \$15,558.61.

13. $P = 800 \left[\dfrac{1 - \left(1 + \dfrac{0.12}{4}\right)^{-28}}{\dfrac{0.12}{4}} \right] \approx 15{,}011.29$, or \$15,011.29.

15. She will have

$$S = 1500 \left[\dfrac{(1 + 0.08)^{25} - 1}{0.08} \right] = 109{,}658.91, \text{ or } \$109{,}658.91.$$

17. On October 31, Linda's account will be worth

$$S = 40 \left[\frac{\left(1 + \frac{0.07}{12}\right)^{11} - 1}{\frac{0.07}{12}} \right] \approx 453.06, \text{ or } \$453.06.$$

One month later, this account will be worth $A = (453.06)\left(1 + \frac{0.07}{12}\right) = 455.70$, or

$455.70.

19. The amount in Collin's employee retirement account is given by

$$S = 100 \left[\frac{\left(1 + \frac{0.07}{12}\right)^{144} - 1}{\frac{0.07}{12}} \right] \approx 22{,}469.50, \text{ or } \$22{,}469.50.$$

The amount in Collin's IRA is given by

$$S = 2000 \left[\frac{(1 + 0.09)^8 - 1}{0.09} \right] \approx 22{,}056.95, \text{ or } \$22{,}056.95.$$

Therefore, the total amount in his retirement fund is given by
    $22{,}469.50 + 22{,}056.95 = 44{,}526.45$, or $44,526.45.

21. To find how much Karen has at age 65, we use formula (6) with $R = 150$,
$i = \frac{r}{m} = \frac{0.05}{12}$, and $n = mt = (12)(40) = 480$, giving

$$S = 150 \left[\frac{\left(1 + \frac{0.05}{12}\right)^{480} - 1}{\frac{0.05}{12}} \right] \approx 228{,}903.0235$$

or $228,903.02. To find how much Matt will have upon attaining the age of 65, use
Formula (6) with $R = 250$, $i = \frac{r}{m} = \frac{0.05}{12}$, and $n = mt = (12)(30) = 360$ giving

$$S = 250 \left[\frac{\left(1 + \frac{0.05}{12}\right)^{360} - 1}{\frac{0.05}{12}} \right] \approx 208{,}064.6588, \text{ or } \$208{,}064.66. \text{ So Karen}$$

will have the bigger nest egg.

23. The equivalent cash payment is given by

$$P = 450 \left[\frac{1 - \left(1 + \frac{0.09}{12}\right)^{-24}}{\frac{0.09}{12}} \right] \approx 9850.12, \text{ or } \$9850.12.$$

25. We use the formula for the present value of an annuity obtaining

$$P = 22 \left[\frac{1 - \left(1 + \frac{0.18}{12}\right)^{-36}}{\frac{0.18}{12}} \right] \approx 608.54, \text{ or } \$608.54.$$

27. With an $1200 monthly payment, the present value of their loan would be

$$P = 1200 \left[\frac{1 - \left(1 + \frac{0.095}{12}\right)^{-360}}{\frac{0.095}{12}} \right] \approx 142{,}712.02, \text{ or } \$142{,}712.02.$$

With a $1500 monthly payment, the present value of their loan would be

$$P = 1500 \left[\frac{1 - \left(1 + \frac{0.095}{12}\right)^{-360}}{\frac{0.095}{12}} \right] \approx 178{,}390.02 \text{ or } \$178{,}390.02.$$

Since they intend to make a $25,000 down payment, the range of homes they should consider is $167,712 to $203,390.

29. The lower limit of their investment is

$$A = 1200 \left[\dfrac{1 - \left(1 + \dfrac{0.09}{12}\right)^{-180}}{\dfrac{0.09}{12}} \right] + 25{,}000 \approx 143{,}312.09$$

or approximately \$143,312. The upper limit of their investment is

$$A = 1500 \left[\dfrac{1 - \left(1 + \dfrac{0.09}{12}\right)^{-180}}{\dfrac{0.09}{12}} \right] + 25{,}000 \approx 172{,}890.11$$

or approximately \$172,890. Therefore, the price range of houses they should consider is \$143,312 to \$172,890.

31. False. This statement would be true only if the interest rate is equal to zero.

USING TECHNOLOGY EXERCISES 5.2, page 278

1. \$59,622.15   3. \$8453.59   5. \$35,607.23   7. \$13,828.

EXERCISES 5.3, page 286

1. The size of each installment is given by

$$R = \dfrac{100{,}000(0.08)}{1 - (1 + 0.08)^{-10}} \approx 14{,}902.95, \text{ or } \$14{,}902.95.$$

3. The size of each installment is given by

$$R = \dfrac{5000(0.01)}{1 - (1 + 0.01)^{-12}} \approx 444.24, \text{ or } \$444.24.$$

5. The size of each installment is given by

$$R = \dfrac{25{,}000(0.0075)}{1 - (1 + 0.0075)^{-48}} \approx 622.13, \text{ or } \$622.13.$$

7. The size of each installment is

$$R = \frac{80{,}000(0.00875)}{1-(1+0.00875)^{-360}} \approx 731.79, \text{ or } \$731.79.$$

9. The periodic payment that is required is
$$R = \frac{20{,}000(0.02)}{(1+0.02)^{12}-1} \approx 1491.19, \text{ or } \$1491.19.$$

11. The periodic payment that is required is
$$R = \frac{100{,}000(0.0075)}{(1+0.0075)^{120}-1} \approx 516.76, \text{ or } \$516.76$$

13. The periodic payment that is required is
$$R = \frac{250{,}000(0.00875)}{(1+0.00875)^{300}-1} \approx 172.95, \text{ or } \$172.95.$$

15. The periodic payment that is required is
$$R = \frac{50000\left(\dfrac{.10}{4}\right)}{\left(1+\dfrac{.10}{4}\right)^{20}-1} = 1957.36, \text{ or } \$1957.36.$$

17. The periodic payment that is required is
$$R = \frac{35000\left(\dfrac{0.075}{2}\right)}{1-\left(1+\dfrac{0.075}{2}\right)^{-13}} = 3450.87, \text{ or } \$3450.87.$$

19. The size of each installment is given by
$$R = \frac{100{,}000(0.10)}{1-(1+0.10)^{-10}} \approx 16{,}274.54, \text{ or } \$16{,}274.54.$$

21. The monthly payment in each case is given by

$$R = \frac{100{,}000\left(\dfrac{r}{12}\right)}{1-\left(1+\dfrac{r}{12}\right)^{-360}}$$

Thus, if $r = 0.08$, then $R = \dfrac{100{,}000\left(\dfrac{0.08}{12}\right)}{1-\left(1+\dfrac{0.08}{12}\right)^{-360}} \approx 733.76$, or \$733.76

If $r = 0.09$, then $R = \dfrac{100{,}000\left(\dfrac{0.09}{12}\right)}{1-\left(1+\dfrac{0.09}{12}\right)^{-360}} \approx 804.62$, or \$804.62

If $r = 0.10$, then $R = \dfrac{100{,}000\left(\dfrac{0.10}{12}\right)}{1-\left(1+\dfrac{0.10}{12}\right)^{-360}} \approx 877.57$, or \$877.57.

If $r = 0.11$, then $R = \dfrac{100{,}000\left(\dfrac{0.11}{12}\right)}{1-\left(1+\dfrac{0.11}{12}\right)^{-360}} \approx \$952.32.$

a. The difference in monthly payments in the two loans is
   \$877.57 - \$665.30 = \$212.27.
b. The monthly mortgage payment on a \$150,000 mortgage would be
   1.5(\$877.57) = \$1316.36.
   The monthly mortgage payment on a \$50,000 mortgage would be
   0.5(\$877.57) = \$438.79.

23. a. The amount of the loan required is 16000 - (0.25)(16000) or 12,000 dollars. If the car is financed over 36 months, the payment will be

$$R = \frac{12{,}000\left(\dfrac{0.10}{12}\right)}{1-\left(1+\dfrac{0.10}{12}\right)^{-36}} \approx 387.21, \text{ or } \$387.21 \text{ per month.}$$

If the car is financed over 48 months, the payment will be

$$R = \frac{12,000\left(\dfrac{0.10}{12}\right)}{1-\left(1+\dfrac{0.10}{12}\right)^{-48}} \approx 304.35, \text{ or } \$304.35 \text{ per month.}$$

b. The interest charges for the 36-month plan are

$$36(387.21) - 12000 = 1939.56,$$

or $1939.56. The interest charges for the 48-month plan are

$$48(304.35) - 12000 = 2608.80, \text{ or } \$2608.80.$$

25. The amount borrowed is $270,000 - 30,000 = 240,000$ dollars. The size of the monthly installment is

$$R = \frac{240,000\left(\dfrac{0.08}{12}\right)}{1-\left(1+\dfrac{0.08}{12}\right)^{-360}} \approx 1761.03, \text{ or } \$1761.03.$$

To find their equity after five years, we compute

$$P = 1761.03\left[\frac{1-\left(1+\dfrac{0.08}{12}\right)^{-300}}{\dfrac{0.08}{12}}\right] \approx 228,167$$

or $228,167, and so their equity is $270,000 - 228,167 = 41,833$, or $41,833. To find their equity after ten years, we compute

$$P = 1761.03\left[\frac{1-\left(1+\dfrac{0.08}{12}\right)^{-240}}{\dfrac{0.08}{12}}\right] \approx 210,539, \text{ or } \$210,539.$$

and their equity is $270,000 - 210,539 = 59,461$, or $59,461. To find their equity after twenty years, we compute

$$P = 1761.03 \left[\frac{1 - \left(1 + \dfrac{0.08}{12}\right)^{-120}}{\dfrac{0.08}{12}} \right] \approx 145{,}147, \text{ or } \$145{,}147,$$

and their equity is 270,000 − 145,147, or $124,853.

27. The amount that must be deposited quarterly into this fund is

$$R = \frac{\left(\dfrac{0.09}{4}\right) 200{,}000}{\left(1 + \dfrac{0.09}{4}\right)^{40} - 1} \approx 3{,}135.48, \text{ or } \$3{,}135.48.$$

29. The size of each quarterly installment is given by

$$R = \frac{\left(\dfrac{0.10}{4}\right) 20{,}000}{\left(1 + \dfrac{0.10}{4}\right)^{12} - 1} \approx 1449.74, \text{ or } \$1449.74.$$

31. The size of each monthly installment is given by

$$R = \frac{\left(\dfrac{0.085}{12}\right) 250{,}000}{\left(1 + \dfrac{0.085}{12}\right)^{300} - 1} \approx 242.23, \text{ or } \$242.23.$$

33. The value of the IRA account after 20 years is

$$S = 375 \left[\frac{\left(1 + \dfrac{0.08}{4}\right)^{80} - 1}{\dfrac{0.08}{4}} \right] \approx 72{,}664.48, \text{ or } \$72{,}664.48.$$

The payment he would receive at the end of each quarter for the next 15 years is given by

$$R = \frac{\left(\dfrac{0.08}{4}\right)72{,}664.48}{1-\left(1+\dfrac{0.08}{4}\right)^{-60}} \approx 2090.41, \text{ or } \$2090.41.$$

If he continues working and makes quarterly payments until age 65, the value of the IRA account would be

$$S = 375\left[\frac{\left(1+\dfrac{0.08}{4}\right)^{100}-1}{\dfrac{0.08}{4}}\right] \approx 117{,}087.11, \text{ or } \$117{,}087.11.$$

The payment he would receive at the end of each quarter for the next 10 years is given by

$$R = \frac{\left(\dfrac{0.08}{4}\right)117{,}087.11}{1-\left(1+\dfrac{0.08}{4}\right)^{-40}} \approx 4280.21, \text{ or } \$4280.21.$$

35. The monthly payment the Sandersons are required to make under the terms of their original loan is given by

$$R = \frac{100{,}000\left(\dfrac{0.10}{12}\right)}{1-\left(1+\dfrac{0.10}{12}\right)^{-240}} \approx 965.02, \text{ or } \$965.02.$$

The monthly payment the Sandersons are required to make under the terms of their new loan is given by

$$R = \frac{100{,}000\left(\dfrac{0.078}{12}\right)}{1-\left(1+\dfrac{0.078}{12}\right)^{-240}} \approx 824.04, \text{ or } \$824.04.$$

The amount of money that the Sandersons can expect to save over the life of the loan by refinancing is given by $240(965.02 - 824.04) = 33{,}835.20$, or $\$33{,}835.20$.

37. Sarah's monthly payment is found using Formula (10) with $P = 200{,}000$,

$i = \dfrac{r}{m} = \dfrac{0.06}{12}$, and $n = (12)(15) = 180$. Thus,

$$R = 200,000 \left[\dfrac{\dfrac{0.06}{12}}{1 - \left(1 + \dfrac{0.06}{12}\right)^{-180}} \right] \approx 1687.7137 . \text{ After 85years, she has paid}$$

(5)(12) or 60 payments. Her outstanding principal is given by the sum of the remaining installments, (180 – 60), or 120. Using Formula 8, we find

$$P = 1687.7137 \left[\dfrac{1 - \left(1 + \dfrac{0.06}{12}\right)^{-120}}{\dfrac{0.06}{12}} \right] \approx 152,018.2012$$

So her outstanding principal is $152,018.20.

39. a. Here $P = 200,000$, $i = \dfrac{r}{m} = \dfrac{0.095}{12}$, and $n = mt = (12)(30) = 360$. Therefore,

$$R = \dfrac{200,000\left(\dfrac{0.095}{12}\right)}{1 - \left(1 + \dfrac{0.095}{12}\right)^{-360}} \approx 1681.7084, \text{ and so her monthly payment is } \$1681.71.$$

b. After $(4)(12) = 48$ monthly payments have been made, her outstanding principal is given by the sum of the present values of the remaining installments (which is $360 – 48 = 312$). Using Formula (8), we find it to be

$$P = 1681.7084 \left[\dfrac{1 - \left(1 + \dfrac{0.095}{12}\right)^{-312}}{\dfrac{0.095}{12}} \right] \approx 194,282.6675,$$

or approximately $194,282.67.

c. Here $P = 194,282.6675$, $i = \dfrac{r}{m} = \dfrac{0.0675}{12}$, and $n = (12)(30) = 360$. So

$$R = \frac{194282.67\left(\dfrac{0.0675}{12}\right)}{1-\left(1+\dfrac{0.0675}{12}\right)^{-360}} \approx 1260.1137$$

and so her new monthly payment is $1260.11.

d. Emily will save $1681.71 - 1260.11$, or $421.60 per month.

41. The amount of the loan the Meyers need to secure is $280,000. Using the bank's financing, the monthly payment would be

$$R = \frac{280,000\left(\dfrac{0.11}{12}\right)}{1-\left(1+\dfrac{0.11}{12}\right)^{-300}} \approx 2744.32, \text{ or } \$2744.32.$$

Using the seller's financing, the monthly payment would be

$$R = \frac{280,000\left(\dfrac{0.098}{12}\right)}{1-\left(1+\dfrac{0.098}{12}\right)^{-300}} \approx 2504.99, \text{ or } \$2504.99.$$

By choosing the seller's financing rather than the bank's, the Meyers would save $(2744.32 - 2504.99)(300) = 71,799$, or $71,799 in interest.

USING TECHNOLOGY EXERCISES 5.3, page 290

1. $3645.40   3. $18,443.75   5. $1863.99   7. $707.96

9. $18,288.92. The amortization schedule follows.

End of Period	Interest charged	Repayment made	Payment toward Principal	Outstanding Principal
0				120,000.00
1	10,200.00	18,288.92	8,088.92	111,911.08
2	9,512.44	18,288.92	8,776.48	103,134.60
3	8,766.44	18,288.92	9,522.48	93,612.12
4	7,957.03	18,288.92	10,331.89	83,280.23
5	7,078.82	18,288.92	11,210.10	72,070.13
6	6,125.96	18,288.92	12,162.96	59,907.17
7	5,092.11	18,288.92	13,196.81	46,710.36
8	3,970.38	18,288.92	14,318.54	32,391.82
9	2,753.30	18,288.92	15,535.62	16,856.20
10	1,432.78	18,288.98	16,856.14	.00

EXERCISES 5.4, page 299

1. $a_9 = 6 + (9 - 1)3 = 30$

3. $a_8 = -15 + (8 - 1)\left(\frac{3}{2}\right) = -\frac{9}{2} = -4.5.$

5. $a_{11} - a_4 = (a_1 + 10d) - (a_1 + 3d) = 7d.$ Also, $a_{11} - a_4 = 107 - 30 = 77.$
   Therefore, $7d = 77$, and $d = 11$. Next,
   $$a_4 = a + 3d = a + 3(11) = a + 33 = 30.$$
   and $a = -3$. Therefore, the first five terms are -3, 8, 19, 30, 41.

7. Here $a = x$, $n = 7$, and $d = y$. Therefore, the required term is
   $$a_7 = x + (7 - 1)y = x + 6y.$$

9. Using the formula for the sum of the terms of an arithmetic progression with $a = 4$,
   $d = 7$ and $n = 15$, we have
   $$S_n = \frac{n}{2}[2a + (n-1)d]]$$
   $$S_{15} = \frac{15}{2}[2(4) + (15-1)7] = \frac{15}{2}(106) = 795$$

11. The common difference is $d = 2$ and the first term is $a = 15$. Using the formula for
    the nth term
    $$a_n = a + (n - 1)d,$$

we have $57 = 15 + (n - 1)(2) = 13 + 2n$

$2n = 44,$ and $n = 22.$

Using the formula for the sum of the terms of an arithmetic progression with $a = 15$, $d = 2$ and $n = 22$, we have

$$S_n = \frac{n}{2}[2a + (n-1)d]]$$

$$S_{22} = \frac{22}{2}[2(15) + (22-1)2] = 11(72) = 792.$$

13. $f(1) + f(2) + f(3) + \cdots + f(20)$

$$= [3(1) - 4] + [3(2) - 4] + [3(3) - 4] + \cdots + [3(20) - 4]$$

$$= 3(1 + 2 + 3 + \cdots + 20) + 20(-4)$$

$$= 3\left(\frac{20}{2}\right)[2(1) + (20-1)1] - 80$$

$$= 550.$$

15. $S_n = \frac{n}{2}[2a_1 + (n-1)d]] = \frac{n}{2}(a_1 + a_1 + (n-1)d]$

$$= \frac{n}{2}(a_1 + a_n)$$

a. $S_{11} = \frac{11}{2}(3 + 47) = 275.$ b. $S_{20} = \frac{20}{2}[5 + (-33)] = -280.$

17. Let n be the number of weeks till she reaches 10 miles. Then

$$a_n = 1 + (n-1)\frac{1}{4} = 1 + \frac{1}{4}n - \frac{1}{4} = \frac{1}{4}n + \frac{3}{4} = 10$$

Therefore, $n + 3 = 40$, and $n = 37$; that is, it will take Karen 37 weeks to meet her goal.

19. To compute the Kunwoo's fare by taxi, take $a = 1$, $d = 0.60$, and $n = 25$. Then the required fare is given by

       $a_{25} = 1 + (25 - 1)0.60 = 15.4,$

or \$15.40. Therefore, by taking the airport limousine, the Kunwoo will save

$$15.40 - 7.50 = 7.90, \text{ or } \$7.90.$$

21. a. Using the formula for the sum of an arithmetic progression, we have

$$S_n = \frac{n}{2}[2a + (n-1)d]]$$

$$= \frac{N}{2}[2(1) + (N-1)(1)] = \frac{N}{2}(N+1).$$

b. $$S_{10} = \frac{10}{2}(10+1) = 5(11) = 55$$

$$D_3 = (C - S)\frac{N-(n-1)}{S_N} = (6000 - 500)\frac{10-(3-1)}{55} = 5500\left(\frac{8}{55}\right)$$

$$= 800, \text{ or } \$800.$$

23. This is a geometric progression with $a = 4$ and $r = 2$. Next, $a_7 = 4(2)^6 = 256$,

and $$S_7 = \frac{4(1-2^7)}{1-2} = 508.$$

25. If we compute the ratios

$$\frac{a_2}{a_1} = \frac{-\frac{3}{8}}{\frac{1}{2}} = -\frac{3}{4} \quad \text{and} \quad \frac{a_3}{a_2} = \frac{\frac{1}{4}}{-\frac{3}{8}} = -\frac{2}{3},$$

we see that the given sequence is not geometric since the ratios are not equal.

27. This is a geometric progression with $a = 243$, and $r = 1/3$.

$$a_7 = 243(\tfrac{1}{3})^6 = \tfrac{1}{3}$$

$$S_7 = \frac{243(1-(\tfrac{1}{3})^7)}{1-\tfrac{1}{3}} = 364\tfrac{1}{3}.$$

29. First, we compute

$$r = \frac{a_2}{a_1} = \frac{3}{-3} = -1.$$

Next, $a_{20} = -3(-1)^{19} = 3$ and so $S_{20} = \dfrac{-3[1-(-1)^{20}]}{[1-(-1)]} = 0$.

31. The population in five years is expected to be

$$200{,}000(1.08)^{6-1} = 200{,}000\,(1.08)^5 \approx 293{,}866.$$

33. The salary of a union member whose salary was \$32,000 six years ago is given by the 7th term of a geometric progression whose first term is 32,000 and whose common ratio is 1.03. Thus

$$a_7 = (32,000)(1.03)^6 = 38,209.67, \text{ or } \$38,209.67.$$

35. With 8 percent raises per year, the employee would make

$$S_4 = 38,000 \left[\frac{1-(1.08)^4}{1-1.08} \right] \approx 171,232.26,$$

or \$171,232.26 over the next four years.
With \$2500 raises per year, the employee would make

$$S_4 = \frac{4}{2}[2(38,000)+(4-1)2500] = 167,000$$

or \$167,000 over the next four years. We conclude that the employee should choose annual raises of 8 percent per year.

37. *a.* During the sixth year, she will receive

$$a_6 = 10,000(1.15)^5 \approx 20,113.57, \text{ or } \$20,113.57.$$

b. The total amount of the six payments will be given by

$$S_6 = \frac{10,000[1-(1.15)^6]}{1-1.15} \approx 87,537.38, \text{ or } \$87,537.38.$$

39. The book value of the office equipment at the end of the eighth year is given by

$$V(8) = 150,000 \left(1 - \frac{2}{10} \right)^8 \approx 25,165.82, \text{ or } \$25,165.82.$$

41. The book value of the restaurant equipment at the end of six years is given by

$$V(6) = 150,000(0.8)^6 \approx 39,321.60,$$

or \$39,321.60. By the end of the sixth year, the equipment will have depreciated by

$$D(n) = 150,000 - 39,321.60 = 110,678.40, \text{ or } \$110,678.40.$$

43. True. Suppose d is the common difference of $a_1, a_2, ..., a_n$ and e is the common difference of $b_1, b_2, ..., b_n$. Then $d + e$ is the common difference of $a_1 + b_1, a_2 + b_2, ..., a_n + b_n$ is obtained by adding the constant $c + d$ to it, we see that it is indeed an arithmetic progression.

CHAPTER 5, REVIEW EXERCISES, page 302

1. a. Here $P = 5000$, $r = 0.1$, and $m = 1$. Thus, $i = r = 0.1$ and $n = 4$. So
    $$A = 5000(1.1)^4 = 7320.5, \text{ or } \$7320.50.$$
    b. Here $m = 2$ so that $i = 0.1/2 = 0.05$ and $n = (4)(2) = 8$. So
    $$A = 5000(1.05)^8 \approx 7387.28 \text{ or } \$7387.28.$$
    c. Here $m = 4$, so that $i = 0.1/4 = 0.025$ and $n = (4)(4) = 16$. So
    $$A = 5000(1.025)^{16} \approx 7,422.53, \text{ or } \$7422.53.$$
    d. Here $m = 12$, so that $i = 0.1/12$ and $n = (4)(12) = 48$. So
    $$A = 5000\left(1 + \frac{0.10}{12}\right)^{48} \approx 7446.77, \text{ or } \$7446.77.$$

3. a. The effective rate of interest is given by
    $$r_{eff} = \left(1 + \frac{r}{m}\right)^m - 1 = (1 + 0.12) - 1 = 0.12, \text{ or } 12 \text{ percent.}$$
    b. The effective rate of interest is given by
    $$r_{eff} = \left(1 + \frac{r}{m}\right)^m - 1 = \left(1 + \frac{0.12}{2}\right)^2 - 1 = 0.1236, \text{ or } 12.36 \text{ percent.}$$
    c. The effective rate of interest is given by
    $$r_{eff} = \left(1 + \frac{r}{m}\right)^m - 1 = \left(1 + \frac{0.12}{4}\right)^4 - 1 \approx 0.125509, \text{ or } 12.5509 \text{ percent.}$$

    d. The effective rate of interest is given by
    $$r_{eff} = \left(1 + \frac{r}{m}\right)^m - 1 = \left(1 + \frac{0.12}{12}\right)^{12} - 1 \approx 0.126825, \text{ or } 12.6825 \text{ percent.}$$

5. The present value is given by
    $$P = 41,413\left(1 + \frac{0.065}{4}\right)^{-20} \approx 30,000.29, \text{ or approximately } \$30,000.$$

7.
$$S = 150 \left[\frac{\left(1 + \dfrac{0.08}{4}\right)^{28} - 1}{\dfrac{0.08}{4}} \right] \approx 5557.68, \text{ or } \$5557.68.$$

9. Using the formula for the present value of an annuity with $R = 250$, $n = 36$,

$i = 0.09/12 = 0.0075$, we have
$$P = 250 \left[\frac{1 - (1.0075)^{-36}}{0.0075} \right] \approx 7861.70, \text{ or } \$7861.70.$$

11. Using the amortization formula with $P = 22{,}000$, $n = 36$, and $i = 0.085/12$, we find

$$R = \frac{22{,}000\left(\dfrac{0.085}{12}\right)}{1 - \left(1 + \dfrac{0.085}{12}\right)^{-36}} \approx 694.49, \text{ or } \$694.49.$$

13. Using the sinking fund formula with $S = 18{,}000$, $n = 48$, and $i = 0.06/12$, we have
$$R = \frac{\left(\dfrac{0.06}{12}\right)18{,}000}{\left(1 + \dfrac{0.06}{12}\right)^{48} - 1} \approx 332.73, \text{ or } \$332.73.$$

15. We are asked to find r such that
$$\left(1 + \frac{r}{365}\right)^{365} = \left(1 + \frac{0.072}{12}\right)^{12} = 1.074424168.$$

Then
$$1 + \frac{r}{365} = (1.074424168)^{1/365}$$

$$\frac{r}{365} = (1.074424168)^{1/365} - 1$$

and $\qquad$ $r = 365[1.074424168)^{1/365} - 1]$
$$\approx 0.071791919, \text{ or approximately } 7.179 \text{ percent.}$$

17. Let a_n denote the sales during the nth year of operation. Then
$$\frac{a_{n+1}}{a_n} = 1.14 = r.$$
Therefore, the sales during the fourth year of operation were
$$a_4 = a_1 r^{n-1} = 1,750,000(1.14)^{4-1} = 2,592,702, \text{ or } \$2,592,702.$$
The total sales over the first four years of operation are given by
$$S_4 = \frac{a(1-r^4)}{1-r} = \frac{1,750,000[1-(1.14)^4]}{1-(1.14)} = 8,612,002,$$
or $8,612,002.

19. Using the present value formula for compound interest, we have
$$P = A\left(1 + \frac{r}{m}\right)^{-mt} = 19,440.31\left(1 + \frac{0.065}{12}\right)^{-12(4)} = 15,000.00, \text{ or } \$15,000.$$

21. Using the sinking fund formula with $S = 40,000$, $n = 120$, and $i = 0.08/12$, we find
$$R = \frac{\left(\dfrac{0.08}{12}\right)40,000}{\left(1 + \dfrac{0.08}{12}\right)^{120} - 1} = 218.64, \text{ or } \$218.64.$$

23. Using the formula for the present value of an annuity, we see that the equivalent cash payment of Maria's auto lease is
$$P = 300\left[\frac{1 - \left(1 + \dfrac{0.05}{12}\right)^{-48}}{\dfrac{0.05}{12}}\right] = 13,026.89, \text{ or } \$13,026.89.$$

25. a. The monthly payment is given by

$$P = \frac{(120{,}000)(0.0075)}{1-(1+0.0075)^{-360}} \approx 965.55, \text{ or } \$965.55.$$

b. We can find the total interest payment by computing

$$360(965.55) - 120{,}000 \approx 227{,}598, \text{ or } \$227{,}598.$$

c. We first compute the present value of their remaining payments. Thus,

$$P = 965.55\left[\frac{1-(1+0.0075)^{-240}}{0.0075}\right] \approx 107{,}316.01.$$

or \$107,316.01. Then their equity is
$$150{,}000 - 107{,}316.01, \text{ or approximately } \$42{,}684.$$

27. Using the sinking fund formula with $S = 500{,}000$, $n = 20$, and $I = 0.10/4$, we find that the amount of each installment should be

$$R = \frac{\left(\dfrac{0.10}{4}\right)500{,}000}{\left(1+\dfrac{0.10}{4}\right)^{20}-1} \approx 19{,}573.56, \text{ or } \$19{,}573.56.$$

29. Using the amortization formula, we find that Bill's monthly payment will be

$$R = \frac{3200\left(\dfrac{0.186}{12}\right)}{1-\left(1+\dfrac{0.186}{12}\right)^{-18}} \approx 205.09, \text{ or } \$205.09.$$

CHAPTER 6

EXERCISES 6.1, page 312

1. {x | x is a gold medalist in the 2002 Winter Olympic Games}

3. {x| x is an integer greater than 2 and less than 8}

5. {2,3,4,5,6}   7. {-2}

9. a. True--the order in which the elements are listed is not important.
   b. False-- A is a set, not an element.

11. a. False. The empty set has no elements. b. False. 0 is an element and $\varnothing$ is a set.

13. True.

15. a. True. 2 belongs to A. b. False. For example, 5 belongs to A but $5 \notin \{2,4,6\}$.

17. a. and b.

19. a. $\varnothing$, {1}, {2}, {1,2}
    b. $\varnothing$, {1}, {2}, {3}, {1,2}, {1,3}, {2,3}, {1,2,3}
    c. $\varnothing$, {1}, {2}, {3}, {4}, {1,2}, {1,3}, {1,4}, {2,3}, {2,4}, {3,4}, {1,2,3},
    {1,2,4}, {2,3,4}, {1,3,4}, {1,2,3,4}

21. {1, 2, 3, 4, 6, 8, 10}   23. {Jill, John, Jack, Susan, Sharon}

25. a.   b.   c.

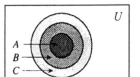
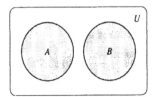
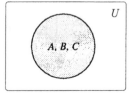

27. a. 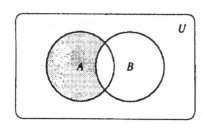   b.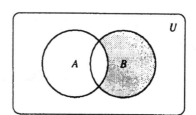

29. a. 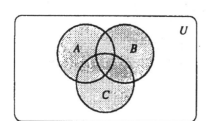   b.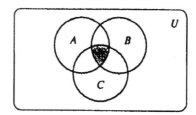

31. a. 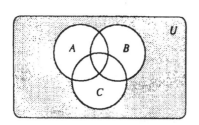   b.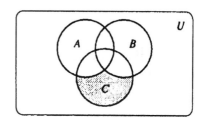

33. a. $A^C = \{2,4,6,8,10\}$
    b. $B \cup C = \{2,4,6,8,10\} \cup \{1,2,4,5,8,9\} = \{1,2,4,5,6,8,9,10\}$
    c. $C \cup C^C = U = \{1,2,3,4,5,6,7,8,9,10\}$

35. a. $(A \cap B) \cup C = C = \{1,2,4,5,8,9\}$
    b. $(A \cup B \cup C)^C = \varnothing$
    c. $(A \cap B \cap C)^C = U = \{1,2,3,4,5,6,7,8,9,10\}$

37. a. The sets are not disjoint. 4 is an element of both sets.
    b. The sets are disjoint as they have no common elements.

39. a. The set of all employees at the Universal Life Insurance Company who do not drink tea.
    b. The set of all employees at the Universal Life Insurance Company who do not drink coffee.

41. a. The set of all employees at the Universal Life Insurance Company who drink tea but not coffee.
    b. The set of all employees at the Universal Life Insurance Company who drink coffee but not tea.

43. a. The set of all employees at the hospital who are not doctors
    b. The set of all employees at the hospital who are not nurses

45. a. The set of all employees at the hospital who are female doctors.
    b. The set of all employees at the hospital who are both doctors and administrators.

47. a. $D \cap F$ b. $R \cap F^C \cap L^C$

49. a. B^C b. $A \cap B$ c. $A \cap B \cap C^C$

51. a. Region 1: $A \cap B \cap C$ is the set of tourists who used all three modes of transportation over a 1-week period in London.
    b. Regions 1 and 4: $A \cap C$ is the set of tourists who have taken the underground and a bus over a 1-week period in London.
    c. Regions 4, 5, 7, and 8: B^c is the set of tourists who have not taken a cab over a 1-week period in London.

53. $A \subset A \cup B$ $B \subset A \cup B$

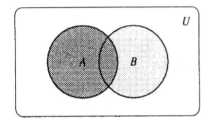

55. $A \cup (B \cup C) = (A \cup B) \cup C$

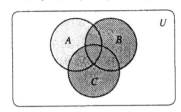
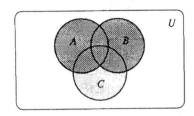

57. $A \cap (B \cup C) = (A \cap B) \cup (A \cap C)$

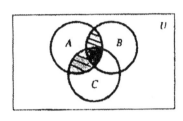
=
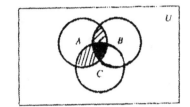

59. a. $A \cup (B \cup C) = \{1,3,5,7,9\} \cup (\{1,2,4,7,8\} \cup \{2,4,6,8\})$
$= \{1\ 3,5,7,9\} \cup \{1,2,4,6,7,8\}$
$= \{1,2,3,4,5,6,7,8,9\}$
$(A \cup B) \cup C = (\{1,3,5,7,9\} \cup (\{1,2,4,7,8\}) \cup \{2,4,6,8\})$
$= \{1,2,3,4,5,7,8,9\} \cup \{2,4,6,8\}$
$= \{1,2,3,4,5,6,7,8,9\}$

b. $A \cap (B \cap C) = \{1,3,5,7,9\} \cap (\{1,2,4,7,8\} \cap \{2,4,6,8\})$
$= \{1,3,5,7,9\} \cap (\{2,4,8\}$
$= \varnothing$
$(A \cap B) \cap C = (\{1,3,5,7,9\} \cap \{1,2,4,7,8\}) \cap \{2,4,6,8\}$
$= \{1,7\} \cap \{2,4,6,8\}$
$= \varnothing.$

61. a. r, u, v, w, x, y b. v, r

63. a. t, y, s b. t, s, w, x, z

65. $A \subset C$

67. False. Since every element in a set A belongs to A, A is a subset of itself.

69. True. If at least one of the sets A or B is nonempty, then $A \cup B \neq \varnothing$.

71. True. $(A \cup A^c)^c = U^c = \varnothing$.

EXERCISES 6.2, page 320

1. $A \cup B = \{a,e,g,h,i,k,l,m,o,u\}$, and so $n(A \cup B) = 10$. Next,
    $n(A) + n(B) = 5 + 5 = 10$.

3. a. $A = \{2,4,6,8\}$ and $n(A) = 4$.   b. $B = \{6,7,8,9,10\}$ and $n(B) = 5$
    c. $A \cup B = \{2,4,6,7,8,9,10\}$ and $n(A \cup B) = 7$.
    d. $A \cap B = \{6,8\}$ and $n(A \cap B) = 2$.

5. Using the results of Exercise 3, we see that $n(A \cup B) = 7$ and
    $n(A) + n(B) - n(A \cap B) = 4 + 5 - 2 = 7$.

7. Since $n(A \cup B) = n(A) + n(B) - n(A \cap B)$
    $\qquad\qquad n(B) = n(A \cup B) + n(A \cap B) - n(A)$
    $\qquad\qquad\qquad = 30 + 5 - 15 = 20$.

9. Refer to the Venn diagram at the right.
    a. $n(A \cup B) = 60 + 40 + 40 = 140$.
    b. $n(A^C) = 40 + 60 = 100$.
    c. $n(A \cap B^C) = 60$.

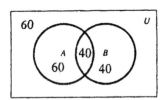

11. $n(A \cup B) = n(A) + n(B) - n(A \cap B) = 6 + 10 - 3 = 13$.

13. $n(A \cup B) = n(A) + n(B) - n(A \cap B)$ so
    $n(A \cap B) = n(A) + n(B) - n(A \cup B) = 4 + 5 - 9 = 0$.

15. $n(A \cap B \cap C)$
    $\qquad = n(A \cup B \cup C) - n(A) - n(B) - n(C) + n(A \cap B) + n(A \cap C) + n(B \cap C)$
    so $n(C) = n(A \cup B \cup C) - n(A \cap B \cap C) - n(A) - n(B)$
    $\qquad\qquad + n(A \cap B) + n(A \cap C) + n(B \cap C$
    $\qquad\qquad = 25 - 2 - 12 - 12 + 5 + 5 + 4 = 13$.

17. Let A denote the set of prisoners in the Wilton County Jail who were accused of a felony and B the set of prisoners in that jail who were accused of a misdemeanor. Then we are given that

$$n(A \cup B) = 190$$

Refer to the diagram at the right. Then the number of prisoners who were accused of both a felony and a misdemeanor is given by

$$(A \cap B) = n(A) + n(B) - n(A \cup B)$$
$$= 130 + 121 - 190 = 61.$$

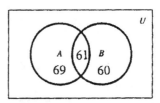

19. Let U denote the set of all customers surveyed, and let

$$A = \{ x \in U \mid x \text{ buys brand } A \}$$
$$B = \{ x \in U \mid x \text{ buys brand } B \}.$$

Then $n(U) = 120$, $n(A) = 80$, $n(B) = 68$, and $n(A \cap B) = 42$.

Refer to the diagram at the right.

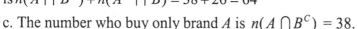

a. The number of customers who buy at least one of these brands is

$$n(A \cup B) = 80 + 68 - 42 = 106.$$

b. The number who buy exactly one of these brands is $n(A \cap B^C) + n(A^C \cap B) = 38 + 26 = 64$

c. The number who buy only brand A is $n(A \cap B^C) = 38$.

d. The number who buy none of these brands is $n[(A \cup B)^C] = 120 - 106 = 14$.

21. Let U denote the set of 200 investors and let

$$A = \{ x \in U \mid x \text{ uses a discount broker} \}$$
$$B = \{ x \in U \mid x \text{ uses a full-service broker} \}.$$

Refer to the diagram at the right.

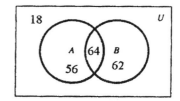

a. The number of investors who use at least one kind of broker is

$$n(A \cup B) = n(A) + n(B) - n(A \cap B) = 120 + 126 - 64 = 182.$$

b. The number of investors who use exactly one kind of broker is

$$n(A \cap B^C) + n(A^C \cap B) = 56 + 62 = 118.$$

c. The number of investors who use only discount brokers is $n(A \cap B^C) = 56$.

d. The number of investors who don't use a broker is

$$n(A \cup B)^C = n(U) - n(A \cup B) = 200 - 182 = 18.$$

23. Let U denote the set of 200 households in the survey and let

$$A = \{x \in U \mid x \text{ owns a desktop computer}\}$$
$$B = \{x \in U \mid x \text{ owns a laptop computer}\}$$

Referring to the figure that follows, we see that the number of households that own both desktop and laptop computers is $n(A \cap B) = 200 - 120 - 10 - 40 = 30$.

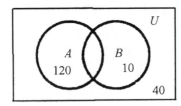

25. a. $n(A \cup B \cup C) = 64$ b. $n(A^C \cap B \cap C) = 10$

27. a. $n(A^C \cap B^C \cap C^C) = n[(A \cup B \cup C)^C] = 36$ b. $n[A^C \cap (B \cup C)] = 36$

29. Let U denote the set of all economists surveyed, and let $A = \{ x \in U \mid x \text{ had lowered his estimate}$ of the consumer inflation rate$\}$ $B = \{x \in U \mid x \text{ had raised his estimate}$ of the *GDP* growth rate$\}$.

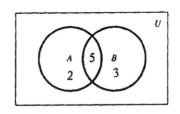

Refer to the diagram at the right. Then $n(U) = 10$, $n(A) = 7$, $n(B) = 8$, and $n(A \cap B^C) = 2$. Then the number of economists who had both lowered their estimate of the consumer inflation rate and raised their estimate of the *GDP* rate is given by $n(A \cap B) = 5$.

31. Let U denote the set of 100 college students who were surveyed and let

$A = \{ x \in U \mid x \text{ is a student who reads } Time$ magazine$\}$
$B = \{x \in U \mid x \text{ is a student who reads}$ *Newsweek* magazine$\}$
and $C = \{x \in U \mid x \text{ is a student who reads } U.S.$ *News and World Report* magazine$\}$

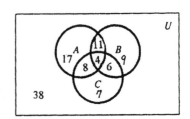

Refer to the diagram at the right.

Then $n(A) = 40$, $n(B) = 30$, $n(C) = 25$, $n(A \cap B) = 15$,

$n(A \cap C) = 12$, $n(B \cap C) = 10$, and $n(A \cap B \cap C) = 4$.

a. The number of students surveyed who read at least one magazine is
$n(A \cup B \cup C) = 17 + 11 + 4 + 8 + 6 + 7 + 9 = 62$

b. The number of students surveyed who read exactly one magazine is

$n(A \cap B^C \cap C^C) + n(A^C \cap B \cap C^C) + n(A^C \cap B^C \cap C)$
$= 17 + 9 + 7 = 33.$

c. The number of students surveyed who read exactly two magazines is

$n(A \cap B \cap C^C) + n(A^C \cap B \cap C) + n(A \cap B^C \cap C)$
$= 11 + 6 + 8 = 25.$

d. The number of students surveyed who did not read any of these magazines is

$n(A \cup B \cup C)^C = 100 - 62 = 38.$

33. Let U denote the set of all customers surveyed, and let

$A = \{ x \in U \mid x \text{ buys brand } A \}$

$B = \{ x \in U \mid x \text{ buys brand } B \}.$

$C = \{ x \in U \mid x \text{ buys brand } C \}.$

Refer to the figure at the right. Then

$n(U) = 120$, $n(A \cap B \cap C^C) = 15,$

$n(A^C \cap B \cap C^C) = 25,$

$n(A^C \cap B^C \cap C) = 26,$

$n(A \cap B \cap C^C) = 15$, $n(A \cap B^C \cap C) = 10,$

$n(A^C \cap B \cap C) = 12$, and $n(A \cap B \cap C) = 8.$

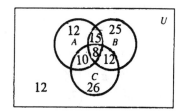

a. The number of customers who buy at least one of these brands is
$n(A \cup B \cup C) = 12 + 15 + 25 + 12 + 8 + 10 + 26 = 108.$

b. The number who buy labels A and B but not C is
$n(A \cap B \cap C^C) = 15$

c. The number who buy brand A is
$n(A) = 12 + 10 + 15 + 8 = 45.$

d. The number who buy none of these brands is
$n[(A \cup B \cup C)^C] = 120 - 108 = 12.$

35. Let U denote the set of 200 employees surveyed, and let

$A = \{x \in U \,|\, x \text{ had investments in stock funds}\}$

$B = \{x \in U \,|\, x \text{ had investments in bond funds}\}$

$C = \{x \in U \,|\, x \text{ had investments in money market funds}\}$

Then

$n(U) = 200, \ n(A) = 141, \ n(B) = 91, \ n(C) = 60, \ n(A \cap B) = 47,$

$n(A \cap C) = 36, \ n(B \cap C) = 36, \text{ and } n(A^c \cap B^c \cap C^c) = n[(A \cup B \cup C)^c] = 5$

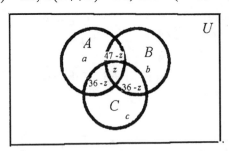

Letting $n(A \cap B \cap C) = z$ and using the fact that $n(A \cap B) = 47, \ n(A \cap C) = 36,$ and $n(B \cap C) = 36,$ leads to the Venn diagram shown. Next, using the fact that $n(A) = 141, \ n(B) = 91,$ and $n(C) = 60$ leads to

$$a + (36 - z) + (47 - z) + z = 141$$
$$b + (47 - z) + (36 - z) + z = 91$$
$$c + (36 - z) + (36 - z) + z = 60$$
$$a + b + c + (36 - z) + (47 - z) + (36 - z) + z + 5 = 200$$

which simplifies to $a - z = 58, \ b - z = 8, \ c - z = -12, \ a + b + c - 2z = 76.$
Solving, we find $a = 80, \ b = 30, \ c = 10,$ and $z = 22.$ Therefore,

a. The number of employees surveyed who had invested in all three vehicles is $n(A \cap B \cap C) = z = 22.$

b. The number who had invested in stock funds only is given by $n(A \cap B^c \cap C^c) = a = 80.$

37. True. $n(A \cup B) = n(A) + n(B) - n(A \cap B).$

39. True. If $A \cap B \neq \varnothing,$ then $n(A \cup B) = n(A) + n(B) - n(A \cap B).$

EXERCISES 6.3, page 327

1. By the multiplication principle, the number of rates is given by $(4)(3) = 12$.

3. By the multiplication principle, the number of ways that a blackjack hand can be dealt is $(4)(16) = 64$.

5. By the multiplication principle, she can create $(2)(4)(3) = 24$ different ensembles.

7. The number of paths is $2 \times 4 \times 3$, or 24.

9. By the multiplication principle, we see that the number of ways a health-care plan can be selected is $(10)(3)(2) = 60$.

11. $10^9 = 1,000,000,000$.

13. The number of different responses is $\underbrace{(5)\,(5)\,...\,(5)}_{50 \text{ terms}} = 5^{50}$.

15. The number of selections is given by $(2)(5)(3)$, or 30 selections.

17. The number of different selections is $(10)(10)(10)(10) - 10 = 10000 - 10 = 9990$.

19 . a. The number of license plate numbers that may be formed is
   $(26)(26)(26)(10)(10)(10)$, or $17,576,000$.
   b. The number of license plate numbers that may be formed is
   $(10)(10)(10)(26)(26)(26) = 17,576,000$.

21. If every question is answered, there are 2^{10}, or 1024, ways. In the second case, there are 3 ways to answer each question, and so we have 3^{10}, or 59,049, ways.

23. The number of ways the first, second and third prizes can be awarded is
   $(15)(14)(13) = 2730$.

25. The number of ways in which the nine symbols on the wheels can appear in the window slot is $(9)(9)(9)$, or 729. The number of ways in which the eight symbols other than the "lucky dollar" can appear in the window slot is $(8)(8)(8)$ or 512. Therefore, the number of ways in which the "lucky dollars" can appear in the window slot is $729 - 512$, or 217.

27. True. There are 4 choices for the digit in the hundreds position, 4 choices in the tens position and 2 choices in the units position, or $4 \cdot 4 \cdot 2$, ore 32 such numbers.

EXERCISES 6.4, page 340

1. $3(5!) = 3(5)(4)(3)(2)(1) = 360.$

3. $\dfrac{5!}{2!3!} = 5(2) = 10.$

5. $P(5,5) = \dfrac{5!}{(5\text{-}5)!} = \dfrac{5!}{0!} = 120$

7. $P(5,2) = \dfrac{5!}{(5-2)!} = \dfrac{5!}{3!} = (5)(4) = 20$

9. $P(n,1) = \dfrac{n!}{(n-1)!} = n$

11. $C(6,6) = \dfrac{6!}{6!0!} = 1$

13. $C(7,4) = \dfrac{7!}{4!3!} = \dfrac{7 \cdot 6 \cdot 5}{3 \cdot 2} = 35$

15. $C(5,0) = \dfrac{5!}{5!0!} = 1$

17. $C(9,6) = \dfrac{9!}{3!6!} = \dfrac{9 \cdot 8 \cdot 7}{3 \cdot 2} = 84$

19. $C(n,2) = \dfrac{n!}{(n-2)!2!} = \dfrac{n(n-1)}{2}$

21. $P(n,n-2) = \dfrac{n!}{(n-(n-2))!} = \dfrac{n!}{(n-n+2)!} = \dfrac{n!}{2}$

23. Order is important here since the word "*glacier*" is different from "*reicalg*", so this is a permutation.

25. Order is not important here. Therefore, we are dealing with a combination. If we consider a sample of three record-o-phones of which one is defective, it does not matter whether the defective record-o-phone is the first member of our sample, the second member of our sample, or the third member of our sample. The net result is a sample of three record-o-phones of which one is defective.

27. The order is important here. Therefore, we are dealing with a permutation. Consider, for example, 9 books on a library shelf. Each of the 9 books would have a call number, and the books would be placed in order of their call numbers; that is, a call

number of 902 would come before a call number of 910.

29. The order is not important here, and consequently we are dealing with a combination. It would not matter if the hand $Q\,Q\,Q\,5\,5$ were dealt or the hand $5\,5\,Q\,Q\,Q$. In each case the hand would consist of three queens and a pair.

31. The number of 4-letter permutations is $P(4,4) = \dfrac{4!}{0!} = 4 \cdot 3 \cdot 2 \cdot 1 = 24$.

33. The number of seating arrangements is $P(4,4) = \dfrac{4!}{0!} = 24$.

35. The number of different batting orders is $P(9,9) = \dfrac{9!}{0!} = 362{,}880$.

37. The number of different ways the 3 candidates can be selected is
$$C(12,3) = \frac{12!}{9!\,3!} = \frac{12 \cdot 11 \cdot 10}{3 \cdot 2 \cdot 1} = 220.$$

39. There are 10 letters in the word *ANTARCTICA*, 3*A*s, 1*N*, 2*T*s, 1*R*, 2*C*s, and 1*I*. Therefore, we use the formula for the permutation of n objects, not all distinct:
$$\frac{n!}{n_1!\,n_2! \cdots n_r!} = \frac{10!}{3!\,2!\,2!} = 151{,}200$$

41. The number of ways the 3 sites can be selected is
$$C(12,3) = \frac{12!}{9!\,3!} = \frac{12 \cdot 11 \cdot 10}{3 \cdot 2 \cdot 1} = 220$$

43. The number of ways in which the sample of 3 transistors can be selected is
$$C(100,3) = \frac{100!}{97!\,3!} = \frac{100 \cdot 99 \cdot 98}{3 \cdot 2 \cdot 1} = 161{,}700.$$

45. In this case order is important, as it makes a difference whether a commercial is shown first, last, or in between. The number of ways that the director can schedule the commercials is given by $P(6,6) = 6! = 720$.

47. The inquiries can be directed in

$$P(12,6) = \frac{12!}{6!} = 12 \cdot 11 \cdot 10 \cdot 9 \cdot 8 \cdot 7 = 665,280, \quad \text{or } 665,280 \text{ ways.}$$

49. a. The ten books can be arranged in
$$P(10,10) = 10! = 3,628,800 \text{ ways.}$$
b. If books on the same subject are placed together, then they can be arranged on the shelf
$$P(3,3) \times P(4,4) \times P(3,3) \times P(3,3) = 5184 \text{ ways.}$$
Here we have computed the number of ways the mathematics books can be arranged times the number of ways the social science books can be arranged times the number of ways the biology books can be arranged times the number of ways the 3 sets of books can be arranged.

51. Notice that order is certainly important here.
a. The number of ways that the 20 featured items can be arranged is given by
$$P(20,20) = 20! = 2.43 \times 10^{18}.$$
b. If items from the same department must appear in the same row, then the number of ways they can be arranged on the page is

Number of ways of arranging the rows	x	Number of ways of arranging the items in each of the 5 rows
$P(5,5)$	•	$P(4,4) \times P(4,4) \times P(4,4) \times P(4,4) \times P(4,4)$

$$= 5! \times (4!)^5 = 955,514,880.$$

53. The number of ways is given by
$$2 \{C(2,2) + [C(3,2) - C(2,2)]\} = 2[1 + (3 - 1)] = 2 \times 3 = 6$$
(number of players)[number of ways to win in exactly 2 sets + number of ways to win in exactly 3 sets]

55. The number of ways the measure can be passed is
$$C(3,3) \times [C(8,6) + C(8,7) + C(8,8)] = 37.$$
Here three of the three permanent members must vote for passage of the bill and this can be done in $C(3,3) = 1$ way. Of the 8 nonpermanent members who are voting 6 can vote for passage of the bill, or 7 can vote for passage, or 8 can vote for passage. Therefore, there are $C(8,6) + C(8,7) + C(8,8) = 37$ ways that the nonpermanent members can vote to ensure passage of the measure. This gives $1 \times 37 = 37$ ways that the members can vote so that the bill is passed.

57. a. If no preference is given to any student, then the number of ways of awarding the 3 teaching assistantships is $C(12,3) = \dfrac{12!}{3!9!} = 220.$

b. If it is stipulated that one particular student receive one of the assistantships, then the remaining two assistantships must be awarded to two of the remaining 11 students. Thus, the number of ways is $C(11,2) = \dfrac{11!}{2!9!} = 55$.

c. If at least one woman is to be awarded one of the assistantships, and the group of students consists of seven men and five women, then the number of ways the assistantships can be awarded is given by

$$C(5,1) \times C(7,2) + C(5,2) \times C(7,1) + C(5,3)$$
$$= \frac{5!}{4!1!} \cdot \frac{7!}{5!2!} + \frac{5!}{3!2!} \cdot \frac{7!}{6!1!} + \frac{5!}{3!2!} = 105 + 70 + 10 = 185.$$

59. The number of ways of awarding the 7 contracts to 3 different firms is given by

$$P(7,3) = \frac{7!}{4!} = 210.$$

The number of ways of awarding the 7 contracts to 2 different firms is

$$C(7,2) \times P(3,2) = 126. \quad \text{(First pick the two firms, and then award the 7 contracts.)}$$

Therefore, the number of ways the contracts can be awarded if no firm is to receive more than 2 contracts is given by $210 + 126 = 336$.

61. The number of different curricula that are available for the student's consideration is given by

$$C(5,1) \times C(3,1) \times C(6,2) \times C(4,1) + C(5,1) \times C(3,1) \times C(6,2) \times C(3,1)$$
$$= \frac{5!}{4!1!} \cdot \frac{3!}{2!1!} \cdot \frac{6!}{4!2!} \cdot \frac{4!}{3!1!} + \frac{5!}{4!1!} \cdot \frac{3!}{2!1!} \cdot \frac{6!}{4!2!} \cdot \frac{3!}{2!1!}$$
$$= (5)(3)(15)(4) + (5)(3)(15)(3) = 900 + 675 = 1575.$$

63. The number of ways is given by $C(10,2) + C(10,1) + C(10,0) = 45 + 10 + 1 = 56$
    or $\quad C(10,8) + C(10,9) + C(10,10) = 56$.

65. The number of ways of dealing a straight flush (5 cards in sequence in the same suit is given by

the number of ways of selecting 5 cards in sequence in the same suit	$\times$	the number of ways of selecting a suit
10	$\bullet$	$C(4,1) = \quad 40.$

67. The number of ways of dealing a flush (5 cards in one suit that are not all in sequence) is given by

the number of ways of selecting - the number of ways of selecting
    5 cards in one suit   5 cards in one suit in sequence

$$4C(13,5) \quad - \quad 4(10)$$
$$= 5148 - 40 = 5108.$$

69. The number of ways of dealing a full house (3 of a kind and a pair) is given by

the number of ways of picking × the number of ways of picking a pair
   3 of a kind from a given rank   from the 12 remaining ranks

$$13C(4,3) \bullet 12C(4,2)$$
$$= 13(4) \bullet (12)(6) = 3744.$$

71. The bus will travel a total of 6 blocks. Each route must include 2 blocks running north and south and 4 blocks running east and west. To compute the total number of possible routes, it suffices to compute the number of ways the 2 blocks running north and south can be selected from the six blocks. Thus,

$$C(6,2) = \frac{6!}{2!4!} = 15.$$

73. The number of ways that the quorum can be formed is given by

$$C(12,6) + C(12,7) + C(12,8) + C(12,9) + C(12,10) + C(12,11) + C(12,12)$$

$$= \frac{12!}{6!6!} + \frac{12!}{7!5!} + \frac{12!}{8!4!} + \frac{12!}{9!3!} + \frac{12!}{10!2!} + \frac{12!}{11!1!} + \frac{12!}{12!0!}$$

$$= 924 + 792 + 495 + 220 + 66 + 12 + 1 = 2510.$$

75. Using the formula given in Exercise 74, we see that the number of ways of seating the 5 commentators at a round table is

$$(5 - 1)! = 4! = 24.$$

77. The number of possible corner points is $C(8,3) = \dfrac{8!}{5!3!} = 56.$

79. True.

81. True. $C(n,r) = \dfrac{n!}{(n-r)!r!}$ and $C(n,n-r) = \dfrac{n!}{[n-(n-r)]!(n-r)!} = \dfrac{n!}{r!(n-r)!}.$

So, $C(n,r) = C(n,n-r).$

USING TECHNOLOGY EXERCISES 6.4, page 343

1. $1.307674368 \times 10^{12}$

3. $2.56094948229 \times 10^{16}$

5. $674,274,182,400$

7. $133,784,560$

9. $4,656,960$

11. Using the multiplication principle, the number of 10-question exams she can set is given by $C(25,3) \times C(40,5) \times C(30,2) = 658,337,004,000$.

CHAPTER 6, REVIEW EXERCISES, page 346

1. $\{3\}$. The set consists of all solutions to the equation $3x - 2 = 7$.

3. $\{4,6,8,10\}$

5. Yes.

7. Yes.

9.

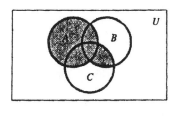

$$A \cup (B \cap C)$$

11.

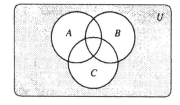

$$A^c \cap B^c \cap C^c$$

13. $A \cup (B \cup C) = \{a,b\} \cup [\{b,c,d\} \cup \{a,d,e\}]$
    $\qquad = \{a,b\} \cup \{a,b,c,d,e\} = \{a,b,c,d,e\}$.
    $(A \cup B) \cup C = [\{a,b\} \cup \{b,c,d\}] \cup \{a,d,e\}$
    $\qquad = \{a,b,c,d\} \cup \{a,d,e\} = \{a,b,c,d,e\}$.

15. $A \cap (B \cup C) = \{a,b\} \cap [\{b,c,d\} \cup \{a,d,e\}] = \{a,b\} \cap \{a,b,c,d,e\} = \{a,b\}$.
    $(A \cap B) \cup (A \cap C) = [\{a,b\} \cap \{b,c,d\}] \cup [\{a,b\} \cap \{a,d,e\}] = \{b\} \cup \{a\} = \{a,b\}$.

17. The set of all participants in a consumer behavior survey who both avoided buying a product because it is not recyclable and boycotted a company's products because of

its record on the environment.

19. The set of all participants in a consumer behavior survey who both did not use cloth diapers rather than disposable diapers and voluntarily recycled their garbage.

In Exercises 21-25, refer to the following Venn diagram.

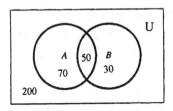

21. $n(A \cup B) = n(A) + n(B) - n(A \cap B) = 120 + 80 - 50 = 150$.

23. $n(B^c) = n(U) - n(B) = 350 - 80 = 270$.

25. $n(A \cap B^c) = n(A) - n(A \cap B) = 120 - 50 = 70$.

27. $C(20,18) = \dfrac{20!}{18!2!} = 190$.

29. $C(5,3) \cdot P(4,2) = \dfrac{5!}{3!2!} \cdot \dfrac{4!}{2!} = 10 \cdot 12 = 120$.

31. Let U denote the set of 5 major cards, and let
   $A = \{x \in U \mid x$ offered cash advances$\}$
   $B = \{x \in U \mid x$ offered extended payments for all goods and services purchased$\}$
   $C = \{x \in U \mid x$ required an annual fee that was less than \$35$\}$
   Thus, $n(A) = 3$, $n(B) = 3$, $n(C) = 2$
   $n(A \cap B) = 2$, $n(B \cap C) = 1$
   and $n(A \cap B \cap C) = 0$
   Using the Venn diagram on the right, we have
   $x + y + 2 = 3$
   $y + 2 = 2$

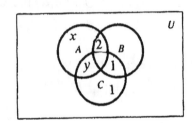

   Solving the system, we find $x = 1$, and $y = 0$.
   Therefore, the number of cards that offer cash advances and have an annual fee that

is less than \$35 is given by $n(A \cap C) = y = 0$.

33. The number of ways the compact discs can be arranged on a shelf is
$$P(6,6) = 6! = 720 \text{ ways.}$$

35. a. Since there is repetition of the letters C, I, and N, we use the formula for the permutation of n objects, not all distinct, with $n = 10$, $n_1 = 2$, $n_2 = 3$, and $n_3 = 3$. Then the number of permutations that can be formed is given by
$$\frac{10!}{2!3!3!} = 50{,}400.$$
b. Here, again, we use the formula for the permutation of n objects, not all distinct, this time with $n = 8$, $n_1 = 2$, $n_2 = 2$, and $n_3 = 2$. Then the number of permutations is given by $\dfrac{8!}{2!2!2!} = 5040$.

37. Let U denote the set comprising Halina's clients, and let
$$A = \{x \in U \,|\, x \text{ owns stocks}\}$$
$$B = \{x \in U \,|\, x \text{ owns bonds}\}$$
$$C = \{x \in U \,|\, x \text{ owns mutual funds}\}$$
Then $n(A) = 300$, $n(B) = 180$, $n(C) = 160$, $n(A \cap B) = 110$, $n(A \cap C) = 120$, $n(B \cap C) = 90$. Let $n(A \cap B \cap C) = z$. Then we are lead to the following Venn diagram.

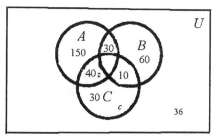

Next, using the fact that $n(A) = 300$, $n(B) = 180$, and $n(C) = 160$ leads to
$$a + (110 - z) + (120 - z) + z = 300$$
$$b + (110 - z) + (90 - z) + z = 180$$
$$c + (120 - z) + (90 - z) + z = 160$$
$$a + b + c + (110 - z) + (120 - z) + (90 - z) + z = 400$$
which simplifies to $a - z = 70$, $b - z = -20$, $c - z = -50$, and $a + b + c - 2z = 80$.

solving, we find $a = 150$, $b = 60$, $c = 30$ and $z = 80$. therefore, the number who own stocks, bonds and mutual funds is $n(A \cap B \cap C) = z = 80$.

39. The number of possible outcomes is $(6)(4)(5)(6) = 720$.

41. a. The number of ways the 7 students can be assigned to seats is $P(7,7) = 7! = 5040$.

    b. The number of ways 2 specified students can be seated next to each other is
            $2(6) = 12$.
    (Think of seven numbered seats. Then the students can be seated in seats 1-2, or 2-3, or 3-4, or 4-5, or 5-6, or 6-7. Since there are 6 such possibilities and the pair of students can be seated in 2 ways, we conclude that there are $2(6)$ possible arrangements.)
    Then the remaining 5 students can be seated in $P(5,5) = 5!$ ways. Therefore, the number of ways the 7 students can be seated if two specified students sit next to each other is $2(6)5! = 1440$.
    Finally, the number of ways the students can be seated if the two students do not sit next to each other is $P(7,7) - 2(6)5! = 5040 - 1440 = 3600$.

43. a. The number of samples that can be selected is $C(15,4) = \dfrac{15!}{4!11!} = 1365$.

    b. There are $C(10,4) = \dfrac{10!}{4!6!} = 210$ ways of selecting 4 balls none of which are white.
    Therefore, there are 1365 - 210, or 1155 ways of selecting 4 balls of which at least one is white.

CHAPTER 7

EXERCISES 7.1, page 356

1. $E \cup F = \{a,b,d,f\}$; $E \cap F = \{a\}$.

3. $F^C = \{b,c,e\}$; $E \cap G^C = \{a,b\} \cap \{a,d,f\} = \{a\}$.

5. Since $E \cap F = \{a\}$ is not a null set, we conclude that E and F are not mutually exclusive.

7. $E \cup F \cup G = \{2,4,6\} \cup \{1,3,5\} \cup \{5,6\} = \{1,2,3,4,5,6\}$.

9. $(E \cup F \cup G)^C = \{1,2,3,4,5,6\}^C = \varnothing$.

11. Yes, $E \cap F = \varnothing$; that is, E and F do not contain any common elements.

13. $E^c = \{2,4,6\}^c = \{1,3,5\} = F$ and so E and F are complementary.

15. $E \cup F$ 17. G^C 19. $(E \cup F \cup G)^C$

21. a. Refer to Example 4, page 353.
$$E = \{(2,1),(3,1),(4,1),(5,1),(6,1),(3,2),(4,2),(5,2),(6,2),$$
$$(4,3),(5,3),(6,3),(5,4),(6,4),(6,5)\}$$
    b. $E = \{(1,2),(2,4),(3,6)\}$

23. $\varnothing$, $\{a\}$, $\{b\}$, $\{c\}$, $\{a,b\}$, $\{a,c\}$, $\{b,c\}$, $\{a,b,c\}$.

25. a. $S = \{R,B\}$ b. $\varnothing$, $\{B\}$, $\{R\}$, $\{B,R\}$

27. a. $S = \{(H,1), (H,2), (H,3), (H,4), (H,5), (H,6), (T,1), (T,2),$
                        $(T,3), (T,4), (T,5), (T,6)\}$
    b. $E = \{(H,2), (H,4), (H,6)\}$

29. a. Here $S = \{1,2,3,4,5,6\}$, $E = \{2\}$, $F = \{2,4,6\}$. Since $E \cap F = \{2\} \neq \varnothing$, we

conclude that E and F are not mutually exclusive.

b. $E^c = \{1,3,4,5,6\} \neq F$ and so E and F are not complementary.

31. $S = \{(d,d,d), (d,d,n), (d,n,d), (n,d,d), (d,n,n), (n,d,n), (n,n,d), (n,n,n)\}$

33. a. $\{ABC, ABD, ABE, ACD, ACE, ADE, BCD, BCE, BDE, CDE\}$;
    b. 6   c. 3   d. 6

35. a. E^C b. $E^C \cap F^C$ c. $E \cup F$ d. $(E \cap F^C) \cup (E^C \cap F)$

37. a. $S = \{t \mid t > 0\}$ b. $E = \{t \mid 0 < t \leq 2\}$ c. $F = \{t \mid t > 2\}$

39. a. $S = \{0,1,2,3,...,10\}$ b. $E = \{0,1,2,3\}$ c. $F = \{5,6,7,8,9,10\}$

41. a. $S = \{0,1,2,...,20\}$ b. $E = \{0,1,2,...,9\}$ c. $F = \{20\}$

43. Let S denote the sample space of the experiment that is the set of 52 cards. Then
    $E = \{x \in S \mid x \text{ is an ace}\}$ and $F = \{x \in S \mid x \text{ is a spade}\}$ and
    $E \cap F = \{x \in S \mid x \text{ is the ace of spades}\}$. Now $n(E) = 4$, $n(F) = 13$, and $n(E \cap F) = 1$.

    Also, $E \cup F = \{x \in S \mid x \text{ is an ace or a spade}\}$ and $n(E \cup F) = 16$, and
    $$n(E) + n(F) - n(E \cap F) = 4 + 13 - 1 = 16 = n(E \cup F).$$

45. $E^C \cap F^C = (E \cup F)^C$ by DeMorgan's Law. Since $(E \cup F) \cap (E \cup F)^C = \emptyset$,
    they are mutually exclusive.

47. False. Let $E = \{1, 2, 3\}$, $F = \{4, 5, 6\}$, and $G = \{4, 5\}$. Then $E \cap F = \emptyset$ and
    $E \cap G = \emptyset$, but $F \cap G = \{4,5\} \neq \emptyset$.

EXERCISES 7.2, page 364

1. $\{(H,H)\}, \{(H,T)\}, \{(T,H)\}, \{(T,T)\}$.

✓ 3. $\{(D,m)\}, \{(D,f)\}, \{(R,m)\}, \{(R,f)\}, \{(I,m)\}, \{(I,f)\}$

5. $\checkmark$ $\{(1,i)\}, \{(1,d)\}, \{(1,s)\}, \{(2,i)\}, \{(2,d)\}, \{(2,s)\}, ..., \{(5,i)\}, \{(5,d)\}, \{(5,s)\}$

7. $\checkmark$ $\{(A,Rh^+)\}, \{(A,Rh^-)\}, \{(B,Rh^+)\}, \{B, Rh^-)\}, \{(AB, Rh^+)\}, \{(AB,Rh^-)\},$
$\{(O,Rh^+)\}, \{(O,Rh^-)\}$

9. The probability distribution associated with this data is

Grade	A	B	C	D	F
Probability	0.10	0.25	0.45	0.15	0.05

11. $\checkmark$ a. $S = \{(0 < x \le 200), (200 < x \le 400), (400 < x \le 600), (600 < x \le 800),$
$(800 < x \le 1000), (x > 1000)\}$

b.

Number of cars (x)	Probability
$0 < x \le 200$	0.075
$200 < x \le 400$	0.1
$400 < x \le 600$	0.175
$600 < x \le 800$	0.35
$800 < x \le 1000$	0.225
$x > 1000$	0.075

13. The probability distribution associated with this data is

Rating	A	B	C	D	E
Probability	0.026	0.199	0.570	0.193	0.012

$\checkmark$
15. The probability distribution is

Number of figures produced (in dozens)	30	31	32	33	34	35	36
Probability	0.125	0	0.1875	0.25	0.1875	0.125	0.125

17. The probability is $\dfrac{84,000,000}{179,000,000} \approx 0.469$.

$\checkmark$
19. a. The probability that a person killed by lightning is a male is $\dfrac{376}{439} \approx 0.856$.

   b. The probability that a person killed by lightning is a female is
   $$\dfrac{439-376}{439} = \dfrac{63}{439} \approx 0.144.$$

21. The probability that the retailer uses electronic tags as antitheft devices is
   $$\dfrac{81}{179} \approx 0.46.$$

23. a. $P(D) = \dfrac{13}{52} = \dfrac{1}{4}$ b. $P(B) = \dfrac{26}{52} = \dfrac{1}{2}$ c. $P(A) = \dfrac{4}{52} = \dfrac{1}{13}$

$\checkmark$
25. The probability of arriving at the traffic light when it is red is
   $$\dfrac{30}{30+5+45} = \dfrac{30}{80} = 0.375.$$

$\checkmark$
27. The required event is $E = \{2,4,\ldots, 36\}$. So
   $$P(E) = \dfrac{1}{38} + \dfrac{1}{38} + \cdots + \dfrac{1}{38} \qquad \text{(eighteen terms)}$$
   $$= \dfrac{18}{38}, \text{ or } \dfrac{9}{19}.$$

29. The probability is
   $$P(600 < x \le 800) + P(800 < x \le 1000) + P(x > 1000)$$
   $$= 0.35 + 0.225 + 0.075 = 0.65.$$

31. There are two ways of getting a 7, one die showing a 3 and the other die showing a 4, and vice versa.

33. No, the outcomes are not equally likely.

35. No. Since the coin is weighted, the outcomes are not equally likely.

✓
37. a. $P(A) = P(s_1) + P(s_3) = \dfrac{1}{12} + \dfrac{1}{12} = \dfrac{1}{6}$

b. $P(B) = P(s_2) + P(s_4) + P(s_5) + P(s_6) = \dfrac{1}{4} + \dfrac{1}{6} + \dfrac{1}{3} + \dfrac{1}{12} = \dfrac{5}{6}$ c. $P(C) = 1.$

✓
39. Let G denote a female birth and let B denote a male birth. Then the eight equally likely outcomes of this experiment are

GGG GGB GBG BGG BGB BBG GBB BBB.

a. The event that there are two girls and a boy in the family is

$E = \{GGB, GBG, BGG\}.$

Since there are three favorable outcomes, $P(E) = 3/8.$

b. The event that the oldest child is a girl is

$F = \{GGG, GGB, GBG, GBB\}.$

Since there are 4 favorable outcomes, $P(F) = 1/2.$

c. The event that the oldest child is a girl and the youngest child is a boy is

$G = \{GGB, GBB\}.$

Since there are two favorable outcomes, $P(F) = 1/4.$

41. The probability that the primary cause of the crash was due to pilot error or bad weather is given by $\dfrac{327 + 22}{327 + 49 + 14 + 22 + 19 + 15} = \dfrac{349}{446} \approx 0.7825.$

43. False. $P(s_1) + P(s_2) + \cdots + P(s_n) = 1$

EXERCISES 7.3, page 373

1. Refer to Example 4, page 373. Let E denote the event of interest. Then

$P(E) = \dfrac{18}{36} = \dfrac{1}{2}$

3. Refer to Example 4, page 353. The event of interest is $E = \{1,1\}$, and $P(E) = 1/36.$

5. Let E denote the event of interest. Then $E = \{(6,2),(6,1),(1,6),(2,6)\}$

and $P(E) = \dfrac{4}{36} = \dfrac{1}{9}.$

7. Let E denote the event that the card drawn is a king, and let F denote the event that

the card drawn is a diamond. Then the required probability is $P(E \cap F) = \dfrac{1}{52}$.

9. Let E denote the event that a face card is drawn. Then $P(E) = \dfrac{12}{52} = \dfrac{3}{13}$.

11. Let E denote the event that an ace is drawn. Then $P(E) = 1/13$. Then E^c is the event that an ace is not drawn and $P(E^C) = 1 - P(E) = \dfrac{12}{13}$.

13. Let E denote the event that a ticket holder will win first prize, then
$$P(E) = \dfrac{1}{500} = 0.002,$$
and the probability of the event that a ticket holder will not win first prize is
$$P(E^c) = 1 - 0.002 = 0.998.$$

15. Property 2 of the laws of probability is violated. The sum of the probabilities must add up to 1. In this case $P(S) = 1.1$, which is not possible.

17. The five events are not mutually exclusive; the probability of winning at least one purse is
$$1 - \text{probability of losing all 5 times} = 1 - \dfrac{9^5}{10^5} = 1 - 0.5905 = 0.4095.$$

19. The two events are not mutually exclusive; hence, the probability of the given event is $\dfrac{1}{6} + \dfrac{1}{6} - \dfrac{1}{36} = \dfrac{11}{36}$.

21. $E^c \cap F^c = \{c,d,e\} \cap \{a,b,e\} = \{e\} \neq \varnothing$.

23. Let G denote the event that a customer purchases a pair of glasses and let C denote the event that the customer purchases a pair of contact lenses. Then
$$P(G \cup C)^C \neq 1 - P(G) - P(C).$$
Mr. Owens has not considered the case in which the customer buy both glasses and contact lenses.

25. a. $P(E \cap F) = 0$ since E and F are mutually exclusive.

b. $P(E \cup F) = P(E) + P(F) - P(E \cap F) = 0.2 + 0.5 = 0.7.$

c. $P(E^c) = 1 - P(E) = 1 - 0.2 = 0.8.$

d. $P(E^C \cap F^C) = P[(E \cup F)^C] = 1 - P(E \cup F) = 1 - 0.7 = 0.3.$

27. a. $P(A) = P(s_1) + P(s_2) = \dfrac{1}{8} + \dfrac{3}{8} = \dfrac{1}{2}; \quad P(B) = P(s_1) + P(s_3) = \dfrac{1}{8} + \dfrac{1}{4} = \dfrac{3}{8}$

  b. $P(A^C) = 1 - P(A) = 1 - \dfrac{1}{2} = \dfrac{1}{2}; \quad P(B^C) = 1 - P(B) = 1 - \dfrac{3}{8} = \dfrac{5}{8}$

  c. $P(A \cap B) = P(s_1) = \dfrac{1}{8}$

  d. $P(A \cup B) = P(A) + P(B) - P(A \cap B) = \dfrac{1}{2} + \dfrac{3}{8} - \dfrac{1}{8} = \dfrac{3}{4}$

29. Referring to the following diagram we see that

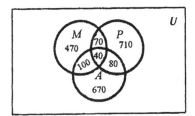

P: lack of parental support
M: malnutrition
A: abused or neglected

the probability that a teacher selected at random from this group said that lack of parental support is the only problem hampering a student's schooling is

$$\frac{710}{2140} = 0.33.$$

31. Let E and F denote the events that the person surveyed learned of the products from *Good Housekeeping* and *The Ladies Home Journal,* respectively. Then

$$P(E) = \frac{140}{500} = \frac{7}{25}, \quad P(F) = \frac{130}{500} = \frac{13}{50}$$

and $P(E \cap F) = \dfrac{80}{500} = 0.16$

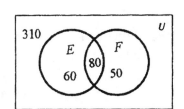

a. $P(E \cap F) = \dfrac{80}{500} = 0.16$ b. $P(E \cup F) = \dfrac{14}{50} + \dfrac{13}{50} - \dfrac{8}{50} = \dfrac{19}{50} = 0.38$

c. $P(E \cap F^C) + P(E^C \cap F) = \dfrac{60}{500} + \dfrac{50}{500} = \dfrac{110}{500} = 0.22$.

33. Let $A = \{t \mid t < 3\}$, $B = \{t \mid t \leq 4\}$, $C = \{t \mid t > 5\}$

a. $D = \{t \mid t \leq 5\}$ and $P(D) = 1 - P(C) = 1 - 0.1 = 0.9$.

b. $E = \{t \mid t > 4\}$ and $P(E) = 1 - P(B) = 1 - 0.6 = 0.4$.

c. $F = \{t \mid 3 \leq t \leq 4\}$ and $P(F) = P(A^C \cap B) = 0.4$.

35. a. The probability that the participant favors tougher gun-control laws is $\dfrac{150}{250} = 0.6$.

b. The probability that the participant owns a handgun is $\dfrac{58 + 25}{250} = 0.332$.

c. The probability that the participant owns a handgun but not a rifle is
$$\dfrac{58}{250} = 0.232.$$
d. The probability that the participant favors tougher gun-control laws and does not own a handgun is $\dfrac{12 + 138}{250} = 0.6$.

37. The probability that Bill will fail to solve the problem is $1 - p_1$, and the probability that Mike will fail to solve the problem is $1 - p_2$. Therefore, the probability that both Bill and Mike will fail to solve the problem is $(1 - p_1)(1 - p_2)$. So, the probability that at least one of them will solve the problem is
$$1 - (1 - p_1)(1 - p_2) = 1 - (1 - p_2 - p_1 + p_1 p_2) = p_1 + p_2 - p_1 p_2$$

39. True. Write $B = A \cup (B - A)$. Since A and $B - A$ are mutually exclusive, we have
$$P(B) = P(A) + P(B - A).$$
Since $P(B) = 0$, we have $P(A) + P(B - A) = 0$. If $P(A) > 0$, then $P(B - A) < 0$ and this is not possible. Therefore, $P(A) = 0$.

41. False. Take $E_1 = \{1, 2\}$ and $E_2 = \{2, 3\}$ where $S = \{1, 2, 3\}$. Then $P(E_1) = \frac{2}{3}$, $P(E_2) = \frac{2}{3}$, but $P(E_1 \cup E_2) = P(S) = 1$.

EXERCISES 7.4, page 381

1. Let E denote the event that the coin lands heads all five times. Then
$$P(E) = \frac{1}{2^5} = \frac{1}{32}.$$

3. Let E denote the event that the coin lands tails all 5 times, then
$$P(E^c) = 1 - P(E) = 1 - \frac{1}{32} = \frac{31}{32},$$
where E^c is the event that the coin lands heads at least once.

5. $P(E) = \dfrac{13 \cdot C(4,2)}{C(52,2)} = \dfrac{78}{1326} \approx 0.0588.$ 7. $P(E) = \dfrac{C(26,2)}{C(52,2)} = \dfrac{325}{1326} \approx 0.2451.$

9. The probability of the event that two of the balls will be white and two will be blue
is $P(E) = \dfrac{n(E)}{n(S)} = \dfrac{C(3,2) \cdot C(5,2)}{C(8,4)} = \dfrac{(3)(10)}{70} = \dfrac{3}{7}.$

11. The probability of the event that exactly three of the balls are blue is
$$P(E) = \frac{n(E)}{n(S)} = \frac{C(5,3)C(3,1)}{C(8,4)} = \frac{30}{70} = \frac{3}{7}.$$

13. $P(E) = \dfrac{C(3,2) \cdot C(1,1)}{8} = \dfrac{3}{8}.$ 15. $P(E) = \dfrac{C(3,3)}{8} = \dfrac{1}{8}.$

17. The number of elements in the sample space is 2^{10}. There are $C(10,6) = \dfrac{10!}{6!4!}$, or 210 ways of answering exactly six questions correctly. Therefore, the required probability is $\dfrac{210}{2^{10}} = \dfrac{210}{1024} \approx 0.205.$

19. a. Let E denote the event that both of the bulbs are defective. Then

$$P(E) = \frac{C(4,2)}{C(24,2)} = \frac{\frac{4!}{2!2!}}{\frac{24!}{22!2!}} = \frac{4 \cdot 3}{24 \cdot 23} = \frac{1}{46} \approx 0.022 \, .$$

b. Let F denote the event that none of the bulbs are defective. Then

$$P(F) = \frac{C(20,2)}{C(24,2)} = \frac{20!}{18!2!} \cdot \frac{22!2!}{24!} = \frac{20}{24} \cdot \frac{19}{23} = 0.6884 \, .$$

Therefore, the probability that at least one of the light bulbs is defective is given by
$1 - P(F) = 1 - 0.6884 = 0.3116$.

21. a. The probability that both of the cartridges are defective is

$$P(E) = \frac{C(6,2)}{C(80,2)} = \frac{15}{3160} = 0.0047.$$

b. Let F denote the event that none of the cartridges are defective. Then

$$P(F) = \frac{C(74,2)}{C(80,2)} = \frac{2701}{3160} = 0.855 \, ,$$

and $P(F^c) = 1 - P(F) = 1 - 0.855 = 0.145$ is the probability that at least 1 of the cartridges is defective.

23. a. The probability that Mary's name will be selected is $P(E) = \dfrac{12}{100} = 0.12$;

The probability that both Mary's and John's names will be selected is

$$P(F) = \frac{C(98,10)}{C(100,12)} = \frac{\frac{98!}{88!10!}}{\frac{100!}{88!12!}} = \frac{12 \cdot 11}{100 \cdot 99} \approx 0.013 \, .$$

b. The probability that Mary's name will be selected is $P(M) = \dfrac{6}{40} = 0.15$.

The probability that both Mary's and John's names will be selected is

$$P(M) \cdot P(J) = \frac{6}{60} \cdot \frac{6}{40} = \frac{36}{2400} = 0.015 \, .$$

25. The probability is given by

$$\frac{C(12,8) \cdot C(8,2)}{C(20,10)} + \frac{C(12,9)C(8,1)}{C(20,10)} + \frac{C(12,10)}{C(20,10)}$$

$$= \frac{(28)(495) + (220)(8) + 66}{184,756} \approx 0.085$$

27. a. The probability that he will select brand B is
$$\frac{C(4,2)}{C(5,3)} = \frac{6}{10} = \frac{3}{5}. \qquad (\frac{\text{the number of selections that include brand } B}{\text{the number of possible selections}})$$

  b. The probability that he will select brands B and C is
$$\frac{C(3,1)}{C(5,3)} = 0.3$$

  c. The probability that he will select at least one of the two brands, B and C is
$$1 - \frac{C(3,3)}{C(5,3)} = 0.9. \qquad \text{(1 - probability that he does not select brands } B \text{ and } C.)$$

29. The probability that the three "Lucky Dollar" symbols will appear in the window of the slot machine is
$$P(E) = \frac{n(E)}{n(S)} = \frac{(1)(1)(1)}{C(9,1)C(9,1)C(9,1)} = \frac{1}{729}.$$

31. The probability of a ticket holder having all four digits in exact order is
$$\frac{1}{C(10,1) \cdot C(10,1) \cdot C(10,1) \cdot C(10,1)} = \frac{1}{10,000} = 0.0001.$$

33. The probability of a ticket holder having a specified digit in exact order is
$$\frac{C(1,1)C(10,1)C(10,1)C(10,1)}{10^4} = 0.10.$$

35. The number of ways of selecting a 5-card hand from 52 cards is given by
$$C(52,5) = 2,598,960.$$
   The number of straight flushes that can be dealt in each suit is 10, so there are 4(10) possible straight flushes. Therefore, the probability of being dealt a straight flush is
$$\frac{4(10)}{C(52,5)} = \frac{40}{2,598,960} = 0.0000154.$$

37. The number of ways of being dealt a flush in one suit is $C(13,5)$, and, since there are four suits, the number of ways of being dealt a flush is $4 \cdot C(13,5)$. Since we wish to exclude the hands that are straight flushes we subtract the number of possible straight flushes from $4 \cdot C(13,5)$. Therefore, the probability of being drawn a flush, but not a straight flush, is
$$\frac{4 \cdot C(13,5) - 40}{C(52,5)} = \frac{5108}{2,598,960} = 0.0019654.$$

39. The total number of ways to select three cards of one rank is $13 \cdot C(4,3)$.
The remaining two cards must form a pair of another rank and there are
$$12 \cdot C(4,2)$$
ways of selecting these pairs. Next, the total number of ways to be dealt a full house
is $13 \cdot C(4,3) \cdot 12 \cdot C(4,2) = 3744$.
Hence, the probability of being dealt a full house is $\dfrac{3,744}{2,598,960} \approx 0.0014406$.

41. Let E denote the event that in a group of 5, no two will have the same sign. Then
$$P(E) = \frac{12 \cdot 11 \cdot 10 \cdot 9 \cdot 8}{12^5} \approx 0.381944.$$
Therefore, the probability that at least two will have the same sign is given by
$$1 - P(E) = 1 - 0.381944 \approx 0.618.$$
b. $P(\text{no Aries}) = \dfrac{11 \cdot 11 \cdot 11 \cdot 11 \cdot 11}{12^5} \approx 0.647228.$

$P(1 \text{ Aries}) = \dfrac{C(5,1) \cdot (1)(11)(11)(11)(11)}{12^5} \approx 0.2941945.$

Therefore, the probability that at least two will have the sign Aries is given by
$$1 - [P(\text{no Aries}) + P(1 \text{ Aries})] = 1 - 0.9414225 \approx 0.059.$$

43. Referring to the table on page 380, we see that in a group of 50 people, the
probability that none of the people will have the same birthday is $1 - 0.970 = 0.03$.

EXERCISES 7.5, page 394

1. a. $P(A|B) = \dfrac{P(A \cap B)}{P(B)} = \dfrac{0.2}{0.5} = \dfrac{2}{5}.$ b. $P(B|A) = \dfrac{P(A \cap B)}{P(A)} = \dfrac{0.2}{0.6} = \dfrac{1}{3}.$

3. $P(A \cap B) = P(A)P(B|A) = (0.6)(0.5) = 0.3.$

5. $P(A) \cdot P(B) = (0.3)(0.6) = 0.18 = P(A \cap B)$. Therefore the events are independent.

7. $P(A \cap B) = P(A) + P(B) - P(A \cup B) = 0.5 + 0.7 - 0.85 = 0.35 = P(A) \cdot P(B)$
so they are independent events.

9. a. $P(A \cap B) = P(A)P(B) = (0.4)(0.6) = 0.24.$
b. $P(A \cup B) = P(A) + P(B) - P(A \cap B) = 0.4 + 0.6 - 0.24 = 0.76.$

11. a. $P(A) = 0.5$ b. $P(E|A) = 0.4$
    c. $P(A \cap E) = P(A)P(E|A) = (0.5)(0.4) = 0.2$
    d. $P(E) = (0.5)(0.4) + (0.5)(0.3) = 0.35.$
    e. No. $P(A \cap E) \ne P(A) \cdot P(E) = (0.5)(0.35)$
    f. A and E are not independent events.

13.

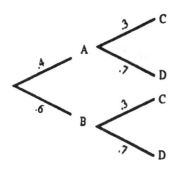

a. $P(A) = 0.4$ b. $P(C|A) = 0.3$ c. $P(A \cap C) = P(A)P(C|A) = (0.4)(0.3) = 0.12$
d. $P(C) = (0.4)(0.3) + (0.6)(0.3) = 0.3$ e. Yes. $P(A \cap C) = 0.12 = P(A)P(C)$ f. Yes.

15. a. Refer to Figure 7.15, page 386. Here $E = \{(5,1),(5,2),(5,3),(5,4),(5,5),(5,6)\}$
    and $F = \{(6,4),(5,5),(4,6)\}$. So $P(F) = \frac{3}{36} = \frac{1}{12}$.
    b. $P(E \cap F) = \frac{1}{36}$ since $E \cap F = \{(5,5)\}$.   c. $P(F|E) = \frac{1}{6}$.   d. $P(E) = \frac{6}{36} = \frac{1}{6}$.
    e. $P(F|E) = \dfrac{P(E \cap F)}{P(E)} = \dfrac{\frac{1}{36}}{\frac{1}{6}} = \dfrac{1}{6} \ne P(F) = \dfrac{1}{12}$ and so the events are not independent.

17. Let A denote the event that the sum of the numbers is less than 9 and let B denote the
    event that one of the numbers is a 6. Then, $P(A|B) = \dfrac{P(A \cap B)}{P(B)} = \dfrac{\frac{4}{36}}{\frac{11}{36}} = \dfrac{4}{11}$.

19. Refer to Figure 7.15, page 386 in the text. Here
    $$E = \{(3,1),(3,2),(3,3),(3,4),(3,5),(3,6)\}$$
    and $F = \{(1,6),(6,1),(2,5),(5,2),(3,4),(4,3)\}$. Then $E \cap F = \{(3,4)\}$.
    Now, $P(E \cap F) = \dfrac{1}{36}$ and this is equal to $P(E) \cdot P(F) = \left(\dfrac{6}{36}\right)\left(\dfrac{6}{36}\right) = \dfrac{1}{36}$.
    So E and F are independent events.

21. $P(E \cap F) = \frac{13}{24} = \frac{1}{4}$, $P(E) = \frac{26}{52} = \frac{1}{2}$, and $P(F) = \frac{13}{52} = \frac{1}{4}$. Now,

$$P(E) \cdot P(F) = (\tfrac{1}{2})(\tfrac{1}{4}) = \tfrac{1}{8} \ne P(E \cap F) = \tfrac{1}{4}.$$

So E and F are not independent events. The knowledge that the card drawn is black increases the probability that it is a spade.

23. Let A denote the event that the battery lasts 10 or more hours and let B denote the event that the battery lasts 15 or more hours.

Then $\qquad P(A) = 0.8, \quad P(B) = 0.15$

and $\qquad P(A \cap B) = 0.15$.

Therefore, the probability that the battery will last 15 hours or more is

$$P(B|A) = \frac{P(A \cap B)}{P(A)} = \frac{0.15}{0.8} = \frac{3}{16} = 0.1875.$$

25. Refer to the following tree diagram:

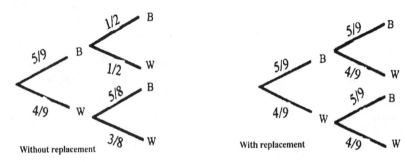

a. The probability that the second ball drawn is a white ball if the second ball is drawn without replacing the first is

$$P(B)P(W|B) + P(W)P(W|W) = (\tfrac{5}{9})(\tfrac{1}{2}) + (\tfrac{4}{9})(\tfrac{3}{8}) = \frac{4}{9}.$$

b. The probability that the second ball drawn is a white ball if the first ball is replaced before the second is drawn is $\quad (\tfrac{5}{9})(\tfrac{4}{9}) + (\tfrac{4}{9})(\tfrac{4}{9}) = \frac{4}{9}$.

27. Refer to the following tree diagram:

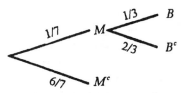

a. The probability that a student selected at random from this medical school is black is $(\frac{1}{7})(\frac{1}{3}) = \frac{1}{21}$

b. The probability that a student selected at random from this medical school is black if it is known that the student is a member of a minority group is $P(B|M) = 1/3$.

29. Let D denote the event that the card drawn is a diamond. Consider the tree diagram that follows.

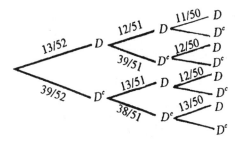

Then the required probability is

$$(\frac{13}{52})(\frac{12}{51})(\frac{11}{50}) + (\frac{13}{52})(\frac{39}{51})(\frac{12}{50}) + (\frac{39}{52})(\frac{13}{51})(\frac{12}{50}) + (\frac{39}{52})(\frac{38}{51})(\frac{13}{50}) = 0.25.$$

31. The sample space for a three-child family is
    $S = \{GGG, GGB, GBG, GBB, BGG, BGB, BBG, BBB\}$.

    Since we know that there is at least one girl in the three-child family we are dealing with a reduced sample space
    $S_1 = \{GGG, GGB, GBG, GBB, BGG, BGB, BBG\}$
    in which there are 7 outcomes. Then the probability that all three children are girls is
    $$P(E) = \frac{n(E)}{n(S)} = \frac{1}{7}.$$

33. a.

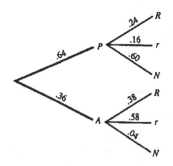

P = Professional
A = Amateur
R = recovered within 48 hrs
r = recovered after 48 hrs
N = never recovered

b. The required probability is 0.24.
c. The required probability is $(0.64)(0.60) + (0.36)(0.04) \approx 0.40$.

35. Refer to the following tree diagram.

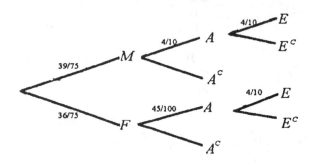

a. $P(A \cap E | M) = (0.4)(0.4) = 0.16$.

b. $P(A) = P(M \cap A) + P(F \cap A) = \dfrac{39}{75} \cdot \dfrac{4}{10} + \dfrac{36}{75} \cdot \dfrac{45}{100} = 0.424$

c. $P(M \cap A \cap E) + P(F \cap A \cap E) = \dfrac{39}{75} \cdot \dfrac{4}{10} \cdot \dfrac{4}{10} + \dfrac{36}{75} \cdot \dfrac{45}{100} \cdot \dfrac{4}{10}$

$$= 0.0832 + 0.0864 = 0.1696.$$

37. a. The probability that none of the dozen eggs is broken is $(0.992)^{12} = 0.908$. Therefore, the probability that at least one egg is broken is $1 - 0.908 = 0.092$.
b. Using the results of (a), we see that the required probability is
$(0.092)(0.092)(0.908) \approx 0.008$.

39. a. $P(A) = \dfrac{1120}{4000} = 0.280$; $P(B) = \dfrac{1560}{4000} = 0.390$;

$P(A \cap B) = \dfrac{720}{4000} = 0.180$;

$P(B|A) = \dfrac{P(A \cap B)}{P(A)} = \dfrac{n(A \cap B)}{n(A)} = \dfrac{720}{1120} \approx 0.643$

$P(B|A^C) = \dfrac{P(A^C \cap B)}{P(A^C)} = \dfrac{n(A^C \cap B)}{n(A^C)} = \dfrac{840}{2880} \approx 0.292$.

b. $P(B|A) \neq P(B)$ so A and B are not independent events.

41. Let C denote the event that a person in the survey was a heavy coffee drinker and Pa denote the event that a person in the survey had cancer of the pancreas. Then

$$P(C) = \dfrac{3200}{10000} = 0.32 \text{ and } P(Pa) = \dfrac{160}{10000} = 0.016$$

$$P(C \cap Pa) = \dfrac{132}{160} = 0.825$$

and $P(C) \cdot P(Pa) = 0.00512 \neq P(C \cap Pa)$. Therefore the events are not independent.

43. The probability that the first test will fail is 0.03, that the second test will fail is 0.015, and the third test will fail is 0.015. Since these are independent events the probability that all three tests will fail is
$$(0.03)(0.015)(0.015) = 0.0000068.$$

45. a. Let $P(A)$, $P(B)$, $P(C)$ denote the probability that the first, second, and third patient suffer a rejection, respectively. Then $P(A) = \dfrac{1}{2}$, $P(B) = \dfrac{1}{3}$, and $P(C) = \dfrac{1}{10}$.

Therefore, the probabilities that each patient does not suffer a rejection are given by $P(A^c) = \dfrac{1}{2}$, $P(B^c) = \dfrac{2}{3}$, and $P(C^c) = \dfrac{9}{10}$. Then, the probability that none of the 3 patients suffers a rejection is given by
$$P(A^c) \cdot P(B^c) \cdot P(C^c) = \dfrac{1}{2} \cdot \dfrac{2}{3} \cdot \dfrac{9}{10} = \dfrac{18}{60} = \dfrac{3}{10}.$$
Therefore, the probability that at least one patient will suffer rejection is
$$1 - P(A^c) \cdot P(B^c) \cdot P(C^c) = 1 - \dfrac{3}{10} = \dfrac{7}{10}.$$

b. The probability that exactly two patients will suffer rejection is

$$P(A)P(B)P(C^c) + P(A)P(B^c)P(C) + P(A^c)P(B)P(C)$$

$$= \frac{1}{2} \cdot \frac{1}{3} \cdot \frac{9}{10} + \frac{1}{2} \cdot \frac{2}{3} \cdot \frac{1}{10} + \frac{1}{2} \cdot \frac{1}{3} \cdot \frac{1}{10} = \frac{9+2+1}{60} = \frac{12}{60} = \frac{1}{5}.$$

47. Let A denote the event that at least one of the floodlights remain functional over the one-year period. Then

$$P(A) = 0.99999 \quad \text{and} \quad P(A^c) = 1 - P(A) = 0.00001.$$

Letting n represent the minimum number of floodlights needed, we have

$$(0.01)^n = 0.00001$$

$$n \log(0.01) = -5$$

$$n(-2) = -5$$

$$n = \frac{5}{2} = 2.5.$$

Therefore, the minimum number of floodlights needed is 3.

49. The probability that the event will not occur in one trial is $1 - p$. Therefore, the probability that it will not occur in n independent trials is $(1 - p)^n$. Therefore, the probability that it will occur at least once in n independent trials is

$$1 - (1 - p)^n.$$

51. $P(E|F) = \dfrac{P(E \cap F)}{P(F)} = \dfrac{P(F)}{P(F)} = 1 \qquad (E \cap F = F \text{ since } F \subset E).$

Interpretation: Since $F \subset E$, an occurrence of F implies an occurrence of E. In other words, given that F has occurred, it is a certainty that E will occur, that is, $P(E|F) = 1$.

53. True. Since A and B are mutually exclusive, $A \cap B = \varnothing$, and

$$P(A|B) = \frac{P(A \cap B)}{P(B)} = \frac{P(\varnothing)}{P(B)} = 0.$$

55. True. This follows from Formula 3, $P(A \cap B) = P(A) \cdot P(B|A)$.

EXERCISES 7.6, page 402

1.

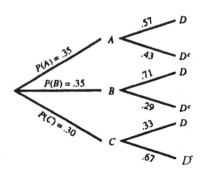

3. a. $P(D^C) = \dfrac{15+10+20}{35+35+30} = 0.45$;   b. $P(B|D^C) = \dfrac{10}{15+10+20} = 0.22$.

5. a. $P(D) = \dfrac{25+20+15}{50+40+35} = 0.48$ b. $P(B|D) = \dfrac{20}{25+20+15} = 0.33$

7. a. $P(A) \cdot P(D|A) = (0.4)(0.2) = 0.08$ b. $P(B) \cdot P(D|B) = (0.6)(0.25) = 0.15$

   c. $P(A|D) = \dfrac{P(A) \cdot P(D|A)}{P(A) \cdot P(D|A) + P(B) \cdot P(D|B)} = \dfrac{(0.4)(0.2)}{0.08+0.15} \approx 0.35$

9. a. $P(A) \cdot P(D|A) = \dfrac{1}{3} \cdot \dfrac{1}{4} = \dfrac{1}{12}$ b. $P(B) \cdot P(D|B) = \dfrac{1}{2} \cdot \dfrac{1}{2} = \dfrac{1}{4}$

   c. $P(C) \cdot P(D|C) = \dfrac{1}{6} \cdot \dfrac{1}{3} = \dfrac{1}{18}$

   d. $P(A|D) = \dfrac{P(A) \cdot P(D|A)}{P(A) \cdot P(D|A) + P(B) \cdot P(D|B) + P(C) \cdot P(C|B)}$

   $= \dfrac{\frac{1}{12}}{\frac{1}{12} + \frac{1}{4} + \frac{1}{18}} = \dfrac{1}{12} \cdot \dfrac{36}{14} = \dfrac{3}{14}$

11. Let A denote the event that the first card drawn is a heart and B the event that the second card drawn is a heart. Then

$$P(A|B) = \frac{P(A) \cdot P(B|A)}{P(A) \cdot P(B|A) + P(A^C) \cdot P(B|A^C)}$$

$$= \frac{\frac{1}{4} \cdot \frac{12}{51}}{\frac{1}{4} \cdot \frac{12}{51} + \frac{3}{4} \cdot \frac{13}{51}} = \frac{4}{17}.$$

13. Using the following tree diagram, we see that

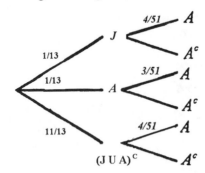

$$P(J|A) = \frac{\frac{1}{13} \cdot \frac{4}{51}}{\frac{1}{13} \cdot \frac{4}{51} + \frac{1}{13} \cdot \frac{3}{51} + \frac{11}{13} \cdot \frac{4}{51}} = \frac{\frac{4}{13 \cdot 51}}{\frac{51}{13 \cdot 51}} = 0.0784 .$$

15. The probabilities associated with this experiment are represented in the following tree diagram.

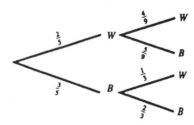

17. Referring to the tree diagram in Exercise 15, we see that the probability that the transferred ball was black given that the second ball was white is

$$P(B|W) = \frac{\frac{3}{5} \cdot \frac{1}{3}}{\frac{2}{5} \cdot \frac{4}{9} + \frac{3}{5} \cdot \frac{1}{3}} = \frac{9}{17}.$$

19. Let D denote the event that a senator selected at random is a Democrat, R denote the event that a senator selected at random is a Republican, and M the event that a

senator has served in the military. From the following tree diagram

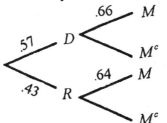

we see that the probability that a senator selected at random who has served in the military is a Republican is

$$P(R|M) = \frac{P(R)P(M|R)}{P(M)} = \frac{(0.64)(0.43)}{(0.66)(0.57) + (0.64)(0.43)}$$

$$\approx 0.422.$$

21. Let H_2 denote the event that the coin tossed is the two-headed coin, H_B denote the event that the coin tossed is the biased coin, and H_F denote the event that the coin tossed is the fair coin. Referring to the following tree diagram, we see that

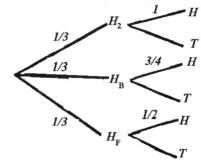

a. $P(H) = \frac{1}{3} \cdot 1 + \frac{1}{3} \cdot \frac{3}{4} + \frac{1}{3} \cdot \frac{1}{2} = \frac{1}{3} + \frac{1}{4} + \frac{1}{6} = \frac{9}{12} = \frac{3}{4}$

b. $P(H_F|H) = \dfrac{\frac{1}{3} \cdot \frac{1}{2}}{\frac{3}{4}} = \dfrac{2}{9}$

23. Let D denote the event that the person has the disease, and let Y denote the event that the test is positive. Referring to the following tree diagram, we see that the required probability is

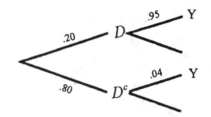

$$P(D|Y) = \frac{P(D) \cdot P(Y|D)}{P(D) \cdot P(Y|D) + P(D^C) \cdot P(Y|D^C)}$$

$$= \frac{(0.2)(0.95)}{(0.2)(0.95) + (0.8)(0.04)} \approx 0.856.$$

25. Let x denote the age of an insured driver, and let A denote the event that an insured driver is in an accident. Using the tree diagram we find,

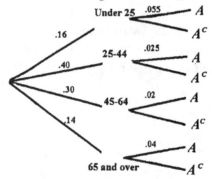

a. $P(A) = (0.16)(0.055) + (0.4)(0.025) + (0.3)(0.02) + (0.14)(0.04)$
    ≈ 0.03.

b. $P(x < 25 | A) = \dfrac{(0.16)(0.055)}{(0.03)} \approx 0.29$

27. Let E, F, and G denote the events that the child selected at random is 12 years old, 13 years old, or 14 years old, respectively; and let C^c denote the event that the child does not have a cavity. Using the tree diagram that follows, we see that

$$P(G|C^c) = \frac{\frac{5}{12}(0.28)}{\frac{1}{4}(0.42) + \frac{1}{3}(0.34) + \frac{5}{12}(0.28)} = 0.348$$

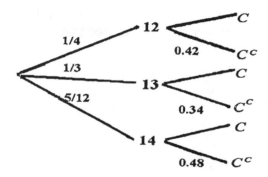

29. Let M and F denote the events that a person arrested for crime in 1988 was male or female, respectively; and let U denote the event that the person was under the age of 18. Using the following tree diagram, we have

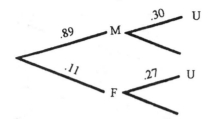

a. $P(U) = (0.89)(0.30) + (0.11)(0.27) = 0.2967$.

b. $P(F|U) = \dfrac{(0.11)(0.27)}{(0.89)(0.30) + (0.11)(0.27)} = 0.1001$.

31. Let I and II denote a customer who purchased the drug in capsule or tablet form, respectively; and let E denote a customer in Group I and II who purchased the extra-strength dosage of the drug. Then using the tree diagram that follows, we see that

$$P(I|E) = \dfrac{(0.57)(0.38)}{(0.57)(0.38) + (0.43)(0.31)} = 0.619$$

33. Let N and D denote the events that a employee was placed by Nancy or Darla,

respectively; and let S denote the event that the employee placed by one of these women was satisfactory. Using the tree diagram that follows, we see that

$$P(D|S^C) = \frac{(0.55)(0.3)}{(0.45)(0.2)+(0.55)(0.3)} = 0.647$$

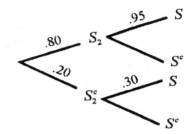

35. Using the tree diagram shown at the right, we see that

$$P(S_2|S) = \frac{(0.95)(0.8)}{(0.95)(0.8)+(0.2)(0.3)} \approx 0.93.$$

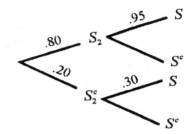

37. Let A, B, C, and D denote the event that the age of a guest is between 21 and 34, between 35 and 44, ..., 55 and over, respectively, and let O denote the event that a man keeps his paper money in order of denomination. Refer to the tree diagram below.

The required probability is

$$P(B|O) = \frac{\left(\frac{3}{8}\right)\left(\frac{61}{100}\right)}{\left(\frac{5}{16}\right)\left(\frac{9}{10}\right)+\left(\frac{3}{8}\right)\left(\frac{61}{100}\right)+\left(\frac{3}{16}\right)\left(\frac{8}{10}\right)+\left(\frac{1}{8}\right)\left(\frac{8}{10}\right)} \approx 0.3010.$$

39. Let D and R denote the event that the respondent is a Democrat or a Republican voter, respectively. Next, let S, O, and R denote the event that the respondent supports, opposes, or either doesn't know or refuses, respectively. Refer to the following diagram

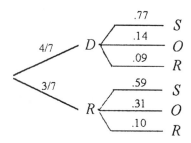

The required probability is

$$P(O|D) = \frac{\left(\frac{4}{7}\right)(0.14)}{\left(\frac{4}{7}\right)(0.14) + \left(\frac{3}{7}\right)(0.31)} = \frac{(0.571428)(0.14)}{(0.571428)(0.14) + (0.428571)(0.31)}$$

$$\approx 0.3758.$$

41. Let A, B, C, D, and E denote the event that the income level is 0-16, 17-33, ..., 96-100 percentile, respectively, and let V denote the event that a person voted in the election . Refer to the following diagram.

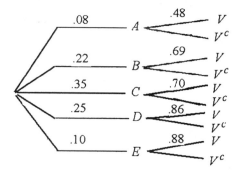

We see that the probability is

$$P(V|B) = \frac{(0.22)(0.69)}{(0.08)(0.48) + (0.22)(0.60) + (0.35)(0.70) + (0.25)(0.86) + (0.10)(0.88)}$$

$$\approx 0.2056.$$

EXERCISES 7.7, page 420

1. Yes. All entries are nonnegative and the sum of the entries in each column is equal to 1.

3. Yes.

5. No. The sum of the entries of the third column is not 1.

7. Yes.

9. No. It is not a square ($n \times n$) matrix.

11. a. The conditional probability that the outcome state 1 will occur given that the outcome state 1 has occurred is 0.3.
    b. 0.7
    c. We compute $X_1 = TX_0 = \begin{bmatrix} .3 & .6 \\ .7 & .4 \end{bmatrix} \begin{bmatrix} .4 \\ .6 \end{bmatrix} = \begin{bmatrix} .48 \\ .52 \end{bmatrix}.$

13. We compute $TX_0 = \begin{bmatrix} .6 & .2 \\ .4 & .8 \end{bmatrix} \begin{bmatrix} .5 \\ .5 \end{bmatrix} = \begin{bmatrix} .4 \\ .6 \end{bmatrix}$

    Thus, after 1 stage of the experiment, the probability of state 1 occurring is 0.4 and the probability of state 2 occurring is 0.6. The tree diagram describing this process follows.

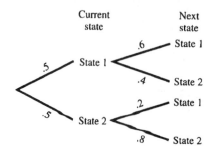

    Using this diagram, we see that the probabilities of state 1 and state 2 occurring in the next stage of the experiment are given by
    $$P(S_1) = (0.5)(0.6) + (0.5)(0.2) = 0.4$$
    $$P(S_2) = (0.5)(0.4) + (0.5)(0.8) = 0.6$$

Observe that these probabilities are precisely those represented in the probability distribution vector TX_0.

15. $$X_1 = TX_0 = \begin{bmatrix} .4 & .8 \\ .6 & .2 \end{bmatrix} \begin{bmatrix} .6 \\ .4 \end{bmatrix} = \begin{bmatrix} .56 \\ .44 \end{bmatrix}.$$

$$X_2 = TX_1 = \begin{bmatrix} .4 & .8 \\ .8 & .2 \end{bmatrix} \begin{bmatrix} .56 \\ .44 \end{bmatrix} = \begin{bmatrix} .576 \\ .424 \end{bmatrix}.$$

17. $$X_1 = TX_0 = \begin{bmatrix} \frac{1}{4} & \frac{1}{4} & \frac{1}{2} \\ \frac{1}{4} & \frac{1}{2} & \frac{1}{2} \\ \frac{1}{2} & \frac{1}{4} & 0 \end{bmatrix} \begin{bmatrix} \frac{1}{4} \\ \frac{1}{2} \\ \frac{1}{4} \end{bmatrix} = \begin{bmatrix} \frac{5}{16} \\ \frac{7}{16} \\ \frac{1}{4} \end{bmatrix}; \quad X_2 = TX_0 = \begin{bmatrix} \frac{1}{4} & \frac{1}{4} & \frac{1}{2} \\ \frac{1}{4} & \frac{1}{2} & \frac{1}{2} \\ \frac{1}{2} & \frac{1}{4} & 0 \end{bmatrix} \begin{bmatrix} \frac{5}{16} \\ \frac{7}{16} \\ \frac{1}{4} \end{bmatrix} = \begin{bmatrix} \frac{5}{16} \\ \frac{27}{64} \\ \frac{17}{64} \end{bmatrix}.$$

19. Since all entires in the matrix are positive, it is regular.

21. $$T^2 = \begin{bmatrix} 1 & .8 \\ 0 & .2 \end{bmatrix} \begin{bmatrix} 1 & .8 \\ 0 & .2 \end{bmatrix} = \begin{bmatrix} 1 & .96 \\ 0 & .04 \end{bmatrix}; \quad T^3 = \begin{bmatrix} 1 & .96 \\ 0 & .04 \end{bmatrix} \begin{bmatrix} 1 & .8 \\ 0 & .2 \end{bmatrix} = \begin{bmatrix} 1 & .992 \\ 0 & .008 \end{bmatrix}$$
and we see that the a_{21} entry will always be zero, so T is not regular.

23. $$T^2 = \begin{bmatrix} \frac{1}{2} & \frac{3}{4} & 0 \\ \frac{1}{2} & 0 & \frac{1}{2} \\ 0 & \frac{1}{4} & \frac{1}{2} \end{bmatrix} \begin{bmatrix} \frac{1}{2} & \frac{3}{4} & 0 \\ \frac{1}{2} & 0 & \frac{1}{2} \\ 0 & \frac{1}{4} & \frac{1}{2} \end{bmatrix} = \begin{bmatrix} \frac{5}{8} & \frac{3}{8} & \frac{3}{8} \\ \frac{1}{4} & \frac{1}{2} & \frac{1}{4} \\ \frac{1}{8} & \frac{1}{8} & \frac{3}{8} \end{bmatrix}$$ and so this matrix is regular.

25. $$T^2 = \begin{bmatrix} .7 & .2 & .3 \\ .3 & .8 & .3 \\ 0 & 0 & .4 \end{bmatrix} \begin{bmatrix} .7 & .2 & .3 \\ .3 & .8 & .3 \\ 0 & 0 & .4 \end{bmatrix} = \begin{bmatrix} .55 & .3 & .39 \\ .45 & .7 & .45 \\ 0 & 0 & .16 \end{bmatrix}$$
and so forth. Continuing, we see that $T^3, T^4, \ldots$, will have the a_{31} and a_{32} entries equal to zero and T is not regular.

27. We solve the matrix equation

$$\begin{bmatrix} \frac{1}{3} & \frac{1}{4} \\ \frac{2}{3} & \frac{3}{4} \end{bmatrix} \begin{bmatrix} x \\ y \end{bmatrix} = \begin{bmatrix} x \\ y \end{bmatrix}$$

7   *Probability*

or equivalently, the system of equations

$$\tfrac{1}{3}x + \tfrac{1}{4}y = x$$
$$\tfrac{2}{3}x + \tfrac{3}{4}y = y$$
$$x + y = 1.$$

Solving this system of equations, we find the required vector to be $\begin{bmatrix} \frac{3}{11} \\ \frac{8}{11} \end{bmatrix}$.

29. We have $TX = X$, that is, $\begin{bmatrix} .5 & .2 \\ .5 & .8 \end{bmatrix}\begin{bmatrix} x \\ y \end{bmatrix} = \begin{bmatrix} x \\ y \end{bmatrix}$, or equivalently, the sytem of equations

$$.5x + .2y = x$$
$$.5x + .8y = y.$$

These two equations are equivalent to the single equation $0.5x - 0.2y = 0$.
We must also have $x + y = 1$. So we have the system

$$.5x - .2y = x$$
$$x + y = 1$$

The second equation gives $y = 1 - x$, which when substituted into the first equation
yields, $0.5x - 0.2(1 - x) = 0$, $0.7x - 0.2 = 0$, or $x = 2/7$. Therefore, $y = 5/7$ and the

steady-state distribution vector is $\begin{bmatrix} \frac{2}{7} \\ \frac{5}{7} \end{bmatrix}$.

31. We solve the system

$$\begin{bmatrix} 0 & \frac{1}{8} & 1 \\ 1 & \frac{5}{8} & 0 \\ 1 & \frac{1}{4} & 0 \end{bmatrix}\begin{bmatrix} x \\ y \\ z \end{bmatrix} = \begin{bmatrix} x \\ y \\ z \end{bmatrix}$$

together with the equation $x + y + z = 1$; that is, the system

$$-x + \tfrac{1}{8}y + z = 0$$
$$x - \tfrac{3}{8}y = 0$$
$$\tfrac{1}{4}y - z = 0$$
$$x + y + z = 1.$$

Using the Gauss-Jordan method, we find that the required steady-state vector is

$$\begin{bmatrix} \frac{3}{13} \\ \frac{8}{13} \\ \frac{2}{13} \end{bmatrix}.$$

33. We solve the system

$$\begin{bmatrix} .2 & 0 & .3 \\ 0 & .6 & .4 \\ .8 & .4 & .3 \end{bmatrix} \begin{bmatrix} x \\ y \\ z \end{bmatrix} = \begin{bmatrix} x \\ y \\ z \end{bmatrix}$$

together with the equation $x + y + z = 1$, or equivalently, the system

$$-0.8x \qquad\quad + 0.3z = 0$$
$$\qquad -0.4y + 0.4z = 0$$
$$0.8x + 0.4y - 0.7z = 0$$
$$x \quad + y \quad + z = 1.$$

Using the Gauss-Jordan method, we find that the required steady-state vector is

$$\begin{bmatrix} \frac{3}{19} \\ \frac{8}{19} \\ \frac{8}{19} \end{bmatrix}.$$

35. a.

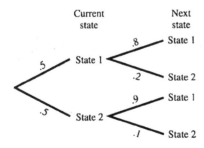

b.
$$\begin{array}{cc} & L \quad R \\ T = \begin{array}{c} L \\ R \end{array} & \begin{bmatrix} .8 & .9 \\ .2 & .1 \end{bmatrix} \end{array}$$

c.
$$X_0 = \begin{array}{c} L \\ R \end{array} \begin{bmatrix} .5 \\ .5 \end{bmatrix}$$

d.
$$X_1 = \begin{array}{c} L \\ R \end{array}\begin{array}{cc} L & R \\ \begin{bmatrix} .8 & .9 \\ .2 & .1 \end{bmatrix} \end{array}\begin{bmatrix} .5 \\ .5 \end{bmatrix} = \begin{array}{c} L \\ R \end{array}\begin{bmatrix} .85 \\ .15 \end{bmatrix}$$ and the probability that the mouse will turn left on the second trial is 0.85.

37. a. The transition matrix for the Markov chain is given by
$$\begin{array}{c} \\ A \\ U \\ N \end{array}\begin{array}{ccc} A & U & N \\ \begin{bmatrix} .85 & 0 & .10 \\ .10 & .95 & .05 \\ .05 & .05 & .85 \end{bmatrix} \end{array}$$

b. The probability vector describing the distribution of land 10 years ago is given by
$$\begin{array}{c} A \\ U \\ N \end{array}\begin{bmatrix} .50 \\ .15 \\ .35 \end{bmatrix}$$

c.
$$TX_0 = \begin{bmatrix} .85 & 0 & .10 \\ .10 & .95 & .05 \\ .05 & .05 & .85 \end{bmatrix}\begin{bmatrix} .50 \\ .15 \\ .35 \end{bmatrix} = \begin{bmatrix} .46 \\ .21 \\ .33 \end{bmatrix}$$

$$TX_1 = \begin{bmatrix} .85 & 0 & .10 \\ .10 & .95 & .05 \\ .05 & .05 & .85 \end{bmatrix}\begin{bmatrix} .47 \\ .37 \\ .16 \end{bmatrix} = \begin{bmatrix} .379 \\ .420 \\ .202 \end{bmatrix}$$

Then the probability vector describing the distribution of land 10 years ago is given by
$$\begin{array}{c} A \\ U \\ N \end{array}\begin{bmatrix} .424 \\ .262 \\ .314 \end{bmatrix}.$$

39. $$X_1 = TX_0 = \begin{bmatrix} .75 & .05 & .05 \\ .15 & .90 & .10 \\ .10 & .05 & .85 \end{bmatrix}\begin{bmatrix} .6 \\ .3 \\ .1 \end{bmatrix} = \begin{bmatrix} .47 \\ .37 \\ .16 \end{bmatrix}$$

$$X_2 = TX_1 = \begin{bmatrix} .75 & .05 & .05 \\ .15 & .90 & .10 \\ .10 & .05 & .85 \end{bmatrix} \begin{bmatrix} .47 \\ .37 \\ .16 \end{bmatrix} = \begin{bmatrix} .379 \\ .420 \\ .202 \end{bmatrix}$$

After 1 year A has 47%, B has 37%, and C has 16%; after 2 years A has 38%, B has 42%, and C has 20%.

41. $X_1 = TX_0 = \begin{bmatrix} .80 & .10 & .20 & .10 \\ .10 & .70 & .10 & .05 \\ .05 & .10 & .60 & .05 \\ .05 & .10 & .10 & .80 \end{bmatrix} \begin{bmatrix} .3 \\ .3 \\ .2 \\ .2 \end{bmatrix} = \begin{bmatrix} .33 \\ .27 \\ .175 \\ .225 \end{bmatrix}$

Similarly

$$X_2 = TX_1 = \begin{bmatrix} .3485 \\ .25075 \\ .15975 \\ .241 \end{bmatrix} \quad \text{and} \quad X_3 = TX_2 = \begin{bmatrix} .3599 \\ .2384 \\ .1504 \\ .2513 \end{bmatrix}$$

Assuming that the present trend continues, 36% of the students in their senior year will major in business, 23.8% will major in the humanities, 15% will major in education, and 25.1% will major in the natural sciences.

43. a. We compute $X_1 = \begin{bmatrix} .72 & .12 \\ .28 & .88 \end{bmatrix} \begin{bmatrix} .48 \\ .52 \end{bmatrix} = \begin{bmatrix} .408 \\ .592 \end{bmatrix}$, and conclude that, ten years from now, there will be 40.8 percent 1-wage-earners and 59.2% 2-wage earners.

b. We solve the system $\begin{bmatrix} .72 & .12 \\ .28 & .88 \end{bmatrix} \begin{bmatrix} x \\ y \end{bmatrix} = \begin{bmatrix} x \\ y \end{bmatrix}$ together with the equation

c.
$$x + y = 1.$$
$$-0.28x + 0.12y = 0$$
$$0.28x - 0.12y = 0$$
$$x + \quad y = 1$$

Solving, we find $x = 0.3$ and $y = 0.7$, and conclude that in the long run, there will be 30% 1-wage earners and 70% 2-wage earners.

45. If this trend continues, the percentage of homeowners in this city who will own single-family homes or condominiums two years from now will be given by $X_2 = TX_1$. Thus,

$$X_1 = TX_0 = \begin{bmatrix} .85 & .35 \\ .15 & .65 \end{bmatrix} \begin{bmatrix} .8 \\ .2 \end{bmatrix} = \begin{bmatrix} .75 \\ .25 \end{bmatrix}$$

$$X_2 = TX_1 = \begin{bmatrix} .85 & .35 \\ .15 & .65 \end{bmatrix} \begin{bmatrix} .75 \\ .25 \end{bmatrix} = \begin{bmatrix} .725 \\ .275 \end{bmatrix}$$

and we conclude that 72.5% will own single-family home and 27.5% will own condominiums at that time. We solve the system

$$\begin{bmatrix} .85 & .35 \\ .15 & .65 \end{bmatrix} \begin{bmatrix} x \\ y \end{bmatrix} = \begin{bmatrix} x \\ y \end{bmatrix}$$

together with the equation $x + y = 1$. Thus,

$$-0.15x + 0.35y = 0$$
$$0.15x - 0.35y = 0$$
$$x + \quad y = 1$$

Solving, we find $x = 0.7$ and $y = 0.3$, and conclude that in the long run 70% will own single family homes and 30% will own condominiums.

47. In the long-run each network will command 33 1/3% of the audience. Observe that the same steady state is reached regardless of the initial state of the system.

49. We wish to solve the system

$$\begin{bmatrix} .75 & .05 & .05 \\ .15 & .90 & .10 \\ .10 & .05 & .85 \end{bmatrix} \begin{bmatrix} x \\ y \\ z \end{bmatrix} = \begin{bmatrix} x \\ y \\ z \end{bmatrix},$$

or, equivalently,

$$.75x + .05y + .05z = x$$
$$.15x + .90y + .10z = y$$
$$.10x + .05y + .85z = z$$
$$x + y + z = 1$$

Using the Gauss-Jordan method to solve this system of equations, we find $\begin{bmatrix} .167 \\ .542 \\ .292 \end{bmatrix}$

and conclude that in the long run manufacturer A will have 16.7% of the market, manufacturer B willl have 54.2% of the market, and manufacturer C will have 29.2% of the market.

51. False. It must add up to exactly one.

53. True

USING TECHNOLOGY EXERCISES 7.7, page 426

1.
$$\begin{bmatrix} .204489 \\ .131869 \\ .261028 \\ .186814 \\ .2158 \end{bmatrix}$$

3. Manufacturer A will have 23.95% of the market, Manufacturer B will have 49.71% of the market share, and manufacturer C will have 26.34 percent of the market share.

5.
$$\begin{bmatrix} 0.2045 \\ 0.1319 \\ 0.2610 \\ 0.1868 \\ 0.2158 \end{bmatrix}$$

CHAPTER 7, REVIEW EXERCISES, page 429

1. a. $P(E \cap F) = 0$ since E and F are mutually exclusive.

    b. $P(E \cup F) = P(E) + P(F) - P(E \cap F)$
    $$= 0.4 + 0.2 = 0.6$$

    c. $P(E^c) = 1 - P(E) = 1 - 0.4 = 0.6$.

    d. $P(E^c \cap F^c) = P(E \cup F)^c = 1 - P(E \cup F) = 1 - 0.6 = 0.4$.

e. $P(E^C \cup F^C) = P(E \cap F)^C = 1 - P(E \cap F) = 1 - 0 = 1$.

3. a. The probability of the number being even is
$$P(2) + P(4) + P(6) = 0.12 + 0.18 + 0.19 = 0.49.$$
    b. The probability that the number is either a 1 or a 6 is
$$P(1) + P(6) = 0.20 + 0.19 = 0.39.$$
    c. The probability that the number is less than 4 is
$$P(1) + P(2) + P(3) = 0.20 + 0.12 + 0.16 = 0.48.$$

5. Let A denote the event that a video-game cartridge has an audio defect and let V denote the event that a video-game cartridge has a video defect. Then using the following Venn diagram, we have

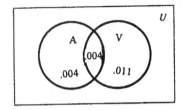

    a. the probability that a cartridge purchased by a customer will have a video or audio defect is
$$P(A \cup V) = P(A \cap V^C) + P(V \cap A^C) + P(A \cap V)$$
$$= 0.004 + 0.004 + 0.011 = 0.019$$
    b. the probability that a cartridge purchased by a customer will have not have a video or audio defect is
$$P(A \cup V)^C = 1 - P(A \cup V) = 1 - 0.019 = 0.981.$$

7. $P(A \cap E) = (0.3)(0.6) = 0.18$.   9. $P(C \cap E) = (0.2)(0.3) = 0.06$.

11. $P(E) = 0.18 + 0.25 + 0.06 = 0.49$

13. a. The probability that none of the pens in the sample are defective is
$$\frac{C(18,3)}{C(20,3)} = \frac{\frac{18!}{15!3!}}{\frac{20!}{17!3!}} = \frac{18!}{15!} \cdot \frac{17!}{20!} = \frac{68}{95} \approx 0.71579.$$
     Therefore, the probability that at least one is defective is given by

$$1 - 0.71579 \approx 0.284.$$

b. The probability that two are defective is given by

$$\frac{C(2,2) \cdot C(18,1)}{C(20,3)} = \frac{18}{\dfrac{20!}{17!3!}} = \frac{6}{380} \approx 0.0158 .$$

Therefore, the probability that no more than 1 is defective is given by

$$1 - \frac{6}{380} = \frac{374}{380} \approx 0.984 .$$

15. Let E denote the event that the sum of the numbers is 8 and let D denote the event that the numbers appearing on the face of the two dice are different. Then

$$P(E|D) = \frac{P(E \cap D)}{P(D)} = \frac{4}{30} = \frac{2}{15} .$$

17. Let E, F, and G denote the events that the first toss of a die results in a 6 being thrown, the second toss results in tails being thrown, and finally that a face card is selected, respectively. Then $P(E) = \frac{1}{6}$, $P(F) = \frac{1}{2}$, and $P(G) = \frac{12}{52} = \frac{3}{13}$. Since the outcomes are independent, the required probability is

$$P(E \cap F \cap G) = P(E)P(F)P(G) = (\tfrac{1}{6})(\tfrac{1}{2})(\tfrac{3}{13}) = \tfrac{1}{52} .$$

19. The probability that all three cards are face cards is $\dfrac{C(12,3)}{C(52,3)} = 0.00995.$

21. Referring to the tree diagram at the right, we see that the probability that the second card is black, given that the first card was red is

$$\frac{26}{51} = 0.510 .$$

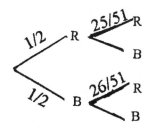

23. Since the entries $a_{21} = -2$ and $a_{22} = -8$ are negative, the given matrix is not stochastic and is hence not a regular stochastic matrix.

7 *Probability*

25. $T^2 = \begin{bmatrix} \frac{1}{2} & 0 & \frac{1}{3} \\ 0 & 0 & \frac{1}{3} \\ \frac{1}{2} & 1 & \frac{1}{3} \end{bmatrix} \begin{bmatrix} \frac{1}{2} & 0 & \frac{1}{3} \\ 0 & 0 & \frac{1}{3} \\ \frac{1}{2} & 1 & \frac{1}{3} \end{bmatrix} = \begin{bmatrix} \frac{5}{12} & \frac{1}{3} & \frac{5}{18} \\ \frac{1}{6} & \frac{1}{3} & \frac{1}{9} \\ \frac{5}{12} & \frac{1}{3} & \frac{11}{18} \end{bmatrix}$

and so the matrix is regular.

27. $X_1 = \begin{bmatrix} 0 & \frac{1}{4} & \frac{3}{5} \\ \frac{2}{5} & \frac{1}{2} & \frac{1}{5} \\ \frac{3}{5} & \frac{1}{4} & \frac{1}{5} \end{bmatrix} \begin{bmatrix} \frac{1}{2} \\ \frac{1}{2} \\ 0 \end{bmatrix} = \begin{bmatrix} \frac{1}{8} \\ \frac{9}{20} \\ \frac{17}{40} \end{bmatrix}$, $X_2 = \begin{bmatrix} 0 & \frac{1}{4} & \frac{3}{5} \\ \frac{2}{5} & \frac{1}{2} & \frac{1}{5} \\ \frac{3}{5} & \frac{1}{4} & \frac{1}{5} \end{bmatrix} \begin{bmatrix} \frac{1}{8} \\ \frac{9}{20} \\ \frac{17}{40} \end{bmatrix} = \begin{bmatrix} \frac{147}{400} \\ \frac{9}{25} \\ \frac{109}{400} \end{bmatrix} = \begin{bmatrix} .3675 \\ .36 \\ .2725 \end{bmatrix}$.

29. We solve the matrix equation

$$\begin{bmatrix} .6 & .3 \\ .4 & .7 \end{bmatrix} \begin{bmatrix} x \\ y \end{bmatrix} = \begin{bmatrix} x \\ y \end{bmatrix}$$

or equivalently, the system of equations

$-.4x + .3y = 0$

$.4x - .3y = 0$

$x + y = 1$

Solving this system of equations, we find the steady-state distribution vector to be

$\begin{bmatrix} \frac{3}{7} \\ \frac{4}{7} \end{bmatrix}$ and the steady-state matrix to be $\begin{bmatrix} \frac{3}{7} & \frac{3}{7} \\ \frac{4}{7} & \frac{4}{7} \end{bmatrix}$.

31. We solve the system

$$\begin{bmatrix} .6 & .4 & .3 \\ .2 & .2 & .2 \\ .2 & .4 & .5 \end{bmatrix} \begin{bmatrix} x \\ y \\ z \end{bmatrix} = \begin{bmatrix} x \\ y \\ z \end{bmatrix}$$

together with the equation $x + y + z = 1$, or equivalently, the system

$.6x + .4y + .3z = x$

$.2x + .2y + .2z = y$

$.2x + .4y + .5z = z$

$x + y + z = 1$

upon solving the system, we find the $x = .457$, $y = .20$, and $z = .343$, and the steady-state distribution vector is given by $\begin{bmatrix} .457 \\ .20 \\ .343 \end{bmatrix}$ and the steady-state matrix

is $\begin{bmatrix} .457 & .457 & .457 \\ .20 & .20 & .20 \\ .343 & .343 & .343 \end{bmatrix}$.

33. Let M denote the event that an employee at the insurance company is a male and let F denote the event that an employee at the insurance company is on flex time. Then

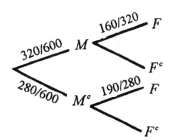

$$P(M|F)=\dfrac{\dfrac{320}{600}\cdot\dfrac{160}{320}}{\dfrac{320}{600}\cdot\dfrac{160}{320}+\dfrac{280}{600}\cdot\dfrac{190}{280}}=0.4571.$$

35. Let E denote the event that an applicant selected at random is eligible for admission, and let Pa denote the event that an applicant selected at random passes the admission exam. Using the tree diagram that follows, we see that

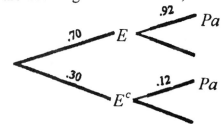

a. $P(Pa) = (0.70)(0.92) + (0.30)(0.12) = 0.68$.

b. $P(E^C|Pa) = \dfrac{(0.30)(0.12)}{(0.70)(0.92) + (0.30)(0.12)} = 0.053$.

37. Let $X = \begin{bmatrix} x \\ y \\ z \end{bmatrix}$ be the steady-state distribution vector associated with the Markov

process under consideration, where x, y, and z are to be determined. Then, the given information translates into the system

$$\begin{bmatrix} .3 & .1 & .1 \\ .3 & .5 & .2 \\ .4 & .4 & .7 \end{bmatrix} \begin{bmatrix} x \\ y \\ z \end{bmatrix} = \begin{bmatrix} x \\ y \\ z \end{bmatrix}$$

together with the equation $x + y + z = 1$, or, equivalently, the system of equations

$$.3x + .1y + .1z = x$$
$$.3x + .5y + .2z = y$$
$$.4x + .4y + .7z = z$$
$$x + y + z = 1$$

Solving the system, we find $x = .125$, $y = .3036$, and $z = .5714$, and the steady-state distribution vector is given by

$$X = \begin{bmatrix} .125 \\ .3036 \\ .5714 \end{bmatrix}.$$

That is, in the long run 12.5% of the cars in this area will be large cars, 30.36% of the cars will be intermediate in size and 57.14% of the cars will be small.

CHAPTER 8

EXERCISES 8.1, page 438

1. a. See part (b).
    b.   c. $\{GGG\}$

Outcome	GGG	GGR	GRG	RGG	GRR	RGR	RRG	RRR
Value	3	2	2	2	1	1	1	0

3. X may assume the values in the set $S = \{1,2,3,...\}$.

5. The event that the sum of the dice is 7 is $E = \{(1,6),(2,5),(3,4),(4,3),(5,2),(6,1)\}$ and $P(E) = \dfrac{6}{36} = \dfrac{1}{6}$.

7. X may assume the value of any positive integer. The random variable is infinite discrete.

9. $\{d \mid d \geq 0\}$. The random variable is continuous.

11. X may assume the value of any positive integer. The random variable is infinite discrete.

13. a. $P(X = -10) = 0.20$
    b. $P(X \geq 5) = 0.1 + 0.25 + 0.1 + 0.15 = 0.60$
    c. $P(-5 \leq X \leq 5) = 0.15 + 0.05 + 0.1 = 0.30$
    d. $P(X \leq 20) = 0.20 + 0.15 + 0.05 + 0.1 + 0.25 + 0.1 + 0.15 = 1$

15.

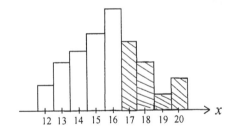

17. a.

x	1	2	3	4	5	6
P(X = x)	$\frac{1}{6}$	$\frac{1}{6}$	$\frac{1}{6}$	$\frac{1}{6}$	$\frac{1}{6}$	$\frac{1}{6}$

y	1	2	3	4	5	6
P(Y = y)	$\frac{1}{6}$	$\frac{1}{6}$	$\frac{1}{6}$	$\frac{1}{6}$	$\frac{1}{6}$	$\frac{1}{6}$

  b.

x + y	2	3	4	5	6	7	8	9	10	11	12
P(X + Y) = x + y	$\frac{1}{36}$	$\frac{2}{36}$	$\frac{3}{36}$	$\frac{4}{36}$	$\frac{5}{36}$	$\frac{6}{36}$	$\frac{5}{36}$	$\frac{4}{36}$	$\frac{3}{36}$	$\frac{2}{36}$	$\frac{1}{36}$

19. a.

x	0	1	2	3	4	5
P(X=x)	0.017	0.067	0.033	0.117	0.233	0.133

x	6	7	8	9	10
P(X=x)	0.167	0.1	0.05	0.067	0.017

b.

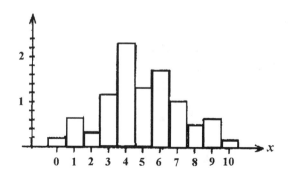

21.

x	1	2	3	4	5
$P(X=x)$	0.007	0.029	0.021	0.079	0.164

x	6	7	8	9	10
$P(X=x)$	0.15	0.20	0.207	0.114	0.029

23.   True. This follows from the definition.

USING TECHNOLOGY EXERCISES 8.1, page 442

1.   Data from Table 8.1, page 434 in   3.
      the text

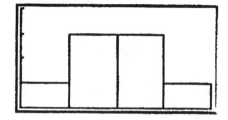

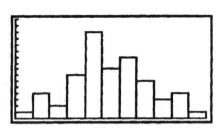

EXERCISES 8.2, page 452

1.   a. The student's grade-point average is given by

$$\frac{(2)(4)(3)+(3)(3)(3)+(4)(2)(3)+(1)(1)(3)}{(10)(3)} \quad \text{or } 2.6.$$

b.

X	0	1	2	3	4
$P(X=x)$	0	0.1	0.4	0.3	0.2

$$E(X) = 1(0.1) + 2(0.4) + 3(0.3) + 4(0.2) = 2.6.$$

3. $E(X) = -5(0.12) + -1(0.16) + 0(0.28) + 1(0.22) + 5(0.12) + 8(0.1)$
   $= 0.86.$

5. $E(X) = 0(0.07) + 25(0.12) + 50(0.17) + 75(0.14) + 100(0.28)$
   $\quad + 125(0.18) + 150(0.04)$
   $= 78.5, \quad$ or $78.50.

7. A customer entering the store is expected to buy
   $E(X) = (0)(0.42) + (1)(0.36) + (2)(0.14) + (3)(0.05) + (4)(0.03)$
   $= 0.91,$ or 0.91 videocassettes.

9. The expected number of accidents is given by
   $E(X) = (0)(0.935) + (1)(0.03) + (2)(0.02) + (3)(0.01) + (4)(0.005) = 0.12$

11. The expected number of machines that will break down on a given day is given by
    $E(X) = (0)(0.43) + (1)(0.19) + (2)(0.12) + (3)(0.09) + (4)(0.04)$
    $\quad + (5)(0.03) + (6)(0.03) + (7)(0.02) + (8)(0.05)$
    $= 1.73.$

13. The associated probabilities are $\frac{3}{50}, \frac{8}{50}, ..., \frac{5}{50}$, respectively. Therefore, the expected interest rate is
    $$(4.9)(\tfrac{3}{50}) + 5(\tfrac{8}{50}) + 5.1(\tfrac{12}{50}) + 5.2(\tfrac{14}{50}) + 5.3(\tfrac{8}{50}) + 5.4(\tfrac{5}{50})$$
    $\approx 5.162, \quad$ or 5.16%

15. The expected net earnings of a person who buys one ticket are
    $-1(0.997) + 24(0.002) + 99(0.0006) + 499(0.0002) + 1999(0.0002)$
    $= -0.39,$ or a loss of $0.39 per ticket.

17. The expected gain of the insurance company is given by
$$E(X) = 0.992(130) - (9870)(0.008) = 50, \text{ or } \$50.$$

19. The company's expected gross profit is given by
$$E(X) = (0.3)(80{,}000) + (0.60)(75{,}000) + (0.10)(70{,}000) - 64{,}000$$
$$= 12{,}000, \text{ or } \$12{,}000.$$

21. City A: $E(X) = (10{,}000{,}000)(0.2) - 250{,}000 = 1{,}750{,}000,$ or \$1.75 million.
City B: $E(X) = (7{,}000{,}000)(0.3) - 200{,}000 = 1{,}900{,}000,$ or \$1.9 million.
We see that the company should bid for the rights in city B.

23. The expected number of houses sold per year at company A is given by
$$E(X) = (12)(.02) + (13)(.03) + (14)(.05) + (15)(.07) + (16)(.07)$$
$$+ (17)(.16) + (18)(.17) + (19)(.13) + (20)(.11)$$
$$+ (21)(.09) + (22)(.06) + (23)(.03) + (24)(.01)$$
$$= 18.09.$$
The expected number of houses sold per year at company B is given by
$$E(X) = (6)(.01) + (7)(.04) + (8)(.07) + (9)(.06) + (10)(.11)$$
$$+ (11)(.12) + (12)(.19) + (13)(.17) + (14)(.13)$$
$$+ (15)(.04) + (16)(.03) + (17)(.02) + (18)(.01)$$
$$= 11.77.$$
Then, Sally's expected commission at company A is given by
$$(0.03)(154{,}000)(18.09) = 83{,}575.80, \text{ or } \$83{,}575.80$$
Her expected commission at company B is given by
$$(0.03)(237{,}000)(11.77) = 83{,}684.70,$$
or \$83,684.70. Based on these expectations, she should accept the job offer with company B.

25. Maria might expect her business to grow at the rate of
$$(5)(0.12) + (4.5)(0.24) + (3)(0.4) + (0)(0.2) + (-0.5)(0.04) = 2.86$$
or 2.86%/year for the upcoming year.

27. The expected value of the winnings on a \$1 bet placed on a split is
$$E(X) = 17 \cdot \frac{2}{38} + (-1) \cdot \frac{36}{38} \approx -0.0526, \text{ or a loss of 5.3 cents.}$$

29. The expected value of a player's winnings are

$$(1)(\frac{18}{37})+(-1)(\frac{19}{37}) = -\frac{1}{37} \approx -0.027, \text{ or a loss of 2.7 cents per bet.}$$

31.   The odds in favor of E occurring are $\dfrac{P(E)}{P(E^C)} = \dfrac{0.4}{0.6}$, or 2 to 3. The odds against E

occurring are 3 to 2.

33.   The probability of E not occurring is given by $P(E) = \dfrac{2}{3+2} = \dfrac{2}{5} = 0.4.$

35.   The probability that she will win her match is $P(E) = \dfrac{7}{7+5} = \dfrac{7}{12} = 0.5833.$

37.   The probability that the business deal will not go through is
$$P(E) = \frac{5}{5+9} = \frac{5}{14} \approx 0.3571.$$

39.   a. The mean is given by
$$\frac{40 + 45 + 2(50) + 55 + 2(60) + 2(75) + 2(80) + 4(85) + 2(90) + 2(95) + 100}{20}$$
$$= 74.$$
   The mode is 85 (the value that appears most frequently).
   The median is 80 (the middle value).
   b. The mode is the least representative of this set of test scores.

41.   True This follows from the definition.

EXERCISES 8.3, page 463

1. $\mu = (1)(0.4) + (2)(0.3) + 3(0.2) + (4)(0.1) = 2.$
   $\text{Var}\,(X) = (0.4)(1 - 2)^2 + (0.3)(2 - 2)^2 + (0.2)(3 - 2)^2 + (0.1)(4 - 2)^2$
   $\qquad = 0.4 + 0 + 0.2 + 0.4 = 1$
   $\sigma = \sqrt{1} = 1.$

3. $\mu = -2(\frac{1}{16}) + -1(\frac{4}{16}) + 0(\frac{6}{16}) + 1(\frac{4}{16}) + 2(\frac{1}{16}) = \frac{0}{16} = 0.$
   $\text{Var}\,(X) = \frac{1}{16}(-2-0)^2 + \frac{4}{16}(-1-0)^2 + \frac{6}{16}(0-0)^2 + \frac{4}{16}(1-0)^2 + \frac{1}{16}(2-0)^2$

$$= 1$$
$$\sigma = \sqrt{1} = 1.$$

5. $\mu = 0.1(430) + (0.2)(480) + (0.4)(520) + (0.2)(565) + (0.1)(580)$
    $= 518.$
    Var $(X) = 0.1(430 - 518)^2 + (0.2)(480 - 518)^2 + (0.4)(520 - 518)^2$
    $\qquad + (0.2)(565 - 518)^2 + (0.1)(580 - 518)^2$
    $= 1891.$
    $\sigma = \sqrt{1891} \approx 43.49.$

7. The mean of the histogram in Figure (b) is more concentrated about its mean than the histogram in Figure (a). Therefore, the histogram in Figure (a) has the larger variance.

9. $E(X) = 1(0.1) + 2(0.2) + 3(0.3) + 4(0.2) + 5(0.2) = 3.2.$
    Var $(X) = (0.1)(1 - 3.2)^2 + (0.2)(2 - 3.2)^2 + (0.3)(3 - 3.2)^2$
    $\qquad + (0.2)(4 - 3.2)^2 + (0.2)(5 - 3.2)^2$
    $= 1.56$

11. $\mu = \dfrac{1+2+3+ \cdots +8}{8} = 4.5$

    $V(X) = \frac{1}{8}(1-4.5)^2 + \frac{1}{8}(2-4.5)^2 + \cdots + \frac{1}{8}(8-4.5)^2 = 5.25$

13. a. Let X be the annual birth rate during the years 1991 - 2000.
    b.

x	14.5	14.6	14.7	14.8	15.2	15.5	15.9	16.3
$P(X = x)$	0.2	0.1	0.2	0.1	0.1	0.1	0.1	0.1

    c. $E(X) = (0.2)(14.5) + (0.1)(14.6) + (0.2)(14.7) + (0.1)(14.8)$
    $\qquad + (0.1)(15.2) + (0.1)(15.5) + (0.1)(15.9) + (0.1)(16.3)$
    $= 15.07.$
    $V(X) = (0.2)(14.5 - 15.07)^2 + (0.1)(14.6 - 15.07)^2$
    $\qquad + (0.2)(14.7 - 15.07)^2 + (0.1)(14.8 - 15.07)^2$
    $\qquad + (0.1)(15.2 - 15.07)^2 + (0.1)(15.5 - 15.07)^2$
    $\qquad + (0.1)(15.9 - 15.07)^2 + (0.1)(16.3 - 15.07)^2$
    $= 0.3621$
    $\sigma = \sqrt{0.3621} \approx 0.602 .$

15. a. Mutual Fund A
$$\mu = (0.2)(-4) + (0.5)(8) + (0.3)(10) = 6.2, \text{ or } \$620.$$
$$V(X) = (0.2)(-4 - 6.2)^2 + (0.5)(8 - 6.2)^2 + (0.3)(10 - 6.2)^2$$
$$= 26.76, \text{ or } \$267,600.$$
   Mutual Fund B
$$\mu = (0.2)(-2) + (0.4)(6) + (0.4)(8)$$
$$= 5.2, \text{ or } \$520.$$
$$V(X) = (0.2)(-2 - 5.2)^2 + (0.4)(6 - 5.2)^2 + (0.4)(8 - 5.2)^2$$
$$= 13.76, \text{ or } \$137,600.$$

   b. Mutual Fund A c. Mutual Fund B

17. $\text{Var } (X) = (0.4)(1)^2 + (0.3)(2)^2 + (0.2)(3)^2 + (0.1)(4)^2 - (2)^2 = 1.$

19. $\mu = [\frac{10}{500}(180) + \frac{20}{500}(190) + \cdots + \frac{5}{500}(350)] = 239.6, \text{ or } \$239,600.$
$$V(X) = [\frac{10}{500}(180-239.6)^2 + \frac{20}{500}(190-239.66)^2 + \cdots + \frac{5}{500}(350-239.66)^2][(1000)^2]$$
$$= 1443.84 \times 10^6 \text{ dollars.}$$
$$\sigma = \sqrt{1443.84 \times 10^6} = 37.998 \times 10^3, \text{ or } \$37,998.$$

21. The mean is given by
$$22\left(\frac{1332}{27127}\right) + 27\left(\frac{4219}{27127}\right) + 32\left(\frac{6345}{27127}\right) + 37\left(\frac{7598}{27127}\right) + 42\left(\frac{7633}{27127}\right) \approx 34.9456, \text{ or } 34.95.$$
$$\text{Var}(x) = \left(\frac{1332}{27127}\right)(22-34.9456)^2 + \left(\frac{4219}{27127}\right)(27-34.9456)^2 + \cdots$$
$$+ \left(\frac{7633}{27127}\right)(42-34.9456)^2 \approx 35.26222.$$
   Therefore, $\sigma = 5.938$, or 5.94.

23. a. Using Chebychev's inequality we have
$$P(\mu - k\sigma \leq X \leq \mu + k\sigma) \geq 1 - 1/k^2.$$
$$\mu - k\sigma = 42 - k(2) = 38, \text{ and } k = 2,$$
   and $P(\mu - k\sigma \leq X \leq \mu + k\sigma) \geq 1 - 1/(2)^2$
$$\geq 1 - 1/4$$
$$\geq 3/4, \text{ or at least } 0.75.$$

   b. Using Chebychev's inequality we have
$$P(\mu - k\sigma \leq X \leq \mu + k\sigma) \geq 1 - 1/k^2.$$
$$\mu - k\sigma = 42 - k(2) = 32, \text{ and } k = 5,$$
   and $P(\mu - k\sigma \leq X \leq \mu + k\sigma) \geq 1 - 1/(5)^2$
$$\geq 1 - 1/25 \geq 24/25, \text{ or at least } 0.96.$$

25. Here $\mu = 50$ and $\sigma = 1.4$. Now, we require that $c = k\sigma$, or $k = \dfrac{c}{1.4}$.

   Next, we solve $0.96 = 1 - \left(\dfrac{1.4}{c}\right)^2$; $\dfrac{1.96}{c^2} = 0.04$; $c^2 = \dfrac{1.96}{0.04} = 49$, or $c = 7$.

27. Using Chebychev's inequality we have $P(\mu - k\sigma \leq X \leq \mu + k\sigma) \geq 1 - 1/k^2$.
   Here, $\mu - k\sigma = 24 - k(3) = 20$, and $k = 4/3$.
   So $P(\mu - k\sigma \leq X \leq \mu + k\sigma) \geq 1 - 1/(4/3)^2 \geq 1 - 9/16 = 7/16$, or at least 0.4375.

29. Using Chebychev's inequality we have
   $$P(\mu - k\sigma \leq X \leq \mu + k\sigma) \geq 1 - 1/k^2.$$
   Here, $\mu - k\sigma = 42{,}000 - k(500) = 40{,}000$, and $k = 4$.
   So $P(\mu - k\sigma \leq X \leq \mu + k\sigma) \geq 1 - 1/(4)^2 \geq 1 - 1/16 \geq 15/16$, or at least 0.9375.

31. True. This follows from the definition.

USING TECHNOLOGY EXERCISES 8.3, page 467

1. a.

3. a.

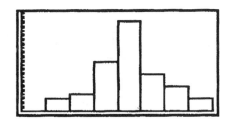

b. $\mu = 4$ and $\sigma \approx 1.40$

b. $\mu = 17.34$ and $\sigma \approx 1.11$

5.   a. Let X denote the random variable that gives the weight of a carton of sugar.
   b. The probability distribution for the random variable X is

x	4.96	4.97	4.98	4.99	5.00	5.01	5.02	5.03	5.04	5.05	5.06
$P(X = x)$	$\frac{3}{30}$	$\frac{4}{30}$	$\frac{4}{30}$	$\frac{1}{30}$	$\frac{1}{30}$	$\frac{5}{30}$	$\frac{3}{30}$	$\frac{3}{30}$	$\frac{4}{30}$	$\frac{1}{30}$	$\frac{1}{30}$

   $\mu = 5.00467 \approx 5.00$; $V(X) = 0.0009$; $\sigma = \sqrt{0.0009} = 0.03$

7. a.

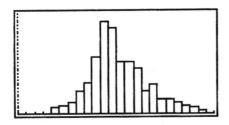

b. $\mu = 65.875$ and $\sigma = 1.73$.

EXERCISES 8.4, page 476

1. Yes. The number of trials is fixed, there are two outcomes of the experiment, the

    probability in each trial is fixed $(p = \frac{1}{6})$, and the trials are independent of each
    other.

3. No. There are more than 2 outcomes in each trial.

5. No. There are more than 2 outcomes in each trial and the probability of success (an
    accident) in each trial is not the same.

7. $C(4,2)(\frac{1}{3})^2(\frac{2}{3})^2 = \dfrac{4!}{2!2!}(\frac{4}{81}) \approx 0.296$.

9. $C(5,3)(0.2)^3(0.8)^2 = (\frac{5!}{2!3!})(0.2)^3(0.8)^2 \approx 0.0512$.

11. The required probability is given by $P(X = 0) = C(5,0)(\frac{1}{3})^0(\frac{2}{3})^5 \approx 0.132$.

13. The required probability is given by
$$P(X \geq 3) = C(6,3)(\tfrac{1}{2})^3(\tfrac{1}{2})^{6-3} + C(6,4)(\tfrac{1}{2})^4(\tfrac{1}{2})^{6-4} + C(6,5)(\tfrac{1}{2})^5(\tfrac{1}{2})^{6-5}$$
$$+ C(6,6)(\tfrac{1}{2})^6(\tfrac{1}{2})^{6-6}$$
$$= \tfrac{6!}{3!3!}(\tfrac{1}{2})^6 + \tfrac{6!}{4!2!}(\tfrac{1}{2})^6 + \tfrac{6!}{5!1!}(\tfrac{1}{2})^6 + \tfrac{6!}{6!0!}(\tfrac{1}{2})^6 = \tfrac{1}{64}(20+15+6+1) = \tfrac{21}{32}.$$

15. The probability of no failures, or, equivalently, the probability of five successes is
$$P(X = 5) = C(5,5)(\tfrac{1}{3})^5(\tfrac{2}{3})^{5-5} = \tfrac{1}{243} \approx 0.00412.$$

17. Here $n = 4$, and $p = 1/6$. Then $P(X = 2) = C(4,2)(\tfrac{1}{6})^2(\tfrac{5}{6})^2 = (\tfrac{25}{216}) \approx 0.116.$

19. Here $n = 5$, $p = 0.4$, and therefore, $q = 1 - 0.4 = 0.6$.
    a. $P(X = 0) = C(5,0)(.4)^0(.6)^5 \approx .078$; $P(X = 1) = C(5,1)(.4)(.6)^4 \approx .259$
    $P(X = 2) = C(5,2)(.4)^2(.6)^3 \approx .346$; $P(X = 3) = C(5,3)(.4)^3(.6)^3 \approx .230$
    $P(X = 4) = C(5,4)(.4)^4(.6)^2 \approx .077$; $P(X = 5) = C(5,5)(.4)^5(.6) \approx .010$
    b.

X	0	1	2	3	4	5
$P(X = x)$	0.078	0.259	0.346	0.230	0.077	0.010

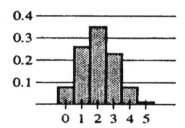

c. $\mu = np = 5(0.4) = 2$; $\sigma = \sqrt{npq} = \sqrt{5(0.4)(0.6)} = \sqrt{1.2} \approx 1.095.$

21. Here $1 - p = 1/50$ or $p = 49/50$. So the probability of obtaining 49 or 50 nondefective fuses is
    $$P(X = 49) + P(X = 50) = C(50,49)(\tfrac{49}{50})^{49}(\tfrac{1}{50}) + C(50,50)(\tfrac{49}{50})^{50}(\tfrac{1}{50})^0 \approx 0.74.$$
    This is also the probability of at most one defective fuse. So the inference is incorrect.

23. The probability that she will serve 0 or 1 ace is given by
    $$P(X = 0) + P(X = 1) = C(5,0)(.15)^0(.85)^5 + C(5,1)(.15)^1(.85)^4$$
    $$= \frac{5!}{0!5!}(.85)^5 + \frac{5!}{1!4!}(.15)(.85)^4 \approx 0.83521.$$

Therefore, the probability that she will serve at least two aces is
$$1 - P(X = 0) - P(X = 1) = 1 - .83521 = .16479, \text{ or approximately } .165.$$

25. The required probability is given by $P(X = 6) = C(6,6)(\frac{1}{4})^6 (\frac{3}{4})^0 \approx 0.0002.$

27. a. The probability that six or more people stated a preference for brand A is
$$P(X \geq 6) = C(10,6)(.6)^6(.4)^4 + C(10,7)(.6)^7(.4)^3$$
$$+ C(10,8)(.6)^8(.4)^2 + C(10,9)(.6)^9(.4)^1$$
$$+ C(10,10)(.6)^{10}(.4)^0$$
$$\approx 0.251 + 0.215 + 0.121 + 0.040 + 0.006 = 0.633.$$
b. The required probability is $1 - 0.633 = 0.367.$

29. a. Let X denote the number of new buildings that are in violation of the building code. Then $p = \frac{1}{3}$ and $p = \frac{2}{3}$. Therefore, the probability that the first 3 new buildings will pass inspection and the remaining 2 will fail the inspection is $(\frac{2}{3})^3 (\frac{1}{3})^2 = \frac{2^3}{3^5}$,
or approximately 0.0329.
b. The probability that just 3 of the new buildings will pass inspection (and so 2 will fail the inspection) is $C(5,2)(\frac{2}{3})^3 (\frac{1}{3})^2 = \frac{5!}{3!2!} \cdot \frac{2^3}{3^5}$, or approximately 0.3292.

31. This is a binomial experiment with $n = 9$, $p = 1/3$, and $q = 2/3$.
a. The probability is given by
$$P(X = 3) = C(9,3)(\frac{1}{3})^3 (\frac{2}{3})^6 = (\frac{9!}{6!3!})(\frac{1}{3})^3 (\frac{2}{3})^6 \approx 0.273.$$
b. The probability is given by
$$P(X = 0) + P(X = 1) + P(X = 2) + P(X = 3)$$
$$= C(9,0)(\frac{1}{3})^0 (\frac{2}{3})^9 + C(9,1)(\frac{1}{3})(\frac{2}{3})^8 + C(9,2)(\frac{1}{3})^2 (\frac{2}{3})^7 + C(9,3)(\frac{1}{3})^3 (\frac{2}{3})^6$$
$$= 0.026 + 0.117 + 0.234 + 0.273 \approx 0.650.$$

33. This is a binomial experiment with $n = 10$, $p = 0.02$, and $q = 0.98$.
a. The probability that the sample contains no defectives is given by
$$P(X = 0) = C(10,0)(0.02)^0(0.98)^{10} \approx 0.817.$$
b. The probability that the sample contains at most 2 defective sets is given by
$$P(X \leq 2) = C(10,0)(0.02)^0(0.98)^{10} + C(10,1)((0.02)(0.98)^9$$
$$+ C(10,2)(0.02)^2(0.98)^8$$
$$\approx 0.817 + 0.167 + 0.015 = 0.999.$$

35. a. The required probability is $P(X=2) = C(10,2)(0.05)^2(0.95)^8 \approx 0.075$.
   b. The required probability is
   $$P(X \geq 2) = 1 - P(X \leq 2)$$
   $$= 1 - [C(10,2)(.05)^2(.95)^8 + C(10,1)(.05)(.95)^9$$
   $$+ C(10,0)(.05)^0(.95)^{10}$$
   $$\approx 0.012.$$

37. a. The required probability is
   $$P(X=0) = C(20,0)(0.1)^0(0.9)^{20} \approx 0.1216.$$
   b. The required probability is
   $$P(X=0) = C(20,0)(0.05)^0(0.95)^{20} \approx 0.3585.$$

39. The required probability is $P(X=0) = C(10,0)(0.1)^0(0.9)^{10} \approx 0.3487$.

41. The mean number of people for whom the drug is effective is
   $$\mu = np = (500)(0.75) = 375.$$
   The standard deviation of the number of people for whom the drug can
   be expected to be effective is $\sigma = \sqrt{npq} = \sqrt{(500)(0.75)(0.25)} \approx 9.68$.

43. False. There are exactly two outcomes.

45. False. Here $p = \frac{1}{4}$ and $q = 1 - \frac{1}{4} = \frac{3}{4}$. The probability that the batter will get a hit if
   he bats four times is
   $$1 - P(X=0) = 1 - C(4,0)(\tfrac{1}{4})^0(\tfrac{3}{4})^4 = 1 - \frac{4!}{4!0!} \cdot 1 \cdot \frac{81}{256} = 1 - \frac{81}{256} = \frac{175}{256} \approx 0.68,$$
   which is far from guaranteeing that he gets a hit.

EXERCISES 8.5, page 486

1. $P(Z < 1.45) = 0.9265$.   3. $P(Z < -1.75) = 0.0401$.

5. $P(-1.32 < Z < 1.74) = P(Z < 1.74) - P(Z < -1.32) = 0.9591 - 0.0934 = 0.8657$.

7. $P(Z < 1.37) = 0.9147$.

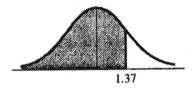

1.37

9. $P(Z < -0.65) = 0.2578.$

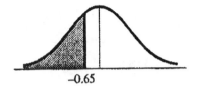

−0.65

11. $P(Z > -1.25) = 1 - P(Z < -1.25) = 1 - 0.1056 = 0.8944$

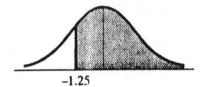

−1.25

13. $P(0.68 < Z < 2.02) = P(Z < 2.02) - P(Z < 0.68)$
$$= 0.9783 - 0.7517 = 0.2266.$$

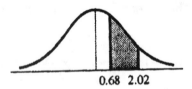

0.68 2.02

15.   a. Referring to Table 2, we see that $P(Z < z) = 0.8907$ implies that $z = 1.23$.
      b. Referring to Table 2, we see that $P(Z < z) = 0.2090$ implies that $z = -0.81$.

17.   a. $P(Z > -z) = 1 - P(Z < -z) = 1 - 0.9713 = 0.0287$ implies $z = 1.9$.
      b. $P(Z < -z) = 0.9713$ implies that $z = -1.9$.

19.   a. $P(X < 60) = P(Z < \dfrac{60-50}{5}) = P(Z < 2) = 0.9772.$

      b. $P(X > 43) = P(Z > \dfrac{43-50}{5}) = P(Z > -1.4) = P(Z < 1.4) = 0.9192.$

c. $P(46 < X < 58) = P(\dfrac{46-50}{5} < Z < \dfrac{58-50}{5}) = P(-0.8 < Z < 1.6)$

$\qquad\qquad = P(Z < 1.6) - P(Z < -0.8) = 0.9452 - 0.2119 = 0.7333.$

EXERCISES 8.6, page 493

1. $\mu = 20$ and $\sigma = 2.6$.

a. $P(X > 22) = P(Z > \dfrac{22-20}{2.6}) = P(Z > 0.77) = P(Z < -0.77) = 0.2206.$

b. $P(X < 18) = P(Z < \dfrac{18-20}{2.6}) = P(Z < -0.77) = 0.2206.$

c. $P(19 < X < 21) = P(\dfrac{19-20}{2.6} < Z < \dfrac{21-20}{2.6}) = P(-0.38 < Z < 0.38)$

$\qquad\qquad = P(Z < 0.38) - P(Z < -0.38) = 0.6480 - 0.3520 = 0.2960.$

3. $\mu = 750$ and $\sigma = 75$.

a. $P(X > 900) = P(Z > \dfrac{900-750}{75}) = P(Z > 2) = P(Z < -2) = 0.0228.$

b. $P(X < 600) = P(Z < \dfrac{600-750}{75}) = P(Z < -2) = 0.0228.$

c. $P(750 < X < 900) = P(Z < \dfrac{750-750}{75} < Z < \dfrac{900-750}{75})$

$\qquad\qquad = P(0 < Z < 2) = P(Z < 2) - P(Z < 0)$

$\qquad\qquad = 0.9772 - 0.5000 = 0.4772.$

d. $P(600 < X < 800) = P(\dfrac{600-750}{75} < Z < \dfrac{800-750}{75})$

$\qquad\qquad = P(-2 < Z < .67) = P(Z < .67) - P(Z < -2)$

$\qquad\qquad = 0.7486 - 0.0228 = 0.7258.$

5. $\mu = 100$ and $\sigma = 15$.

a. $P(X > 140) = P(Z > \dfrac{140-100}{15}) = P(Z > 2.67) = P(Z < -2.67) = 0.0038.$

b. $P(X > 120) = P(Z > \dfrac{120-100}{15}) = P(Z > 1.33) = P(Z < -1.33) = 0.0918.$

c. $P(100 < X < 120) = P(\dfrac{100-100}{15} < Z < \dfrac{120-100}{15}) = P(0 < Z < 1.33)$

$$= P(Z < 0) - P(Z < 1.33) = 0.9082 - 0.5000 = 0.4082.$$

d. $P(X < 90) = P(Z < \dfrac{90 - 100}{15}) = P(Z < -0.67) = 0.2514.$

7. Here $\mu = 575$ and $\sigma = 50$.

$P(550 < X < 650) = P(\dfrac{550 - 575}{50} < Z < \dfrac{650 - 575}{50}) = P(-0.5 < Z < 1.5)$

$$= P(Z < 1.5) - P(Z < -0.5) = 0.9332 - 0.3085 = 0.6247.$$

9. Here $\mu = 22$ and $\sigma = 4$.

$P(X < 12) = P(Z < \dfrac{12 - 22}{4}) = P(Z < -2.5) = 0.0062,$ or 0.62 percent.

11. $\mu = 70$ and $\sigma = 10$.
To find the cut-off point for an A, we solve $P(Y < y) = 0.85$ for y. Now

$P(Y < y) = P\left(Z < \dfrac{y - 70}{10}\right) = 0.85$ implies $\dfrac{y - 70}{10} = 1.04$, or $y = 80.4 \approx 80.$

For a B: $P(Y < y) = P\left(Z < \dfrac{y - 70}{10}\right) = 0.75$ implies $\dfrac{y - 70}{10} = 0.67,$ or $y \approx 77.$

For a C: $P(Y < y) = P\left(Z < \dfrac{y - 70}{10}\right) = 0.60$ implies $\dfrac{y - 70}{10} = 0.25,$ or $y \approx 73.$

For a D: $P\left(Z \leq \dfrac{y - 70}{10}\right) = 0.2$ implies $\dfrac{y - 70}{10} = -0.84$ or $y \approx 62.$

For a F: $P\left(Z < \dfrac{y - 70}{10}\right) = 0.05$ implies $\dfrac{y - 70}{10} = -1.65,$ or $y \approx 54.$

13. Let X denote the number of heads in 25 tosses of the coin. Then X is a binomial random variable. Also, $n = 25$, $p = 0.4$, and $q = 0.6$. So

$$\mu = (25)(0.4) = 10; \quad \sigma = \sqrt{(25)(0.4)(0.6)} \approx 2.45.$$

Approximating the binomial distribution by a normal distribution with a mean of 10 and a standard deviation of 2.45, we find upon letting Y denote the associated normal random variable,

a. $P(X < 10) \approx P(Y < 9.5) = P\left(Z < \dfrac{9.5 - 10}{2.45}\right) = P(Z < -0.20) = 0.4207.$

b. $P(10 \le X \le 12) \approx P(9.5 < Y < 12.5)$

$$= P\left(\frac{9.5-10}{2.45} < Z < \frac{12.5-10}{2.45}\right) = P(Z < 1.02) - P(Z < -0.20)$$

$$= P(Z < 1.02) - P(Z < -0.20) = 0.8461 - 0.4207 = 0.4254.$$

c. $P(X > 15) \approx P(Y \ge 15)$

$$= P(Z > \frac{15.5-10}{2.45}) = P(Z > 2.24) = P(Z < -2.24) = 0.0125.$$

15. Let X denote the number of times the marksman hits his target. Then X has a binomial distribution with $n = 30$, $p = 0.6$ and $q = 0.4$. Therefore,

$$\mu = (30)(0.6) = 18, \ \sigma = \sqrt{(30)(0.6)(0.4)} = 2.68.$$

a. $P(X \ge 20) \approx P(Y \ge 19.5)$

$$= P\left(Z > \frac{19.5-18}{2.68}\right) = P(Z > 0.56) = P(Z < -0.56) = 0.2877.$$

b. $P(X < 10) \approx P(Y < 9.5) = P\left(Z < \frac{9.5-18}{2.68}\right) = P(Z < -3.17) = 0.0008.$

c. $P(15 \le X \le 20) \approx P(14.5 < Y < 20.5) = P\left(\frac{14.5-18}{2.68} < Z < \frac{20.5-18}{2.68}\right)$

$$= P(Z < 0.93) - P(Z < -1.31) = 0.8238 - 0.0951 = 0.7287.$$

17. Let X denote the number of "seconds." Then X has a binomial distribution with $n = 200$, $p = 0.03$, and $q = 0.97$. Then

$$\mu = (200)(0.03) = 6; \ \sigma = \sqrt{(200)(0.03)(0.97)} \approx 2.41,$$

and $P(X < 10) \approx P(Y < 9.5) = P\left(Z < \frac{9.5-6}{2.41}\right) = P(Z < 1.45) = 0.9265.$

19. Let X denote the number of workers who meet with an accident during a 1-year period. Then $\mu = (800)(0.1) = 80$; $\sigma = \sqrt{(800)(0.1)(0.9)} \approx 8.49$,

and $\qquad P(X > 70) \approx P(Y > 70.5)$

$$= P\left(Z > \frac{70.5-80}{8.49}\right) = P(Z > -1.12) = P(Z < 1.12) = 0.8686.$$

21. a. Let X denote the number of mice that recovered from the disease. Then X has a binomial distribution with $n = 50$, $p = 0.5$, and $q = 0.5$, so

$$\mu = (50)(0.5) = 25; \ \sigma = \sqrt{(50)(0.5)(0.5)} \approx 3.54,$$

Approximating the binomial distribution by a normal distribution with a mean of 25 and a standard deviation of 3.54, we find that the probability that 35 or more of the mice would recover from the disease without benefit of the drug is

$$P(X \geq 35) \approx P(Y > 34.5)$$

$$= P(Z > \frac{34.5 - 25}{3.54}) = P(Z > 2.68) = P(Z < -2.68) = 0.0037.$$

b. The drug is very effective.

23. Let n denote the number of reservations the company should accept. Then we need to find

$$P(X \geq 2000) \approx P(Y > 1999.5) = 0.01$$

or equivalently,

$$P(Z \geq \frac{1999.5 - np}{\sqrt{npq}}) = 0.01 \qquad [\text{Here p = 0.92 and q = 0.08.}]$$

or $\qquad P(Z \leq \frac{np - 1999.5}{\sqrt{npq}}) = 0.01$. Next, $\frac{.92n - 1999.5}{\sqrt{0.0736n}} = -2.33$

$$(0.92n - 1999.5)^2 = (-2.33)^2(0.0736n)$$
$$0.8464n^2 - 3679.08n + 3,998,000.25 = 0.39956704n,$$

or $\qquad 0.8464n^2 - 3679.479567n + 3,998,000.25 = 0.$

Using the quadratic formula, we obtain $\ n = \dfrac{3679.479567 \pm \sqrt{2940.2376}}{1.6928} \approx 2142,$

or 2142. [You can verify that 2206 is not a root of the original equation (before squaring).] Therefore, the company should accept no more than 2142 reservations.

CHAPTER 8, REVIEW EXERCISES, page 497

1. a. $S = \{WWW, WWB, WBW, WBB, BWW, BWB, BBW, BBB\}$

b.

Outcome	WWW	BWW	WBW	WWB	BBW	BWB	WBB	BBB
Value	0	1	1	1	2	2	2	3

c.

x	0	1	2	3
$P(X = x)$	$\frac{1}{35}$	$\frac{12}{35}$	$\frac{18}{35}$	$\frac{4}{35}$

d.

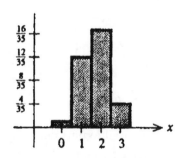

3. a. $P(1 \leq X \leq 4) = 0.1 + 0.2 + 0.3 + 0.2 = 0.8.$
    b. $\mu = 0(0.1) + 1(0.1) + 2(0.2) + 3(0.3) + 4(0.2) + 5(0.1) = 2.7.$
    $V(X) = 0.1(0 - 2.7)^2 + 0.1(1 - 2.7)^2 + 0.2(2 - 2.7)^2 + 0.3(3 - 2.7)^2$
    $$+ 0.2(4 - 2.7)^2 + 0.1(5 - 2.7)^2$$
    $$= 2.01$$
    $$\sigma = \sqrt{2.01} \approx 1.418.$$

5. $P(Z < 0.5) = 0.6915.$

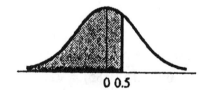

7. $P(-0.75 < Z < 0.5) = P(Z < 0.5) - P(Z < -0.75) = 0.6915 - 0.2266 = 0.4649.$

9. If $P(Z < z) = 0.9922,$ then $z = 2.42.$

11. If $P(Z > z) = 0.9788,$ then $P(Z < -z) = 0.9788,$ and $-z = 2.03,$
    or $z = -2.03.$

13. $P(X < 11) = P\left(Z < \dfrac{11-10}{2}\right) = P(Z < 0.5) = 0.6915.$

15. $P(7 < X < 9) = P\left(\dfrac{7-10}{2} < Z < \dfrac{9-10}{2}\right) = P(-1.5 < Z < -0.5)$

$$= P(Z < -0.5) - P(Z < -1.5) = 0.3085 - 0.0668 = 0.2417.$$

17. This is a binomial experiment with $p = 0.7$, and so $q = 0.3$. The probability that he will get exactly two strikes in four attempts is given by
$$P(X = 2) = C(4,2)(0.7)^2(0.3)^2 \approx 0.2646.$$
The probability that he will get at least two strikes in four attempts is given by
$$P(X = 2) + P(X = 3) + P(X = 4)$$
$$= C(4,2)(0.7)^2(0.3)^2 + C(4,3)(0.7)^3(0.3) + C(4,4)(0.7)^4(0.3)^0$$
$$= 0.2646 + 0.4116 + 0.2401 \approx 0.9163.$$

19. Here $\mu = 64.5$ and $\sigma = 2.5$. Next, $64.5 - 2.5k = 59.5$ and $64.5 + 2.5k = 69.5$ and $k = 2$. Therefore, the required probability is given by
$$P(59.5 \le X \le 69.5) \ge 1 - \dfrac{1}{2^2} = 0.75.$$

21. Let the random variable X be the number of people for whom the drug is effective. Then $\mu = (0.15)(800) = 120$ and $\sigma = \sqrt{(800)(0.15)(0.85)} = \sqrt{102} \approx 10.1$.

23. Here $\mu = (0.6)(100) = 60$ and $\sigma = \sqrt{100(0.6)(0.4)} = 4.899.$
Then $P(X > 50) \approx P(Y > 50.5) = P\left(Z > \dfrac{50.5-60}{4.899}\right) = P(Z > -1.94)$
$$= P(Z < 1.94) = 0.9738.$$

CHAPTER 9

EXERCISES 9.1, page 502

1. $27^{2/3} = (3^3)^{2/3} = 3^2 = 9$.

3. $\left(\dfrac{1}{\sqrt{3}}\right)^0 = 1$. Recall that any number raised to the zero power is 1.

5. $\left[\left(\dfrac{1}{8}\right)^{1/3}\right]^{-2} = \left(\dfrac{1}{2}\right)^{-2} = (2^2) = 4$.

7. $\left(\dfrac{7^{-5} \cdot 7^2}{7^{-2}}\right)^{-1} = (7^{-5+2+2})^{-1} = (7^{-1})^{-1} = 7^1 = 7$.

9. $(125^{2/3})^{-1/2} = 125^{(2/3)(-1/2)} = 125^{-1/3} = \dfrac{1}{125^{1/3}} = \dfrac{1}{5}$.

11. $\dfrac{\sqrt{32}}{\sqrt{8}} = \sqrt{\dfrac{32}{8}} = \sqrt{4} = 2$.   13. $\dfrac{16^{5/8}16^{1/2}}{16^{7/8}} = 16^{(5/8+1/2-7/8)} = 16^{1/4} = 2$.

15. $16^{1/4} \cdot 8^{-1/3} = 2 \cdot \left(\dfrac{1}{8}\right)^{1/3} = 2 \cdot \dfrac{1}{2} = 1$.   17. True.

19. False. $x^3 \times 2x^2 = 2x^{3+2} = 2x^5 \neq 2x^6$.

21. False. $\dfrac{2^{4x}}{1^{3x}} = \dfrac{2^{4x}}{1} = 2^{4x}$.   23. False. $\dfrac{1}{4^{-3}} = 4^3 = 64$.

25. False. $(1.2^{1/2})^{-1/2} = (1.2)^{-1/4} \neq 1$.   27. $(xy)^{-2} = \dfrac{1}{(xy)^2}$.

29. $\dfrac{x^{-1/3}}{x^{1/2}} = x^{(-1/3)-(1/2)} = x^{-5/6} = \dfrac{1}{x^{5/6}}$.

31. $12^0(s+t)^{-3} = 1 \cdot \dfrac{1}{(s+t)^3} = \dfrac{1}{(s+t)^3}$.

33. $\dfrac{x^{7/3}}{x^{-2}} = x^{(7/3)+2} = x^{(7/3)+(6/3)} = x^{13/3}$.

35. $(x^2 y^{-3})(x^{-5} y^3) = (x^{2-5} y^{-3+3}) = x^{-3} y^0 = x^{-3} = \dfrac{1}{x^3}$.

37. $\dfrac{x^{3/4}}{x^{-1/4}} = x^{(3/4)-(-1/4)} = x^{4/4} = x$.

39. $\left(\dfrac{x^3}{-27 y^{-6}}\right)^{-2/3} = x^{3(-2/3)}\left(-\dfrac{1}{27}\right)^{-2/3} y^{6(-2/3)} = x^{-2}\left(-\dfrac{1}{3}\right)^{-2} y^{-4} = \dfrac{9}{x^2 y^4}$.

41. $\left(\dfrac{x^{-3}}{y^{-2}}\right)^2 \left(\dfrac{y}{x}\right)^4 = \dfrac{x^{-3(2)} y^4}{y^{-2(2)} x^4} = \left(\dfrac{y^{4+4}}{x^{4+6}}\right) = \dfrac{y^8}{x^{10}}$.

43. $\sqrt[3]{x^{-2}} \cdot \sqrt{4x^5} = x^{-2/3} \cdot 4^{1/2} \cdot x^{5/2} = x^{-(2/3)+(5/2)} \cdot 2 = 2x^{11/6}$.

45. $-\sqrt[4]{16 x^4 y^8} = -(16^{1/4} \cdot x^{4/4} \cdot y^{8/4}) = -2xy^2$.

47. $\sqrt[6]{64 x^8 y^3} = (64)^{1/6} \cdot x^{8/6} y^{3/6} = 2x^{4/3} y^{1/2}$.

49. $2^{3/2} = (2)(2^{1/2}) = 2(1.414) = 2.828$.

51. $9^{3/4} = (3^2)^{3/4} = 3^{6/4} = 3^{3/2} = 3 \cdot 3^{1/2} = 3(1.732) = 5.196$.

53. $10^{3/2} = 10^{1/2} \cdot 10 = (3.162)(10) = 31.62$.

55. $10^{2.5} = 10^2 \cdot 10^{1/2} = 100(3.162) = 316.2$.

57. $\dfrac{3}{2\sqrt{x}} \cdot \dfrac{\sqrt{x}}{\sqrt{x}} = \dfrac{3\sqrt{x}}{2x}$.

59. $\dfrac{2y}{\sqrt{3y}} \cdot \dfrac{\sqrt{3y}}{\sqrt{3y}} = \dfrac{2y\sqrt{3y}}{3y} = \dfrac{2}{3}\sqrt{3y}$.

61. $\dfrac{1}{\sqrt[3]{x}} \cdot \dfrac{\sqrt[3]{x^2}}{\sqrt[3]{x^2}} = \dfrac{\sqrt[3]{x^2}}{\sqrt[3]{x^3}} = \dfrac{\sqrt[3]{x^2}}{x}$.

63. $\dfrac{2\sqrt{x}}{3} \cdot \dfrac{\sqrt{x}}{\sqrt{x}} = \dfrac{2x}{3\sqrt{x}}$.

65. $\sqrt{\dfrac{2y}{x}} = \dfrac{\sqrt{2y}}{\sqrt{x}} \cdot \dfrac{\sqrt{2y}}{\sqrt{2y}} = \dfrac{2y}{\sqrt{2xy}}$.

67. $\dfrac{\sqrt[3]{x^2z}}{y} \cdot \dfrac{\sqrt[3]{xz^2}}{\sqrt[3]{xz^2}} = \dfrac{\sqrt[3]{x^3z^3}}{y\sqrt[3]{xz^2}} = \dfrac{xz}{y\sqrt[3]{xz^2}}$.

EXERCISES 9.2, page 511

1. $(7x^2 - 2x + 5) + (2x^2 + 5x - 4) = 7x^2 - 2x + 5 + 2x^2 + 5x - 4 = 9x^2 + 3x + 1$.

3. $(5y^2 - 2y + 1) - (y^2 - 3y - 7) = 5y^2 - 2y + 1 - y^2 + 3y + 7 = 4y^2 + y + 8$.

5. $x - \{2x - [-x - (1 - x)]\} = x - \{2x - [-x - 1 + x]\} = x - \{2x + 1\} = x - 2x - 1 = -x - 1$.

7. $(\dfrac{1}{3} - 1 + e) - (-\dfrac{1}{3} - 1 + e^{-1}) = \dfrac{1}{3} - 1 + e + \dfrac{1}{3} + 1 - \dfrac{1}{e} = \dfrac{2}{3} + e - \dfrac{1}{e} = \dfrac{3e^2 + 2e - 3}{3e}$.

9. $3\sqrt{8} + 8 - 2\sqrt{y} + \dfrac{1}{2}\sqrt{x} - \dfrac{3}{4}\sqrt{y} = 3\sqrt{4 \cdot 2} + 8 + \dfrac{1}{2}\sqrt{x} - \dfrac{11}{4}\sqrt{y}$

$$= 6\sqrt{2} + 8 + \dfrac{1}{2}\sqrt{x} - \dfrac{11}{4}\sqrt{y}.$$

11. $(x + 8)(x - 2) = x(x - 2) + 8(x - 2) = x^2 - 2x + 8x - 16 = x^2 + 6x - 16$.

13. $(a + 5)^2 = (a + 5)(a + 5) = a(a + 5) + 5(a + 5) = a^2 + 5a + 5a + 25$
    $= a^2 + 10a + 25$.

15. $(x + 2y)^2 = (x + 2y)(x + 2y) = x(x + 2y) + 2y(x + 2y)$
    $= x^2 + 2xy + 2yx + 4y^2 = x^2 + 4xy + 4y^2$.

17. $(2x + y)(2x - y) = 2x(2x - y) + y(2x - y) = 4x^2 - 2xy + 2xy - y^2 = 4x^2 - y^2$.

19. $(x^2 - 1)(2x) - x^2(2x) = 2x^3 - 2x - 2x^3 = -2x.$

21. $2(t + \sqrt{t})^2 - 2t^2 = 2(t + \sqrt{t})(t + \sqrt{t}) - 2t^2 = 2(t^2 + 2t\sqrt{t} + t) - 2t^2$
$$= 2t^2 + 4t\sqrt{t} + 2t - 2t^2 = 4t\sqrt{t} + 2t = 2t(2\sqrt{t} + 1).$$

23. $4x^5 - 12x^4 - 6x^3 = 2x^3(2x^2 - 6x - 3).$

25. $7a^4 - 42a^2b^2 + 49a^3b = 7a^2(a^2 - 6b^2 + 7ab).$

27. $e^{-x} - xe^{-x} = e^{-x}(1 - x).$

29. $2x^{-5/2} - \frac{3}{2}x^{-3/2} = \frac{1}{2}x^{-5/2}(4 - 3x).$

31. $6ac + 3bc - 4ad - 2bd = 3c(2a + b) - 2d(2a + b) = (2a + b)(3c - 2d).$

33. $4a^2 - b^2 = (2a + b)(2a - b).$ (Difference of two squares)

35. $10 - 14x - 12x^2 = -2(6x^2 + 7x - 5) = -2(3x + 5)(2x - 1).$

37. $3x^2 - 6x - 24 = 3(x^2 - 2x - 8) = 3(x - 4)(x + 2).$

39. $12x^2 - 2x - 30 = 2(6x^2 - x - 15) = 2(3x - 5)(2x + 3).$

41. $9x^2 - 16y^2 = (3x)^2 - (4y)^2 = (3x - 4y)(3x + 4y).$

43. $x^6 + 125 = (x^2)^3 + (5)^3 = (x^2 + 5)(x^4 - 5x^2 + 25).$

45. $(x^2 + y^2)x - xy(2y) = x^3 + xy^2 - 2xy^2 = x^3 - xy^2.$

47. $2(x - 1)(2x + 2)^3[4(x - 1) + (2x + 2)]$
$$= 2(x - 1)(2x + 2)^3[4x - 4 + 2x + 2]$$
$$= 2(x - 1)(2x + 2)^3[6x - 2]$$
$$= 4(x - 1)(3x - 1)(2x + 2)^3.$$

49. $4(x - 1)^2(2x + 2)^3(2) + (2x + 2)^4(2)(x - 1)$
$$= 2(x - 1)(2x + 2)^3[4(x - 1) + (2x + 2)] = 2(x - 1)(2x + 2)^3(6x - 2)$$
$$= 4(x - 1)(3x - 1)(2x + 2)^3.$$

51. $(x^2 + 2)^2[5(x^2 + 2)^2 - 3](2x) = (x^2 + 2)^2[5(x^4 + 4x^2 + 4) - 3](2x)$
$$= (2x)(x^2 + 2)^2(5x^4 + 20x^2 + 17).$$

53. $x^2 + x - 12 = 0$, or $(x + 4)(x - 3) = 0$, so that $x = -4$ or $x = 3$. We conclude that the roots are $x = -4$ and $x = 3$.

55. $4t^2 + 2t - 2 = (2t - 1)(2t + 2) = 0$. Thus, $t = 1/2$ and $t = -1$ are the roots.

57. $\frac{1}{4}x^2 - x + 1 = (\frac{1}{2}x - 1)(\frac{1}{2}x - 1) = 0$. Thus $\frac{1}{2}x = 1$, and $x = 2$ is a double root of the equation.

59. Here we use the quadratic formula to solve the equation $4x^2 + 5x - 6 = 0$. Then, $a = 4$, $b = 5$, and $c = -6$. Therefore,
$$x = \frac{-b \pm \sqrt{b^2 - 4ac}}{2a} = \frac{-(5) \pm \sqrt{(5)^2 - 4(4)(-6)}}{2(4)} = \frac{-5 \pm \sqrt{121}}{8} = \frac{-5 \pm 11}{8}.$$
Thus, $x = -\frac{16}{8} = -2$ and $x = \frac{6}{8} = \frac{3}{4}$ are the roots of the equation.

61. We use the quadratic formula to solve the equation $8x^2 - 8x - 3 = 0$. Here $a = 8$, $b = -8$, and $c = -3$. Therefore,
$$x = \frac{-b \pm \sqrt{b^2 - 4ac}}{2a} = \frac{-(-8) \pm \sqrt{(-8)^2 - 4(8)(-3)}}{2(8)} = \frac{8 \pm \sqrt{160}}{16}$$
$$= \frac{8 \pm 4\sqrt{10}}{16} = \frac{2 \pm \sqrt{10}}{4}.$$
Thus, $x = \frac{1}{2} + \frac{1}{4}\sqrt{10}$ and $x = \frac{1}{2} - \frac{1}{4}\sqrt{10}$ are the roots of the equation.

63. We use the quadratic formula to solve $2x^2 + 4x - 3 = 0$. Here, $a = 2$, $b = 4$, and $c = -3$. Therefore
$$x = \frac{-b \pm \sqrt{b^2 - 4ac}}{2a} = \frac{-(4) \pm \sqrt{(4)^2 - 4(2)(-3)}}{2(2)} = \frac{-4 \pm \sqrt{40}}{4}$$
$$= \frac{-4 \pm 2\sqrt{10}}{4} = \frac{-2 \pm \sqrt{10}}{2}.$$
Thus, $x = -1 + \frac{1}{2}\sqrt{10}$ and $x = -1 - \frac{1}{2}\sqrt{10}$ are the roots of the equation.

EXERCISES 9.3, page 518

1. $\dfrac{x^2+x-2}{x^2-4} = \dfrac{(x+2)(x-1)}{(x+2)(x-2)} = \dfrac{x-1}{x-2}.$

3. $\dfrac{12t^2+12t+3}{4t^2-1} = \dfrac{3(4t^2+4t+1)}{4t^2-1} = \dfrac{3(2t+1)(2t+1)}{(2t+1)(2t-1)} = \dfrac{3(2t+1)}{2t-1}.$

5. $\dfrac{(4x-1)(3)-(3x+1)(4)}{(4x-1)^2} = \dfrac{12x-3-12x-4}{(4x-1)^2} = -\dfrac{7}{(4x-1)^2}.$

7. $\dfrac{(2x+3)(1)-(x+1)(2)}{(2x+3)^2} = \dfrac{2x+3-2x-2}{(2x+3)^2} = \dfrac{1}{(2x+3)^2}.$

9. $\dfrac{(e^x+1)e^x-e^x(2)(e^x+1)(e^x)}{(e^x+1)^4} = \dfrac{e^x(e^x+1)(1-2e^x)}{(e^x+1)^4} = \dfrac{e^x(1-2e^x)}{(e^x+1)^3}.$

11. $\dfrac{2a^2-2b^2}{b-a} \cdot \dfrac{4a+4b}{a^2+2ab+b^2} = \dfrac{2(a+b)(a-b)4(a+b)}{-(a-b)(a+b)(a+b)} = -8.$

13. $\dfrac{3x^2+2x-1}{2x+6} \div \dfrac{x^2-1}{x^2+2x-3} = \dfrac{(3x-1)(x+1)}{2(x+3)} \cdot \dfrac{(x+3)(x-1)}{(x+1)(x-1)} = \dfrac{3x-1}{2}.$

15. $\dfrac{58}{3(3t+2)} + \dfrac{1}{3} = \dfrac{58+3t+2}{3(3t+2)} = \dfrac{3t+60}{3(3t+2)} = \dfrac{t+20}{3t+2}.$

17. $\dfrac{2x}{2x-1} - \dfrac{3x}{2x+5} = \dfrac{2x(2x+5)-3x(2x-1)}{(2x-1)(2x+5)} = \dfrac{4x^2+10x-6x^2+3x}{(2x-1)(2x+5)}$

$= \dfrac{-2x^2+13x}{(2x-1)(2x+5)} = -\dfrac{x(2x-13)}{(2x-1)(2x+5)}.$

19. $(x+\dfrac{1}{x})(x^2-1) = x^3+x-x-\dfrac{1}{x} = \dfrac{x^4-1}{x}.$

21. $\dfrac{4}{x^2-9}-\dfrac{5}{x^2-6x+9}=\dfrac{4}{(x+3)(x-3)}-\dfrac{5}{(x-3)^2}$

$$=\dfrac{4(x-3)-5(x+3)}{(x-3)^2(x+3)}=-\dfrac{x+27}{(x-3)^2(x+3)}.$$

23. $2+\dfrac{1}{a+2}-\dfrac{2a}{a-2}=\dfrac{2(a+2)(a-2)+a-2-2a(a+2)}{(a+2)(a-2)}$

$$=\dfrac{2a^2-8+a-2-2a^2-4a}{(a+2)(a-2)}=-\dfrac{3a+10}{(a+2)(a-2)}.$$

25. $\dfrac{1+\dfrac{1}{x}}{1-\dfrac{1}{x}}=\dfrac{\dfrac{x+1}{x}}{\dfrac{x-1}{x}}=\dfrac{x+1}{x}\cdot\dfrac{x}{x-1}=\dfrac{x+1}{x-1}.$

27. $\dfrac{x^{-2}-y^{-2}}{x+y}=\dfrac{\dfrac{1}{x^2}-\dfrac{1}{y^2}}{x+y}=\dfrac{\dfrac{y^2-x^2}{x^2y^2}}{x+y}=\dfrac{(y+x)(y-x)}{x^2y^2}\cdot\dfrac{1}{x+y}=\dfrac{y-x}{x^2y^2}.$

29. $\dfrac{4x^2}{2\sqrt{2x^2+7}}+\sqrt{2x^2+7}=\dfrac{4x^2+2\sqrt{2x^2+7}\sqrt{2x^2+7}}{2\sqrt{2x^2+7}}=\dfrac{4x^2+4x^2+14}{2\sqrt{2x^2+7}}$

$$=\dfrac{4x^2+7}{\sqrt{2x^2+7}}.$$

31. $\dfrac{2x(x+1)^{-1/2}-(x+1)^{1/2}}{x^2}=\dfrac{(x+1)^{-1/2}(2x-x-1)}{x^2}=\dfrac{(x+1)^{-1/2}(x-1)}{x^2}$

$$=\dfrac{x-1}{x^2\sqrt{x+1}}.$$

33. $\dfrac{(2x+1)^{1/2}-(x+2)(2x+1)^{-1/2}}{2x+1}=\dfrac{(2x+1)^{-1/2}(2x+1-x-2)}{2x+1}$

$$= \frac{(2x+1)^{-1/2}(x-1)}{2x+1} = \frac{x-1}{(2x+1)^{3/2}}.$$

35. (b) is equivalent to $\dfrac{a}{1+\dfrac{b}{x}} = \dfrac{a}{\dfrac{x+b}{x}} = \dfrac{ax}{x+b}$ which is (a).

37. (b) is equivalent to $\dfrac{x+\dfrac{1}{x^3}}{1-\dfrac{1}{x^3}} = \dfrac{x^4+1}{x^3} \cdot \dfrac{x^3}{x^3-1} = \dfrac{x^4+1}{x^3-1}$ which is (a).

39. $\dfrac{1}{\sqrt{3}-1} \cdot \dfrac{\sqrt{3}+1}{\sqrt{3}+1} = \dfrac{\sqrt{3}+1}{3-1} = \dfrac{\sqrt{3}+1}{2}.$

41. $\dfrac{1}{\sqrt{x}-\sqrt{y}} \cdot \dfrac{\sqrt{x}+\sqrt{y}}{\sqrt{x}+\sqrt{y}} = \dfrac{\sqrt{x}+\sqrt{y}}{x-y}.$

43. $\dfrac{\sqrt{a}+\sqrt{b}}{\sqrt{a}-\sqrt{b}} \cdot \dfrac{\sqrt{a}+\sqrt{b}}{\sqrt{a}+\sqrt{b}} = \dfrac{(\sqrt{a}+\sqrt{b})^2}{a-b}.$

45. $\dfrac{\sqrt{x}}{3} \cdot \dfrac{\sqrt{x}}{\sqrt{x}} = \dfrac{x}{3\sqrt{x}}.$

47. $\dfrac{1-\sqrt{3}}{3} \cdot \dfrac{1+\sqrt{3}}{1+\sqrt{3}} = \dfrac{1^2-(\sqrt{3})^2}{3(1+\sqrt{3})} = -\dfrac{2}{3(1+\sqrt{3})}.$

49. $\dfrac{1+\sqrt{x+2}}{\sqrt{x+2}} \cdot \dfrac{1-\sqrt{x+2}}{1-\sqrt{x+2}} = \dfrac{1-(x+2)}{\sqrt{x+2}(1-\sqrt{x+2})} = -\dfrac{x+1}{\sqrt{x+2}(1-\sqrt{x+2})}.$

EXERCISES 9.4, page 524

1. The statement is false because -3 is greater than -20. (See the number line that follows).

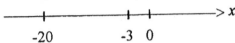

3. The statement is false because 2/3 [which is equal to (4/6)] is less than 5/6.

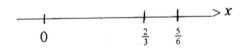

5. The interval (3,6) is shown on the number line that follows. Note that this is an open interval indicated by (and)

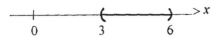

7. The interval [-1,4) is shown on the number line that follows. Note that this is a half-open interval indicated by [(closed) and) (open).

9. The infinite interval (0,∞) is shown on the number line that follows.

11. First, $2x + 4 < 8$ (Add -4 to each side of the inequality.)
    Next, $2x < 4$ (Multiply each side of the inequality by 1/2)
    and $x < 2$.

13. We are given the inequality $-4x \geq 20$.
    Then $x \leq -5$.   (Multiply both sides of the inequality by -1/4
                                        and reverse the sign of the inequality.)
    We write this in interval notation as $(-\infty, -5]$.

15. We are given the inequality $-6 < x - 2 < 4$.
    First $-6 + 2 < x < 4 + 2$ (Add +2 to each member of the inequality.)
    and $-4 < x < 6$,
    so the solution set is the open interval $(-4,6)$.

17. We want to find the values of x that satisfy the inequalities $x + 1 > 4$ or $x + 2 < -1$.
    Adding -1 to both sides of the first inequality, we obtain $x + 1 - 1 > 4 - 1$, or $x > 3$. Similarly, adding -2 to both sides of the second inequality, we obtain $x + 2 - 2 < -1 - 2$, or $x < -3$. Therefore, the solution set is $(-\infty, -3) \cup (3, \infty)$.

19. We want to find the values of x that satisfy the inequalities
                $x + 3 > 1$ and $x - 2 < 1$.
    Adding -3 to both sides of the first inequality, we obtain
                $x + 3 - 3 > 1 - 3$, or $x > -2$.

277

Similarly, adding 2 to each side of the second inequality, we obtain $x < 3$, and the solution set is (-2,3).

21. We want to find the values of x that satisfy the inequalities $(x + 3)(x - 5) \leq 0$. From the sign diagram

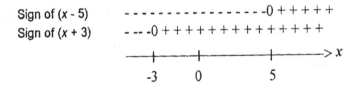

we see that the given inequality is satisfied when $-3 \leq x \leq 5$, that is, when the signs of the two factors are different or when one of the factors is equal to zero.

23. We want to find the values of x that satisfy the inequalities $(2x - 3)(x - 1) \geq 0$. From the sign diagram

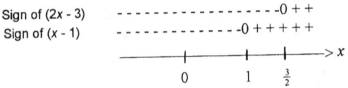

we see that the given inequality is satisfied when $x \leq 1$ or $x \geq \frac{3}{2}$; that is, when the signs of both factors are the same, or one of the factors is equal to zero.

25. We want to find the values of x that satisfy the inequalities $\dfrac{x+3}{x-2} \geq 0$. From the sign diagram we see that the given inequality is satisfied when $x \leq -3$ or $x > 2$, that is,

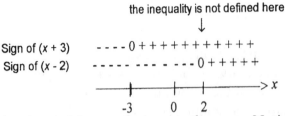

when the signs of the two factors are the same. Notice that $x = 2$ is not included because the inequality is not defined at that value of x.

27. We want to find the values of x that satisfy the inequalities $\dfrac{x-2}{x-1} \leq 2$. Subtracting

from each side of the given inequality gives
$$\frac{x-2}{x-1}-2\le 0,\quad \frac{x-2-2(x-1)}{x-1}\le 0,\text{ or }\quad -\frac{x}{x-1}\le 0.$$
From the sign diagram

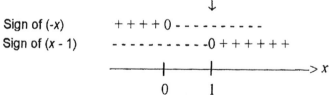

the inequality is not defined here

Sign of (-x)

Sign of (x - 1)

we see that the given inequality is satisfied when $x\le 0$ or $x>1$; that is, when the signs of the two factors differ. Notice that $x=1$ is not included because the inequality is undefined at that value of x.

29. $\left|-6+2\right|=4.$

31. $\dfrac{\left|-12+4\right|}{\left|16-12\right|}=\dfrac{\left|-8\right|}{4}=2.$

33. $\sqrt{3}\left|-2\right|+3\left|-\sqrt{3}\right|=\sqrt{3}(2)+3\sqrt{3}=5\sqrt{3}.$ 35. $\left|\pi-1\right|+2=\pi-1+2=\pi+1.$

37. $\left|\sqrt{2}-1\right|+\left|3-\sqrt{2}\right|=\sqrt{2}-1+3-\sqrt{2}=2.$

39. False. If $a>b$, then $-a<-b$, $-a+b<-b+b$, and $b-a<0$.

41. False. Let $a=-2$ and $b=-3$. Then $a^2=4$ and $b^2=9$, and $4<9$. Note that we only need to provide a counterexample to show that the statement is not always true.

43. True. There are three possible cases.
   Case 1 If $a>0$, $b>0$, then $a^3>b^3$, since $a^3-b^3=(a-b)(a^2+ab+b^2)>0.$
   Case 2 If $a>0$, $b<0$, then $a^3>0$ and $b^3<0$ and it follows that $a^3>b^3.$
   Case 3 If $a<0$ and $b<0$, then $a^3-b^3=(a-b)(a^2+ab+b^2)>0$, and we see that $a^3>b^3.$ (Note that $(a-b)>0$ and $ab>0.$)

45. False. Take $a=-2$, then $\left|-a\right|=\left|-(-2)\right|=\left|2\right|=2\ne a.$

47. True. If $a-4<0$, then $\left|a-4\right|=4-a=\left|4-a\right|.$ If $a-4>0$, then
   $$\left|4-a\right|=a-4=\left|a-4\right|.$$

49. False. Take $a = 3$, $b = -1$. Then $|a + b| = |3 - 1| = 2 \neq |a| + |b| = 3 + 1 = 4$.

51. If the car is driven in the city, then it can be expected to cover
$$(18.1)(20) = 362 \qquad \text{(miles/gal} \cdot \text{gal)}$$
or 362 miles on a full tank. If the car is driven on the highway, then it can be expected to cover
$$(18.1)(27) = 488.7 \qquad \text{(miles/gal} \cdot \text{gal)}$$
or 488.7 miles on a full tank. Thus, the driving range of the car may be described by the interval [362, 488.7].

53. $6(P - 2500) \leq 4(P + 2400)$
$6P - 15000 \leq 4P + 9600$
$2P \leq 24600$, or $P \leq 12300$.
Therefore, the maximum profit is $12,300.

55. Let x represent the salesman's monthly sales in dollars. Then
$$0.15(x - 12000) \geq 3000$$
$$15(x - 12000) \geq 300000$$
$$15x - 180000 \geq 300000$$
$$15x \geq 480000$$
$$x \geq 32,000.$$
We conclude that the salesman must attain sales of at least $32,000 to reach his goal.

57. The rod is acceptable if $0.49 < 0.51$ or $-0.01 < x - 0.5 < 0.01$. This gives the required inequality $|x - 0.5| < 0.01$.

59. We want to solve the inequality
$$-6x^2 + 30x - 10 \geq 14. \qquad \text{(Remember } x \text{ is expressed in thousands.)}$$
Adding -14 to both sides of this inequality, we have
$$-6x^2 + 30x - 10 - 14 \geq 14 - 14,$$
or $\qquad - 6x^2 + 30x - 24 \geq 0$.
Dividing both sides of the inequality by -6 (which reverses the sign of the inequality), we have $x^2 - 5x + 4 \leq 0$. Factoring this last expression, we have
$$(x - 4)(x - 1) \leq 0.$$
From the following sign diagram,

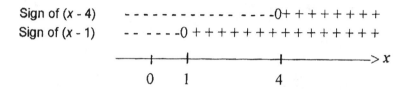

we see that x must lie between 1 and 4. (The inequality is only satisfied when the two factors have opposite signs.) Since x is expressed in thousands of units, we see that the manufacturer must produce between 1000 and 4000 units of the commodity.

CHAPTER 9 REVIEW EXERCISES, page 526

1. $\left(\dfrac{9}{4}\right)^{3/2} = \dfrac{9^{3/2}}{4^{3/2}} = \dfrac{27}{8}.$

3. $(3 \cdot 4)^{-2} = 12^{-2} = \dfrac{1}{12^2} = \dfrac{1}{144}.$

5. $\dfrac{(3 \cdot 2^{-3})(4 \cdot 3^5)}{2 \cdot 9^3} = \dfrac{3 \cdot 2^{-3} \cdot 2^2 \cdot 3^5}{2 \cdot (3^2)^3} = \dfrac{2^{-1} \cdot 3^6}{2 \cdot 3^6} = \dfrac{1}{4}.$

7. $\dfrac{4(x^2 + y)^3}{x^2 + y} = 4(x^2 + y)^2.$

9. $\dfrac{\sqrt[4]{16x^5 yz}}{\sqrt[4]{81xyz^5}} = \dfrac{(2^4 x^5 yz)^{1/4}}{(3^4 xyz^5)^{1/4}} = \dfrac{2x^{5/4} y^{1/4} z^{1/4}}{3x^{1/4} y^{1/4} z^{5/4}} = \dfrac{2x}{3z}.$

11. $\left(\dfrac{3xy^2}{4x^3 y}\right)^{-2} \left(\dfrac{3xy^3}{2x^2}\right)^3 = \left(\dfrac{3y}{4x^2}\right)^{-2} \left(\dfrac{3y^3}{2x}\right)^3 = \left(\dfrac{4x^2}{3y}\right)^2 \left(\dfrac{3y^3}{2x}\right)^3 = \dfrac{(16x^4)(27y^9)}{(9y^2)(8x^3)} = 6xy^7.$

13. $\sqrt[3]{81x^5 y^{10}} \cdot \sqrt[3]{9xy^2} = 3^{4/3} x^{5/3} y^{10/3} \cdot 3^{2/3} x^{1/3} y^{2/3} = 3^2 x^2 y^4 = 9x^2 y^4.$

15. $-2\pi^2 r^3 + 100\pi r^2 = -2\pi r^2(\pi r - 50).$

17. $16 - x^2 = 4^2 - x^2 = (4 - x)(4 + x).$

19. $-2x^2 - 4x + 6 = (-2x - 6)(x - 1) = -2(x + 3)(x - 1).$

21. $\dfrac{(t+6)(60)-(60t+180)}{(t+6)^2} = \dfrac{60t+360-60t-180}{(t+6)^2} = \dfrac{180}{(t+6)^2}.$

23. $\dfrac{2}{3}\left(\dfrac{4x}{2x^2-1}\right)+3\left(\dfrac{3}{3x-1}\right) = \dfrac{8x}{3(2x^2-1)}+\dfrac{9}{3x-1} = \dfrac{8x(3x-1)+27(2x^2-1)}{3(2x^2-1)(3x-1)}$

$= \dfrac{78x^2-8x-27}{3(2x^2-1)(3x-1)}.$

25. $8x^2 + 2x - 3 = (4x + 3)(2x - 1) = 0$ and $x = -3/4$ and $x = 1/2$ are the roots of the equation.

27. $-x^3 - 2x^2 + 3x = -x(x^2 + 2x - 3) = -x(x + 3)(x - 1) = 0$ and the roots of the equation are $x = 0$, $x = -3$, and $x = 1$.

29. Adding x to both sides yields $3 \le 3x + 9$ or $3x \ge -6$, $\Rightarrow x \ge -2$. We conclude that the solution set is $[-2,\infty)$.

31. The inequalities imply $x > 5$ or $x < -4$. So the solution is $(-\infty,-4) \cup (5,\infty)$.

33. $|-5+7|+|-2| = |2|+|-2| = 2+2 = 4.$

35. $|2\pi - 6| - \pi = 2\pi - 6 - \pi = \pi - 6.$

37. Factoring the given expression, we have $(2x - 1)(x + 2) \le 0$. From the sign diagram we conclude that the given inequality is satisfied when $-2 \le x \le \tfrac{1}{2}$.

Sign of (2x - 1) - - - - - - - - -- - - - - -- 0 + + + + +
Sign of (x + 2)   - - - - - - -0 + ++ + + + + + + + + +

$\qquad\qquad\qquad$ -2 $\qquad$ 0 $\qquad$ $\tfrac{1}{2}$

39. The given inequality is equivalent to $|2x - 3| < 5$ or $-5 < 2x - 3 < 5$.
Thus, $-2 < 2x < 8$, or $-1 < x < 4$.

41. $\dfrac{\sqrt{x}-1}{x-1} = \dfrac{\sqrt{x}-1}{x-1}\cdot\dfrac{\sqrt{x}+1}{\sqrt{x}+1} = \dfrac{(\sqrt{x})^2-1}{(x-1)(\sqrt{x}+1)} = \dfrac{x-1}{(x-1)(\sqrt{x}+1)} = \dfrac{1}{\sqrt{x}+1}.$

43. Here we use the quadratic formula to solve the equation $x^2 - 2x - 5$. Then $a = 1$, $b = -2$, and $c = -5$. Thus,

$$x = \frac{-b \pm \sqrt{b^2 - 4ac}}{2a} = \frac{-(-2) \pm \sqrt{(-2)^2 - 4(1)(-5)}}{2(1)} = \frac{2 \pm \sqrt{24}}{2} = 1 \pm \sqrt{6}.$$

45. $2(1.5C + 80) \leq 2(2.5C - 20) \Rightarrow 1.5C + 80 \leq 2.5C - 20$, so $C \geq 100$ and the minimum cost is $100.

CHAPTER 10

EXERCISES 10.1, page 537

1. $f(x) = 5x + 6$. Therefore

   $f(3) = 5(3) + 6 = 21$

   $f(-3) = 5(-3) + 6 = -9$

   $f(a) = 5(a) + 6 = 5a + 6$

   $f(-a) = 5(-a) + 6 = -5a + 6$

   $f(a + 3) = 5(a + 3) + 6 = 5a + 15 + 6 = 5a + 21.$

3. $g(x) = 3x^2 - 6x - 3;$ $g(0) = 3(0) - 6(0) - 3 = -3$

   $g(-1) = 3(-1)^2 - 6(-1) - 3 = 3 + 6 - 3 = 6$

   $g(a) = 3(a)^2 - 6(a) - 3 = 3a^2 - 6a - 3$

   $g(-a) = 3(-a)^2 - 6(-a) - 3 = 3a^2 + 6a - 3$

   $g(x + 1) = 3(x + 1)^2 - 6(x + 1) - 3 = 3(x^2 + 2x + 1) - 6x - 6 - 3$

   $\qquad = 3x^2 + 6x + 3 - 6x - 9 = 3x^2 - 6.$

5. $f(x) = 2x + 5.$ $f(a + h) = 2(a + h) + 5 = 2a + 2h + 5$

   $f(-a) = 2(-a) + 5 = -2a + 5$

   $f(a^2) = 2(a^2) + 5 = 2a^2 + 5$

   $f(a - 2h) = 2(a - 2h) + 5 = 2a - 4h + 5$

   $f(2a - h) = 2(2a - h) + 5 = 4a - 2h + 5$

7. $s(t) = \dfrac{2t}{t^2 - 1}.$ Therefore,

   $s(4) = \dfrac{2(4)}{(4)^2 - 1} = \dfrac{8}{15}.$ $s(0) = \dfrac{2(0)}{0^2 - 1} = 0$

$$s(a) = \frac{2(a)}{a^2 - 1} = \frac{2a}{a^2 - 1}$$

$$s(2+a) = \frac{2(2+a)}{(2+a)^2 - 1} = \frac{2(2+a)}{a^2 + 4a + 4 - 1} = \frac{2(2+a)}{a^2 + 4a + 3}$$

$$s(t+1) = \frac{2(t+1)}{(t+1)^2 - 1} = \frac{2(t+1)}{t^2 + 2t + 1 - 1} = \frac{2(t+1)}{t(t+2)}.$$

9. $f(t) = \dfrac{2t^2}{\sqrt{t-1}}$. Therefore, $f(2) = \dfrac{2(2^2)}{\sqrt{2-1}} = 8$

$$f(a) = \frac{2a^2}{\sqrt{a-1}}; \qquad f(x+1) = \frac{2(x+1)^2}{\sqrt{(x+1)-1}} = \frac{2(x+1)^2}{\sqrt{x}}$$

$$f(x-1) = \frac{2(x-1)^2}{\sqrt{(x-1)-1}} = \frac{2(x-1)^2}{\sqrt{x-2}}.$$

11. Since $x = -2 \le 0$, we see that $f(-2) = (-2)^2 + 1 = 4 + 1 = 5$

Since $x = 0 \le 0$, we see that $f(0) = (0)^2 + 1 = 1$

Since $x = 1 > 0$, we see that $f(1) = \sqrt{1} = 1.$

13. Since $x = -1 < 1$, $f(-1) = -\frac{1}{2}(-1)^2 + 3 = \frac{5}{2}.$

Since $x = 0 < 1$, $f(0) = -\frac{1}{2}(0)^2 + 3 = 3.$

Since $x = 1 \ge 1$, $f(1) = 2(1^2) + 1 = 3.$

Since $x = 2 \ge 1, f(2) = 2(2^2) + 1 = 9.$

15. a. $f(0) = -2;$
b. (i) $f(x) = 3$ when $x \approx 2$ (ii) $f(x) = 0$ when $x = 1$
c. $[0,6]$ d. $[-2, 6]$

17. $g(2) = \sqrt{2^2 - 1} = \sqrt{3}$ and the point $(2, \sqrt{3})$ lies on the graph of g.

19. $f(-2) = \dfrac{|-2 - 1|}{-2 + 1} = \dfrac{|-3|}{-1} = -3$ and the point $(-2, -3)$ does lie on the graph of f.

21. Since $f(x)$ is a real number for any value of x, the domain of f is $(-\infty, \infty)$.

23. $f(x)$ is not defined at $x = 0$ and so the domain of f is $(-\infty, 0) \cup (0, \infty)$.

25. $f(x)$ is a real number for all values of x. Note that $x^2 + 1 \geq 1$ for all x. Therefore, the domain of f is $(-\infty, \infty)$.

27. Since the square root of a number is defined for all real numbers greater than or equal to zero, we have
$$5 - x \geq 0, \text{ or } \quad -x \geq -5$$
and $\quad x \leq 5$. (Recall that multiplying by -1 reverses the sign of an inequality.) Therefore, the domain of g is $(-\infty, 5]$.

29. The denominator of f is zero when $x^2 - 1 = 0$ or $x = \pm 1$.
Therefore, the domain of f is $(-\infty, -1) \cup (-1, 1) \cup (1, \infty)$.

31. f is defined when $x + 3 \geq 0$, that is, when $x \geq -3$. Therefore, the domain of f is $[-3, \infty)$.

33. The numerator is defined when $1 - x \geq 0$, $\quad -x \geq -1 \quad$ or $\quad x \leq 1$.
Furthermore, the denominator is zero when $x = \pm 2$. Therefore, the domain is the set of all real numbers in $(-\infty, -2) \cup (-2, 1]$.

35. a. The domain of f is the set of all real numbers.
   b. $f(x) = x^2 - x - 6$. Therefore,
   $f(-3) = (-3)^2 - (-3) - 6 = 9 + 3 - 6 = 6; \quad f(-2) = (-2)^2 - (-2) - 6 = 4 + 2 - 6 = 0.$
   $f(-1) = (-1)^2 - (-1) - 6 = 1 + 1 - 6 = -4; \quad f(0) = (0)^2 - (0) - 6 = -6.$
   $f\left(\tfrac{1}{2}\right) = \left(\tfrac{1}{2}\right)^2 - \left(\tfrac{1}{2}\right) - 6 = \tfrac{1}{4} - \tfrac{2}{4} - \tfrac{24}{4} = -\tfrac{25}{4}; \quad f(1) = (1)^2 - 1 - 6 = -6.$

$$f(2) = (2)^2 - 2 - 6 = 4 - 2 - 6 = -4; \quad f(3) = (3)^2 - 3 - 6 = 9 - 3 - 6 = 0.$$

c.

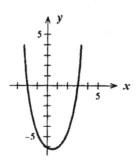

37. $f(x) = 2x^2 + 1$

x	-3	-2	-1	0	1	2	3
f(x)	19	9	3	1	3	9	19

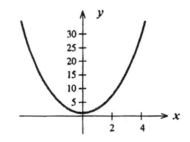

$(-\infty, \infty); \quad [1, \infty)$

39. $f(x) = 2 + \sqrt{x}$

x	0	1	2	4	9	16
f(x)	2	3	3.41	4	5	6

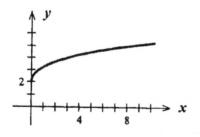

$[0, \infty); \quad [2, \infty)$

41. $f(x) = \sqrt{1-x}$

x	0	-1	-3	-8	-15
$f(x)$	1	1.4	2	3	4

$(-\infty, 1]; [0, \infty)$

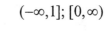

43. $f(x) = |x| - 1$

x	-3	-2	-1	0	1	2	3
$f(x)$	2	1	0	-1	0	1	2

$(-\infty, \infty); [-1, \infty)$

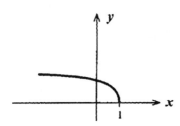

45. $f(x) = \begin{cases} x & \text{if } x < 0 \\ 2x+1 & \text{if } x \ge 0 \end{cases}$

x	-3	-2	-1	0	1	2	3
$f(x)$	-3	-2	-1	1	3	5	7

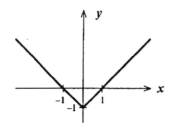

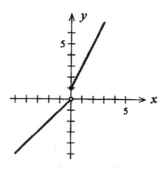

$(-\infty,\infty)$; $(-\infty,0)\cup[1,\infty)$

47. If $x \le 1$, the graph of f is the half-line $y = -x + 1$. For $x > 1$, use the table

x	2	3	4
$f(x)$	3	8	15

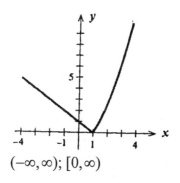

$(-\infty,\infty)$; $[0,\infty)$

49. Each vertical line cuts the given graph at exactly one point, and so the graph represents y as a function of x.

51. Since there is a vertical line that intersects the graph at three points, the graph does not represent y as a function of x.

53. Each vertical line intersects the graph of f at exactly one point, and so the graph represents y as a function of x.

55. Each vertical line intersects the graph of f at exactly one point, and so the graph represents y as a function of x.

57. The circumference of a circle with a 5-inch radius is given by
$$C(5) = 2\pi(5) = 10\pi, \text{ or } 10\pi \text{ inches.}$$

59. $\dfrac{4}{3}(\pi)(2r)^3 = \dfrac{4}{3}\pi 8r^3 = 8(\dfrac{4}{3}\pi r^3)$. Therefore, the volume of the tumor is increased by a factor of 8.

61. a. From $t = 0$ to $t = 5$, the graph for cassettes lies above that for CDs so from 1985 to 1990, sales of prerecorded cassettes were greater than that of CDs.
b. Sales of prerecorded CDs were greater than that of prerecorded cassettes from 1990 on.
c. The graphs intersect at the point with coordinates $x = 5$ and $y \approx 3.5$, and this tells us that the sales of the two formats were the same in 1990 with the level of sales at approximately $3.5 billion.

63. a. The slope of the straight line passing through the points (0, 0.58) and (20, 0.95) is
$$m = \dfrac{0.95 - 0.58}{20 - 0} = 0.0185,$$
and so an equation of the straight line passing through these two points is
$$y - 0.58 = 0.0185(t - 0) \text{ or } y = 0.0185t + 0.58$$
Next, the slope of the straight line passing through the points (20, 0.95) and (30, 1.1) is
$$m = \dfrac{1.1 - 0.95}{30 - 20} = 0.015$$
and so an equation of the straight line passing through the two points is
$$y - 0.95 = 0.015(t - 20) \text{ or } y = 0.015t + 0.65.$$
Therefore, the rule for f is
$$f(t) = \begin{cases} 0.0185t + 0.58 & 0 \le t \le 20 \\ 0.015t + 0.65 & 20 < t \le 30 \end{cases}$$
b. The ratios were changing at the rates of 0.0185/yr and 0.015/yr from 1960 through 1980, and from 1980 through 1990, respectively.
c. The ratio was 1 when $t \approx 23.3$. This shows that the number of bachelor's degrees earned by women equaled the number earned by men for the first time around 1983.

65. a. $T(x) = 0.06x$

  b. $T(200) = 0.06(200) = 12$, or $12.00.

    $T(5.65) = 0.06(5.65) = 0.34$, or $0.34.

67. The child should receive $D(4) = \frac{2}{25}(500)(4) = 160$, or 160 mg.

69. a. The daily cost of leasing from Ace is $C_1(x) = 30 + 0.45x$,
    while the daily cost of leasing from Acme is $C_2(x) = 25 + 0.50x$,
    where x is the number of miles driven.
    b.

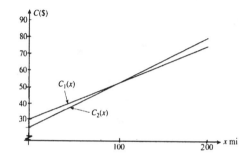

  c. The costs will be the same when $C_1(x) = C_2(x)$, that is, when
    $$30 + 0.45x = 25 + 0.50x \text{ or } -0.05x = -5, \text{ or } x = 100.$$
    Since $\qquad C_1(70) = 30 + 0.45(70) = 61.5$ (Ace)
    and $\qquad C_2(70) = 25 + 0.50(70) = 60$, (Acme)
    and the customer plans to drive less than 70 miles, she should rent from Acme.

71. Here $V = -20{,}000n + 1{,}000{,}000$.
    The book value in 1999 will be $V = -20{,}000(15) + 1{,}000{,}000$, or $700,000.
    The book value in 2003 will be $V = -20{,}000(19) + 1{,}000{,}000$, or $620,000.
    The book value in 2007 will be $V = -20{,}000(23) + 1{,}000{,}000$, or $540,000.

73. a. We require that $0.04 - r^2 \geq 0$ and $r \geq 0$. This is true if $0 \leq r \leq 0.2$. Therefore, the
    domain of v is $[0, 0.2]$.
    b. Compute
    $$v(0) = 1000[0.04 - (0)^2] = 1000(0.04) = 40.$$
    $$v(0.1) = 1000[0.04 - (0.1)^2] = 1000(0.04 - .01)$$

$$= 1000(0.03) = 30.$$
$$v(0.2) = 1000[0.04 - (0.2)^2] = 1000(0.04 - 0.04) = 0.$$

c.

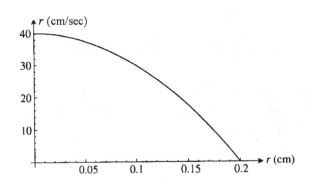

d. As the distance r increases, the velocity of the blood decreases.

75. Between 8 A.M. and 9 A.M., the average worker can be expected to assemble
$$N(1) - N(0) = (-1 + 6 + 15) - 0 = 20,$$
or 20 walkie-talkies. Between 9 A.M. and 10 A.M., we can expect
$$N(2) - N(1) = [-2^3 + 6(2^2) + 15(2)] - (-1 + 6 + 15)$$
$$= 46 - 20 = 26,$$
or 26 walkie-talkies can be assembled by the average worker.

77. The percentage at age 65 that are expected to have Alzheimer's disease is given by
$$P(0) = 0.0726(0)^2 + 0.7902(0) + 4.9623 = 4.9623, \quad \text{or } 4.96\%.$$
The percentage at age 90 that are expected to have Alzheimer's disease is given by
$$P(5) = 0.0726(25)^2 + 0.7902(25) + 4.9623 = 70.09, \quad \text{or } 70.09\%.$$

79. a. The amount of solids discharged in 1989 ($t = 0$) was 130 tons/day; in 1992
($t = 3$), it was 100 tons/day; and in 1996 ($t = 7$), it was

$$f(7) = 1.25(7)^2 - 26.25(7) + 162.5 = 40, \text{ or 40 tons/day.}$$

b.

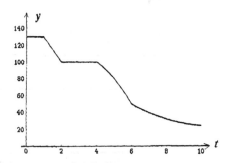

81. True, by definition the of a function .

83. False. Let $f(x) = x^2$, then take $a = 1$, and $b = 2$. Then $f(a) = f(1) = 1$ and
$f(b) = f(2) = 4$ and $f(a) + f(b) = 1 + 4 \neq f(a+b) = f(3) = 9$.

USING TECHNOLOGY EXERCISES 10.1, page 545

1.

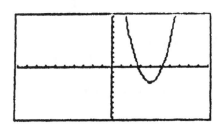

3.

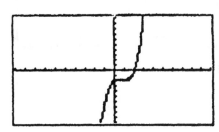

5.

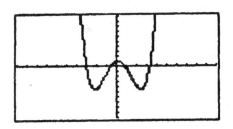

7.

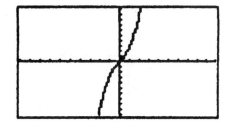

9. a.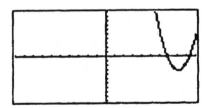

b.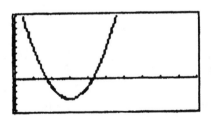

11. a.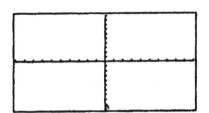

b.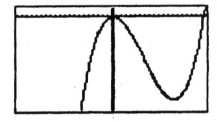

13. a.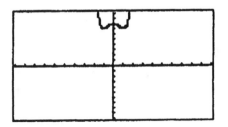

b.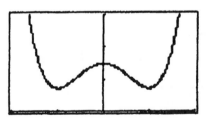

15. a.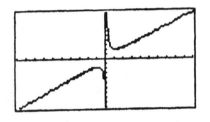

b.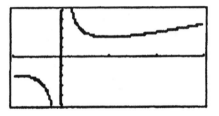

17. a.

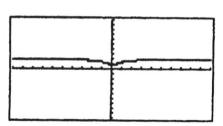

b.

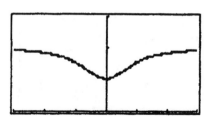

19. a.

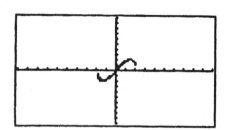

b.

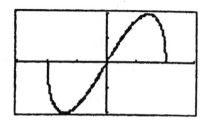

21.

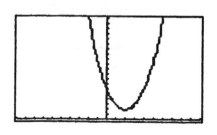

23.

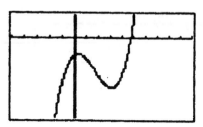

25.

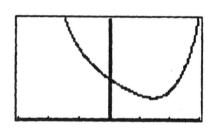

27.

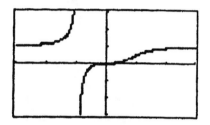

29.

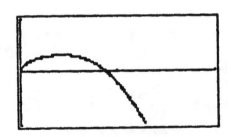

31. $18; f(-1) = -3(-1)^3 + 5(-1)^2 - 2(-1) + 8 = 3 + 5 + 2 + 8 = 18.$

33. $2; f(1) = \dfrac{(1)^4 - 3(1)^2}{1-2} = \dfrac{1-3}{-1} = 2.$

35. $f(2.145) \approx 18.5505$

37. $f(1.28) \approx 17.3850$

39. $f(2.41) \approx 4.1616$

41. $f(0.62) \approx 1.7214$

43. a.

b. $f(2) \approx 9.4066$, or approximately 9.4066%/yr
$f(4) \approx 8.7062$, or approximately 8.7062 %/yr.

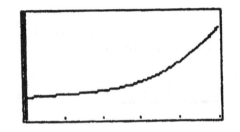

45. a.

b. $f(6) = 44.7;$

$f(8) = 52.7;$

$f(11) = 129.2.$

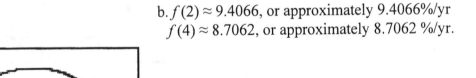

1. $(f+g)(x) = f(x) + g(x) = (x^3 + 5) + (x^2 - 2) = x^3 + x^2 + 3.$

3. $fg(x) = f(x)g(x) = (x^3 + 5)(x^2 - 2) = x^5 - 2x^3 + 5x^2 - 10.$

5. $\dfrac{f}{g}(x) = \dfrac{f(x)}{g(x)} = \dfrac{x^3 + 5}{x^2 - 2}.$

7. $\dfrac{fg}{h}(x) = \dfrac{f(x)g(x)}{h(x)} = \dfrac{(x^3 + 5)(x^2 - 2)}{2x + 4} = \dfrac{x^5 - 2x^3 + 5x^2 - 10}{2x + 4}$

9. $(f+g)(x) = x - 1 + \sqrt{x + 1}.$

11. $(fg)(x) = (x-1)\sqrt{x+1}$

13. $\dfrac{g}{h}(x) = \dfrac{g(x)}{h(x)} = \dfrac{\sqrt{x+1}}{2x^3 - 1}.$

15. $\dfrac{fg}{h}(x) = \dfrac{(x-1)(\sqrt{x+1})}{2x^3 - 1}$

17. $\dfrac{f-h}{g}(x) = \dfrac{x - 1 - (2x^3 - 1)}{\sqrt{x+1}} = \dfrac{x - 2x^3}{\sqrt{x+1}}.$

19. $(f+g)(x) = x^2 + 5 + \sqrt{x} - 2 = x^2 + \sqrt{x} + 3.$

$(f-g)(x) = x^2 + 5 - (\sqrt{x} - 2) = x^2 - \sqrt{x} + 7.$

$(fg)(x) = (x^2 + 5)(\sqrt{x} - 2); \ (\dfrac{f}{g})(x) = \dfrac{x^2 + 5}{\sqrt{x} - 2}.$

21. $(f+g)(x) = \sqrt{x+3} + \dfrac{1}{x-1} = \dfrac{(x-1)\sqrt{x+3} + 1}{x-1}.$

$(f-g)(x) = \sqrt{x+3} - \dfrac{1}{x-1} = \dfrac{(x-1)\sqrt{x+3} - 1}{x-1}.$

$$(f \, g)(x) = \sqrt{x+3} \left(\frac{1}{x-1} \right) = \frac{\sqrt{x+3}}{x-1}. \qquad (\frac{f}{g}) = \sqrt{x+3}(x-1).$$

23. $(f + g)(x) = \dfrac{x+1}{x-1} + \dfrac{x+2}{x-2} = \dfrac{(x+1)(x-2)+(x+2)(x-1)}{(x-1)(x-2)}$

$$= \frac{x^2 - x - 2 + x^2 + x - 2}{(x-1)(x-2)} = \frac{2x^2 - 4}{(x-1)(x-2)} = \frac{2(x^2 - 2)}{(x-1)(x-2)}.$$

$(f - g)(x) = \dfrac{x+1}{x-1} - \dfrac{x+2}{x-2} = \dfrac{(x+1)(x-2)-(x+2)(x-1)}{(x-1)(x-2)}$

$$= \frac{x^2 - x - 2 - x^2 - x + 2}{(x-1)(x-2)} = \frac{-2x}{(x-1)(x-2)}.$$

$(f g)(x) = \dfrac{(x+1)(x+2)}{(x-1)(x-2)}; \quad (\dfrac{f}{g}) = \dfrac{(x+1)(x-2)}{(x-1)(x+2)}.$

25. $(f \circ g)(x) = f(g(x)) = f(x^2) = (x^2)^2 + x^2 + 1 = x^4 + x^2 + 1.$
$(g \circ f)(x) = g(f(x)) = g(x^2 + x + 1) = (x^2 + x + 1)^2.$

27. $(f \circ g)(x) = f(g(x)) = f(x^2 - 1) = \sqrt{x^2 - 1} + 1.$
$(g \circ f)(x) = g(f(x)) = g(\sqrt{x} + 1) = (\sqrt{x} + 1)^2 - 1 = x + 2\sqrt{x} + 1 - 1 = x + 2\sqrt{x}.$

29. $(f \circ g)(x) = f(g(x)) = f\left(\dfrac{1}{x}\right) = \dfrac{1}{x} \div \left(\dfrac{1}{x^2} + 1\right) = \dfrac{1}{x} \cdot \dfrac{x^2}{x^2 + 1} = \dfrac{x}{x^2 + 1}.$

$(g \circ f)(x) = g(f(x)) = g\left(\dfrac{x}{x^2 + 1}\right) = \dfrac{x^2 + 1}{x}.$

31. $h(2) = g[f(2)]$. But $f(2) = 4 + 2 + 1 = 7$, so $h(2) = g(7) = 49$.

33. $h(2) = g[f(2)]$. But $f(2) = \dfrac{1}{2(2)+1} = \dfrac{1}{5}$, so $h(2) = g(\dfrac{1}{5}) = \dfrac{1}{\sqrt{5}} = \dfrac{\sqrt{5}}{5}.$

35. $f(x) = 2x^3 + x^2 + 1, \, g(x) = x^5.$

37. $f(x) = x^2 - 1, g(x) = \sqrt{x}.$

39. $f(x) = x^2 - 1, g(x) = \dfrac{1}{x}.$

41. $f(x) = 3x^2 + 2, g(x) = \dfrac{1}{x^{3/2}}.$

43. $f(a+h) - f(a) = [3(a+h)+4] - (3a+4) = 3a + 3h + 4 - 3a - 4 = 3h.$

45. $f(a+h) - f(a) = 4 - (a+h)^2 - (4-a^2)$
$$= 4 - a^2 - 2ah - h^2 - 4 + a^2 = -2ah - h^2 = -h(2a+h).$$

47. $\dfrac{f(a+h) - f(a)}{h} = \dfrac{[(a+h)^2 + 1] - (a^2 + 1)}{h} = \dfrac{a^2 + 2ah + h^2 + 1 - a^2 - 1}{h} = \dfrac{2ah + h^2}{h}$
$$= \dfrac{h(2a+h)}{h} = 2a + h$$

49. $\dfrac{f(a+h) - f(a)}{h} = \dfrac{[(a+h)^3 - (a+h)] - (a^3 - a)}{h}$
$$= \dfrac{a^3 + 3a^2h + 3ah^2 + h^3 - a - h - a^3 + a}{h}$$
$$= \dfrac{3a^2h + 3ah^2 + h^3 - h}{h} = 3a^2 + 3ah + h^2 - 1$$

51. $\dfrac{f(a+h) - f(a)}{h} = \dfrac{\dfrac{1}{a+h} - \dfrac{1}{a}}{h} = \dfrac{\dfrac{a - (a+h)}{a(a+h)}}{h} = -\dfrac{1}{a(a+h)}.$

53. $F(t)$ represents the total revenue for the two restaurants at time t.

55. $f(t)g(t)$ represents the (dollar) value of Nancy's holdings at time t.

57. $g \circ f$ is the function giving the amount of carbon monoxide pollution at time t.

59. $C(x) = 0.6x + 12,100$.

61. a. $P(x) = R(x) - C(x)$
$$= -0.1x^2 + 500x - (0.000003x^3 - 0.03x^2 + 200x + 100,000)$$
$$= -0.000003x^3 - 0.07x^2 + 300x - 100,000.$$
   b. $P(1500) = -0.000003(1500)^3 - 0.07(1500)^2 + 300(1500) - 100,000$
$$= 182,375 \quad \text{or } \$182,375.$$

63. a. The gap is
$$G(t) - C(t) = (3.5t^2 + 26.7t + 436.2) - (24.3t + 365)$$
$$= 3.5t^2 + 2.4t + 71.2.$$
   b. At the beginning of 1983, the gap was
$$G(0) = 3.5(0)^2 + 2.4(0) + 71.2 = 71.2, \text{ or } \$71,200.$$
   At the beginning of 1986, the gap was
$$G(3) = 3.5(3)^2 + 2.4(3) + 71.2 = 109.9, \text{ or } \$109,900.$$

65. a. The occupancy rate at the beginning of January is
$$r(0) = \frac{10}{81}(0)^3 - \frac{10}{3}(0)^2 + \frac{200}{9}(0) + 55 = 55, \text{ or } 55 \text{ percent.}$$
$$r(5) = \frac{10}{81}(5)^3 - \frac{10}{3}(5)^2 + \frac{200}{9}(5) + 55 = 98.2, \text{ or } 98.2 \text{ percent.}$$
   b. The monthly revenue at the beginning of January is
$$R(55) = -\frac{3}{5000}(55)^3 + \frac{9}{50}(55)^2 = 444.68, \text{ or } \$444,700.$$
   The monthly revenue at the beginning of June is
$$R(98.2) = -\frac{3}{5000}(98.2)^3 + \frac{9}{50}(98.2)^2 = 1167.6, \text{ or } \$1,167,600.$$

67. True. $(f + g)(x) = f(x) + g(x) = g(x) + f(x) = (g + f)(x)$.

69. False. Take $f(x) = \sqrt{x}$ and $g(x) = x + 1$. Then $(g \circ f)(x) = \sqrt{x} + 1$, but $(f \circ g)(x) = \sqrt{x + 1}$.

EXERCISES 10.3, page 566

1. f is a polynomial function in x of degree 6.

3. Expanding $G(x) = 2(x^2 - 3)^3$, we have $3G(x) = 2x^6 - 18x^4 + 54x^2 - 54$, and we conclude that G is a polynomial function in x of degree 6.

5. f is neither a polynomial nor a rational function.

7. The individual's disposable income is

$$D = (1 - 0.28)40,000 = 28,800, \text{ or } \$28,800.$$

9. The child should receive

$$D(4) = \left(\frac{4+1}{24}\right)(500) = 104.17, \text{ or } 104 \text{ mg.}$$

11. When 1000 units are produced,

$$R(1000) = -0.1(1000)^2 + 500(1000) = 400,000, \text{ or } \$400,000.$$

13. a.

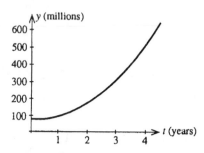

b. $f(4) = 38.57(4^2) - 24.29(4) + 79.14 = 599.1$, or 599,100,000.

15. a. The given data implies that $R(40) = 50$, that is,

$$\frac{100(40)}{b+40} = 50$$

$$50(b+40) = 4000, \text{ or } b = 40.$$

Therefore, the required response function is $R(x) = \dfrac{100x}{40+x}$.

b. The response will be $R(60) = \dfrac{100(60)}{40+60} = 60$, or approximately 60 percent.

17. Using the formula given in Problem 16, we have

$$V(2) = 100,000 - \frac{(100,000 - 30,000)}{5}(2) = 100,000 - \frac{70,000}{5}(2)$$

$$= 72,000, \text{ or } \$72,000.$$

19. $f(t) = 0.1714t^2 + 0.6657t + 0.7143$

a. $f(0) = 0.7143$ or $714,300$.

b. $f(5) = 0.1714(25) + 0.6657(5) + 0.7143 = 8.3278$, or 8.33 million.

21. a.

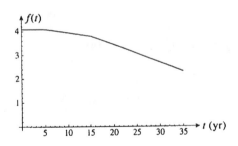

b. At the beginning of 2005, the ratio will be $f(10) = -0.03(10) + 4.25 = 3.95$
   At the beginning of 2020, the ratio will be $f(25) = -0.075(25) + 4.925 = 3.05$.

c. The ratio is constant from 1995 to 2000.

d. The decline of the ratio is greatest from 2010 through 2030. It is
$$\frac{f(35) - f(15)}{35 - 15} = \frac{2.3 - 3.8}{20} = -0.075.$$

23. $h(t) = f(t) - g(t) = \dfrac{110}{\frac{1}{2}t + 1} - 26(\frac{1}{4}t^2 - 1)^2 - 52.$

$h(0) = f(0) - g(0) = \dfrac{110}{\frac{1}{2}(0) + 1} - 26\left[\frac{1}{4}(0)^2 - 1\right]^2 - 52 = 110 - 26 - 52 = 32$, or $\$32$.

$$h(1) = f(1) - g(1) = \frac{110}{\frac{1}{2}(1)+1} - 26\left[\frac{1}{4}(1)^2 - 1\right]^2 - 52 = 6.71, \text{ or } \$6.71.$$

$$h(2) = f(2) - g(2) = \frac{110}{\frac{1}{2}(2)+1} - 26\left[\frac{1}{4}(2)^2 - 1\right]^2 - 52 = 3, \text{ or } \$3.$$

We conclude that the price gap was narrowing.

25. a.

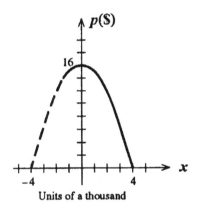

Units of a thousand

b. If $p = 7$, we have $7 = -x^2 + 16$, or $x^2 = 9$, so that $x = \pm 3$. Therefore, the quantity demanded when the unit price is \$7 is 3000 units.

27. a.

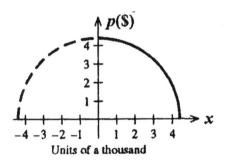

Units of a thousand

b. If $p = 3$, then $3 = \sqrt{18 - x^2}$, and $9 = 18 - x^2$, so that $x^2 = 9$ and $x = \pm 3$. Therefore, the quantity demanded when the unit price is \$3 is 3000 units.

29. a.

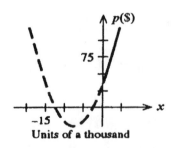

Units of a thousand

b. If $x = 2$, then $p = 2^2 + 16(2) + 40 = 76$, or $76.

31. a.

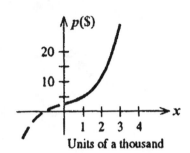

Units of a thousand

b. $p = 2^3 + 2(2) + 3 = 15$, or $15.

33. Substituting $x = 6$ and $p = 8$ into the given equation gives

$$8 = \sqrt{-36a + b}, \quad \text{or } -36a + b = 64.$$

Next, substituting $x = 8$ and $p = 6$ into the equation gives

$$6 = \sqrt{-64a + b}, \quad \text{or } -64a + b = 36.$$

Solving the system

$$-36a + b = 64$$

$$-64a + b = 36$$

for a and b, we find $a = 1$ and $b = 100$. Therefore the demand equation is $p = \sqrt{-x^2 + 100}$. When the unit price is set at $7.50, we have $7.5 = \sqrt{-x^2 + 100}$, or $56.25 = -x^2 + 100$ from which we deduce that $x = \pm 6.614$. So, the quantity demanded is 6614 units.

35. Substituting $x = 10{,}000$ and $p = 20$ into the given equation yields
$$20 = a\sqrt{10{,}000} + b = 100a + b.$$
Next, substituting $x = 62{,}500$ and $p = 35$ into the equation yields
$$35 = a\sqrt{62{,}500} + b = 250a + b.$$
Subtracting the first equation from the second yields
$$15 = 150a, \text{ or } a = \tfrac{1}{10}.$$
Substituting this value of a into the first equation gives $b = 10$. Therefore, the required equation is $p = \tfrac{1}{10}\sqrt{x} + 10$. The graph of the supply function follows.

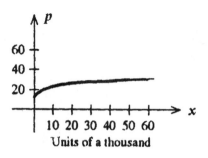

Units of a thousand

Substituting $x = 40{,}000$ into the supply equation yields
$$p = \tfrac{1}{10}\sqrt{40{,}000} + 10 = 30, \text{ or } \$30.$$

37. We solve the system of equations $p = -x^2 - 2x + 100$ and $p = 8x + 25$.
Thus, $-x^2 - 2x + 100 = 8x + 25$, or $x^2 + 10x - 75 = 0$.
Factoring this equation, we have $(x + 15)(x - 5) = 0$, or $x = -15$ and $x = 5$. Rejecting the negative root, we have $x = 5$ and the corresponding value of p is
$$p = 8(5) + 25 = 65.$$
We conclude that the equilibrium quantity is 5000 and the equilibrium price is \$65.

39. We solve the system $\quad p = 60 - 2x^2$
and $\quad p = x^2 + 9x + 30.$
Equating these two equations, we have
$$x^2 + 9x + 30 = 60 - 2x^2$$
$$3x^2 + 9x - 30 = 0$$
$$x^2 + 3x - 10 = 0$$
$$(x + 5)(x - 2) = 0$$

and $x = -5$ or $x = 2$. We take $x = 2$. The corresponding value of p is 5. Therefore, the equilibrium quantity is 2000 and the equilibrium price is $52.

41. Equating the two equations, we have

$$144 - x^2 = 48 + \tfrac{1}{2}x^2$$
$$288 - 2x^2 = 96 + x^2$$
$$3x^2 = 192; \quad x^2 = 64,$$

or $x = \pm 8$. We take $x = 8$, and the corresponding value of p is $144 - 8^2 = 80$. We conclude that the equilibrium quantity is 8000 tires and the equilibrium price is $80.

43. Since there is 80 feet of fencing available,
$$2x + 2y = 80; \quad x + y = 40 \text{ and } y = 40 - x.$$
Then the area of the garden is given by $f = xy = x(40 - x) = 40x - x^2$.
The domain of f is $[0, 40]$.

45. The volume of the box is given by the (area of the base) $\times$ the height of the box.
Thus, $V = f(x) = (15 - 2x)(8 - 2x)x$.

47. Since the perimeter of a circle is $2\pi r$, we know that the perimeter of the semicircle

is πx. Next, the perimeter of the rectangular portion of the window is given by $2y + 2x$, so the perimeter of the Norman window is $\pi x + 2y + 2x$ and
$$\pi x + 2y + 2x = 28, \text{ or } y = \tfrac{1}{2}(28 - \pi x - 2x)$$
Since the area of the window is given by $2xy + \tfrac{1}{2}\pi x^2$, we see that
$$A = 2xy + \tfrac{1}{2}\pi x^2.$$
Substituting the value of y found earlier, we see that

$$A = f(x) = x\left(28 - \pi x - 2x\right) + \tfrac{1}{2}\pi x^2$$
$$= \tfrac{1}{2}\pi x^2 + 28x - \pi x^2 - 2x^2$$
$$= 28x - \frac{\pi}{2}x^2 - 2x^2.$$

49. $xy = 50$ and so $y = \dfrac{50}{x}$. The area of the printed page is
$$A = (x-1)(y-2) = (x-1)(\tfrac{50}{x} - 2) = -2x + 52 - \tfrac{50}{x}.$$
So the required function is $f(x) = -2x + 52 - \tfrac{50}{x}$. We must have $x > 0$, $x - 1 \ge 0$,
And $\tfrac{50}{x} - 2 \ge 2$; The last inequality is solved as follows:
$$\frac{50}{x} \ge 4; \quad \frac{x}{50} \le \frac{1}{4}; \quad x \le \frac{50}{4} = \frac{25}{2}.$$
So the domain is $[1, \tfrac{25}{2}]$.

51. a. Let x denote the number of people beyond 20 who sign up for the cruise. Then the revenue is
$$R(x) = (20 + x)(600 - 4x)$$
$$= -4x^2 + 520x + 12,000$$

b. $\quad R(40) = -4(40^2) + 520(40) + 12000 = 26,400$, or 26,400 passengers.
$\quad R(60) = -4(60^2) + 520(60) + 12,000 = 28,800$ or 28,800 passengers.

53. a. $V = \dfrac{4}{3}\pi r^3$, $\quad r^3 = \dfrac{3V}{4\pi}$, $\quad$ or $\quad r = f(V) = \sqrt[3]{\dfrac{3V}{4\pi}}$ $\qquad$ b. $g(t) = \dfrac{9}{2}\pi t$

$\quad$ c. $h(t) = (f \circ g)(t) = f(g(t)) = \left(\dfrac{3g(t)}{4\pi}\right)^{1/3} = \left[\dfrac{3(9)\pi}{4\pi(2)}\right]^{1/3} = \dfrac{3}{2}\sqrt[3]{t}$

$\quad$ d. $h(8) = \dfrac{3}{2}\sqrt[3]{8} = 3$, or 3 ft.

55. True. If $P(x)$ is a polynomial function, then $P(x) = \dfrac{P(x)}{1}$ and so it is a rational

function. The converse if false. For example, $R(x) = \dfrac{x+1}{x-1}$ is a rational function

that is not a polynomial.

57. False. A power function has the form x^r, where r is a real number.

1. (-3.0414, 0.1503); (3.0414, 7.4497)   3.   (-2.3371, 2.4117); (6.0514, -2.5015)

5. (-1.0219, -6.3461); (1.2414, -1.5931), and (5.7805, 7.9391)

7. a.   b.   438 wall clocks; $40.92

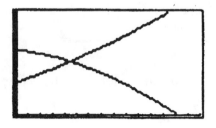

9. a. $y = 0.1375t^2 + 0.675t + 3.1$
    b.

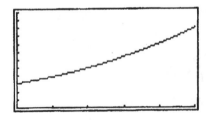

    c. 3.1; 3.9; 5; 6.4; 8; 9.9

11. a. $y = -0.02028t^3 + 0.31393t^2 + 0.40873t + 0.66024$
    b.

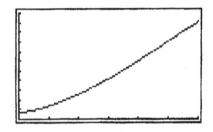

c. 0.66; 1.36; 2.57; 4.16; 6.02; 8.02; 10.03

13. a. $f(t) = 0.0012t^3 - 0.053t^2 + 0.497t + 2.55$ $(0 \le t \le 20)$

b.

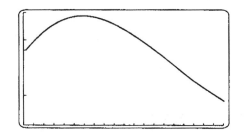

c.

t	0	7	10	20
$f(t)$	2.55	3.84	3.42	0.89

15. a. $y = 0.05833t^3 - 0.325t^2 + 1.8881t + 5.07143$

b.

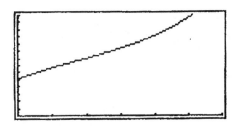

c. 6.7; 8.0; 9.4; 11.2; 13.7

17. a. $y = 0.0125t^4 - 0.01389t^3 + 0.55417t^2 + 0.53294t + 4.95238$ $(0 \le t \le 5)$

b.

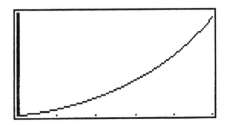

c. 5.0; 6.0; 8.3; 12.2; 18.3; 27.5

1. $\lim_{x \to -2} f(x) = 3.$ 3. $\lim_{x \to 3} f(x) = 3.$ 5. $\lim_{x \to -2} f(x) = 3.$

7. The limit does not exist. If we consider any value of x to the right of $x = -2$, $f(x) \le 2$. If we consider values of x to the left of $x = -2$, $f(x) \ge -2$. Since $f(x)$ does not approach any one number as x approaches $x = -2$, we conclude that the limit does not exist.

9. $\lim_{x \to 2} (x^2 + 1) = 5.$

x	1.9	1.99	1.999	2.001	2.01	2.1
$f(x)$	4.61	4.9601	4.9960	5.004	5.0401	5.41

11.

x	-0.1	-0.01	-0.001	0.001	0.01	0.1
$f(x)$	-1	-1	-1	1	1	1

The limit does not exist.

13.

x	0.9	0.99	0.999	1.001	1.01	1.1
$f(x)$	100	10,000	1,000,000	1,000,000	10,000	100

The limit does not exist.

15.

x	0.9	0.99	0.999	1.001	1.01	1.1
$f(x)$	2.9	2.99	2.999	3.001	3.01	3.1

$$\lim_{x \to 1} \frac{x^2 + x - 2}{x - 1} = 3.$$

17.

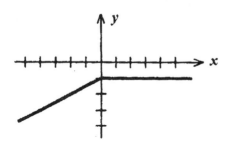

$$\lim_{x \to 0} f(x) = -1$$

19.

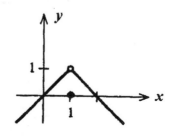

$$\lim_{x \to 1} f(x) = 1$$

21.

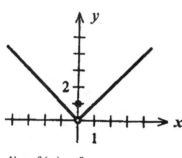

$$\lim_{x \to 0} f(x) = 0$$

23. $\lim_{x \to 2} 3 = 3$

25. $\lim_{x \to 3} x = 3$

27. $\lim_{x \to 1} (1 - 2x^2) = 1 - 2(1)^2 = -1$

29. $\lim_{x \to 1} (2x^3 - 3x^2 + x + 2) = 2(1)^3 - 3(1)^2 + 1 + 2 = 2.$

31. $\lim_{s \to 0} (2s^2 - 1)(2s + 4) = (-1)(4) = -4.$

33. $\displaystyle\lim_{x\to 2}\frac{2x+1}{x+2}=\frac{2(2)+1}{2+2}=\frac{5}{4}$.

35. $\displaystyle\lim_{x\to 2}\sqrt{x+2}=\sqrt{2+2}=2$.

37. $\displaystyle\lim_{x\to -3}\sqrt{2x^4+x^2}=\sqrt{2(-3)^4+(-3)^2}=\sqrt{162+9}=\sqrt{171}=3\sqrt{19}$.

39. $\displaystyle\lim_{x\to -1}\frac{\sqrt{x^2+8}}{2x+4}=\frac{\sqrt{(-1)^2+8}}{2(-1)+4}=\frac{\sqrt{9}}{2}=\frac{3}{2}$.

41. $\displaystyle\lim_{x\to a}[f(x)-g(x)]=\lim_{x\to a}f(x)-\lim_{x\to a}g(x)=3-4=-1$.

43. $\displaystyle\lim_{x\to a}[2f(x)-3g(x)]=\lim_{x\to a}2f(x)-\lim_{x\to a}3g(x)=2(3)-3(4)=-6$.

45. $\displaystyle\lim_{x\to a}\sqrt{g(x)}=\lim_{x\to a}\sqrt{4}=2$.

47. $\displaystyle\lim_{x\to a}\frac{2f(x)-g(x)}{f(x)g(x)}=\frac{2(3)-(4)}{(3)(4)}=\frac{2}{12}=\frac{1}{6}$.

49. $\displaystyle\lim_{x\to 1}\frac{x^2-1}{x-1}=\lim_{x\to 1}\frac{(x-1)(x+1)}{x-1}=\lim_{x\to 1}(x+1)=1+1=2$.

51. $\displaystyle\lim_{x\to 0}\frac{x^2-x}{x}=\lim_{x\to 0}\frac{x(x-1)}{x}=\lim_{x\to 0}(x-1)=0-1=-1$.

53. $\displaystyle\lim_{x\to -5}\frac{x^2-25}{x+5}=\lim_{x\to -5}\frac{(x+5)(x-5)}{x+5}=\lim_{x\to -5}(x-5)=-10$.

55. $\displaystyle\lim_{x\to 1}\frac{x}{x-1}$ does not exist.

57. $\displaystyle\lim_{x\to -2}\frac{x^2-x-6}{x^2+x-2}=\lim_{x\to -2}\frac{(x-3)(x+2)}{(x+2)(x-1)}=\lim_{x\to -2}\frac{x-3}{x-1}=\frac{-2-3}{-2-1}=\frac{5}{3}$.

59. $\lim\limits_{x\to 1}\dfrac{\sqrt{x}-1}{x-1}=\lim\limits_{x\to 1}\dfrac{\sqrt{x}-1}{x-1}\cdot\dfrac{\sqrt{x}+1}{\sqrt{x}+1}=\lim\limits_{x\to 1}\dfrac{x-1}{(x-1)(\sqrt{x}+1)}=\lim\limits_{x\to 1}\dfrac{1}{\sqrt{x}+1}=\dfrac{1}{2}.$

61. $\lim\limits_{x\to 1}\dfrac{x-1}{x^3+x^2-2x}=\lim\limits_{x\to 1}\dfrac{x-1}{x(x-1)(x+2)}=\lim\limits_{x\to 1}\dfrac{1}{x(x+2)}=\dfrac{1}{3}.$

63. $\lim\limits_{x\to\infty} f(x)=\infty$ (does not exist) and $\lim\limits_{x\to-\infty} f(x)=\infty$ (does not exist).

65. $\lim\limits_{x\to\infty} f(x)=0$ and $\lim\limits_{x\to-\infty} f(x)=0.$

67. $\lim\limits_{x\to\infty} f(x)=-\infty$ (does not exist) and $\lim\limits_{x\to-\infty} f(x)=-\infty$ (does not exist).

69.

x	1	10	100	1000
$f(x)$	0.5	0.009901	0.0001	0.000001

x	-1	-10	-100	-1000
$f(x)$	0.5	0.009901	0.0001	0.000001

$\lim\limits_{x\to\infty} f(x)=0$ and $\lim\limits_{x\to-\infty} f(x)=0$

71.

x	1	5	10	100	1000
$f(x)$	12	360	2910	2.99×10^6	2.999×10^9

x	-1	-5	-10	-100	-1000
$f(x)$	6	-390	-3090	-3.01×10^6	-3.0×10^9

$$\lim_{x \to \infty} f(x) = \infty \text{ (does not exist) and } \lim_{x \to -\infty} f(x) = -\infty \text{ (does not exist)}.$$

73. $\displaystyle\lim_{x \to \infty} \frac{3x+2}{x-5} = \lim_{x \to \infty} \frac{3+\dfrac{2}{x}}{1-\dfrac{5}{x}} = \frac{3}{1} = 3.$

75. $\displaystyle\lim_{x \to -\infty} \frac{3x^3+x^2+1}{x^3+1} = \lim_{x \to -\infty} \frac{3+\dfrac{1}{x}+\dfrac{1}{x^3}}{1+\dfrac{1}{x^3}} = 3.$

77. $\displaystyle\lim_{x \to -\infty} \frac{x^4+1}{x^3-1} = \lim_{x \to -\infty} \frac{x+\dfrac{1}{x^3}}{1-\dfrac{1}{x^3}} = -\infty$; that is, the limit does not exist.

79. $\displaystyle\lim_{x \to \infty} \frac{x^5-x^3+x-1}{x^6+2x^2+1} = \lim_{x \to \infty} \frac{\dfrac{1}{x}-\dfrac{1}{x^3}+\dfrac{1}{x^5}-\dfrac{1}{x^6}}{1+\dfrac{2}{x^4}+\dfrac{1}{x^6}} = 0.$

81. a. The cost of removing 50 percent of the pollutant is
$$C(50) = \frac{0.5(50)}{100-50} = 0.5 \text{, or } \$500,000.$$
Similarly, we find that the cost of removing 60, 70, 80, 90, and 95 percent of the pollutants is $750,000; $1,166,667; $2,000,000; $4,500,000, and $9,500,000, respectively.

b. $\displaystyle\lim_{x \to 100} \frac{0.5x}{100-x} = \infty,$

which means that the cost of removing the pollutant increases astronomically if we wish to remove almost all of the pollutant.

83. $\displaystyle\lim_{x \to \infty} \overline{C}(x) = \lim_{x \to \infty} 2.2 + \frac{2500}{x} = 2.2 \text{, or } \$2.20 \text{ per video disc.}$

In the long-run, the average cost of producing x video discs will approach $2.20/disc.

85. a. $T(1) = \dfrac{120}{1+4} = 24$, or \$24 million. $\qquad T(2) = \dfrac{120(4)}{8} = 60$, \$60 million.

$T(3) = \dfrac{120(9)}{13} = 83.1$, or \$83.1 million.

b. In the long run, the movie will gross

$$\lim_{x \to \infty} \frac{120x^2}{x^2 + 4} = \lim_{x \to \infty} \frac{120}{1 + \dfrac{4}{x^2}} = 120, \text{ or } \$120 \text{ million.}$$

87. a. The average cost of driving 5000 miles per year is

$$C(5) = \frac{2010}{5^{2.2}} + 17.80 = 76.07,$$

or 76.1 cents per mile. Similarly, we see that the average cost of driving 10,000 miles per year; 15,000 miles per year; 20,000 miles per year; and 25,000 miles per year is 30.5, 23; 20.6, and 19.5 cents per mile, respectively.

b.

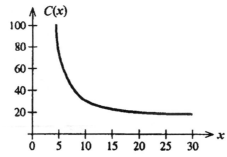

c. It approaches 17.80 cents per mile.

89. False. Let $f(x) = \begin{cases} -1 & \text{if } x < 0 \\ 1 & \text{if } x > 0 \end{cases}$. Then $\lim_{x \to 0} f(x) = 1$, but $f(1)$ is not defined.

91. True. Division by zero is not permitted.

93. True. Each limit in the sum exists. Therefore,

$$\lim_{x \to 2} \left(\frac{x}{x+1} + \frac{3}{x-1} \right) = \lim_{x \to 2} \frac{x}{x+1} + \lim_{x \to 2} \frac{3}{x-1} = \frac{2}{3} + \frac{3}{1} = \frac{11}{3}.$$

95. $\lim\limits_{x\to\infty}\dfrac{ax}{x+b}=\lim\limits_{x\to\infty}\dfrac{a}{1+\frac{b}{x}}=a.$ As the amount of substrate becomes very large, the initial

speed approaches the constant a moles per liter per second.

97. Consider the functions $f(x)=\begin{cases}-1 & \text{if } x<0 \\ 1 & \text{if } x\geq 0\end{cases}$ and $g(x)=\begin{cases}1 & \text{if } x<0 \\ -1 & \text{if } x\geq 0\end{cases}$.

Then $\lim\limits_{x\to 0}f(x)$ and $\lim\limits_{x\to 0}g(x)$ do not exist, but $\lim\limits_{x\to 0}[f(x)g(x)]=\lim\limits_{x\to 0}(-1)=-1.$

This example does not contradict Theorem 1 because the hypothesis of Theorem 1 says that if $\lim\limits_{x\to 0}f(x)$ and $\lim\limits_{x\to 0}g(x)$ both exist, then the limit of the product of f and g also exists. It does not say that if the former do not exist, then the latter might not exist.

USING TECHNOLOGY EXERCISES 10.4, page 596

1. 5   3. 3   5. $\dfrac{2}{3}$ 7. $\dfrac{1}{2}$ 9. e^2, or 7.38906

11. From the graph we see that $f(x)$ does not
    approach any finite number as x approaches 3.

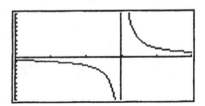

13. a.

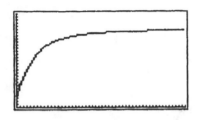

    b. $\lim\limits_{t\to\infty}\dfrac{25t^2+125t+200}{t^2+5t+40}=25$, so in the

    long run the population will approach
    25,000.

1. $\lim\limits_{x\to 2^-} f(x) = 3$, $\lim\limits_{x\to 2^+} f(x) = 2$, $\lim\limits_{x\to 2} f(x)$ does not exist.

3. $\lim\limits_{x\to -1^-} f(x) = \infty$, $\lim\limits_{x\to -1^+} f(x) = 2$. Therefore $\lim\limits_{x\to -1} f(x)$ does not exist.

5. $\lim\limits_{x\to 1^-} f(x) = 0$, $\lim\limits_{x\to 1^+} f(x) = 2$, $\lim\limits_{x\to 1} f(x)$ does not exist.

7. $\lim\limits_{x\to 0^-} f(x) = -2$, $\lim\limits_{x\to 0^+} f(x) = 2$, $\lim\limits_{x\to 0} f(x)$ does not exist.

9. True   11. True   13. False   15. True   17. False   19. True

21. $\lim\limits_{x\to 1^+} (2x+4) = 6$.

23. $\lim\limits_{x\to 2^-} \dfrac{x-3}{x+2} = \dfrac{2-3}{2+2} = -\dfrac{1}{4}$.

25. $\lim\limits_{x\to 0^+} \dfrac{1}{x}$ does not exist because $1/x \to \infty$ as $x \to 0$ from the right..

27. $\lim\limits_{x\to 0^+} \dfrac{x-1}{x^2+1} = \dfrac{-1}{1} = -1$.

29. $\lim\limits_{x\to 0^+} \sqrt{x} = \sqrt{\lim\limits_{x\to 0^+} x} = 0$.

31. $\lim\limits_{x\to -2^+} (2x+\sqrt{2+x}) = \lim\limits_{x\to -2^+} 2x + \lim\limits_{x\to -2^+} \sqrt{2+x} = -4+0 = -4$.

33. $\lim\limits_{x\to 1^-} \dfrac{1+x}{1-x} = \infty$, that is, the limit does not exist.

35. $\lim\limits_{x\to 2^-} \dfrac{x^2-4}{x-2} = \lim\limits_{x\to 2^-} \dfrac{(x+2)(x-2)}{x-2} = \lim\limits_{x\to 2^-} (x+2) = 4$.

37. $\lim\limits_{x\to 3^+} \dfrac{x^2-9}{x+3} = \dfrac{9-9}{3+3} = 0$.

39. $\lim_{x \to 0^+} f(x) = \lim_{x \to 0^+} x^2 = 0, \; \lim_{x \to 0^-} f(x) = \lim_{x \to 0^-} 2x = 0$

41. $\lim_{x \to 1^+} f(x) = \lim_{x \to 1^+} \sqrt{x+3} = \sqrt{4} = 2.$

$\lim_{x \to 1^-} f(x) = \lim_{x \to 1^-} (2+\sqrt{x}) = 2+\sqrt{x} = 2+\sqrt{1} = 3.$

43. The function is discontinuous at $x = 0$. Conditions 2 and 3 are violated.

45. The function is continuous everywhere.

47. The function is discontinuous at $x = 0$. Condition 3 is violated.

49. The function is discontinuous at $x = 0$. Condition 3 is violated.

51. f is continuous for all values of x.

53. f is continuous for all values of x. Note that $x^2 + 1 \geq 1 > 0.$

55. f is discontinuous at $x = 1/2$, where the denominator is 0.

57. Observe that $x^2 + x - 2 = (x + 2)(x - 1) = 0$ if $x = -2$ or $x = 1$. So, f is discontinuous at these values of x.

59. f is continuous everywhere since all three conditions are satisfied.

61. f is continuous everywhere because all three conditions are satisfied.

63. f is continuous everywhere since all three conditions are satisfied. Observe that

$$\lim_{x \to 1} f(x) = \lim_{x \to 1} \frac{x^2 - 1}{x - 1} = \lim_{x \to 1} \frac{(x-1)(x+1)}{x-1} = \lim_{x \to 1}(x+1) = 2 = f(1).$$

65. f is continuous everywhere since all three conditions are satisfied.

67. Since the denominator $x^2 - 1 = (x - 1)(x + 1) = 0$ if $x = -1$ or 1, we see that f is discontinuous at these points.

69. Since $x^2 - 3x + 2 = (x - 2)(x - 1) = 0$ if $x = 1$ or 2, we see that the denominator is zero at these points and so f is discontinuous at these points.

71. The function f is discontinuous at $x = 1, 2, 3, ..., 11$ because the limit of f does not exist at these points.

73. Having made steady progress up to $x = x_1$, Michael's progress came to a standstill. Then at $x = x_2$ a sudden break-through occurs and he then continues to successfully complete the solution to the problem.

75. Conditions 2 and 3 are not satisfied at each of these points.

77. The graph of f follows.

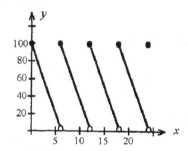

f is discontinuous at $x = 6, 12, 18, 24$.

79.

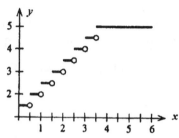

f is discontinuous at $x = \frac{1}{2}, 1, 1\frac{1}{2}, ..., 4$.

81. a. $\lim\limits_{t \to 0^+} S(t) = \lim\limits_{t \to 0^+} \dfrac{a}{t} + b = \infty$. As the time taken to excite the tissue is made smaller

and smaller, the strength of the electric current gets stronger and stronger.

b. $\lim\limits_{t \to \infty} \dfrac{a}{t} + b = b$. As the time taken to excite the tissue is made larger and larger,

the strength of the electric current gets smaller and smaller and approaches b.

83.   We require that $f(1) = 1 + 2 = 3 = \lim\limits_{x \to 1^+} kx^2 = k$, or $k = 3$.

85. a. Yes, because if $f + g$ were continuous at a, then $g = (f + g) - f$
would be continuous (the difference of two continuous functions is continuous),
and this would imply that g is continuous, a contradiction.
b. No. Consider the functions f and g defined by

$$f(x) = \begin{cases} -1 & \text{if } x < 0 \\ 1 & \text{if } x \geq 0 \end{cases} \quad \text{and } g(x) = \begin{cases} 1 & \text{if } x < 0 \\ -1 & \text{if } x \geq 0 \end{cases}.$$

Both f and g are discontinuous at $x = 0$, but $f + g$ is continuous everywhere.

87. a. f is a polynomial of degree 2 and is therefore continuous everywhere, and in
particular in $[1,3]$.
b. $f(1) = 3$ and $f(3) = -1$ and so f must have at least one zero in $(1,3)$.

89. f is a polynomial and is therefore continuous on $[-1,1]$.
$$f(-1) = (-1)^3 - 2(-1)^2 + 3(-1) + 2 = -1 - 2 - 3 + 2 = -4.$$
$$f(1) = 1 - 2 + 3 + 2 = 4.$$
Since $f(-1)$ and $f(1)$ have opposite signs, we see that f has at least one zero in
$(-1,1)$.

91. $f(0) = 6$ and $f(3) = 3$ and f is continuous on $[0,3]$. So the Intermediate Value
Theorem guarantees that there is at least one value of x for which $f(x) = 4$. Solving
$f(x) = x^2 - 4x + 6 = 4$, we find $x^2 - 4x + 2 = 0$. Using the quadratic formula, we
find that $x = 2 \pm \sqrt{2}$. Since $2 \pm \sqrt{2}$ does not lie in $[0,3]$, we see that
$x = 2 - \sqrt{2} \approx 0.59$.

93. $x^5 + 2x - 7 = 0$

Step	Root of f(x) = 0 lies in
1	(1,2)
2	(1,1.5)
3	(1.25,1.5)
4	(1.25,1.375)
5	(1.3125,1.375)
6	(1.3125,1.34375)
7	(1.328125,1.34375)
8	(1.3359375,1.34375)
9	(1.33984375,1.34375)

We see that the required root is approximately 1.34.

95. a. $h(t) = 4 + 64(0) - 16(0) = 4$, and $h(2) = 4 + 64(2) - 16(4) = 68$.

b. The function h is continuous on [0,2]. Furthermore, the number 32 lies between 4 and 68. Therefore, the Intermediate Value Theorem guarantees that there is at least one value of t such that $h(t) = 32$, that is, Joan must see the ball at least once during the time the ball is in the air.

c. We solve

$$h(t) = 4 + 64t - 16t^2 = 32$$
or
$$16t^2 - 64t + 28 = 0$$
$$4t^2 - 16t + 7 = 0$$
$$(2t - 1)(2t - 7) = 0$$

giving $t = \frac{1}{2}$ or $t = \frac{7}{2}$. Joan sees the ball on its way up half a second after it was thrown and again $3\frac{1}{2}$ seconds later when it is on its way down.

97. False. Take
$$f(x) = \begin{cases} -1 & \text{if } x < 2 \\ 4 & \text{if } x = 2 \\ 1 & \text{if } x > 2 \end{cases}$$

Then $f(2) = 4$ but $\lim_{x \to 2}$ does not exist.

99. False. Consider the function $f(x) = x^2 - 1$ on the interval [-2, 2]. Here, $f(-2) = f(2) = 3$, but f has zeros at $x = -1$ and $x = 1$.

321   *10 Functions, Limits, and the Derivative*

101. False. Let $f(x) = \begin{cases} x & \text{if } x \neq 0 \\ 1 & \text{if } x = 0 \end{cases}$. Then $\lim_{x \to 0^+} f(x) = \lim_{x \to 0^-} f(x)$, but $f(0) = 1$.

103. False. Take $f(x) = \begin{cases} \frac{1}{x} & \text{if } x \neq 0 \\ 0 & \text{if } x = 0 \end{cases}$. Then f is continuous for all $x \neq 0$ but

$\lim_{x \to 0} f(x)$ does not exist.

105. True. Since the number 2 lies between $f(-2) = 3$ and $f(3) = 1$, the intermediate value theorem guarantees that there exists at least one number $-2 \leq c \leq 3$ such that $f(c) = 2$.

107. a. Both $g(x) = x$ and $h(x) = \sqrt{1 - x^2}$ are continuous on $[-1,1]$ and so
   $f(x) = x - \sqrt{1 - x^2}$ is continuous on $[-1,1]$.
   b. $f(-1) = -1$ and $f(1) = 1$ and so f has at least one zero in $(-1,1)$.
   c. Solving $f(x) = 0$, we have $x = \sqrt{1 - x^2}$, $x^2 = 1 - x^2$, $2x^2 = 1$, or $x = \frac{\pm\sqrt{2}}{2}$.

109. a. (i). Repeated use of property 3 shows that $g(x) = x^n = x \cdot x \cdots x$ (n times) is a continuous function since $f(x) = x$ is continuous by Property 1.
   (ii). Properties 1 and 5 combine to show that $c \cdot x^n$ is continuous using the results of (a).
   (iii). Each of the terms of $p(x) = a_0 x^n + a_1 x^{n-1} + \cdots + a_n$ is continuous and so Property 4 implies that p is continuous.
   b. Property 6 now shows that $R(x) = \dfrac{p(x)}{q(x)}$ is continuous if $q(a) \neq 0$ since p

   and q are continuous at $x = a$.

USING TECHNOLOGY EXERCISES 10.5, page 612

1. $x = 0, 1$ 3. $x = 2$ 5. $x = 0, \frac{1}{2}$

7. $x = -\frac{1}{2}, 2$ 9. $x = -2, 1$

11.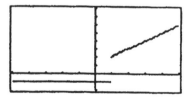

13.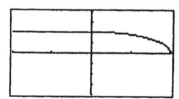

15.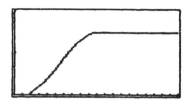

EXERCISES 10.6, page 627

1. The rate of change of the average infant's weight when $t = 3$ is $(7.5)/5$, or 1.5 lb/month. The rate of change of the average infant's weight when $t = 18$ is $(3.5)/6$, or approximately 0.6 lb/month. The average rate of change over the infant's first year of life is $(22.5 - 7.5)/(12)$, or 1.25 lb/month.

3. The rate of change of the percentage of households watching television at 4 P.M. is $(12.3)/4$, or approximately 3.1 percent per hour. The rate at 11 P.M. is $(-42.3)/2 = -21.15$; that is, it is dropping off at the rate of 21.15 percent per hour.

5. a. Car A is travelling faster than Car B at t_1 because the slope of the tangent line to the graph of f is greater than the slope of the tangent line to the graph of g at t_1.
   b. Their speed is the same because the slope of the tangent lines are the same at t_2.
   c. Car B is travelling faster than Car A.
   d. They have both covered the same distance and are once again side by side at t_3.

7. a. P_2 is decreasing faster at t_1 because the slope of the tangent line to the graph of g at t_1 is greater than the slope of the tangent line to the graph of f at t_1.
   b. P_1 is decreasing faster than P_2 at t_2.

c. Bactericide B is more effective in the short run, but bactericide A is more effective in the long run.

9. $f(x) = 13$

Step 1 $f(x + h) = 13$

Step 2 $f(x + h) - f(x) = 13 - 13 = 0$

Step 3 $\dfrac{f(x+h) - f(x)}{h} = \dfrac{0}{h} = 0$

Step 4 $f'(x) = \lim\limits_{h \to 0} \dfrac{f(x+h) - f(x)}{h} = \lim\limits_{h \to 0} 0 = 0$

11. $f(x) = 2x + 7$

Step 1 $f(x + h) = 2(x + h) + 7$

Step 2 $f(x + h) - f(x) = 2(x + h) + 7 - (2x + 7) = 2h$

Step 3 $\dfrac{f(x+h) - f(x)}{h} = \dfrac{2h}{h} = 2$

Step 4 $f'(x) = \lim\limits_{h \to 0} \dfrac{f(x+h) - f(x)}{h} = \lim\limits_{h \to 0} 2 = 2$

13. $f(x) = 3x^2$

Step 1 $f(x + h) = 3(x + h)^2 = 3x^2 + 6xh + 3h^2$

Step 2 $f(x + h) - f(x) = (3x^2 + 6xh + 3h^2) - 3x^2 = 6xh + 3h^2 = h(6x + 3h)$

Step 3 $\dfrac{f(x+h) - f(x)}{h} = \dfrac{h(6x + 3h)}{h} = 6x + 3h$

Step 4 $f'(x) = \lim\limits_{h \to 0} \dfrac{f(x+h) - f(x)}{h} = \lim\limits_{h \to 0} (6x + 3h) = 6x.$

15. $f(x) = -x^2 + 3x$

Step 1 $f(x + h) = -(x + h)^2 + 3(x + h) = -x^2 - 2xh - h^2 + 3x + 3h$

Step 2 $f(x + h) - f(x) = (-x^2 - 2xh - h^2 + 3x + 3h) - (-x^2 + 3x)$

$$= -2xh - h^2 + 3h = h(-2x - h + 3)$$

Step 3 $\dfrac{f(x+h) - f(x)}{h} = \dfrac{h(-2x - h + 3)}{h} = -2x - h + 3$

Step 4 $f'(x) = \lim\limits_{h \to 0} \dfrac{f(x+h) - f(x)}{h} = \lim\limits_{h \to 0} (-2x - h + 3) = -2x + 3.$

17. $f(x) = 2x + 7$. Using the four-step process,

Step 1 $f(x + h) = 2(x + h) + 7 = 2x + 2h + 7$

Step 2 $f(x + h) - f(x) = 2x + 2h + 7 - 2x - 7 = 2h$

Step 3 $\dfrac{f(x+h) - f(x)}{h} = \dfrac{2h}{h} = 2$

Step 4 $f'(x) = \lim\limits_{h \to 0} \dfrac{f(x+h) - f(x)}{h} = \lim\limits_{h \to 0} 2 = 2$

we find that $f'(x) = 2$. In particular, the slope at $x = 2$ is also 2. Therefore, a required equation is $y - 11 = 2(x - 2)$ or $y = 2x + 7$.

19. $f(x) = 3x^2$. We first compute $f'(x) = 6x$ (see Problem 13). Since the slope of the tangent line is $f'(1) = 6$, we use the point-slope form of the equation of a line and find that a required equation is $y - 3 = 6(x - 1)$, or $y = 6x - 3$.

21. $f(x) = -1/x$. We first compute $f'(x)$ using the four-step process.

Step 1 $f(x + h) = -\dfrac{1}{x + h}$

Step 2 $f(x + h) - f(x) = -\dfrac{1}{x + h} + \dfrac{1}{x} = \dfrac{-x + (x + h)}{x(x + h)} = \dfrac{h}{x(x + h)}$

Step 3 $\dfrac{f(x+h) - f(x)}{h} = \dfrac{\dfrac{h}{x(x+h)}}{h} = \dfrac{1}{x(x + h)}$

Step 4 $f'(x) = \lim\limits_{h \to 0} \dfrac{f(x+h) - f(x)}{h} = \lim\limits_{h \to 0} \dfrac{1}{x(x + h)} = \dfrac{1}{x^2}.$

The slope of the tangent line is $f'(3) = 1/9$. Therefore, a required equation is
$$y - (-\tfrac{1}{3}) = \tfrac{1}{9}(x - 3) \quad \text{or} \quad y = \tfrac{1}{9}x - \tfrac{2}{3}.$$

23. a. $f(x) = 2x^2 + 1$. We use the four-step process.

Step 1 $f(x + h) = 2(x + h)^2 + 1 = 2x^2 + 4xh + 2h^2 + 1$

Step 2 $f(x + h) - f(x) = (2x^2 + 4xh + 2h^2 + 1) - (2x^2 + 1) = 4xh + 2h^2$

$$= h(4x + 2h)$$

Step 3 $\dfrac{f(x+h)-f(x)}{h} = \dfrac{h(4x+2h)}{h} = 4x+2h$

Step 4 $f'(x) = \lim\limits_{h \to 0} \dfrac{f(x+h)-f(x)}{h} = \lim\limits_{h \to 0}\ (4x+2h) = 4x$

b. The slope of the tangent line is $f'(1) = 4(1) = 4$. Therefore, an equation is $y - 3 = 4(x-1)$ or $y = 4x - 1$.

c.

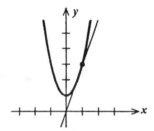

25. a. $f(x) = x^2 - 2x + 1$. We use the four-step process:

Step 1 $f(x+h) = (x+h)^2 - 2(x+h) + 1 = x^2 + 2xh + h^2 - 2x - 2h + 1$

Step 2 $f(x+h) - f(x) = (x^2 + 2xh + h^2 - 2x - 2h + 1) - (x^2 - 2x + 1)]$
$$= 2xh + h^2 - 2h = h(2x + h - 2)$$

Step 3 $\dfrac{f(x+h)-f(x)}{h} = \dfrac{h(2x+h-2)}{h} = 2x + h - 2$

Step 4 $f'(x) = \lim\limits_{h \to 0} \dfrac{f(x+h)-f(x)}{h} = \lim\limits_{h \to 0}\ (2x+h-2) = 2x - 2.$

b. At a point on the graph of f where the tangent line to the curve is horizontal, $f'(x) = 0$. Then $2x - 2 = 0$, or $x = 1$. Since $f(1) = 1 - 2 + 1 = 0$, we see that the required point is $(1,0)$.

c.

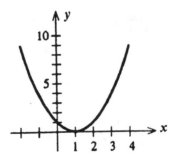

d. It is changing at the rate of 0 units per unit change in x.

27. a. $f(x) = x^2 + x$

$$\frac{f(3) - f(2)}{3 - 2} = \frac{(3^2 + 3) - (2^2 + 2)}{1} = 6$$

$$\frac{f(2.5) - f(2)}{2.5 - 2} = \frac{(2.5^2 + 2.5) - (2^2 + 2)}{0.5} = 5.5$$

$$\frac{f(2.1) - f(2)}{2.1 - 2} = \frac{(2.1^2 + 2.1) - (2^2 + 2)}{0.1} = 5.1$$

b. We first compute $f'(x)$ using the four-step process.

Step 1 $f(x + h) = (x + h)^2 + (x + h) = x^2 + 2xh + h^2 + x + h$

Step 2 $f(x + h) - f(x) = (x^2 + 2xh + h^2 + x + h) - (x^2 + x)]$
$$= 2xh + h^2 + h = h(2x + h + 1)$$

Step 3 $\dfrac{f(x + h) - f(x)}{h} = \dfrac{h(2x + h + 1)}{h} = 2x + h + 1$

Step 4 $f'(x) = \lim\limits_{h \to 0} \dfrac{f(x + h) - f(x)}{h} = \lim\limits_{h \to 0} (2x + h + 1) = 2x + 1.$

The instantaneous rate of change of y at $x = 2$ is $f'(2) = 5$ or 5 units per unit change in x.

c. The results in (a) suggest that the average rates of change of f at $x = 2$ approach 5 as the interval $[2, 2+h]$ gets smaller and smaller ($h = 1, 0.5,$ and 0.1). This number is the instantaneous rate of change of f at $x = 2$ as computed in (b).

29. a. $f(t) = 2t^2 + 48t$. The average velocity of the car over $[20,21]$ is

$$\frac{f(21) - f(20)}{21 - 20} = \frac{[2(21)^2 + 48(21)] - [2(20)^2 + 48(20)]}{1} = 130 \text{ ft / sec}$$

Its average velocity over $[20, 20.1]$ is

$$\frac{f(20.1) - f(20)}{20.1 - 20} = \frac{[2(20.1)^2 + 48(20.1)] - [2(20)^2 + 48(20)]}{0.1} = 128.2 \text{ ft / sec}$$

Its average velocity over $[20, 20.01]$

$$\frac{f(20.01) - f(20)}{20.01 - 20} = \frac{[2(20.01)^2 + 48(20.01)] - [2(20)^2 + 48(20)]}{0.01} = 128.02 \text{ ft / sec}$$

b. We first compute $f'(t)$ using the four-step process.

Step 1 $f(t + h) = 2(t + h)^2 + 48(t + h) = 2t^2 + 4th + 2h^2 + 48t + 48h$

Step 2 $f(t + h) - f(t) = (2t^2 + 4th + 2h^2 + 48t + 48h) - (2t^2 + 48t)]$

$$= 4th + 2h^2 + 48h = h(4t + 2h + 48).$$

Step 3 $\dfrac{f(t + h) - f(t)}{h} = \dfrac{h(4t + 2h + 48)}{h} = 4t + 2h + 48$

Step 4 $f'(t) = \lim\limits_{t \to 0} \dfrac{f(t + h) - f(t)}{h} = \lim\limits_{t \to 0} 4t + 2h + 48 = 4t + 48$

The instantaneous velocity of the car at $t = 20$ is $f'(20) = 4(20) + 48$, or 128 ft/sec.

c. Our results shows that the average velocities do approach the instantaneous velocity as the intervals over which they are computed decreases.

31. a. We solve the equation $16t^2 = 400$ obtaining $t = 5$ which is the time it takes the screw driver to reach the ground.

b. The average velocity over the time $[0,5]$ is

$$\dfrac{f(5) - f(0)}{5 - 0} = \dfrac{16(25) - 0}{5} = 80, \text{ or 80 ft/sec.} \quad [\text{Let s = f(t) = } 16t^2.]$$

c. The velocity of the screwdriver at time t is

$$v(t) = \lim\limits_{h \to 0} \dfrac{f(t + h) - f(t)}{h} = \lim\limits_{h \to 0} \dfrac{16(t + h)^2 - 16t^2}{h}$$

$$= \lim\limits_{h \to 0} \dfrac{16t^2 + 32th + 16h^2 - 16t^2}{h} = \lim\limits_{h \to 0} \dfrac{(32t + 16h)h}{h} = 32t .$$

In particular, the velocity of the screwdriver when it hits the ground (at $t = 5$) is
$$v(5) = 32(5) = 160, \text{ or 160 ft/sec.}$$

33. a. $V = \dfrac{1}{p}$. The average rate of change of V is

$$\dfrac{f(3) - f(2)}{3 - 2} = \dfrac{\frac{1}{3} - \frac{1}{2}}{1} = -\dfrac{1}{6}, \quad [\text{Write } V = f(p) = \dfrac{1}{p}.]$$

or a decrease of $\frac{1}{6}$ liter/atmosphere.

b. $\quad V'(t) = \lim\limits_{h \to 0} \dfrac{f(p + h) - f(p)}{h} = \lim\limits_{h \to 0} \dfrac{\frac{1}{p + h} - \frac{1}{p}}{h}$

$$= \lim\limits_{h \to 0} \dfrac{p - (p + h)}{hp(p + h)} = \lim\limits_{h \to 0} - \dfrac{1}{p(p + h)} = -\dfrac{1}{p^2} .$$

In particular, the rate of change of V when $p = 2$ is

$$V'(2) = -\frac{1}{2^2}, \text{ or a decrease of } \tfrac{1}{4} \text{ liter/atmosphere}$$

35. a. Using the four-step process, we find that

$$P'(x) = \lim_{h \to 0} \frac{P(x+h) - P(x)}{h}$$

$$= \lim_{h \to 0} \frac{-\tfrac{1}{3}(x^2 + 2xh + h^2) + 7x + 7h + 30 - (-\tfrac{1}{3}x^2 + 7x + 30)}{h}$$

$$= \lim_{h \to 0} \frac{-\tfrac{2}{3}xh - \tfrac{1}{3}h + 7h}{h} = \lim_{h \to 0} (-\tfrac{2}{3}x - \tfrac{1}{3}h + 7) = -\tfrac{2}{3}x + 7.$$

   b. $P'(10) = -\tfrac{2}{3}(10) + 7 \approx 0.333$, or \$333 per quarter.

   $P'(30) = -\tfrac{2}{3}(30) + 7 \approx -13$, or a decrease of \$13,000 per quarter.

37. $N(t) = t^2 + 2t + 50$. We first compute $N'(t)$ using the four–step process.

   Step 1 $N(t + h) = (t + h)^2 + 2(t + h) + 50$
   $$= t^2 + 2th + h^2 + 2t + 2h + 50$$

   Step 2 $N(t + h) - N(t)$
   $$= (t^2 + 2th + h^2 + 2t + 2h + 50) - (t^2 + 2t + 50)$$
   $$= 2th + h^2 + 2h = h(2t + h + 2).$$

   Step 3 $\dfrac{N(t+h) - N(t)}{h} = 2t + h + 2.$

   Step 4 $N'(t) = \lim_{h \to 0}(2t + h + 2) = 2t + 2.$

The rate of change of the country's GNP two years from now will be $N'(2) = 6$, or \$6 billion/yr. The rate of change four years from now will be $N'(4) = 10$, or \$10 billion/yr.

39. $\dfrac{f(a+h) - f(a)}{h}$ gives the average rate of change of the seal population over the

time interval $[a, a + h]$.

$\displaystyle \lim_{h \to 0} \frac{f(a+h) - f(a)}{h}$ gives the instantaneous rate of change of the seal population

at $x = a$.

41. $\dfrac{f(a+h)-f(a)}{h}$ gives the average rate of change of the country's industrial production over the time interval $[a, a + h]$.

$\displaystyle\lim_{h\to 0}\dfrac{f(a+h)-f(a)}{h}$ gives the instantaneous rate of change of the country's industrial production at $x = a$.

43. $\dfrac{f(a+h)-f(a)}{h}$ gives the average rate of change of the atmospheric pressure over the altitudes $[a, a + h]$.

$\displaystyle\lim_{h\to 0}\dfrac{f(a+h)-f(a)}{h}$ gives the instantaneous rate of change of the atmospheric pressure at $x = a$.

45. a. f has a limit at $x = a$.
    b. f is not continuous at $x = a$ because $f(a)$ is not defined.
    c. f is not differentiable at $x = a$ because it is not continuous there.

47. a. f has a limit at $x = a$. b. f is continuous at $x = a$.
    c. f is not differentiable at $x = a$ because f has a kink at the point $x = a$.

49. a. f does not have a limit at $x = a$ because it is unbounded in the neighborhood of a.
    b. f is not continuous at $x = a$.
    c. f is not differentiable at $x = a$ because it is not continuous there.

51. Our computations yield the following results:
    32.1, 30.939, 30.814, 30.8014, 30.8001, 30.8000.
    The motorcycle's instantaneous velocity at $t = 2$ is approximately 30.8 ft/sec.

53. False. Let $f(x) = |x|$. Then f is continuous at $x = 0$, but is not differentiable there.

55. Observe that the graph of f has a kink at $x = -1$. We have

$$\dfrac{f(-1+h)-f(-1)}{h} = 1 \text{ if } h > 0, \text{ and } -1 \text{ if } h < 0,$$

so that $\lim\limits_{h\to 0}\dfrac{f(-1+h)-f(-1)}{h}$ does not exist.

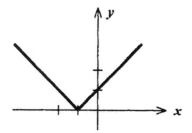

57. For continuity, we require that $f(1) = 1 = \lim\limits_{x\to 1^+}(ax+b) = a+b$, or $a+b = 1$.

In order that the derivative exist at $x = 1$, we require that $\lim\limits_{x\to 1^-}2x = \lim\limits_{x\to 1^+}a$, or $2 = a$.

Therefore, $b = -1$ and so $f(x) = \begin{cases} x^2 & \text{if } x \le 1 \\ 2x-1 & \text{if } x > 1 \end{cases}$. The graph of f follows.

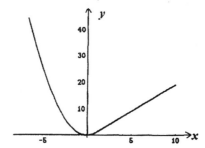

59. We have $f(x) = x$ if $x > 0$ and $f(x) = -x$ if $x < 0$. Therefore, when $x > 0$

$$f'(x) = \lim_{h\to 0}\frac{f(x+h)-f(x)}{h} = \lim_{h\to 0}\frac{x+h-x}{h} = \lim_{h\to 0}\frac{h}{h} = 1,$$

and when $x < 0$

$$f'(x) = \lim_{h\to 0}\frac{f(x+h)-f(x)}{h} = \lim_{h\to 0}\frac{-x-h-(-x)}{h} = \lim_{h\to 0}\frac{-h}{h} = -1.$$

Since the right–hand limit does not equal the left–hand limit, we conclude that $\lim\limits_{h\to 0}f(x)$ does not exist.

1. a. $y = 4x - 3$
    b.

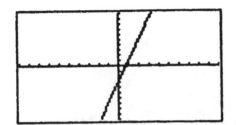

3. a. $y = -7x - 8$
    b.

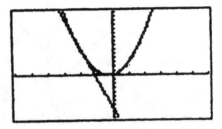

5. a. $y = 9x - 11$ b.

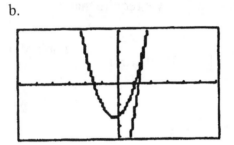

7. a. $y = 2$
    b.

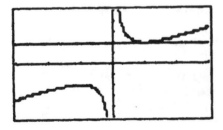

9. a. $y = \frac{1}{4}x + 1$
    b.

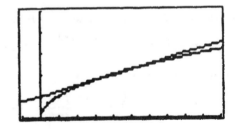

11. a. 4   b. $y = 4x - 1$

   c.

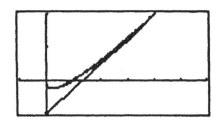

13. a. 20   b. $y = 20x - 35$

   c.

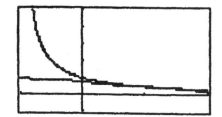

15. a. $\frac{3}{4}$ b. $y = \frac{3}{4}x - 1$

   c.

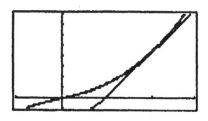

17. a. $-\frac{1}{4}$ b. $y = -\frac{1}{4}x + \frac{3}{4}$

   c.

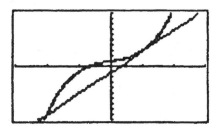

19. a. 4.02   b. $y = 4.02x - 3.57$

   c.

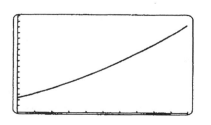

21. a.

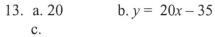

b. 41.22 cents/mile   c. 1.22 cents/mile

1. a. $9 - x \geq 0$ gives $x \leq 9$ and the domain is $(-\infty, 9]$.
    b. $2x^2 - x - 3 = (2x - 3)(x + 1)$, and $x = 3/2$ or $x = -1$.
    Since the denominator of the given expression is zero at these points, we see that
    the domain of f cannot include these points and so the domain of f is
    $(-\infty, -1) \cup (-1, \tfrac{3}{2}) \cup (\tfrac{3}{2}, \infty)$.

3. a.

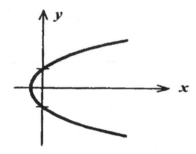

    b. For each value of $x > 0$, there are two values of y. We conclude that y is not a
    function of x. Equivalently, the function fails the vertical line test.
    c. Yes. For each value of y, there is only 1 value of x.

5. a. $f(x)g(x) = \dfrac{2x + 3}{x}$

    b. $\dfrac{f(x)}{g(x)} = \dfrac{1}{x(2x + 3)}$

    c. $f(g(x)) = \dfrac{1}{2x + 3}$.

    d. $g(f(x)) = 2\left(\dfrac{1}{x}\right) + 3 = \dfrac{2}{x} + 3$.

7. $\lim_{x \to 1}(x^2 + 1) = [(1)^2 + 1] = 1 + 1 = 2$.

9. $\lim_{x \to 3}\dfrac{x - 3}{x + 4} = \dfrac{3 - 3}{3 + 4} = 0$.

11. $\lim_{x \to -2}\dfrac{x^2 - 2x - 3}{x^2 + 5x + 6}$ does not exist. (The denominator is 0 at $x = -2$.)

13. $\lim\limits_{x\to 3}\dfrac{4x-3}{\sqrt{x+1}}=\dfrac{12-3}{\sqrt{4}}=\dfrac{9}{2}$.

15. $\lim\limits_{x\to 1^-}\dfrac{\sqrt{x}-1}{x-1}=\lim\limits_{x\to 1^-}\dfrac{(\sqrt{x}-1)(\sqrt{x}+1)}{(x-1)(\sqrt{x}+1)}=\lim\limits_{x\to 1^-}\dfrac{x-1}{(x-1)(\sqrt{x}+1)}=\lim\limits_{x\to 1^-}\dfrac{1}{\sqrt{x}+1}=\dfrac{1}{2}$.

17. $\lim\limits_{x\to -\infty}\dfrac{x+1}{x}=\lim\limits_{x\to -\infty}\left(1+\dfrac{1}{x}\right)=1$.

19. $\lim\limits_{x\to -\infty}\dfrac{x^2}{x+1}=\lim\limits_{x\to -\infty}x\cdot\dfrac{1}{1+\dfrac{1}{x}}=-\infty$, so the limit does not exist.

21. $\lim\limits_{x\to 2^+}f(x)=\lim\limits_{x\to 2^+}(x+2)=4;$

$\lim\limits_{x\to 2^-}f(x)=\lim\limits_{x\to 2^-}(4-x)=2.$

Therefore, $\lim\limits_{x\to 2}f(x)$ does not exist.

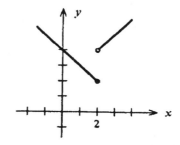

23. Since the denominator
$$4x^2-2x-2=2(2x^2-x-1)=2(2x+1)(x-1)=0$$
if $x=-1/2$ or 1, we see that f is discontinuous at these points.

25. The function is discontinuous at $x=0$.

27. $f(x)=3x+5$. Using the four-step process, we find

Step 1 $\quad f(x+h)=3(x+h)+5=3x+3h+5$

Step 2 $\quad f(x+h)-f(x)=3x+3h+5-3x-5=3h$

Step 3 $\quad \dfrac{f(x+h)-f(x)}{h}=\dfrac{3h}{h}=3.$

Step 4 $\quad f'(x)=\lim\limits_{h\to 0}\dfrac{f(x+h)-f(x)}{h}=\lim\limits_{h\to 0}(3)=3.$

29. $f(x) = 3x + 5$. We use the four-step process to obtain

Step 1 $f(x + h) = \frac{3}{2}(x + h) + 5 = \frac{3}{2}x + \frac{3}{2}h + 5.$

Step 2 $f(x + h) - f(x) = \frac{3}{2}x + \frac{3}{2}h + 5 - \frac{3}{2}x - 5 = \frac{3}{2}h.$

Step 3 $\dfrac{f(x + h) - f(x)}{h} = \dfrac{3}{2}.$

Step 4 $f'(x) = \lim\limits_{h \to 0} \dfrac{f(x + h) - f(x)}{h} = \lim\limits_{h \to 0} \dfrac{3}{2} = \dfrac{3}{2}.$

Therefore, the slope of the tangent line to the graph of the function f at the point (-2,2) is 3/2. To find the equation of the tangent line to the curve at the point (-2,2), we use the point–slope form of the equation of a line obtaining
$$y - 2 = \tfrac{3}{2}[x - (-2)] \quad \text{or} \quad y = \tfrac{3}{2}x + 5.$$

31. a. f is continuous at $x = a$ because the three conditions for continuity are satisfied at $x = a$; that is,

    i. $f(x)$ is defined   ii. $\lim\limits_{x \to a} f(x)$ exists   iii. $\lim\limits_{x \to a} f(x) = f(a)$

   b. f is not differentiable at $x = a$ because the graph of f has a kink at $x = a$.

33. a. The line passes through (0, 2.4) and (5, 7.4) and has slope $m = \dfrac{7.4 - 2.4}{5 - 0} = 1.$

Letting y denote the sales, we see that an equation of the line is
$$y - 2.4 = 1(t - 0), \text{ or } y = t + 2.4.$$

We can also write this in the form $S(t) = t + 2.4.$
b. The sales in 2002 were $S(3) = 3 + 2.4 = 5.4$, or $5.4 million.

35. Substituting the first equation into the second yields
$$3x - 2(\tfrac{3}{4}x + 6) + 3 = 0 \quad \text{or} \quad \tfrac{3}{2}x - 12 + 3 = 0$$
or $x = 6$. Substituting this value of x into the first equation then gives $y = 21/2$, so the point of intersection is $(6, \tfrac{21}{2})$.

37. We solve the system $3x + p - 40 = 0$
                                   $2x - p + 10 = 0.$

Adding these two equations, we obtain $5x - 30 = 0$, or $x = 6$. So,
$$p = 2x + 10 = 12 + 10 = 22.$$
Therefore, the equilibrium quantity is 6000 and the equilibrium price is $22.

39. $R(30) = -\frac{1}{2}(30)^2 + 30(30) = 450$, or $45,000.

41. $T = f(n) = 4n\sqrt{n-4}$.

$\quad f(4) = 0, \ f(5) = 20\sqrt{1} = 20, \ f(6) = 24\sqrt{2} \approx 33.9, \ f(7) = 28\sqrt{3} \approx 48.5,$

$\quad f(8) = 32\sqrt{4} = 64, \ f(9) = 36\sqrt{5} \approx 80.5, \ f(10) = 40\sqrt{6} \approx 98,$

$\quad f(11) = 44\sqrt{7} \approx 116 \quad \text{and} \quad f(12) = 48\sqrt{8} \approx 135.8.$

The graph of f follows:

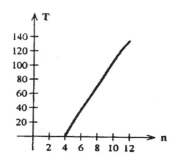

43.

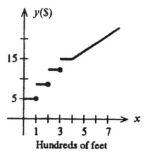

The function is discontinuous at $x = 100$, 200, and 300.

CHAPTER 11

1. $f'(x) = \dfrac{d}{dx}(-3) = 0.$

3. $f'(x) = \dfrac{d}{dx}(x^5) = 5x^4.$

5. $f'(x) = \dfrac{d}{dx}(x^{2.1}) = 2.1x^{1.1}.$

7. $f'(x) = \dfrac{d}{dx}(3x^2) = 6x.$

9. $f'(r) = \dfrac{d}{dr}(\pi r^2) = 2\pi r.$

11. $f'(x) = \dfrac{d}{dx}(9x^{1/3}) = \dfrac{1}{3}(9)x^{(1/3-1)} = 3x^{-2/3}.$

13. $f'(x) = \dfrac{d}{dx}(3\sqrt{x}) = \dfrac{d}{dx}(3x^{1/2}) = \dfrac{1}{2}(3)x^{-1/2} = \dfrac{3}{2}x^{-1/2} = \dfrac{3}{2\sqrt{x}}.$

15. $f'(x) = \dfrac{d}{dx}(7x^{-12}) = (-12)(7)x^{(-12-1)} = -84x^{-13}.$

17. $f'(x) = \dfrac{d}{dx}(5x^2 - 3x + 7) = 10x - 3.$

19. $f'(x) = \dfrac{d}{dx}(-x^3 + 2x^2 - 6) = -3x^2 + 4x.$

21. $f'(x) = \dfrac{d}{dx}(0.03x^2 - 0.4x + 10) = 0.06x - 0.4.$

23. If $f(x) = \dfrac{x^3 - 4x^2 + 3}{x} = x^2 - 4x + \dfrac{3}{x},$

    then $f'(x) = \dfrac{d}{dx}(x^2 - 4x + 3x^{-1}) = 2x - 4 - \dfrac{3}{x^2}.$

25. $f'(x) = \dfrac{d}{dx}\left(4x^4 - 3x^{5/2} + 2\right) = 16x^3 - \tfrac{15}{2}x^{3/2}$.

27. $f'(x) = \dfrac{d}{dx}\left(3x^{-1} + 4x^{-2}\right) = -3x^{-2} - 8x^{-3}$.

29. $f'(t) = \dfrac{d}{dt}\left(4t^{-4} - 3t^{-3} + 2t^{-1}\right) = -16t^{-5} + 9t^{-4} - 2t^{-2}$.

31. $f'(x) = \dfrac{d}{dx}\left(2x - 5x^{1/2}\right) = 2 - \dfrac{5}{2}x^{-1/2} = 2 - \dfrac{5}{2\sqrt{x}}$.

33. $f'(x) = \dfrac{d}{dx}\left(2x^{-2} - 3x^{-1/3}\right) = -4x^{-3} + x^{-4/3} = -\dfrac{4}{x^3} + \dfrac{1}{x^{4/3}}$.

35. a. $f'(x) = \dfrac{d}{dx}\left(2x^3 - 4x\right) = 6x^2 - 4.$ $f'(-2) = 6(-2)^2 - 4 = 20.$

  b. $f'(0) = 6(0) - 4 = -4.$ c. $f'(2) = 6(2)^2 - 4 = 20.$

37. The given limit is $f'(1)$ where $f(x) = x^3$. Since $f'(x) = 3x^2$, we have

$$\lim_{h \to 0} \frac{(1+h)^3 - 1}{h} = f'(1) = 3$$

39. Let $f(x) = 3x^2 - x$. Then

$$\lim_{h \to 0} \frac{3(2+h)^2 - (2+h) - 10}{h} = \lim_{h \to 0} \frac{f(2+h) - f(2)}{h}$$

because $f(2 + h) - f(2) = 3(2 + h)^2 - (2 + h) - [3(4) - 2]$
$$= 3(2 + h)^2 - (2 + h) - 10.$$

But the last limit is $f'(2)$. Since $f'(x) = 6x - 1$, we have $f'(2) = 11$.

Therefore, $\displaystyle\lim_{h \to 0} \frac{3(2+h)^2 - (2+h) - 10}{h} = 11.$

41. $f(x) = 2x^2 - 3x + 4$. The slope of the tangent line at any point $(x, f(x))$ on the graph of f is $f'(x) = 4x - 3$. In particular, the slope of the tangent line at the point $(2,6)$ is $f'(2) = 4(2) - 3 = 5$. An equation of the required tangent line is

$$y - 6 = 5(x - 2) \qquad \text{or} \qquad y = 5x - 4.$$

43. $f(x) = x^4 - 3x^3 + 2x^2 - x + 1$. $f'(x) = 4x^3 - 9x^2 + 4x - 1$.
The slope is $f'(1) = 4 - 9 + 4 - 1 = -2$. An equation
of the tangent line is $y - 0 = -2(x - 1)$ or $y = -2x + 2$.

45. a. $f'(x) = 3x^2$. At a point where the tangent line is horizontal,
$f'(x) = 0$, or $3x^2 = 0$ giving $x = 0$. Therefore, the point is $(0,0)$.
b.

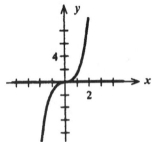

47. a. $f(x) = x^3 + 1$. The slope of the tangent line at any point $(x, f(x))$ on the graph
of f is $f'(x) = 3x^2$. At the point(s) where the slope is 12, we have
$3x^2 = 12$, or $x = \pm 2$. The required points are $(-2,-7)$ and $(2,9)$.
b. The tangent line at $(-2,-7)$ has equation
$$y - (-7) = 12[x - (-2)], \qquad \text{or} \qquad y = 12x + 17,$$
and the tangent line at $(2,9)$ has equation
$$y - 9 = 12(x - 2), \qquad \text{or} \qquad y = 12x - 15.$$
c.

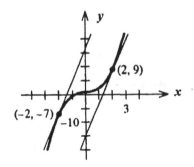

49. If $f(x) = \frac{1}{4}x^4 - \frac{1}{3}x^3 - x^2$, then $f'(x) = x^3 - x^2 - 2x$.
a. $f'(x) = x^3 - x^2 - 2x = -2x$

$$x^3 - x^2 = 0$$
$$x^2(x - 1) = 0 \qquad \text{and} \qquad x = 0 \text{ or } x = 1.$$
$$f(1) = \tfrac{1}{4}(1)^4 - \tfrac{1}{3}(1)^3 - (1)^2 = -\tfrac{13}{12}.$$
$$f(0) = \tfrac{1}{4}(0)^4 - \tfrac{1}{3}(0)^3 - (0)^2 = 0.$$

We conclude that the corresponding points on the graph are $(1, -\tfrac{13}{12})$ and $(0,0)$.

b. $\quad f'(x) = x^3 - x^2 - 2x = 0$
$$x(x^2 - x - 2) = 0$$
$$x(x - 2)(x + 1) = 0 \quad \text{and} \quad x = 0, 2, \text{ or } -1.$$
$$f(0) = 0$$
$$f(2) = \frac{1}{4}(2)^4 - \frac{1}{3}(2)^3 - (2)^2 = 4 - \frac{8}{3} - 4 = -\frac{8}{3}.$$
$$f(-1) = \frac{1}{4}(-1)^4 - \frac{1}{3}(-1)^3 - (-1)^2 = \frac{1}{4} + \frac{1}{3} - 1 = -\frac{5}{12}.$$

We conclude that the corresponding points are $(0,0)$, $(2, -\tfrac{8}{3})$ and $(-1, -\tfrac{5}{12})$.

c. $\quad f'(x) = x^3 - x^2 - 2x = 10x$
$$x^3 - x^2 - 12x = 0$$
$$x(x^2 - x - 12) = 0$$
$$x(x - 4)(x + 3) = 0$$
and $x = 0, 4,$ or -3.
$$f(0) = 0$$
$$f(4) = \tfrac{1}{4}(4)^4 - \tfrac{1}{3}(4)^3 - (4)^2 = 48 - \tfrac{64}{3} = \tfrac{80}{3}.$$
$$f(-3) = \tfrac{1}{4}(-3)^4 - \tfrac{1}{3}(-3)^3 - (-3)^2 = \tfrac{81}{4} + 9 - 9 = \tfrac{81}{4}.$$

We conclude that the corresponding points are $(0,0)$, $(4, \tfrac{80}{3})$ and $(-3, \tfrac{81}{4})$.

51. $\quad V(r) = \tfrac{4}{3}\pi r^3$. $\quad V'(r) = 4\pi r^2$.

a. $V'(\tfrac{2}{3}) = 4\pi(\tfrac{4}{9}) = \tfrac{16}{9}\pi$ cm^3/cm. b. $V'(\tfrac{5}{4}) = 4\pi(\tfrac{25}{16}) = \tfrac{25}{4}\pi$ cm^3/cm.

53. $\quad \dfrac{dA}{dx} = 26.5\dfrac{d}{dx}(x^{-0.45}) = 26.5(-0.45)x^{-1.45} = -\dfrac{11.925}{x^{1.45}}.$

Therefore, $\left.\dfrac{dA}{dx}\right|_{x=0.25} = -\dfrac{11.925}{(0.25)^{1.45}} \approx -89.01$ and $\left.\dfrac{dA}{dx}\right|_{x=2} = -\dfrac{11.925}{(2)^{1.45}} \approx -4.36$

Our computations reveal that if you make 0.25 stops per mile, your average speed will decrease at the rate of approximately 89.01 mph per stop per mile. If you make 2 stops per mile, your average speed will decrease at the rate of approximately 4.36 mph per stop per mile.

55. $I'(t) = -0.6t^2 + 6t$.
   a. In 1999, it was changing at a rate of $I'(5) = -0.6(25) + 6(5)$, or 15 points/yr. In 2001, it was $I'(7) = -0.6(49) + 6(7)$, or 12.6 pts/yr. In 2004, it was $I'(10) = -0.6(100) + 6(10)$, or 0 pts/yr.
   b. The average rate of increase of the CPI over the period from 1999 to 2004 was
   $$\frac{I(10) - I(5)}{5} = \frac{[-0.2(1000) + 3(100) + 100] - [-0.2(125) + 3(25) + 100]}{5}$$
   $$= \frac{200 - 150}{5} = 10, \text{ or } 10 \text{ pts/yr.}$$

57. $P(t) = 50,000 + 30t^{3/2} + 20t$. The rate at which the population will be increasing at any time t is $P'(t) = 45t^{1/2} + 20$. Nine months from now the population will be increasing at the rate of $P'(9) = 45(3) + 20$, or 155 people/month. Sixteen months from now the population will be increasing at the rate of
   $P'(16) = 45(4) + 20$, or 200 people/month.

59. $N(t) = 2t^3 + 3t^2 - 4t + 1000$. $N'(t) = 6t^2 + 6t - 4$.
   $N'(2) = 6(4) + 6(2) - 4 = 32$, or 32 turtles/yr.
   $N'(8) = 6(64) + 6(8) - 4 = 428$, or 428 turtles/yr.
   The population ten years after implementation of the conservation measures will be $N(10) = 2(10^3) + 3(10^2) - 4(10) + 1000$, or 3260 turtles.

61. a. $f(t) = 120t - 15t^2$. $v = f'(t) = 120 - 30t$ b. $v(0) = 120$ ft/sec
   c. Setting $v = 0$ gives $120 - 30t = 0$, or $t = 4$. Therefore, the stopping distance is
   $f(4) = 120(4) - 15(16)$ or 240 ft.

63. a. The number of temporary workers at the beginning of 1994 ($t = 3$) was
   $$N(3) = 0.025(3^2) + 0.255(3) + 1.505 = 2.495 \text{ million.}$$
   b. $N(t) = 0.025t^2 + 0.255t + 1.505$ $N'(t) = 0.05t + 0.255$.
   So, at the beginning of 1994 ($t = 3$), the number of temporary workers was growing at the rate of $N'(3) = 0.05(3) + 0.255 = 0.405$, or 405,000 per year.

65. a. $f'(x) = \frac{d}{dx}\left[0.0001x^{5/4} + 10\right] = \frac{5}{4}(0.0001x^{1/4}) = 0.000125x^{1/4}$
   b. $f'(10,000) = 0.000125(10,000)^{1/4} = 0.00125$, or \$0.00125/radio.

67. a. $f(t) = 20t - 40\sqrt{t} + 50$. $f'(t) = 20 - 40\left(\dfrac{1}{2}\right)t^{-1/2} = 20\left(1 - \dfrac{1}{\sqrt{t}}\right)$.

b. $f(0) = 20(0) - 40\sqrt{0} + 50 = 50$; $f(1) = 20(1) - 40\sqrt{1} + 50 = 30$

$f(2) = 20(2) - 40\sqrt{2} + 50 \approx 33.43$.

The average velocity at 6, 7, and 8 A.M. is 50 mph, 30 mph, and 33.43 mph, respectively.

c. $f'(\tfrac{1}{2}) = 20 - 20(\tfrac{1}{2})^{-1/2} \approx -8.28$. $f'(1) = 20 - 20(1)^{-1/2} \approx 0$.

$f'(2) = 20 - 20(2)^{-1/2} \approx 5.86$.

At 6:30 A.M. the average velocity is decreasing at the rate of 8.28 mph/hr; at 7 A.M., it is unchanged, and at 8 A.M., it is increasing at the rate of 5.86 mph.

69. a. $\dfrac{d}{dx}[0.075t^3 + 0.025t^2 + 2.45t + 2.4] = 0.225t^2 + 0.05t + 2.45$

b. $f'(3) = 0.225(3)^2 + 0.05(3) + 2.45 = 4.625$, or \$4.625 billion/yr.

c. $f(3) = 0.075(3)^3 + 0.025(3)^2 + 2.45(3) + 2.4 = 12$, or \$12 billion/yr.

71. True. $\dfrac{d}{dx}[2f(x) - 5g(x)] = \dfrac{d}{dx}[2f(x)] - \dfrac{d}{dx}[5g(x)] = 2f'(x) - 5g'(x)$.

73. $\dfrac{d}{dx}(x^3) = \lim_{h \to 0}\dfrac{(x+h)^3 - x^3}{h} = \lim_{h \to 0}\dfrac{x^3 + 3x^2h + 3xh^2 + h^3 - x^3}{h}$

$= \lim_{h \to 0}\dfrac{h(3x^2 + 3xh + h^2)}{h} = \lim_{h \to 0}(3x^2 + 3xh + h^2) = 3x^2$.

USING TECHNOL0GY EXERCISES 11.1, page 651

1. 1   3. 0.4226   5.   0.1613

7.   a.   9. a.

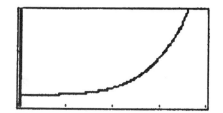   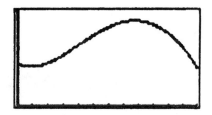

b. 3.4295 parts/million;
105.4332 parts/million

b. decreasing at the rate of 9 days/yr
increasing at the rate of 13 days/yr

11.   a.

b. Increasing at the rate of 1.1557%/yr
decreasing at the rate of 0.2116%/yr

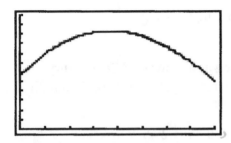

EXERCISES 11.2, page 661

1. $f(x) = 2x(x^2 + 1)$.
$$f'(x) = 2x\frac{d}{dx}(x^2 + 1) + (x^2 + 1)\frac{d}{dx}(2x)$$
$$= 2x(2x) + (x^2 + 1)(2) = 6x^2 + 2.$$

3. $f(t) = (t - 1)(2t + 1)$
$$f'(t) = (t - 1)\frac{d}{dt}(2t + 1) + (2t + 1)\frac{d}{dt}(t - 1)$$
$$= (t - 1)(2) + (2t + 1)(1) = 4t - 1$$

5. $f(x) = (3x + 1)(x^2 - 2)$
$$f'(x) = (3x + 1)\frac{d}{dx}(x^2 - 2) + (x^2 - 2)\frac{d}{dx}(3x + 1)$$
$$= (3x + 1)(2x) + (x^2 - 2)(3) = 9x^2 + 2x - 6.$$

7. $f(x) = (x^3 - 1)(x + 1)$.
$$f'(x) = (x^3 - 1)\frac{d}{dx}(x + 1) + (x + 1)\frac{d}{dx}(x^3 - 1)$$
$$= (x^3 - 1)(1) + (x + 1)(3x^2) = 4x^3 + 3x^2 - 1.$$

9. $f(w) = (w^3 - w^2 + w - 1)(w^2 + 2)$.

$$f'(w) = (w^3 - w^2 + w - 1)\frac{d}{dw}(w^2 + 2) + (w^2 + 2)\frac{d}{dw}(w^3 - w^2 + w - 1)$$

$$= (w^3 - w^2 + w - 1)(2w) + (w^2 + 2)(3w^2 - 2w + 1)$$

$$= 2w^4 - 2w^3 + 2w^2 - 2w + 3w^4 - 2w^3 + w^2 + 6w^2 - 4w + 2$$

$$= 5w^4 - 4w^3 + 9w^2 - 6w + 2.$$

11. $f(x) = (5x^2 + 1)(2\sqrt{x} - 1)$

$$f'(x) = (5x^2 + 1)\frac{d}{dx}(2x^{1/2} - 1) + (2x^{1/2} - 1)\frac{d}{dx}(5x^2 + 1)$$

$$= (5x^2 + 1)(x^{-1/2}) + (2x^{1/2} - 1)(10x)$$

$$= 5x^{3/2} + x^{-1/2} + 20x^{3/2} - 10x \ = \frac{25x^2 - 10x\sqrt{x} + 1}{\sqrt{x}}.$$

13. $f(x) = (x^2 - 5x + 2)(x - \frac{2}{x})$

$$f'(x) = (x^2 - 5x + 2)\frac{d}{dx}(x - \frac{2}{x}) + (x - \frac{2}{x})\frac{d}{dx}(x^2 - 5x + 2)$$

$$= \frac{(x^2 - 5x + 2)(x^2 + 2)}{x^2} + \frac{(x^2 - 2)(2x - 5)}{x}$$

$$= \frac{(x^2 - 5x + 2)(x^2 + 2) + x(x^2 - 2)(2x - 5)}{x^2}$$

$$= \frac{x^4 + 2x^2 - 5x^3 - 10x + 2x^2 + 4 + 2x^4 - 5x^3 - 4x^2 + 10x}{x^2}$$

$$= \frac{3x^4 - 10x^3 + 4}{x^2}.$$

15. $f(x) = \frac{1}{x-2}.$ $\quad f'(x) = \dfrac{(x-2)\frac{d}{dx}(1) - (1)\frac{d}{dx}(x-2)}{(x-2)^2} = \dfrac{0 - 1(1)}{(x-2)^2} = -\dfrac{1}{(x-2)^2}.$

17. $f(x) = \dfrac{x-1}{2x+1}.$

$$f'(x) = \frac{(2x+1)\dfrac{d}{dx}(x-1)-(x-1)\dfrac{d}{dx}(2x+1)}{(2x+1)^2}$$

$$= \frac{2x+1-(x-1)(2)}{(2x+1)^2} = \frac{3}{(2x+1)^2}.$$

19. $f(x) = \dfrac{1}{x^2+1}.$

$$f'(x) = \frac{(x^2+1)\dfrac{d}{dx}(1)-(1)\dfrac{d}{dx}(x^2+1)}{(x^2+1)^2}$$

$$= \frac{(x^2+1)(0)-1(2x)}{(x^2+1)^2} = -\frac{2x}{(x^2+1)^2}.$$

21. $f(s) = \dfrac{s^2-4}{s+1}.$

$$f'(s) = \frac{(s+1)\dfrac{d}{ds}(s^2-4)-(s^2-4)\dfrac{d}{ds}(s+1)}{(s+1)^2}$$

$$= \frac{(s+1)(2s)-(s^2-4)(1)}{(s+1)^2} = \frac{s^2+2s+4}{(s+1)^2}.$$

23. $f(x) = \dfrac{\sqrt{x}}{x^2+1}.$

$$f'(x) = \frac{(x^2+1)\dfrac{d}{dx}(x^{1/2})-(x^{1/2})\dfrac{d}{dx}(x^2+1)}{(x^2+1)^2} = \frac{(x^2+1)(\tfrac{1}{2}x^{-1/2})-(x^{1/2})(2x)}{(x^2+1)^2}$$

$$= \frac{(\tfrac{1}{2}x^{-1/2})[(x^2+1)-4x^2]}{(x^2+1)^2} = \frac{1-3x^2}{2\sqrt{x}(x^2+1)^2}.$$

25. $f(x) = \dfrac{x^2+2}{x^2+x+1}.$

$$f'(x) = \frac{(x^2+x+1)\dfrac{d}{dx}(x^2+2)-(x^2+2)\dfrac{d}{dx}(x^2+x+1)}{(x^2+x+1)^2}$$

$$= \frac{(x^2+x+1)(2x)-(x^2+2)(2x+1)}{(x^2+x+1)^2}$$

$$= \frac{2x^3+2x^2+2x-2x^3-x^2-4x-2}{(x^2+x+1)^2} = \frac{x^2-2x-2}{(x^2+x+1)^2}.$$

27. $f(x) = \dfrac{(x+1)(x^2+1)}{x-2} = \dfrac{(x^3+x^2+x+1)}{x-2}.$

$$f'(x) = \frac{(x-2)\dfrac{d}{dx}(x^3+x^2+x+1)-(x^3+x^2+x+1)\dfrac{d}{dx}(x-2)}{(x-2)^2}$$

$$= \frac{(x-2)(3x^2+2x+1)-(x^3+x^2+x+1)}{(x-2)^2}$$

$$= \frac{3x^3+2x^2+x-6x^2-4x-2-x^3-x^2-x-1}{(x-2)^2} = \frac{2x^3-5x^2-4x-3}{(x-2)^2}.$$

29. $f(x) = \dfrac{x}{x^2-4} - \dfrac{x-1}{x^2+4} = \dfrac{x(x^2+4)-(x-1)(x^2-4)}{(x^2-4)(x^2+4)} = \dfrac{x^2+8x-4}{(x^2-4)(x^2+4)}.$

$$f'(x) = \frac{(x^2-4)(x^2+4)\dfrac{d}{dx}(x^2+8x-4)-(x^2+8x-4)\dfrac{d}{dx}(x^4-16)}{(x^2-4)^2(x^2+4)^2}$$

$$= \frac{(x^2-4)(x^2+4)(2x+8)-(x^2+8x-4)(4x^3)}{(x^2-4)^2(x^2+4)^2}$$

$$= \frac{2x^5+8x^4-32x-128-4x^5-32x^4+16x^3}{(x^2-4)^2(x^2+4)^2}$$

$$= \frac{-2x^5-24x^4+16x^3-32x-128}{(x^2-4)^2(x^2+4)^2}.$$

31. $h'(x) = f(x)g'(x)+f'(x)g(x)$, by the Product Rule. Therefore,
$h'(1) = f(1)g'(1)+f'(1)g(1) = (2)(3)+(-1)(-2) = 8.$

33. Using the Quotient Rule followed by the Product Rule, we have

$$h'(x) = \frac{[x+g(x)]\frac{d}{dx}[xf(x)] - xf(x)\frac{d}{dx}[x+g(x)]}{[x+g(x)]^2}$$

$$= \frac{[x+g(x)][xf'(x)+f(x)] - xf(x)[1+g'(x)]}{[x+g(x)]^2}$$

Therefore, $h'(1) = \dfrac{[1+g(1)][f'(1)+f(1)] - f(1)[1+g'(1)]}{[1+g(1)]^2}$

$$= \frac{(1-2)(-1+2) - 2(1+3)}{(1-2)^2} = \frac{-1-8}{1} = -9.$$

35. $f(x) = (2x-1)(x^2+3)$

$$f'(x) = (2x-1)\frac{d}{dx}(x^2+3) + (x^2+3)\frac{d}{dx}(2x-1)$$

$$= (2x-1)(2x) + (x^2+3)(2) = 6x^2 - 2x + 6 = 2(3x^2 - x + 3).$$

At $x = 1, f'(1) = 2[3(1)^2 - (1) + 3] = 2(5) = 10.$

37. $f(x) = \dfrac{x}{x^4 - 2x^2 - 1}.$

$$f'(x) = \frac{(x^4 - 2x^2 - 1)\frac{d}{dx}(x) - x\frac{d}{dx}(x^4 - 2x^2 - 1)}{(x^4 - 2x^2 - 1)^2}$$

$$= \frac{(x^4 - 2x^2 - 1)(1) - x(4x^3 - 4x)}{(x^4 - 2x^2 - 1)^2} = \frac{-3x^4 + 2x^2 - 1}{(x^4 - 2x^2 - 1)^2}.$$

Therefore, $f'(-1) = \dfrac{-3+2-1}{(1-2-1)^2} = -\dfrac{2}{4} = -\dfrac{1}{2}.$

39. $f(x) = (x^3+1)(x^2-2).$

$$f'(x) = (x^3+1)\frac{d}{dx}(x^2-2) + (x^2-2)\frac{d}{dx}(x^3+1)$$

$$= (x^3+1)(2x) + (x^2-2)(3x^2).$$

The slope of the tangent line at $(2,18)$ is $f'(2) = (8+1)(4) + (4-2)(12) = 60.$
An equation of the tangent line is $y - 18 = 60(x - 2),$ or $y = 60x - 102.$

41. $f(x) = \dfrac{x+1}{x^2+1}.$

$$f'(x) = \frac{(x^2+1)\dfrac{d}{dx}(x+1)-(x+1)\dfrac{d}{dx}(x^2+1)}{(x^2+1)^2}$$

$$= \frac{(x^2+1)(1)-(x+1)(2x)}{(x^2+1)^2} = \frac{-x^2-2x+1}{(x^2+1)^2}.$$

At $x = 1$, $f'(1) = \dfrac{-1-2+1}{4} = -\dfrac{1}{2}$. Therefore, the slope of the tangent line at $x = 1$ is -1/2. Then an equation of the tangent line is
$$y-1 = -\tfrac{1}{2}(x-1) \quad \text{or} \quad y = -\tfrac{1}{2}x + \tfrac{3}{2}.$$

43. $f(x) = (x^3+1)(3x^2-4x+2)$

$$f'(x) = (x^3+1)\frac{d}{dx}(3x^2-4x+2)+(3x^2-4x+2)\frac{d}{dx}(x^3+1)$$

$$= (x^3+1)(6x-4)+(3x^2-4x+2)(3x^2)$$

$$= 6x^4+6x-4x^3-4+9x^4-12x^3+6x^2$$

$$= 15x^4-16x^3+6x^2+6x-4.$$

At $x = 1$, $f'(1) = 15(1)^4 - 16(1)^3 + 6(1) + 6(1) - 4 = 7$. The slope of the tangent line at the point $x = 1$ is 7. The equation of the tangent line is
$$y - 2 = 7(x - 1), \quad \text{or} \quad y = 7x - 5.$$

45. $f(x) = (x^2+1)(2-x)$

$$f'(x) = (x^2+1)\frac{d}{dx}(2-x)+(2-x)\frac{d}{dx}(x^2+1)$$

$$= (x^2+1)(-1)+(2-x)(2x) = -3x^2+4x-1.$$

At a point where the tangent line is horizontal, we have
$$f'(x) = -3x^2+4x-1 = 0$$

or $3x^2 - 4x + 1 = (3x-1)(x-1) = 0$, giving $x = 1/3$ or $x = 1$.

Since $f(\tfrac{1}{3}) = (\tfrac{1}{9}+1)(2-\tfrac{1}{3}) = \tfrac{50}{27}$, and $f(1) = 2(2-1) = 2$, we see that the required points are $(\tfrac{1}{3}, \tfrac{50}{27})$ and $(1, 2)$.

47. $f(x) = (x^2+6)(x-5)$

$$f'(x) = (x^2+6)\frac{d}{dx}(x-5)+(x-5)\frac{d}{dx}(x^2+6)$$

$$= (x^2+6)(1) + (x-5)(2x) = x^2 + 6 + 2x^2 - 10x = 3x^2 - 10x + 6.$$

At a point where the slope of the tangent line is -2, we have

$f'(x) = 3x^2 - 10x + 6 = -2.$

This gives $3x^2 - 10x + 8 = (3x - 4)(x - 2) = 0.$ So $x = \frac{4}{3}$ or $x = 2.$

Since $f(\frac{4}{3}) = (\frac{16}{9} + 6)(\frac{4}{3} - 5) = -\frac{770}{27}$ and $f(2) = (4+6)(2-5) = -30,$

the required points are $(\frac{4}{3}, -\frac{770}{27})$ and $(2, -30).$

49. $y = \dfrac{1}{1+x^2}.$ $y' = \dfrac{(1+x^2)\dfrac{d}{dx}(1) - (1)\dfrac{d}{dx}(1+x^2)}{(1+x^2)^2} = \dfrac{-2x}{(1+x^2)^2}.$

So, the slope of the tangent line at $(1, \frac{1}{2})$ is

$$y'\Big|_{x=1} = \dfrac{-2x}{(1+x^2)^2}\Big|_{x=1} = \dfrac{-2}{4} = -\dfrac{1}{2}$$

and the equation of the tangent line is $y - \frac{1}{2} = -\frac{1}{2}(x-1),$ or $y = -\frac{1}{2}x + 1.$

Next, the slope of the required normal line is 2 and its equation is

$$y - \tfrac{1}{2} = 2(x-1), \quad \text{or} \quad y = 2x - \tfrac{3}{2}.$$

51. $C(x) = \dfrac{0.5x}{100-x}.$ $C'(x) = \dfrac{(100-x)(0.5) - 0.5x(-1)}{(100-x)^2} = \dfrac{50}{(100-x)^2}.$

$C'(80) = \dfrac{50}{20^2} = 0.125;$ $\qquad C'(90) = \dfrac{50}{10^2} = 0.5,$

$C'(95) = \dfrac{50}{5^2} = 2;$ $\qquad C'(99) = \dfrac{50}{1} = 50.$

The rates of change of the cost in removing 80%, 90%, and 99% of the toxic waste are 0.125, 0.5, 2, and 50 million dollars per 1% more of the waste to be removed, respectively. It is too costly to remove *all* of the pollutant.

53. $N(t) = \dfrac{10,000}{1+t^2} + 2000$

$N'(t) = \dfrac{d}{dt}[10,000(1+t^2)^{-1} + 2000] = -\dfrac{10,000}{(1+t^2)^2}(2t) = -\dfrac{20,000t}{(1+t^2)^2}.$

The rate of change after 1 minute and after 2 minutes is

$N'(1) = -\dfrac{20,000}{(1+1^2)^2} = -5000;$ $N'(2) = -\dfrac{20,000(2)}{(1+2^2)^2} = -1600.$

The population of bacteria after one minute is $N(1) = \dfrac{10{,}000}{1+1} + 2000 = 7000$.

The population after two minutes is $N(2) = \dfrac{10{,}000}{1+4} + 2000 = 4000$.

55.   a. $N(t) = \dfrac{60t + 180}{t + 6}$.

$$N'(t) = \dfrac{(t+6)\dfrac{d}{dt}(60t+180) - (60t+180)\dfrac{d}{dt}(t+6)}{(t+6)^2}$$

$$= \dfrac{(t+6)(60) - (60t+180)(1)}{(t+6)^2} = \dfrac{180}{(t+6)^2}.$$

b. $N'(1) = \dfrac{180}{(1+6)^2} = 3.7, \ N'(3) = \dfrac{180}{(3+6)^2} = 2.2, \ N'(4) = \dfrac{180}{(4+6)^2} = 1.8,$

$$N'(7) = \dfrac{180}{(7+6)^2} = 1.1$$

We conclude that the rate at which the average student is increasing his or her speed one week, three weeks, four weeks, and seven weeks into the course is 3.7, 2.2, 1.8, and 1.1 words per minute, respectively.

c. Yes

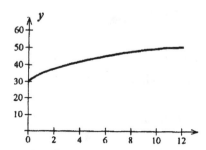

d. $N(12) = \dfrac{60(12) + 180}{12 + 6} = 50$, or 50 words/minute.

57. $f(t) = \dfrac{0.055t + 0.26}{t + 2}; \ f'(t) = \dfrac{(t+2)(0.055) - (0.055t + 0.26)(1)}{(t+2)^2} = -\dfrac{0.15}{(t+2)^2}.$

$\dfrac{d}{dx}[f(x)g(x)] = \dfrac{d}{dx}(x^2) = 2x \neq f'(x)g'(x) = 1.$

At the beginning, the formaldehyde level is changing at the rate of

$$f'(0) = -\frac{0.15}{4} = -0.0375;$$

that is, it is dropping at the rate of 0.0375 parts per million per year. Next,

$$f'(3) = -\frac{0.15}{5^2} = -0.006,$$

and so the level is dropping at the rate of 0.006 parts per million per year at the beginning of the fourth year ($t = 3$).

59.   False. Take $f(x) = x$ and $g(x) = x$. Then $f(x)g(x) = x^2$. So

60. True. Using the Product Rule,

$$\frac{d}{dx}\left[x\,f(x)\right] = f(x)\frac{d}{dx}(x) + x\frac{d}{dx}\left[f(x)\right] = f(x)(1) + xf'(x).$$

61.   False. Let $f(x) = x^3$. Then

$$\frac{d}{dx}\left[\frac{f(x)}{x^2}\right] = \frac{d}{dx}\left(\frac{x^3}{x^2}\right) = \frac{d}{dx}(x) = 1 \neq \frac{f'(x)}{2x} = \frac{3x^2}{2x} = \frac{3}{2}x.$$

63.   Let $f(x) = u(x)v(x)$ and $g(x) = w(x)$. Then $h(x) = f(x)g(x)$. Therefore,
      $$h'(x) = f'(x)g(x) + f(x)g'(x).$$
      But $f'(x) = u(x)v'(x) + u'(x)v(x).$
      Therefore, $h'(x) = [u(x)v'(x) + u'(x)v(x)]g(x) + u(x)v(x)w'(x)$
      $$= u(x)v(x)w'(x) + u(x)v'(x)w(x) + u'(x)v(x)w(x).$$

USING TECHNOLOGY EXERCISES 11.2, page 665

1. 0.8750   3. 0.0774   5. -0.5000   7. 87,322 per year

EXERCISES 11.3, page 674

1. $f(x) = (2x-1)^4.\ f'(x) = 4(2x-1)^3\dfrac{d}{dx}(2x-1) = 4(2x-1)^3(2) = 8(2x-1)^3.$

3. $f(x) = (x^2+2)^5.\ f'(x) = 5(x^2+2)^4(2x) = 10x(x^2+2)^4.$

5. $f(x) = (2x - x^2)^3$.

$$f'(x) = 3(2x - x^2)^2 \frac{d}{dx}(2x - x^2) = 3(2x - x^2)^2(2 - 2x) = 6x^2(1 - x)(2 - x)^2.$$

7. $f(x) = (2x + 1)^{-2}$.

$$f'(x) = -2(2x + 1)^{-3} \frac{d}{dx}(2x + 1) = -2(2x + 1)^{-3}(2) = -4(2x + 1)^{-3}.$$

9. $f(x) = (x^2 - 4)^{3/2}$.

$$f'(x) = \tfrac{3}{2}(x^2 - 4)^{1/2} \frac{d}{dx}(x^2 - 4) = \tfrac{3}{2}(x^2 - 4)^{1/2}(2x) = 3x(x^2 - 4)^{1/2}.$$

11. $f(x) = \sqrt{3x - 2} = (3x - 2)^{1/2}$.

$$f'(x) = \frac{1}{2}(3x - 2)^{-1/2}(3) = \frac{3}{2}(3x - 2)^{-1/2} = \frac{3}{2\sqrt{3x - 2}}.$$

13. $f(x) = \sqrt[3]{1 - x^2}$.

$$f'(x) = \frac{d}{dx}(1 - x^2)^{1/3} = \frac{1}{3}(1 - x^2)^{-2/3} \frac{d}{dx}(1 - x^2)$$

$$= \frac{1}{3}(1 - x^2)^{-2/3}(-2x) = -\frac{2}{3}x(1 - x^2)^{-2/3} = \frac{-2x}{3(1 - x^2)^{2/3}}.$$

15. $f(x) = \dfrac{1}{(2x + 3)^3} = (2x + 3)^{-3}$.

$$f'(x) = -3(2x + 3)^{-4}(2) = -6(2x + 3)^{-4} = -\frac{6}{(2x + 3)^4}.$$

17. $f(t) = \dfrac{1}{\sqrt{2t - 3}}$.

$$f'(t) = \frac{d}{dt}(2t - 3)^{-1/2} = -\frac{1}{2}(2t - 3)^{-3/2}(2) = -(2t - 3)^{-3/2} = -\frac{1}{(2t - 3)^{3/2}}.$$

19. $y = \dfrac{1}{(4x^4 + x)^{3/2}}$.

$$\frac{dy}{dx} = \frac{d}{dx}(4x^4 + x)^{-3/2} = -\frac{3}{2}(4x^4 + x)^{-5/2}(16x^3 + 1) = -\frac{3}{2}(16x^3 + 1)(4x^4 + x)^{-5/2}.$$

21. $f(x) = (3x^2 + 2x + 1)^{-2}$.

$$f'(x) = -2(3x^2 + 2x + 1)^{-3}\frac{d}{dx}(3x^2 + 2x + 1)$$

$$= -2(3x^2 + 2x + 1)^{-3}(6x + 2) = -4(3x + 1)(3x^2 + 2x + 1)^{-3}.$$

23. $f(x) = (x^2 + 1)^3 - (x^3 + 1)^2$.

$$f'(x) = 3(x^2 + 1)^2 \frac{d}{dx}(x^2 + 1) - 2(x^3 + 1)\frac{d}{dx}(x^3 + 1)$$

$$= 3(x^2 + 1)^2(2x) - 2(x^3 + 1)(3x^2)$$

$$= 6x[(x^2 + 1)^2 - x(x^3 + 1)] = 6x(2x^2 - x + 1).$$

25. $f(t) = (t^{-1} - t^{-2})^3$. $f'(t) = 3(t^{-1} - t^{-2})^2 \frac{d}{dt}(t^{-1} - t^{-2}) = 3(t^{-1} - t^{-2})^2(-t^{-2} + 2t^{-3})$.

27. $f(x) = \sqrt{x+1} + \sqrt{x-1} = (x+1)^{1/2} + (x-1)^{1/2}$.

$$f'(x) = \tfrac{1}{2}(x+1)^{-1/2}(1) + \tfrac{1}{2}(x-1)^{-1/2}(1) = \tfrac{1}{2}[(x+1)^{-1/2} + (x-1)^{-1/2}].$$

29. $f(x) = 2x^2(3 - 4x)^4$.

$$f'(x) = 2x^2(4)(3 - 4x)^3(-4) + (3 - 4x)^4(4x) = 4x(3 - 4x)^3(-8x + 3 - 4x)$$

$$= 4x(3 - 4x)^3(-12x + 3) = (-12x)(4x - 1)(3 - 4x)^3.$$

31. $f(x) = (x - 1)^2(2x + 1)^4$.

$$f'(x) = (x - 1)^2 \frac{d}{dx}(2x + 1)^4 + (2x + 1)^4 \frac{d}{dx}(x - 1)^2 \quad \text{[Product Rule]}$$

$$= (x - 1)^2(4)(2x + 1)^3 \frac{d}{dx}(2x + 1) + (2x + 1)^4(2)(x - 1)\frac{d}{dx}(x - 1)$$

$$= 8(x - 1)^2(2x + 1)^3 + 2(x - 1)(2x + 1)^4$$

$$= 2(x - 1)(2x + 1)^3(4x - 4 + 2x + 1) = 6(x - 1)(2x - 1)(2x + 1)^3.$$

33. $f(x) = \left(\dfrac{x+3}{x-2}\right)^3.$

$$f'(x) = 3\left(\frac{x+3}{x-2}\right)^2 \frac{d}{dx}\left(\frac{x-3}{x-2}\right) = 3\left(\frac{x+3}{x-2}\right)^2 \left[\frac{(x-2)(1)-(x+3)(1)}{(x-2)^2}\right]$$

$$= 3\left(\frac{x+3}{x-2}\right)^2\left[-\frac{5}{(x-2)^2}\right] = -\frac{15(x+3)^2}{(x-2)^4}.$$

35. $s(t) = \left(\dfrac{t}{2t+1}\right)^{3/2}.$

$$s'(t) = \frac{3}{2}\left(\frac{t}{2t+1}\right)^{1/2} \frac{d}{dt}\left(\frac{t}{2t+1}\right) = \frac{3}{2}\left(\frac{t}{2t+1}\right)^{1/2}\left[\frac{(2t+1)(1)-t(2)}{(2t+1)^2}\right]$$

$$= \frac{3}{2}\left(\frac{t}{2t+1}\right)^{1/2}\left[\frac{1}{(2t+1)^2}\right] = \frac{3t^{1/2}}{2(2t+1)^{5/2}}.$$

37. $g(u) = \left(\dfrac{u+1}{3u+2}\right)^{1/2}.$

$$g'(u) = \frac{1}{2}\left(\frac{u+1}{3u+2}\right)^{-1/2}\frac{d}{du}\left(\frac{u+1}{3u+2}\right)$$

$$= \frac{1}{2}\left(\frac{u+1}{3u+2}\right)^{-1/2}\left[\frac{(3u+2)(1)-(u+1)(3)}{(3u+2)^2}\right] = -\frac{1}{2\sqrt{u+1}\,(3u+2)^{3/2}}.$$

39. $f(x) = \dfrac{x^2}{(x^2-1)^4}.$

$$f'(x) = \frac{(x^2-1)^4\dfrac{d}{dx}(x^2) - (x^2)\dfrac{d}{dx}(x^2-1)^4}{\left[(x^2-1)^4\right]^2}$$

$$= \frac{(x^2-1)^4(2x) - x^2(4)(x^2-1)^3(2x)}{(x^2-1)^8}$$

$$= \frac{(x^2-1)^3(2x)(x^2-1-4x^2)}{(x^2-1)^8} = \frac{(-2x)(3x^2+1)}{(x^2-1)^5}.$$

41. $h(x) = \dfrac{(3x^2 + 1)^3}{(x^2 - 1)^4}.$

$$h'(x) = \dfrac{(x^2-1)^4(3)(3x^2+1)^2(6x) - (3x^2+1)^3(4)(x^2-1)^3(2x)}{(x^2-1)^8}$$

$$= \dfrac{2x(x^2-1)^3(3x^2+1)^2[9(x^2-1) - 4(3x^2+1)]}{(x^2-1)^8}$$

$$= -\dfrac{2x(3x^2+13)(3x^2+1)^2}{(x^2-1)^5}.$$

43. $f(x) = \dfrac{\sqrt{2x+1}}{x^2-1}.$

$$f'(x) = \dfrac{(x^2-1)(\frac{1}{2})(2x+1)^{-1/2}(2) - (2x+1)^{1/2}(2x)}{(x^2-1)^2}$$

$$= \dfrac{(2x+1)^{-1/2}[(x^2-1)-(2x+1)(2x)]}{(x^2-1)^2} = -\dfrac{3x^2+2x+1}{\sqrt{2x+1}(x^2-1)^2}.$$

45. $g(t) = \dfrac{(t+1)^{1/2}}{(t^2+1)^{1/2}}.$

$$g'(t) = \dfrac{(t^2+1)^{1/2}\dfrac{d}{dt}(t+1)^{1/2} - (t+1)^{1/2}\dfrac{d}{dt}(t^2+1)^{1/2}}{t^2+1}$$

$$= \dfrac{(t^2+1)^{1/2}(\frac{1}{2})(t+1)^{-1/2}(1) - (t+1)^{1/2}(\frac{1}{2})(t^2+1)^{-1/2}(2t)}{t^2+1}$$

$$= \dfrac{\frac{1}{2}(t+1)^{-1/2}(t^2+1)^{-1/2}[(t^2+1) - 2t(t+1)]}{t^2+1} = -\dfrac{t^2+2t-1}{2\sqrt{t+1}(t^2+1)^{3/2}}.$$

47. $f(x) = (3x+1)^4(x^2-x+1)^3$

$$f'(x) = (3x+1)^4 \cdot \dfrac{d}{dx}(x^2-x+1)^3 + (x^2-x+1)^3 \dfrac{d}{dx}(3x+1)^4$$

$$= (3x+1)^4 \cdot 3(x^2-x+1)^2(2x-1) + (x^2-x+1)^3 \cdot 4(3x+1)^3 \cdot 3$$

$$= 3(3x+1)^3(x^2-x+1)^2[(3x+1)(2x-1) + 4(x^2-x+1)]$$

$$= 3(3x+1)^3(x^2-x+1)^2(6x^2-3x+2x-1+4x^2-4x+4)$$

$$= 3(3x+1)^3 (x^2 - x + 1)^2 (10x^2 - 5x + 3)$$

49. $y = g(u) = u^{4/3}$ and $\dfrac{dy}{du} = \dfrac{4}{3} u^{1/3}$, $u = f(x) = 3x^2 - 1$, and $\dfrac{du}{dx} = 6x$.

So $\dfrac{dy}{dx} = \dfrac{dy}{du} \cdot \dfrac{du}{dx} = \frac{4}{3} u^{1/3} (6x) = \frac{4}{3} (3x^2 - 1)^{1/3} 6x = 8x(3x^2 - 1)^{1/3}$.

51. $\dfrac{dy}{du} = -\dfrac{2}{3} u^{-5/3} = -\dfrac{2}{3u^{5/3}}$, $\dfrac{du}{dx} = 6x^2 - 1$.

$\dfrac{dy}{dx} = \dfrac{dy}{du} \cdot \dfrac{du}{dx} = -\dfrac{2(6x^2 - 1)}{3u^{5/3}} = -\dfrac{2(6x^2 - 1)}{3(2x^3 - x + 1)^{5/3}}$.

53. $\dfrac{dy}{du} = \frac{1}{2} u^{-1/2} - \frac{1}{2} u^{-3/2}$, $\dfrac{du}{dx} = 3x^2 - 1$.

$\dfrac{dy}{dx} = \dfrac{dy}{du} \cdot \dfrac{du}{dx} = \left[\dfrac{1}{2\sqrt{x^3 - x}} - \dfrac{1}{2(x^3 - x)^{3/2}} \right] (3x^2 - 1)$

$= \dfrac{(3x^2 - 1)(x^3 - x - 1)}{2(x^3 - x)^{3/2}}$.

55. $F(x) = g(f(x))$; $F'(x) = g'(f(x)) f'(x)$ and $F'(2) = g'(3)(-3) = (4)(-3) = -12$

57. Let $g(x) = x^2 + 1$, then $F(x) = f(g(x))$. Next, $F'(x) = f'(g(x))g'(x)$
and $F'(1) = f'(2)(2x) = (3)(2) = 6$.

59. No. Suppose $h = g(f(x))$. Let $f(x) = x$ and $g(x) = x^2$. Then
$h = g(f(x)) = g(x) = x^2$ and $h'(x) = 2x \neq g'(f'(x)) = g'(1) = 2(1) = 2$.

61. $f(x) = (1 - x)(x^2 - 1)^2$.

$f'(x) = (1 - x)2(x^2 - 1)(2x) + (-1)(x^2 - 1)^2$

$= (x^2 - 1)(4x - 4x^2 - x^2 + 1) = (x^2 - 1)(-5x^2 + 4x + 1)$.

Therefore, the slope of the tangent line at $(2, -9)$ is

$f'(2) = [(2)^2 - 1][-5(2)^2 + 4(2) + 1] = -33$.

Then the required equation is $y + 9 = -33(x - 2)$, or $y = -33x + 57$.

63. $f(x) = x\sqrt{2x^2 + 7}$. $f'(x) = \sqrt{2x^2 + 7} + x(\frac{1}{2})(2x^2 + 7)^{-1/2}(4x)$.

The slope of the tangent line is $f'(3) = \sqrt{25} + (\frac{3}{2})(25)^{-1/2}(12) = \frac{43}{5}$.

An equation of the tangent line is $y - 15 = \frac{43}{5}(x - 3)$ or $y = \frac{43}{5}x - \frac{54}{5}$.

65. $N(x) = (60 + 2x)^{2/3}$. $N'(x) = \frac{2}{3}(60 + 2x)^{-1/3} \dfrac{d}{dx}(60 + 2x) = \frac{4}{3}(60 + 2x)^{-1/3}$.

The rate of increase at the end of the second week is
$$N'(2) = \frac{4}{3}(64)^{-1/3} = \frac{1}{3}, \text{ or } \frac{1}{3} \text{ million/week}$$

At the end of the 12th week, $N'(12) = \frac{4}{3}(84)^{-1/3} \approx 0.3$ million/wk. The number of viewers in the 2nd and 24th week are $N(2) = (60 + 4)^{2/3} = 16$ million and $N(24) = (60 + 48)^{2/3} = 22.7$ million, respectively.

67. $C(t) = 0.01(0.2t^2 + 4t + 64)^{2/3}$.

a. $C'(t) = 0.01(\frac{2}{3})(0.2t^2 + 4t + 64)^{-1/3} \dfrac{d}{dt}(0.2t^2 + 4t + 64)$

$= (0.01)(0.667)(0.4t + 4)(0.2t^2 + 4t + 4)^{-1/3}$

$= 0.027(0.1t + 1)(0.2t^2 + 4t + 64)^{-1/3}$.

b. $C'(5) = 0.007[0.4(5) + 4][0.2(25) + 4(5) + 64]^{-1/3} \approx 0.009$, or 0.009 parts per million per year.

69. $P(t) = 33.55(t + 5)^{0.205}$. $P'(t) = 33.55(0.205)(t + 5)^{-0.795}(1) = 6.87775(t + 5)^{-0.795}$
The rate of change at the beginning of 2000 is
$$P'(20) = 6.87775(25)^{-0.795} \approx 0.5322 \text{ or } 0.53\%/\text{yr.}$$
The percent of these mothers was $P(20) = 33.55(25)^{0.205} \approx 64.90$, or 64.9%.

71. a. $A(t) = 0.03t^3(t - 7)^4 + 60.2$
$A'(t) = 0.03[3t^2(t - 7)^4 + t^3(4)(t - 7)^3] = 0.03t^2(t - 7)^3[3(t - 7) + 4t]$
$= 0.21t^2(t - 3)(t - 7)^3$.
b. $A'(1) = 0.21(-2)(-6)^3 = 90.72$; $A'(3) = 0$. $A'(4) = 0.21(16)(1)(-3)^3 = -90.72$.
The amount of pollutant is increasing at the rate of 90.72 units/hr at 8 A.M. Its rate of change is 0 units/hr at 10 A.M.; its rate of change is -90.72 units/hr at 11 A.M.

73. $P(t) = \dfrac{300\sqrt{\frac{1}{2}t^2 + 2t + 25}}{t + 25} = \dfrac{300(\frac{1}{2}t^2 + 2t + 25)^{1/2}}{t + 25}$.

$P'(t) = 300\left[\dfrac{(t+25)\frac{1}{2}(\frac{1}{2}t^2 + 2t + 25)^{-1/2}(t+2) - (\frac{1}{2}t^2 + 2t + 25)^{1/2}(1)}{(t+25)^2}\right]$

$\qquad = 300\left[\dfrac{(\frac{1}{2}t^2 + 2t + 25)^{-1/2}[(t+25)(t+2) - 2(\frac{1}{2}t^2 + 2t + 25)]}{(t+25)^2}\right]$

$\qquad = \dfrac{3450t}{(t+25)^2\sqrt{\frac{1}{2}t^2 + 2t + 25}}$.

Ten seconds into the run, the athlete's pulse rate is increasing at

$P'(10) = \dfrac{3450(10)}{(35)^2\sqrt{50 + 20 + 25}} \approx 2.9$, or approximately 2.9 beats per minute per

minute. Sixty seconds into the run, it is increasing at

$P'(60) = \dfrac{3450(60)}{(85)^2\sqrt{1800 + 120 + 25}} \approx 0.65$, or approximately 0.7 beats per minute per

minute. Two minutes into the run, it is increasing at

$P'(120) = \dfrac{3450(120)}{(145)^2\sqrt{7200 + 240 + 25}} \approx 0.23$, or approximately 0.2 beats per minute

per minute. The pulse rate two minutes into the run is given by

$P(120) = \dfrac{300\sqrt{7200 + 240 + 25}}{120 + 25} \approx 178.8$, or approximately 179 beats per minute.

75. The area is given by $A = \pi r^2$. The rate at which the area is increasing is given by

dA/dt, that is, $\dfrac{dA}{dt} = \dfrac{d}{dt}(\pi r^2) = \dfrac{d}{dt}(\pi r^2)\dfrac{dr}{dt} = 2\pi r\dfrac{dr}{dt}$.

If $r = 40$ and $dr/dt = 2$, then $\dfrac{dA}{dt} = 2\pi(40)(2) = 160\pi$, that is, it is increasing at the

rate of 160π, or approximately 503, sq ft/sec.

77. $f(t) = 6.25t^2 + 19.75t + 74.75$. $g(x) = -0.00075x^2 + 67.5$.

$\dfrac{dS}{dt} = g'(x)f'(t) = (-0.0015x)(12.5t + 19.75)$.

When $t = 4$, we have $x = f(4) = 6.25(16) + 19.75(4) + 74.75 = 253.75$

and $\dfrac{dS}{dt}\Big|_{t=4} = (-0.0015)(253.75)[12.5(4) + 19.75] \approx -26.55;$

that is, the average speed will be dropping at the rate of approximately 27 mph per decade. The average speed of traffic flow at that time will be

$S = g(f(4)) = -0.00075(253.75^2) + 67.5 = 19.2$, or approximately 19 mph.

79. $N(x) = 1.42x$ and $x(t) = \dfrac{7t^2 + 140t + 700}{3t^2 + 80t + 550}$. The number of construction jobs as a function of time is $n(t) = N[x(t)]$. Using the Chain Rule,

$$n'(t) = \dfrac{dN}{dx}\cdot\dfrac{dx}{dt} = 1.42\dfrac{dx}{dt}$$

$$= (1.42)\left[\dfrac{(3t^2 + 80t + 550)(14t + 140) - (7t^2 + 140t + 700)(6t + 80)}{(3t^2 + 80t + 550)^2}\right]$$

$$= \dfrac{1.42(140t^2 + 3500t + 21000)}{(3t^2 + 80t + 550)^2}.$$

$n'(1) = \dfrac{1.42(140 + 3500 + 21000)}{(3 + 80 + 550)^2} \approx 0.0873216$, or approximately 87,322 jobs/year.

81. $x = f(p) = 10\sqrt{\dfrac{50 - p}{p}}$;

$$\dfrac{dx}{dp} = \dfrac{d}{dp}\left[10\left(\dfrac{50 - p}{p}\right)^{1/2}\right] = (10)(\tfrac{1}{2})\left(\dfrac{50 - p}{p}\right)^{-1/2}\dfrac{d}{dp}\left(\dfrac{50 - p}{p}\right)$$

$$= 5\left(\dfrac{50 - p}{p}\right)^{-1/2}\cdot\dfrac{d}{dp}\left(\dfrac{50}{p} - 1\right)$$

$$= 5\left(\dfrac{50 - p}{p}\right)^{-1/2}\left(-\dfrac{50}{p^2}\right) = -\dfrac{250}{p^2\left(\dfrac{50 - p}{p}\right)^{1/2}}$$

$$\dfrac{dx}{dp}\Big|_{p=25} = -\dfrac{250}{p^2\left(\dfrac{50 - p}{p}\right)^{1/2}} = -\dfrac{250}{(625)\left(\dfrac{25}{25}\right)^{1/2}} = -0.4$$

So the quantity demanded is falling at the rate of 0.4(1000) or 400 wristwatches per dollar increase in price.

83. True. This is just the statement of the Chain Rule.

85. True. $\dfrac{d}{dx}\sqrt{f(x)} = \dfrac{d}{dx}[f(x)]^{1/2} = \dfrac{1}{2}[f(x)]^{-1/2}f'(x) = \dfrac{f'(x)}{2\sqrt{f(x)}}$.

87. Let $f(x) = x^{1/n}$ so that $[f(x)]^{n} = x$.
Differentiating both sides with respect to x, we get
$$n[f(x)]^{n-1}f'(x) = 1$$
$$f'(x) = \dfrac{1}{n[f(x)]^{n-1}} = \dfrac{1}{n[x^{1/n}]^{n-1}} = \dfrac{1}{nx^{1-(1/n)}} = \dfrac{1}{n}x^{(1/n)-1}.$$
as was to be shown.

USING TECHNOLOGY EXERCISES 11.3 page 678

1. 0.5774   3. 0.9390   5. –4.9498

7. a. 10,146,200/decade   b. 7,810,520/decade

EXERCISES 11.4, page 691

1. a. $C(x)$ is always increasing because as x, the number of units produced, increases, the greater the amount of money that must be spent on production.
b. This occurs at $x = 4$, or a production level of 4000. You can see this by looking at the slopes of the tangent lines for x less than, equal to, and a little larger then $x = 4$.

3. a. The actual cost incurred in the production of the 1001st record is given by
$$C(1001) - C(1000) = [2000 + 2(1001) - 0.0001(1001)^{2}]$$
$$-[2000 + 2(1000) - 0.0001(1000)^{2}]$$
$$= 3901.7999 - 3900 = 1.7999,$$
or \$1.80. The actual cost incurred in the production of the 2001st record is given by $\;C(2001) - C(2000) = [2000 + 2(2001) - 0.0001(2001)^{2}]$
$$-[2000 + 2(2000) - 0.0001(2000)^{2}]$$

$$= 5601.5999 - 5600 = 1.5999, \text{ or } \$1.60.$$

b. The marginal cost is $C'(x) = 2 - 0.0002x$. In particular

$$C'(1000) = 2 - 0.0002(1000) = 1.80$$

and $\qquad C'(2000) = 2 - 0.0002(2000) = 1.60.$

5. a. $\overline{C}(x) = \dfrac{C(x)}{x} = \dfrac{100x + 200,000}{x} = 100 + \dfrac{200,000}{x}.$

   b. $\overline{C}'(x) = \dfrac{d}{dx}(100) + \dfrac{d}{dx}(200,000x^{-1}) = -200,000x^{-2} = -\dfrac{200,000}{x^2}.$

   c. $\displaystyle\lim_{x \to \infty} \overline{C}(x) = \lim_{x \to \infty}\left[100 + \dfrac{200,000}{x}\right] = 100$

   and this says that the average cost approaches \$100 per unit if the production level is very high.

7. $\overline{C}(x) = \dfrac{C(x)}{x} = \dfrac{2000 + 2x - 0.0001x^2}{x} = \dfrac{2000}{x} + 2 - 0.0001x.$

   $\overline{C}'(x) = -\dfrac{2000}{x^2} + 0 - 0.0001 = -\dfrac{2000}{x^2} - 0.0001.$

9. a. $R'(x) = \dfrac{d}{dx}(8000x - 100x^2) = 8000 - 200x.$

   b. $R'(39) = 8000 - 200(39) = 200. \quad R'(40) = 8000 - 200(40) = 0$
   $R'(41) = 8000 - 200(41) = -200$
   This suggests the total revenue is maximized if the price charged/ passenger is \$40.

11. a. $P(x) = R(x) - C(x) = (-0.04x^2 + 800x) - (200x + 300,000)$
   $\qquad = -0.04x^2 + 600x - 300,000.$
   b. $P'(x) = -0.08x + 600$
   c. $P'(5000) = -0.08(5000) + 600 = 200 \quad P'(8000) = -0.08(8000) + 600 = -40.$
   d.

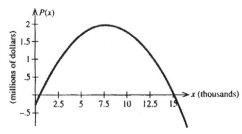

The profit realized by the company increases as production increases, peaking at a level of production of 7500 units. Beyond this level of production, the profit begins to fall.

13. a. The revenue function is $R(x) = px = (600 - 0.05x)x = 600x - 0.05x^2$
and the profit function is
$$P(x) = R(x) - C(x)$$
$$= (600x - 0.05x^2) - (0.000002x^3 - 0.03x^2 + 400x + 80,000)$$
$$= -0.000002x^3 - 0.02x^2 + 200x - 80,000.$$

b. $C'(x) = \dfrac{d}{dx}(0.000002x^3 - 0.03x^2 + 400x + 80,000) = 0.000006x^2 - 0.06x + 400.$

$R'(x) = \dfrac{d}{dx}(600x - 0.05x^2) = 600 - 0.1x.$

$P'(x) = \dfrac{d}{dx}(-0.000002x^3 - 0.02x^2 + 200x - 80,000) = -0.000006x^2 - 0.04x + 200.$

c. $C'(2000) = 0.000006(2000)^2 - 0.06(2000) + 400 = 304$, and this says that at a level of production of 2000 units, the cost for producing the 2001st unit is \$304. $R'(2000) = 600 - 0.1(2000) = 400$ and this says that the revenue realized in selling the 2001st unit is \$400. $P'(2000) = R'(2000) - C'(2000) = 400 - 304 = 96$, and this says that the revenue realized in selling the 2001st unit is \$96.

d.

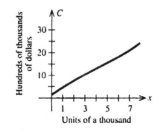

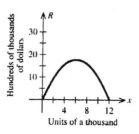

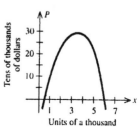

15. $\overline{C}(x) = \dfrac{C(x)}{x} = \dfrac{0.000002x^3 - 0.03x^2 + 400x + 80,000}{x}$
$$= 0.000002x^2 - 0.03x + 400 + \dfrac{80,000}{x}.$$

a. $\overline{C}'(x) = 0.000004x - 0.03 - \dfrac{80,000}{x^2}.$

b. $\overline{C}'(5000) = 0.000004(5000) - 0.03 - \dfrac{80,000}{5000^2} \approx -0.0132$, and this says that, at a

11 Differentiation

level of production of 5000 units, the average cost of
production is dropping at the rate of approximately a penny per unit.

$$\bar{C}'(10{,}000) = 0.000004(10000) - 0.03 - \frac{80{,}000}{10{,}000^2} \approx 0.0092,$$

and this says that, at a level of production of 10,000 units, the average cost of
production is increasing at the rate of approximately a penny per unit.

c.

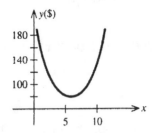

17. a. $R(x) = px = \dfrac{50x}{0.01x^2 + 1}$. b. $R'(x) = \dfrac{(0.01x^2 + 1)50 - 50x(0.02x)}{(0.01x^2 + 1)^2} = \dfrac{50 - 0.5x^2}{(0.01x^2 + 1)^2}$

c. $R'(2) = \dfrac{50 - 0.5(4)}{[0.01(4) + 1]^2} \approx 44.379.$

This result says that at a level of sale of 2000 units, the revenue increases at the
rate of approximately $44,379 per sales of 1000 units.

19. $C(x) = 0.873x^{1.1} + 20.34;\ \ C'(x) = 0.873(1.1)x^{0.1}$
$C'(10) = 0.873(1.1)(10)^{0.1} = 1.21$, or $1.21 billion per billion dollars.

21. The consumption function is given by $C(x) = 0.712x + 95.05$. The marginal
propensity to consume is given by $\dfrac{dC}{dx} = 0.712$. The marginal propensity to save is
given by $\dfrac{dS}{dx} = 1 - \dfrac{dC}{dx} = 1 - 0.712 = 0.288$, or $0.288 billion per billion dollars.

23. Here $x = f(p) = -\tfrac{5}{4}p + 20$ and so $f'(p) = -\tfrac{5}{4}$. Therefore,
$$E(p) = -\frac{pf'(p)}{f(p)} = -\frac{p(-\tfrac{5}{4})}{-\tfrac{5}{4}p + 20} = \frac{5p}{80 - 5p}.$$

$$E(10) = \frac{5(10)}{80 - 5(10)} = \frac{50}{30} = \frac{5}{3} > 1, \text{ and so the demand is elastic.}$$

25. $f(p) = -\frac{1}{3}p + 20;\ f'(p) = -\frac{1}{3}.$

Then the elasticity of demand is given by

$$E(p) = -\frac{p(-\frac{1}{3})}{-\frac{1}{3}p + 20},$$

and $E(30) = -\dfrac{30(-\frac{1}{3})}{-\frac{1}{3}(30) + 20} = 1,$

and we conclude that the demand is unitary at this price.

27. $x^2 = 169 - p$ and $f(p) = (169 - p)^{1/2}.$
   Next, $f'(p) = \frac{1}{2}(169 - p)^{-1/2}(-1) = -\frac{1}{2}(169 - p)^{-1/2}.$
   Then the elasticity of demand is given by

$$E(p) = -\frac{pf'(p)}{f(p)} = -\frac{p(-\frac{1}{2})(169 - p)^{-1/2}}{(169 - p)^{1/2}} = \frac{\frac{1}{2}p}{169 - p}.$$

Therefore, when $p = 29,$
$$E(p) = \frac{\frac{1}{2}(29)}{169 - 29} = \frac{14.5}{140} = 0.104.$$

Since $E(p) < 1,$ we conclude that demand is inelastic at this price.

29. $f(p) = \frac{1}{5}(225 - p^2);\ f'(p) = \frac{1}{5}(-2p) = -\frac{2}{5}p.$
   Then the elasticity of demand is given by
$$E(p) = -\frac{pf'(p)}{f(p)} = -\frac{p(-\frac{2}{5}p)}{\frac{1}{5}(225 - p^2)} = \frac{2p^2}{225 - p^2}.$$

a. When $p = 8,\ E(8) = \dfrac{2(64)}{225 - 64} = 0.8 < 1$ and the demand is inelastic. When $p = 10,$

$$E(10) = \frac{2(100)}{225 - 100} = 1.6 > 1$$

and the demand is elastic.

b. The demand is unitary when $E = 1$. Solving $\dfrac{2p^2}{225 - p^2} = 1$ we find $2p^2 = 225 - p^2$,

$3p^2 = 225$, and $p = 8.66$. So the demand is unitary when $p = 8.66$.

c. Since demand is elastic when $p = 10$, lowering the unit price will cause the revenue to increase.

d. Since the demand is inelastic at $p = 8$, a slight increase in the unit price will cause the revenue to increase.

31. $f(p) = \frac{2}{3}(36 - p^2)^{1/2}$

$f'(p) = \frac{2}{3}(\frac{1}{2})(36 - p^2)^{-1/2}(-2p) = -\frac{2}{3}p(36 - p^2)^{-1/2}$.

Then the elasticity of demand is given by

$$E(p) = -\frac{pf'(p)}{f(p)} = -\frac{-\frac{2}{3}p(36 - p^2)^{-1/2}\, p}{\frac{2}{3}(36 - p^2)^{1/2}} = \frac{p^2}{36 - p^2}.$$

When $p = 2$, $E(2) = \dfrac{4}{36 - 4} = \dfrac{1}{8} < 1$, and we conclude that the demand is inelastic.

b. Since the demand is inelastic, the revenue will increase when the rental price is increased.

33. We first solve the demand equation for x in terms of p. Thus,

$$p = \sqrt{9 - 0.02x}$$
$$p^2 = 9 - 0.02x$$

or $x = -50p^2 + 450$. With $f(p) = -50p^2 + 450$, we find

$$E(p) = -\frac{pf'(p)}{f(p)} = -\frac{p(-100p)}{-50p^2 + 450} = \frac{2p^2}{9 - p^2}.$$

Setting $E(p) = 1$ gives $2p^2 = 9 - p^2$, so $p = \sqrt{3}$. So the demand is inelastic in $[0, \sqrt{3}\,]$, unitary when $p = \sqrt{3}$, and elastic in $(\sqrt{3}, 3)$.

35. True. $\bar{C}'(x) = \dfrac{d}{dx}\left[\dfrac{C(x)}{x}\right] = \dfrac{xC'(x) - C(x)\dfrac{d}{dx}(x)}{x^2} = \dfrac{xC'(x) - C(x)}{x^2}$.

EXERCISES 11.5, page 698

1. $f(x) = 4x^2 - 2x + 1$; $f'(x) = 8x - 2$; $f''(x) = 8$.

3. $f(x) = 2x^3 - 3x^2 + 1;\ f'(x) = 6x^2 - 6x;\ f''(x) = 12x - 6 = 6(2x-1).$

5. $h(t) = t^4 - 2t^3 + 6t^2 - 3t + 10;\ h'(t) = 4t^3 - 6t^2 + 12t - 3$
    $h''(t) = 12t^2 - 12t + 12 = 12(t^2 - t + 1).$

7. $f(x) = (x^2 + 2)^5;\ f'(x) = 5(x^2 + 2)^4(2x) = 10x(x^2 + 2)^4$ and
    $f''(x) = 10(x^2 + 2)^4 + 10x(x^2 + 2)^3(2x)$
    $\quad = 10(x^2 + 2)^3[(x^2 + 2) + 8x^2] = 10(9x^2 + 2)(x^2 + 2)^3.$

9. $g(t) = (2t^2 - 1)^2(3t^2);$
    $g'(t) = 2(2t^2 - 1)(4t)(3t^2) - (2t^2 - 1)^2(6t)$
    $\quad = 6t(2t^2 - 1)[4t^2 + (2t^2 - 1)] = 6t(2t^2 - 1)(6t^2 - 1)$
    $\quad = 6t(12t^4 - 8t^2 + 1) = 72t^5 - 48t^3 + 6t.$
    $g''(t) = 360t^4 - 144t^2 + 6 = 6(60t^4 - 24t^2 + 1)$

11. $f(x) = (2x^2 + 2)^{7/2};\ f'(x) = \frac{7}{2}(2x^2 + 2)^{5/2}(4x) = 14x(2x^2 + 2)^{5/2};$
    $f''(x) = 14(2x^2 + 2)^{5/2} + 14x(\frac{5}{2})(2x^2 + 2)^{3/2}(4x)$
    $\quad = 14(2x^2 + 2)^{3/2}[(2x^2 + 2) + 10x^2] = 28(6x^2 + 1)(2x^2 + 2)^{3/2}.$

13. $f(x) = x(x^2 + 1)^2;$
    $f'(x) = (x^2 + 1)^2 + x(2)(x^2 + 1)(2x)$
    $\quad = (x^2 + 1)[(x^2 + 1) + 4x^2] = (x^2 + 1)(5x^2 + 1);$
    $f''(x) = 2x(5x^2 + 1) + (x^2 + 1)(10x) = 2x(5x^2 + 1 + 5x^2 + 5)$
    $\quad = 4x(5x^2 + 3).$

15. $f(x) = \dfrac{x}{2x+1};\ f'(x) = \dfrac{(2x+1)(1) - x(2)}{(2x+1)^2} = \dfrac{1}{(2x+1)^2};$
    $f''(x) = \dfrac{d}{dx}(2x+1)^{-2} = -2(2x+1)^{-3}(2) = -\dfrac{4}{(2x+1)^3}.$

17. $f(s) = \dfrac{s-1}{s+1}$; $f'(s) = \dfrac{(s+1)(1)-(s-1)(1)}{(s+1)^2} = \dfrac{2}{(s+1)^2}$.

$f''(s) = 2\dfrac{d}{ds}(s+1)^{-2} = -4(s+1)^{-3} = -\dfrac{4}{(s+1)^3}$.

19. $f(u) = \sqrt{4-3u} = (4-3u)^{1/2}$. $f'(u) = \frac{1}{2}(4-3u)^{-1/2}(-3) = -\dfrac{3}{2\sqrt{4-3u}}$.

$f''(u) = -\dfrac{3}{2}\cdot\dfrac{d}{du}(4-3u)^{-1/2} = -\dfrac{3}{2}\left(-\dfrac{1}{2}\right)(4-3u)^{-3/2}(-3) = -\dfrac{9}{4(4-3u)^{3/2}}$.

21. $f(x) = 3x^4 - 4x^3$; $f'(x) = 12x^3 - 12x^2$; $f''(x) = 36x^2 - 24x$; $f'''(x) = 72x - 24$.

23. $f(x) = \dfrac{1}{x}$; $f'(x) = \dfrac{d}{dx}(x^{-1}) = -x^{-2}$; $f''(x) = 2x^{-3}$; $f'''(x) = -6x^{-4} = -\dfrac{6}{x^4}$.

25. $g(s) = (3s-2)^{1/2}$; $g'(s) = \dfrac{1}{2}(3s-2)^{-1/2}(3) = \dfrac{3}{2(3s-2)^{1/2}}$;

$g''(s) = \dfrac{3}{2}\left(-\dfrac{1}{2}\right)(3s-2)^{-3/2}(3) = -\dfrac{9}{4}(3s-2)^{-3/2} = -\dfrac{9}{4(3s-2)^{3/2}}$;

$g'''(s) = \dfrac{27}{8}(3s-2)^{-5/2}(3) = \dfrac{81}{8}(3s-2)^{-5/2} = \dfrac{81}{8(3s-2)^{5/2}}$.

27. $f(x) = (2x-3)^4$; $f'(x) = 4(2x-3)^3(2) = 8(2x-3)^3$

$f''(x) = 24(2x-3)^2(2) = 48(2x-3)^2$; $f'''(x) = 96(2x-3)(2) = 192(2x-3)$.

29. Its velocity at any time t is $v(t) = \dfrac{d}{dt}(16t^2) = 32t$. The hammer strikes the ground when $16t^2 = 256$ or $t = 4$ (we reject the negative root). Therefore, its velocity at the instant it strikes the ground is $v(4) = 32(4) = 128$ ft/sec. Its acceleration at time t is $a(t) = \dfrac{d}{dt}(32t) = 32$. In particular, its acceleration at $t = 4$ is 32 ft/sec^2.

31. $N(t) = -0.1t^3 + 1.5t^2 + 100$.

a. $N'(t) = -0.3t^2 + 3t = 0.3t(10 - t)$. Since $N'(t) > 0$ for $t = 0, 1, 2, ..., 7$, it is evident that $N(t)$ (and therefore the crime rate) was increasing from 1988 through 1995.

b. $N''(t) = -0.6t + 3 = 0.6(5 - t)$. Now $N''(4) = 0.6 > 0$, $N''(5) = 0$, $N''(6) = -0.6 < 0$ and $N''(7) = -1.2 < 0$. This shows that the rate of the rate of change was decreasing beyond $t = 5$ (1990). This shows that the program was working.

33. $N(t) = 0.00037t^3 - 0.0242t^2 + 0.52t + 5.3$ $(0 \le t \le 10)$

$N'(t) = 0.00111t^2 - 0.0484t + 0.52$

$N''(t) = 0.00222t - 0.0484$

So $N(8) = 0.00037(8)^3 - 0.0242(8)^2 + 5.3 = 8.1$

$N'(8) = 0.00111(8)^2 - 0.0484(8) + 0.52 \approx 0.204$.

$N''(8) = 0.00222(8) - 0.0484 = -0.031$.

We conclude that at the beginning of 1998, there were 8.1 million persons receiving disability benefits, the number is increasing at the rate of 0.2 million/yr, and the rate of the rate of change of the number of persons is decreasing at the rate of 0.03 million persons/yr^2.

35. $N(t) = -0.00233t^4 + 0.00633t^3 - 0.05417t^2 + 1.3467t + 25$

$N'(t) = -0.00932t^3 + 0.01899t^2 - 0.10834t + 1.3467$

$N''(t) = -0.02796t^2 + 0.03798t - 0.10834$

So $N'(10) = -7.158$ and $N''(10) = -2.5245$.

Our computations show that at the beginning of the year 2000, the number of Americans aged 45 to 54 was decreasing at the rate of 7 million people per year, and the number decreases at a rate of approximately 2.5 million people per year per year.

37. $f(t) = 10.72(0.9t + 10)^{0.3}$.

$f'(t) = 10.72(0.3)(0.9t + 10)^{-0.7}(0.9) = 2.8944(0.9t + 10)^{-0.7}$

$f''(t) = 2.8944(-0.7)(0.9t + 10)^{-1.7}(0.9) = -1.823472(0.9t + 10)^{-1.7}$

So $f''(10) = -1.823472(19)^{-1.7} \approx -0.01222$. And this says that the rate of the rate of change of the population is decreasing at the rate of 0.01%/yr^2.

39. False. If f has derivatives of order two at $x = a$, then $f''(a) = [f'(a)]^2$.

41. True. If $f(x)$ is a polynomial function of degree n, then $f^{(n+1)}(x) = 0$.

43. True. Using the chain rule, $h'(x) = f'(2x) \cdot \dfrac{d}{dx}(2x) = f'(x) \cdot 2 = 2f'(2x)$

Using the chain rule again, $h''(x) = 2f''(2x) \cdot 2 = 4f''(2x)$.

45. Consider the function $f(x) = x^{(2n+1)/2} = x^{n+(1/2)}$.

Then $\quad f'(x) = (n + \tfrac{1}{2})x^{n-(1/2)}$

$\qquad f''(x) = (n + \tfrac{1}{2})(n - \tfrac{1}{2})x^{n-(3/2)}$

$\qquad$..

$\qquad f^{(n)}(x) = (n + \tfrac{1}{2})(n - \tfrac{1}{2}) \; \cdots \; \tfrac{3}{2}x^{1/2}$

$\qquad f^{(n+1)}(x) = (n + \tfrac{1}{2})(n - \tfrac{1}{2}) \; \cdots \; \tfrac{1}{2}x^{-1/2}$.

The first n derivatives exist at $x = 0$, but the $(n + 1)$st derivative fails to be defined there.

USING TECHNOLOGY EXERCISES 11.5, page 701

1. -18 　　　　　　　 3. 15.2762 　　　　　　　 5. -0.6255 　　　　　　　 7. 0.1973

9. $f''(6) = -68.46214$ and it tells us that at the beginning of 1988, the rate of the rate of the rate at which banks were failing was 68 banks per year per year per year.

EXERCISES 11.6, page 710

1. a. Solving for y in terms of x, we have $y = -\tfrac{1}{2}x + \tfrac{5}{2}$. Therefore, $y' = -\tfrac{1}{2}$.

　 b. Next, differentiating $x + 2y = 5$ implicitly, we have $1 + 2y' = 0$, or $y' = -\tfrac{1}{2}$.

3. a. $xy = 1$, $y = \dfrac{1}{x}$, and $\dfrac{dy}{dx} = -\dfrac{1}{x^2}$.

　 b. $\qquad x\dfrac{dy}{dx} + y = 0$

$\qquad\qquad x\dfrac{dy}{dx} = -y$

$$\frac{dy}{dx} = -\frac{y}{x} = \frac{-\frac{1}{x}}{x} = -\frac{1}{x^2}.$$

5. $x^3 - x^2 - xy = 4.$

a. $-xy = 4 - x^3 + x^2$

$$y = -\frac{4}{x} + x^2 - x \quad \text{and} \quad y' = \frac{4}{x^2} + 2x - 1.$$

b. $x^3 - x^2 - xy = 4$

$$-x\frac{dy}{dx} = -3x^2 + 2x + y$$

$$\frac{dy}{dx} = 3x - 2 - \frac{y}{x}$$

$$= 3x - 2 - \frac{1}{x}(-\frac{4}{x} + x^2 - x) = 3x - 2 + \frac{4}{x^2} - x + 1$$

$$= \frac{4}{x^2} + 2x - 1.$$

7. a. $\dfrac{x}{y} - x^2 = 1$ is equivalent to $\dfrac{x}{y} = x^2 + 1,$ or $y = \dfrac{x}{x^2 + 1}.$ Therefore,

$$y' = \frac{(x^2 + 1) - x(2x)}{(x^2 + 1)^2} = \frac{1 - x^2}{(x^2 + 1)^2}.$$

b. Next, differentiating the equation $x - x^2 y = y$ implicitly, we obtain

$$1 - 2xy - x^2 y' = y', \; y'(1 + x^2) = 1 - 2xy, \text{ or } \; y' = \frac{1 - 2xy}{(1 + x^2)}.$$

(This may also be written in the form $-2y^2 + \dfrac{y}{x}$.) To show that this is equivalent to
the results obtained earlier, use the value of y obtained before, to get

$$y' = \frac{1 - 2x\left(\dfrac{x}{x^2 + 1}\right)}{1 + x^2} = \frac{x^2 + 1 - 2x^2}{(1 + x^2)^2} = \frac{1 - x^2}{(1 + x^2)^2}.$$

9. $x^2 + y^2 = 16.$ Differentiating both sides of the equation implicitly, we obtain

$$2x + 2yy' = 0 \text{ and so } y' = -\frac{x}{y}.$$

11. $x^2 - 2y^2 = 16$. Differentiating implicitly with respect to x, we have

$$2x - 4y\frac{dy}{dx} = 0 \text{ and } \frac{dy}{dx} = \frac{x}{2y}.$$

13. $x^2 - 2xy = 6$. Differentiating both sides of the equation implicitly, we obtain

$$2x - 2y - 2xy' = 0 \text{ and so } y' = \frac{x-y}{x} = 1 - \frac{y}{x}.$$

15. $x^2y^2 - xy = 8$. Differentiating both sides of the equation implicitly, we obtain

$$2xy^2 + 2x^2yy' - y - xy' = 0, \; 2xy^2 - y + y'(2x^2y - x) = 0$$

and so
$$y' = \frac{y(1-2xy)}{x(2xy-1)} = -\frac{y}{x}.$$

17. $x^{1/2} + y^{1/2} = 1$. Differentiating implicitly with respect to x, we have

$$\tfrac{1}{2}x^{-1/2} + \tfrac{1}{2}y^{-1/2}\frac{dy}{dx} = 0. \text{ Therefore, } \frac{dy}{dx} = -\frac{x^{-1/2}}{y^{-1/2}} = -\frac{\sqrt{y}}{\sqrt{x}}.$$

19. $\sqrt{x+y} = x$. Differentiating both sides of the equation implicitly, we obtain

$$\tfrac{1}{2}(x+y)^{-1/2}(1+y') = 1, \; 1+y' = 2(x+y)^{1/2},$$

or
$$y' = 2\sqrt{x+y} - 1.$$

21. $\dfrac{1}{x^2} + \dfrac{1}{y^2} = 1$. Differentiating both sides of the equation implicitly, we obtain

$$-\frac{2}{x^3} - \frac{2}{y^3}y' = 0, \text{ or } y' = -\frac{y^3}{x^3}.$$

23. $\sqrt{xy} = x + y$. Differentiating both sides of the equation implicitly, we obtain

$$\tfrac{1}{2}(xy)^{-1/2}(xy'+y) = 1 + y'$$

$$xy' + y = 2\sqrt{xy}(1+y')$$

$$y'(x - 2\sqrt{xy}) = 2\sqrt{xy} - y$$

or
$$y' = -\frac{(2\sqrt{xy} - y)}{(2\sqrt{xy} - x)} = \frac{2\sqrt{xy} - y}{x - 2\sqrt{xy}}.$$

25. $\dfrac{x+y}{x-y} = 3x$, or $x+y = 3x^2 - 3xy$. Differentiating both sides of the equation

implicitly, we obtain $\quad 1 + y' = 6x - 3xy' - 3y$ or $y' = \dfrac{6x - 3y - 1}{3x + 1}$.

27. $xy^{3/2} = x^2 + y^2$. Differentiating implicitly with respect to x, we obtain

$$y^{3/2} + x\left(\tfrac{3}{2}\right)y^{1/2}\dfrac{dy}{dx} = 2x + 2y\dfrac{dy}{dx}$$

$$2y^{3/2} + 3xy^{1/2}\dfrac{dy}{dx} = 4x + 4y\dfrac{dy}{dx} \qquad \text{(Multiplying by 2.)}$$

$$(3xy^{1/2} - 4y)\dfrac{dy}{dx} = 4x - 2y^{3/2}$$

$$\dfrac{dy}{dx} = \dfrac{2(2x - y^{3/2})}{3xy^{1/2} - 4y}.$$

29. $(x + y)^3 + x^3 + y^3 = 0$. Differentiating implicitly with respect to x, we obtain

$$3(x + y)^2\left(1 + \dfrac{dy}{dx}\right) + 3x^2 + 3y^2\dfrac{dy}{dx} = 0$$

$$(x + y)^2 + (x + y)^2\dfrac{dy}{dx} + x^2 + y^2\dfrac{dy}{dx} = 0$$

$$[(x + y)^2 + y^2]\dfrac{dy}{dx} = -[(x + y)^2 + x^2]$$

$$\dfrac{dy}{dx} = -\dfrac{2x^2 + 2xy + y^2}{x^2 + 2xy + 2y^2}.$$

31. $4x^2 + 9y^2 = 36$. Differentiating the equation implicitly, we obtain
$$8x + 18yy' = 0.$$
At the point $(0,2)$, we have $0 + 36y' = 0$ and the slope of the tangent line is 0.
Therefore, an equation of the tangent line is $y = 2$.

33. $x^2y^3 - y^2 + xy - 1 = 0$. Differentiating implicitly with respect to x, we have
$$2xy^3 + 3x^2y^2\dfrac{dy}{dx} - 2y\dfrac{dy}{dx} + y + x\dfrac{dy}{dx} = 0.$$

11 Differentiation

At (1,1), $2 + 3\dfrac{dy}{dx} - 2\dfrac{dy}{dx} + 1 + \dfrac{dy}{dx} = 0$, and

$2\dfrac{dy}{dx} = -3$ and $\dfrac{dy}{dx} = -\dfrac{3}{2}$.

Using the point-slope form of an equation of a line, we have

$$y - 1 = -\tfrac{3}{2}(x - 1)$$

and the equation of the tangent line to the graph of the function f at $(1,1)$ is

$$y = -\tfrac{3}{2}x + \tfrac{5}{2}.$$

35. $xy = 1$. Differentiating implicitly, we have $xy' + y = 0$, or $y' = -\dfrac{y}{x}$.

Differentiating implicitly once again, we have $xy'' + y' + y' = 0$.

Therefore, $y'' = -\dfrac{2y'}{x} = \dfrac{2\left(\dfrac{y}{x}\right)}{x} = \dfrac{2y}{x^2}$.

37. $y^2 - xy = 8$. Differentiating implicitly we have $2yy' - y - xy' = 0$

and $y' = \dfrac{y}{2y - x}$. Differentiating implicitly again, we have

$$2(y')^2 + 2yy'' - y' - y' - xy'' = 0, \quad \text{or} \quad y'' = \dfrac{2y' - 2(y')^2}{2y - x}.$$

Then $y'' = \dfrac{2\left(\dfrac{y}{2y - x}\right)\left(1 - \dfrac{y}{2y - x}\right)}{2y - x} = \dfrac{2y(2y - x - y)}{(2y - x)^3} = \dfrac{2y(y - x)}{(2y - x)^3}$.

39. a. Differentiating the given equation with respect to t, we obtain

$$\dfrac{dV}{dt} = \pi r^2 \dfrac{dh}{dt} + 2\pi r h \dfrac{dr}{dt} = \pi r \left(r \dfrac{dh}{dt} + 2h \dfrac{dr}{dt} \right).$$

b. Substituting $r = 2$, $h = 6$, $\dfrac{dr}{dt} = 0.1$ and $\dfrac{dh}{dt} = 0.3$ into the expression for $\dfrac{dV}{dt}$

we obtain $\dfrac{dV}{dt} = \pi(2)[2(0.3) + 2(6)(0.1)] = 3.6\pi$

and so the volume is increasing at the rate of 3.6π cu in/sec.

41. We are given $\dfrac{dp}{dt} = 2$ and are required to find $\dfrac{dx}{dt}$ when $x = 9$ and $p = 63$.

Differentiating the equation $p + x^2 = 144$ with respect to t, we obtain

$$\frac{dp}{dt} + 2x\frac{dx}{dt} = 0.$$

When $x = 9$, $p = 63$, and $\dfrac{dp}{dt} = 2$,

$$2 + 2(9)\frac{dx}{dt} = 0$$

and

$$\frac{dx}{dt} = -\frac{1}{9} \approx -0.111,$$

or the quantity demanded is decreasing at the rate of 111 tires per week.

43. $100x^2 + 9p^2 = 3600$. Differentiating the given equation implicitly with respect to t, we have $200x\dfrac{dx}{dt} + 18p\dfrac{dp}{dt} = 0$. Next, when $p = 14$, the given equation yields

$$100x^2 + 9(14)^2 = 3600$$
$$100x^2 = 1836,$$

or $x = 4.2849$. When $p = 14$, $\dfrac{dp}{dt} = -0.15$, and $x = 4.2849$, we have

$$200(4.2849)\frac{dx}{dt} + 18(14)(-0.15) = 0$$

$$\frac{dx}{dt} = 0.0441.$$

So the quantity demanded is increasing at the rate of 44 ten–packs per week.

45. From the results of Problem 44, we have

$$1250p\frac{dp}{dt} - 2x\frac{dx}{dt} = 0.$$

When $p = 1.0770$, $x = 25$, and $\dfrac{dx}{dt} = -1$, we find that

$$1250(1.077)\frac{dp}{dt} - 2(25)(-1) = 0,$$

and
$$\frac{dp}{dt} = -\frac{50}{1250(1.077)} = -0.037.$$

We conclude that the price is decreasing at the rate of 3.7 cents per carton.

47. $p = -0.01x^2 - 0.2x + 8$. Differentiating the given equation implicitly with respect to p, we have

$$1 = -0.02x\frac{dx}{dp} - 0.2\frac{dx}{dp} = [0.02x + 0.2]\frac{dx}{dp}$$

or
$$\frac{dx}{dp} = -\frac{1}{0.02x + 0.2}.$$

When $x = 15$, $p = -0.01(15)^2 - 0.2(15) + 8 = 2.75$

and
$$\frac{dx}{dp} = -\frac{1}{0.02(15) + 0.2} = -2.$$

Therefore, $E(p) = -\frac{pf'(p)}{f(p)} = -\frac{(2.75)(-2)}{15} = 0.37 < 1,$

and the demand is inelastic.

49. $A = \pi r^2$. Differentiating with respect to t, we obtain

$$\frac{dA}{dt} = 2\pi r\frac{dr}{dt}.$$

When the radius of the circle is 40 ft and increasing at the rate of 2 ft/sec,

$$\frac{dA}{dt} = 2\pi(40)(2) = 160\pi \ \text{ft}^2 / \text{sec}.$$

51. Let D denote the distance between the two cars, x the distance traveled by the car heading east, and y the distance traveled by the car heading north as shown in the

diagram at the right. Then
$D^2 = x^2 + y^2$. Differentiating with respect
to t, we have

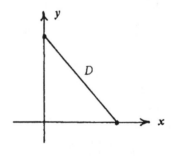

$$2D\frac{dD}{dt} = 2x\frac{dx}{dt} + 2y\frac{dy}{dt},$$

or
$$\frac{dD}{dt} = \frac{x\frac{dx}{dt} + y\frac{dy}{dt}}{D}$$

When $t = 5$, $x = 30$, $y = 40$, $\dfrac{dx}{dt} = 2(5) + 1 = 11$, and $\dfrac{dy}{dt} = 2(5) + 3 = 13$.

Therefore, $\dfrac{dD}{dt} = \dfrac{(30)(11) + (40)(13)}{\sqrt{900 + 1600}} = 17$ ft/sec.

53. Referring to the diagram at the right, we see that
$$D^2 = 120^2 + x^2.$$
Differentiating this last equation with respect to t, we have

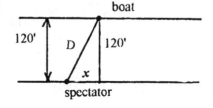

$$2D\dfrac{dD}{dt} = 2x\dfrac{dx}{dt} \quad \text{and} \quad \dfrac{dD}{dt} = \dfrac{x\dfrac{dx}{dt}}{D}.$$

When $x = 50$, $D = \sqrt{120^2 + 50^2} = 130$ and
$$\dfrac{dD}{dt} = \dfrac{(20)(50)}{130} \approx 7.69, \text{ or } 7.69 \text{ ft/sec.}$$

55. Let V and S denote its volume and surface area. Then we are given that
$\dfrac{dV}{dt} = -kS$, where k is the constant of proportionality. But from $V = \left(\dfrac{4}{3}\right)\pi r^3$,
we find, upon differentiating both sides with respect to t, that
$$\dfrac{dV}{dt} = \left(\dfrac{4}{3}\right)\pi(3\pi r^2)\dfrac{dr}{dt} = 4\pi^2 r^2 \dfrac{dr}{dt}$$
and using the fact stated earlier, $\dfrac{dV}{dt} = 4\pi^2 r^2 \dfrac{dr}{dt} = -kS = -k(4\pi r^2)$.

Therefore, $\dfrac{dr}{dt} = -\dfrac{k(4\pi r^2)}{4\pi^2 r^2} = -\dfrac{k}{\pi}$ and this proves that the radius is decreasing at the constant rate of (k/π) units/unit time.

57. Refer to the figure at the right.
We are given that $\dfrac{dx}{dt} = 264$. Using the
Pythagorean Theorem,

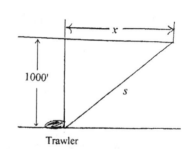

$$s^2 = x^2 + 1000^2 = x^2 + 1000000.$$

We want to find $\dfrac{ds}{dt}$ when $s = 1500$. Differentiating both sides of the equation with respect to t, we have

$$2s\frac{ds}{dt} = 2x\frac{dx}{dt} \quad \text{and so} \quad \frac{ds}{dt} = \frac{x\dfrac{dx}{dt}}{s}.$$

Now, when $s = 1500$, we have

$$1500^2 = x^2 + 10000 \quad \text{or} \quad x = \sqrt{1250000}.$$

Therefore, $\qquad \dfrac{ds}{dt} = \dfrac{\sqrt{1250000} \cdot (264)}{1500} \approx 196.8,$

that is, the aircraft is receding from the trawler at the speed of approximately 196.8 ft/sec.

59. Refer to the diagram at the right.

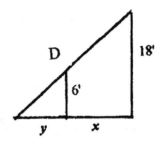

$\dfrac{y}{6} = \dfrac{y+x}{18}, \quad 18y = 6(y+x)$

$3y = y + x, \quad 2y = x, \quad y = \tfrac{1}{2}x.$

Then $D = y + x = \tfrac{3}{2}x$. Differentiating

implicitly, we have $\qquad \dfrac{dD}{dt} = \dfrac{3}{2} \cdot \dfrac{dx}{dt}$

and when $\dfrac{dx}{dt} = 6, \quad \dfrac{dD}{dt} = \dfrac{3}{2}(6) = 9,$ or 9 ft / sec.

61. Differentiating $x^2 + y^2 = 13^2 = 169$ with respect to t gives

$$2x\frac{dx}{dt} + 2y\frac{dy}{dt} = 0.$$

When $x = 12$, we have $144 + y^2 = 169$ or $y = 5$.

Therefore, with $x = 12$, $y = 5$, and $\dfrac{dx}{dt} = 8$, we find $2(12)(8) + 2(5)\dfrac{dy}{dt} = 0$

or $\dfrac{dy}{dt} = -19.2$, that is, the top of the ladder is sliding down the wall at 19.2 ft/sec.

63. True. Differentiating both sides of the equation with respect to x, we have

11 Differentiation

$$\frac{d}{dx}[f(x)g(y)] = \frac{d}{dx}(0)$$

$$f(x)g'(y)\frac{dy}{dx} + f'(x)g(y) = 0$$

$$\frac{dy}{dx} = -\frac{f'(x)g(y)}{f(x)g'(y)}$$

provided $f(x) \neq 0$ and $g'(y) \neq 0$.

EXERCISES 11.7, page 719

1. $f(x) = 2x^2$ and $dy = 4x\, dx$.   3. $f(x) = x^3 - x$ and $dy = (3x^2 - 1)\, dx$.

5. $f(x) = \sqrt{x+1} = (x+1)^{1/2}$ and $dy = \frac{1}{2}(x+1)^{-1/2}\, dx = \frac{dx}{2\sqrt{x+1}}$.

7. $f(x) = 2x^{3/2} + x^{1/2}$ and $dy = (3x^{1/2} + \frac{1}{2}x^{-1/2})\, dx = \frac{1}{2}x^{-1/2}(6x + 1)dx = \frac{6x+1}{2\sqrt{x}}\, dx$.

9. $f(x) = x + \dfrac{2}{x}$ and $dy = \left(1 - \dfrac{2}{x^2}\right)dx = \dfrac{x^2 - 2}{x^2}\, dx$.

11. $f(x) = \dfrac{x-1}{x^2+1}$ and $dy = \dfrac{x^2+1-(x-1)2x}{(x^2+1)^2}\, dx = \dfrac{-x^2+2x+1}{(x^2+1)^2}\, dx$.

13. $f(x) = \sqrt{3x^2 - x} = (3x^2 - x)^{1/2}$ and

$$dy = \frac{1}{2}(3x^2 - x)^{-1/2}(6x - 1)dx = \frac{6x-1}{2\sqrt{3x^2 - x}}\, dx.$$

15. $f(x) = x^2 - 1$.
    a. $dy = 2x\, dx$.   b. $dy \approx 2(1)(0.02) = 0.04$.
    c. $\Delta y = [(1.02)^2 - 1] - [1 - 1] = 0.0404$.

17. $f(x) = \dfrac{1}{x}$.

a. $dy = -\dfrac{dx}{x^2}$.
b. $dy \approx -0.05$
c. $\Delta y = \dfrac{1}{-0.95} - \dfrac{1}{-1} = -0.05263$.

19. $y = \sqrt{x}$ and $dy = \dfrac{dx}{2\sqrt{x}}$. Therefore, $\sqrt{10} = 3 + \dfrac{1}{2 \cdot \sqrt{9}} = 3.167$.

21. $y = \sqrt{x}$ and $dy = \dfrac{dx}{2\sqrt{x}}$. Therefore, $\sqrt{49.5} = 7 + \dfrac{0.5}{2 \cdot 7} = 7.0358$.

23. $y = x^{1/3}$ and $dy = \tfrac{1}{3}x^{-2/3}\, dx$. Therefore, $\sqrt[3]{7.8} = 2 - \dfrac{0.2}{3 \cdot 4} = 1.983$.

25. $y = \sqrt{x}$ and $dy = \dfrac{dx}{2\sqrt{x}}$. Therefore, $\sqrt{0.089} = \tfrac{1}{10}\sqrt{8.9} = \tfrac{1}{10}\left[3 - \dfrac{0.1}{2 \cdot 3}\right] \approx 0.298$.

27. $y = f(x) = \sqrt{x} + \dfrac{1}{\sqrt{x}} = x^{1/2} + x^{-1/2}$. Therefore,

$$\dfrac{dy}{dx} = \dfrac{1}{2}x^{-1/2} - \dfrac{1}{2}x^{-3/2}$$

$$dy = \left(\dfrac{1}{2x^{1/2}} - \dfrac{1}{2x^{3/2}}\right)dx.$$

Letting $x = 4$ and $dx = 0.02$, we find

$$\sqrt{4.02} + \dfrac{1}{\sqrt{4.02}} - f(4) = f(4.02) - f(4) = \Delta y \approx dy$$

$$\sqrt{4.02} + \dfrac{1}{\sqrt{4.02}} = f(4) + dy\big|_{\substack{x=4 \\ dx=0.02}}$$

$$\approx 2 + \dfrac{1}{2} + \left(\dfrac{1}{2 \cdot 2} - \dfrac{1}{2 \cdot 2\sqrt{2}}\right)(0.02) \approx 2.50146.$$

29. The volume of the cube is given by $V = x^3$. Then $dV = 3x^2\, dx$ and when $x = 12$ and $dx = 0.02$, $dV = 3(144)(\pm 0.02) = \pm 8.64$, and the possible error that might occur in calculating the volume is ± 8.64 cm^3.

31. The volume of the hemisphere is given by $V = \tfrac{2}{3}\pi r^3$. The amount of rust-proofer

needed is

$$\Delta V = \frac{2}{3}\pi(r+\Delta r)^3 - \frac{2}{3}\pi r^3$$

$$\approx dV = \left(\frac{2}{3}\right)(3\pi r^2)dr.$$

So, with $r = 60$, and $dr = \frac{1}{12}(0.01)$, we have

$$\Delta V \approx 2\pi(60^2)\left(\frac{1}{12}\right)(0.01) \approx 18.85.$$

So we need approximately 18.85 ft^3 of rust-proofer.

33. $dR = \frac{d}{dr}(k\ell r^{-4})dr = -4k\ell r^{-5}\,dr.$ With $\frac{dr}{r} = 0.1$, we find

$$\frac{dR}{R} = -\frac{4k\ell r^{-5}}{k\ell r^{-4}}\,dr = -4\frac{dr}{r} = -4(0.1) = -0.4.$$

In other words, the resistance will drop by 40%.

35. $f(n) = 4n\sqrt{n-4} = 4n(n-4)^{1/2}.$

Then $df = 4[(n-4)^{1/2} + \frac{1}{2}n(n-4)^{-1/2}]dn$

When $n = 85$ and $dn = 5$, $df = 4[9 + \frac{85}{2.9}]5 \approx 274$ seconds.

37. $N(r) = \frac{7}{1+0.02r^2}$ and $dN = -\frac{0.28r}{(1+0.02r^2)^2}\,dr.$ To estimate the decrease in the

number of housing starts when the mortgage rate is increased from 12 to 12.5 percent, we compute

$$dN = -\frac{(0.28)(12)(0.5)}{(3.88)^2} \approx -0.111595 \quad (r = 12, \; dr = 0.5)$$

or 111,595 fewer housing starts.

39. $p = \frac{30}{0.02x^2 + 1}$ and $dp = -\frac{(1.2x)}{(0.02x^2 + 1)^2}\,dx.$ To estimate the change in the price p

when the quantity demanded changed from 5000 to 5500 units ($x = 5$ to $x = 5.5$) per

week, we compute $dp = \frac{(-1.2)(5)(0.5)}{[0.02(25)+1]^2} \approx -1.33$, or a decrease of $1.33.

41. $P(x) = -0.000032x^3 + 6x - 100$ and $dP = (-0.000096x^2 + 6)\,dx$. To determine the error in the estimate of Trappee's profits corresponding to a maximum error in the forecast of 15 percent $[dx = \pm 0.15(200)]$, we compute
$$dP = [(-0.000096)(200)^2 + 6]\,(\pm 30) = (2.16)(30) = \pm 64.80$$
or $\pm 64,800$.

43. $N(x) = \dfrac{500(400 + 20x)^{1/2}}{(5 + 0.2x)^2}$ and

$$N'(x) = \frac{(5 + 0.2x)^2 250(400 + 20x)^{-1/2}(20) - 500(400 + 20x)^{1/2}(2)(5 + 0.2x)(0.2)}{(5 + 0.2x)^4}\,dx.$$

To estimate the change in the number of crimes if the level of reinvestment changes from 20 cents per dollars deposited to 22 cents per dollar deposited, we compute

$$dN = \frac{(5 + 4)^2(250)(800)^{-1/2}(20) - 500(400 + 400)^{1/2}(2)(9)(0.2)}{(5 + 4)^4}(2) \quad (2)$$

$$= \frac{(14318.91 - 50911.69)}{9^4}(2) \approx -11$$

or a decrease of approximately 11 crimes per year.

45. $A = 10,000\left(1 + \dfrac{r}{12}\right)^{120}$.

a. $dA = 10,000(120)\left(1 + \dfrac{r}{12}\right)^{119}\left(\dfrac{1}{12}\right)dr = 100,000\left(1 + \dfrac{r}{12}\right)^{119}dr$.

b. At 8.1%, it will be worth $100,000\left(1 + \dfrac{0.08}{12}\right)^{119}(0.001)$, or \$220.49 more.

At 8.2%, it will be worth $100,000\left(1 + \dfrac{0.08}{12}\right)^{119}(0.002)$, or \$440.99 more.

At 8.3%, it will be worth $100,000\left(1 + \dfrac{0.08}{12}\right)^{119}(0.003)$, or \$661.48 more.

47. True. $dy = f'(x)dx = \dfrac{d}{dx}(ax + b)dx = a\,dx$. On the other hand,

$$\Delta y = f(x + \Delta x) - f(x) = [a(x + \Delta x) + b] - (ax + b) = a\Delta x = a\,dx.$$

USING TECHNOLOGY EXERCISES 11.7, page 721

1. $dy = f'(3) \, dx = 757.87(0.01) \approx 7.5787.$

3. $dy = f'(1) \, dx = 1.04067285926(0.03) \approx 0.031220185778.$

5. $dy = f'(4)(0.1) = -0.198761598(0.1) = -0.0198761598.$

7. If the interest rate changes from 10% to 10.3% per year, the monthly payment will increase by
$$dP = f'(0.1)(0.003) \approx 26.60279,$$
or approximately \$26.60 per month. If the interest rate changes from 10% to 10.4% per year, it will be \$35.47 per month. If the interest rate changes from 10% to 10.5% per year, it will be \$44.34 per month.

9. $dx = f'(40)(2) \approx -0.625.$ That is, the quantity demanded will decrease by 625 watches per week.

CHAPTER 11 REVIEW, page 725

1. $f'(x) = \dfrac{d}{dx}(3x^5 - 2x^4 + 3x^2 - 2x + 1) = 15x^4 - 8x^3 + 6x - 2.$

3. $g'(x) = \dfrac{d}{dx}(-2x^{-3} + 3x^{-1} + 2) = 6x^{-4} - 3x^{-2}.$

5. $g'(t) = \dfrac{d}{dt}(2t^{-1/2} + 4t^{-3/2} + 2) = -t^{-3/2} - 6t^{-5/2}.$

7. $f'(t) = \dfrac{d}{dt}(t + 2t^{-1} + 3t^{-2}) = 1 - 2t^{-2} - 6t^{-3} = 1 - \dfrac{2}{t^2} - \dfrac{6}{t^3}.$

9. $h'(x) = \dfrac{d}{dx}(x^2 - 2x^{-3/2}) = 2x + 3x^{-5/2} = 2x + \dfrac{3}{x^{5/2}}.$

11. $g(t) = \dfrac{t^2}{2t^2 + 1}.$

$$g'(t) = \frac{(2t^2 + 1)\dfrac{d}{dt}(t^2) - t^2\dfrac{d}{dt}(2t^2 + 1)}{(2t^2 + 1)^2}$$

$$= \frac{(2t^2 + 1)(2t) - t^2(4t)}{(2t^2 + 1)^2} = \frac{2t}{(2t^2 + 1)^2}.$$

13. $f(x) = \dfrac{\sqrt{x}-1}{\sqrt{x}+1} = \dfrac{x^{1/2}-1}{x^{1/2}+1}$.

$$f'(x) = \frac{(x^{1/2}+1)(\frac{1}{2}x^{-1/2})-(x^{1/2}-1)(\frac{1}{2}x^{-1/2})}{(x^{1/2}+1)^2}$$

$$= \frac{\frac{1}{2}+\frac{1}{2}x^{-1/2}-\frac{1}{2}+\frac{1}{2}x^{-1/2}}{(x^{1/2}+1)^2} = \frac{x^{-1/2}}{(x^{1/2}+1)^2} = \frac{1}{\sqrt{x}(\sqrt{x}+1)^2}.$$

15. $f(x) = \dfrac{x^2(x^2+1)}{x^2-1}$.

$$f'(x) = \frac{(x^2-1)\dfrac{d}{dx}(x^4+x^2)-(x^4+x^2)\dfrac{d}{dx}(x^2-1)}{(x^2-1)^2}$$

$$= \frac{(x^2-1)(4x^3+2x)-(x^4+x^2)(2x)}{(x^2-1)^2}$$

$$= \frac{4x^5+2x^3-4x^3-2x-2x^5-2x^3}{(x^2-1)^2}$$

$$= \frac{2x^5-4x^3-2x}{(x^2-1)^2} = \frac{2x(x^4-2x^2-1)}{(x^2-1)^2}.$$

17. $f(x) = (3x^3-2)^8; f'(x) = 8(3x^3-2)^7(9x^2) = 72x^2(3x^3-2)^7.$

19. $f'(t) = \dfrac{d}{dt}(2t^2+1)^{1/2} = \dfrac{1}{2}(2t^2+1)^{-1/2}\dfrac{d}{dt}(2t^2+1)$

$$= \frac{1}{2}(2t^2+1)^{-1/2}(4t) = \frac{2t}{\sqrt{2t^2+1}}.$$

21. $s(t) = (3t^2-2t+5)^{-2}$

$s'(t) = -2(3t^2-2t+5)^{-3}(6t-2) = -4(3t^2-2t+5)^{-3}(3t-1)$

$$= -\frac{4(3t-1)}{(3t^2-2t+5)^3}.$$

23. $h(x) = \left(x + \dfrac{1}{x}\right)^2 = (x + x^{-1})^2.$

$\quad h'(x) = 2(x + x^{-1})(1 - x^{-2}) = 2\left(x + \dfrac{1}{x}\right)\left(1 - \dfrac{1}{x^2}\right)$

$\quad\quad = 2\left(\dfrac{x^2 + 1}{x}\right)\left(\dfrac{x^2 - 1}{x^2}\right) = \dfrac{2(x^2 + 1)(x^2 - 1)}{x^3}.$

25. $h'(t) = (t^2 + t)^4 \dfrac{d}{dt}(2t^2) + 2t^2 \dfrac{d}{dt}(t^2 + t)^4$

$\quad\quad = (t^2 + t)^4 (4t) + 2t^2 \cdot 4(t^2 + t)^3 (2t + 1)$

$\quad\quad = 4t(t^2 + t)^3 [(t^2 + t) + 4t^2 + 2t] = 4t^2(5t + 3)(t^2 + t)^3.$

27. $g(x) = x^{1/2}(x^2 - 1)^3.$

$\quad g'(x) = \dfrac{d}{dx}[x^{1/2}(x^2 - 1)^3] = x^{1/2} \cdot 3(x^2 - 1)^2 (2x) + (x^2 - 1)^3 \cdot \tfrac{1}{2}x^{-1/2}$

$\quad\quad = \tfrac{1}{2}x^{-1/2}(x^2 - 1)^2[12x^2 + (x^2 - 1)]$

$\quad\quad = \dfrac{(13x^2 - 1)(x^2 - 1)^2}{2\sqrt{x}}.$

29. $h(x) = \dfrac{(3x + 2)^{1/2}}{4x - 3}.$

$\quad h'(x) = \dfrac{(4x - 3)\tfrac{1}{2}(3x + 2)^{-1/2}(3) - (3x + 2)^{1/2}(4)}{(4x - 3)^2}$

$\quad\quad = \dfrac{\tfrac{1}{2}(3x + 2)^{-1/2}[3(4x - 3) - 8(3x + 2)]}{(4x - 3)^2} = -\dfrac{12x + 25}{2\sqrt{3x + 2}\,(4x - 3)^2}.$

31. $f(x) = 2x^4 - 3x^3 + 2x^2 + x + 4.$

$\quad f'(x) = \dfrac{d}{dx}(2x^4 - 3x^3 + 2x^2 + x + 4) = 8x^3 - 9x^2 + 4x + 1.$

$\quad f''(x) = \dfrac{d}{dx}(8x^3 - 9x^2 + 4x + 1) = 24x^2 - 18x + 4 = 2(12x^2 - 9x + 2).$

33. $h(t) = \dfrac{t}{t^2+4}$. $h'(t) = \dfrac{(t^2+4)(1)-t(2t)}{(t^2+4)^2} = \dfrac{4-t^2}{(t^2+4)^2}$.

$h''(t) = \dfrac{(t^2+4)^2(-2t)-(4-t^2)2(t^2+4)(2t)}{(t^2+4)^4}$

$\qquad = \dfrac{-2t(t^2+4)[(t^2+4)+2(4-t^2)]}{(t^2+4)^4} = \dfrac{2t(t^2-12)}{(t^2+4)^3}$.

35. $f'(x) = \dfrac{d}{dx}(2x^2+1)^{1/2} = \dfrac{1}{2}(2x^2+1)^{-1/2}(4x) = 2x(2x^2+1)^{-1/2}$.

$f''(x) = 2(2x^2+1)^{-1/2} + 2x\cdot(-\tfrac{1}{2})(2x^2+1)^{-3/2}(4x)$

$\qquad = 2(2x^2+1)^{-3/2}[(2x^2+1)-2x^2] = \dfrac{2}{(2x^2+1)^{3/2}}$.

37. $6x^2 - 3y^2 = 9$ so $12x - 6y\dfrac{dy}{dx} = 0$ and $-6y\dfrac{dy}{dx} = -12x$.

Therefore, $\dfrac{dy}{dx} = \dfrac{-12x}{-6y} = \dfrac{2x}{y}$.

39. $y^3 + 3x^2 = 3y$, so $3y^2y' + 6x = 3y'$, $3y^2y' - 3y' = -6x$,

and $y'(3y^2-3) = -6x$. Therefore, $y' = -\dfrac{6x}{3(y^2-1)} = -\dfrac{2x}{y^2-1}$.

41. $x^2 - 4xy - y^2 = 12$ so $2x - 4xy' - 4y - 2yy' = 0$ and $y'(-4x-2y) = -2x+4y$.

So $y' = \dfrac{-2(x-2y)}{-2(2x+y)} = \dfrac{x-2y}{2x+y}$.

43. $df = f'(x)dx = (2x - 2x^{-3})dx = \left(2x - \dfrac{2}{x^3}\right)dx = \dfrac{2(x^4-1)}{x^3}dx$

45. a. $df = f'(x)dx = \dfrac{d}{dx}(2x^2+4)^{1/2}\,dx = \dfrac{1}{2}(2x^2+4)^{-1/2}(4x) = \dfrac{2x}{\sqrt{2x^2+4}}\,dx$

b. $\Delta f \approx df\big|_{\substack{x=4 \\ dx=0.1}} = \dfrac{2(4)(0.1)}{\sqrt{2(16)+4}} = \dfrac{0.8}{6} = \dfrac{8}{60} = \dfrac{2}{15}.$

c. $\Delta f = f(4.1) - f(4) = \sqrt{2(4.1)^2 + 4} - \sqrt{2(16)+4} = 0.1335$

From (b), $\Delta f \approx \dfrac{2}{15} \approx 0.1333.$

47. $f(x) = 2x^3 - 3x^2 - 16x + 3$ and $f'(x) = 6x^2 - 6x - 16.$

a. To find the point(s) on the graph of f where the slope of the tangent line is equal to –4, we solve
$$6x^2 - 6x - 16 = -4,\ 6x^2 - 6x - 12 = 0,\ 6(x^2 - x - 2) = 0$$
$$6(x - 2)(x + 1) = 0$$
and $x = 2$ or $x = -1$. Then $f(2) = 2(2)^3 - 3(2)^2 - 16(2) + 3 = -25$ and $f(-1) = 2(-1)^3 - 3(-1)^2 - 16(-1) + 3 = 14$ and the points are $(2,-25)$ and $(-1,14)$.

b. Using the point-slope form of the equation of a line, we find
that $\quad y - (-25) = -4(x - 2),\ y + 25 = -4x + 8,\ \text{or } y = -4x - 17$
and $\quad y - 14 = -4(x + 1),\ \text{or } y = -4x + 10$
are the equations of the tangent lines at $(2,-25)$ and $(-1,14)$.

49. $y = (4 - x^2)^{1/2}.$ $y' = \tfrac{1}{2}(4 - x^2)^{-1/2}(-2x) = -\dfrac{x}{\sqrt{4 - x^2}}.$

The slope of the tangent line is obtained by letting $x = 1$, giving
$$m = -\dfrac{1}{\sqrt{3}} = -\dfrac{\sqrt{3}}{3}.$$
Therefore, an equation of the tangent line is

$$y - \sqrt{3} = -\dfrac{\sqrt{3}}{3}(x - 1),\ \text{ or } y = -\dfrac{\sqrt{3}}{3}x + \dfrac{4\sqrt{3}}{3}.$$

51. $f(x) = (2x - 1)^{-1};\ f'(x) = -2(2x - 1)^{-2},\ f''(x) = 8(2x - 1)^{-3} = \dfrac{8}{(2x - 1)^3}.$

$f'''(x) = -48(2x - 1)^{-4} = -\dfrac{48}{(2x - 1)^4}.$

Since $(2x - 1)^4 = 0$ when $x = 1/2$, we see that the domain of f''' is
$(-\infty, \tfrac{1}{2}) \cup (\tfrac{1}{2}, \infty).$

53. $x = \dfrac{25}{\sqrt{p}} - 1$; $f'(p) = -\dfrac{25}{2p^{3/2}}$; $E(p) = -\dfrac{p\left(-\frac{25}{2p^{3/2}}\right)}{\frac{25}{p^{1/2}} - 1} = \dfrac{\frac{25}{2p^{1/2}}}{\frac{25 - p^{1/2}}{p^{1/2}}} = \dfrac{25}{2(25 - p^{1/2})}$

Since $E(p) = 1$,

$$2(25 - p^{1/2}) = 25,$$

$$25 - p^{1/2} = \tfrac{25}{2}, \quad p^{1/2} = \tfrac{25}{2}, \text{ and } p = \tfrac{625}{4}.$$

$E(p) > 1$ and demand is elastic if $p > 156.25$; $E(p) = 1$ and demand is unitary if $p = 156.25$; and $E(p) < 1$ and demand is inelastic, if $p < 156.25$.

55. $p = 9\sqrt[3]{1000 - x}$; $\sqrt[3]{1000 - x} = \dfrac{p}{9}$; $1000 - x = \dfrac{p^3}{729}$; $x = 1000 - \dfrac{p^3}{729}$

Therefore, $x = f(p) = \dfrac{729,000 - p^3}{729}$ and $f'(x) = -\dfrac{3p^2}{729} = -\dfrac{p^2}{243}$.

Then $E(p) = -\dfrac{p\left(-\frac{p^2}{243}\right)}{\frac{729,000 - p^3}{729}} = \dfrac{3p^3}{729,000 - p^3}$.

So $E(60) = \dfrac{3(60)^3}{729,000 - 60^3} = \dfrac{648,000}{513,000} = \dfrac{648}{513} > 1$, and so demand is elastic.

Therefore, raising the price slightly will cause the revenue to decrease.

57. $N(x) = 1000(1 + 2x)^{1/2}$. $N'(x) = 1000(\tfrac{1}{2})(1 + 2x)^{-1/2}(2) = \dfrac{1000}{\sqrt{1 + 2x}}$.

The rate of increase at the end of the twelfth week is $N'(12) = \dfrac{1000}{\sqrt{25}} = 200$,

or 200 subscribers/week.

59. He can expect to live $f(100) = 46.9[1 + 1.09(100)]^{0.1} \approx 75.0433$, or approximately

75.04 years. $f'(t) = 46.9(0.1)(1 + 1.09t)^{-0.9}(1.09) = 5.1121(1 + 1.09t)^{-0.9}$

So the required rate of change is $f'(100) = 5.1121(1 + 1.09)^{-0.9} = 0.074$, or

approximately 0.07 yr/yr.

61. a. $R(x) = px = (-0.02x + 600)x = -0.02x^2 + 600x$

  b. $R'(x) = -0.04x + 600$

  c. $R'(10,000) = -0.04(10,000) + 600 = 200$ and this says that the sale of the
   10,001st phone will bring a revenue of $200.

CHAPTER 12

EXERCISES 12.1, page 740

1. f is decreasing on $(-\infty,0)$ and increasing on $(0, \infty)$.

3. f is increasing on $(-\infty,-1) \cup (1,\infty)$, and decreasing on $(-1,1)$.

5. f is increasing on $(0,2)$ and decreasing on $(-\infty,0) \cup (2,\infty)$.

7. f is decreasing on $(-\infty,-1) \cup (1,\infty)$ and increasing on $(-1,1)$.

9. Increasing on $(20.2, 20.6) \cup (21.7, 21.8)$, constant on $(19.6, 20.2) \cup (20.6, 21.1)$, and decreasing on $(21.1, 21.7) \cup (21.8, 22.7)$.

11. $f(x) = 3x + 5$; $f'(x) = 3 > 0$ for all x and so f is increasing on $(-\infty,\infty)$.

13. $f(x) = x^2 - 3x$. $f'(x) = 2x - 3$ is continuous everywhere and is equal to zero when $x = 3/2$. From the following sign diagram

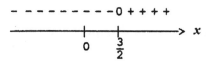

we see that f is decreasing on $(-\infty, \frac{3}{2})$ and increasing on $(\frac{3}{2}, \infty)$.

15. $g(x) = x - x^3$. $g'(x) = 1 - 3x^2$ is continuous everywhere and is equal to zero when $1 - 3x^2 = 0$, or $x = \pm\frac{\sqrt{3}}{3}$. From the following sign diagram

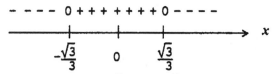

we see that f is decreasing on $(-\infty, -\frac{\sqrt{3}}{3}) \cup (\frac{\sqrt{3}}{3}, \infty)$ and increasing on $(-\frac{\sqrt{3}}{3}, \frac{\sqrt{3}}{3})$.

17. $g(x) = x^3 + 3x^2 + 1$; $g'(x) = 3x^2 + 6x = 3x(x + 2)$.
From the following sign diagram

$$+ + + + + + 0 - - 0 + + + + +$$

$$\xrightarrow{\qquad \overset{|}{-2} \quad \overset{|}{0} \qquad} x$$

we see that g is increasing on $(-\infty,-2) \cup (0,\infty)$ and decreasing on $(-2,0)$.

19. $f(x) = \frac{1}{3}x^3 - 3x^2 + 9x + 20$; $f'(x) = x^2 - 6x + 9 = (x - 3)^2 > 0$ for all x except $x = 3$, at which point $f'(3) = 0$. Therefore, f is increasing on $(-\infty,3) \cup (3,\infty)$.

21. $h(x) = x^4 - 4x^3 + 10$; $h'(x) = 4x^3 - 12x^2 = 4x^2(x - 3)$ if $x = 0$ or 3. From the sign diagram of h',

$$- - - - - - 0 - - - - - -0 + + + + + +$$

$$\xrightarrow{\qquad \overset{|}{0} \qquad\quad \overset{|}{3} \qquad} x$$

we see that h is increasing on $(3,\infty)$ and decreasing on $(-\infty,0) \cup (0,3)$.

23. $f(x) = \dfrac{1}{x-2} = (x-2)^{-1}$. $f'(x) = -1(x-2)^{-2}(1) = -\dfrac{1}{(x-2)^2}$ is discontinuous at

$x = 2$ and is continuous everywhere else. From the sign diagram

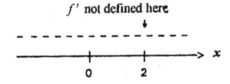

f' not defined here

we see that f is decreasing on $(-\infty,2) \cup (2,\infty)$.

25. $h(t) = \dfrac{t}{t-1}$. $h'(t) = \dfrac{(t-1)(1) - t(1)}{(t-1)^2} = -\dfrac{1}{(t-1)^2}$.
From the following sign diagram,

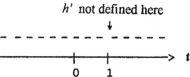

we see that $h'(t) < 0$ whenever it is defined. We conclude that h is decreasing on $(-\infty,1) \cup (1,\infty)$.

27. $f(x) = x^{3/5}$. $f'(x) = \dfrac{3}{5}x^{-2/5} = \dfrac{3}{5x^{2/5}}$. Observe that $f'(x)$ is not defined at $x = 0$, but is positive everywhere else and therefore increasing on $(-\infty,0) \cup (0,\infty)$.

29. $f(x) = \sqrt{x+1}$. $f'(x) = \dfrac{d}{dx}(x+1)^{1/2} = \dfrac{1}{2}(x+1)^{-1/2} = \dfrac{1}{2\sqrt{x+1}}$ and we see that $f'(x) > 0$ if $x > -1$. Therefore, f is increasing on $(-1, \infty)$.

31. $f(x) = \sqrt{16 - x^2} = (16 - x^2)^{1/2}$. $f'(x) = \dfrac{1}{2}(16 - x^2)^{-1/2}(-2x) = -\dfrac{x}{\sqrt{16 - x^2}}$.

Since the domain of f is $[-4,4]$, we consider the sign diagram for f' on this interval. Thus,

f' not defined here

$$+ + + \ 0 \ - - - -$$

and we see that f is increasing on $(-4,0)$ and decreasing on $(0,4)$.

33. $f'(x) = \dfrac{d}{dx}(x - x^{-1}) = 1 + \dfrac{1}{x^2} = \dfrac{x^2 + 1}{x^2}$ and so $f'(x) > 0$ for all $x \ne 0$.

Therefore, f is increasing on $(-\infty,0) \cup (0,\infty)$.

35. $f'(x) = \dfrac{d}{dx}(x-1)^{-2} = -2(x-1)^{-3} = -\dfrac{2}{(x-1)^3}$. From the sign diagram of f'

f' not defined here

$$+ + + + + + + + + + \ - - - -$$

we see that f is increasing on $(-\infty,1)$ and decreasing on $(1,\infty)$.

37. f has a relative maximum of $f(0) = 1$ and relative minima of $f(-1) = 0$ and $f(1) = 0$.

39. f has a relative maximum of $f(-1) = 2$ and a relative minimum of $f(1) = -2$.

41. f has a relative maximum of $f(1) = 3$ and a relative minimum of $f(2) = 2$.

43. f has a relative minimum at $(0,2)$.

45. a   47. d

49. $f(x) = x^2 - 4x$. $f'(x) = 2x - 4 = 2(x - 2)$ has a critical point at $x = 2$. From the following sign diagram

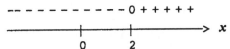

we see that $f(2) = -4$ is a relative minimum by the First Derivative Test.

51. $h(t) = -t^2 + 6t + 6$; $h'(t) = -2t + 6 = -2(t - 3) = 0$ if $t = 3$, a critical point. The sign diagram

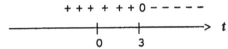

and the First Derivative Test imply that h has a relative maximum at 3 with value $f(3) = -9 + 18 + 6 = 15$.

53. $f(x) = x^{5/3}$. $f'(x) = \frac{5}{3}x^{2/3}$ giving $x = 0$ as the critical point of f.
From the sign diagram

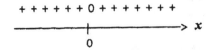

we see that f' does not change sign as we move across $x = 0$, and conclude that f has no relative extremum.

55. $g(x) = x^3 - 3x^2 + 4$. $g'(x) = 3x^2 - 6x = 3x(x - 2) = 0$ if $x = 0$ or 2. From the sign

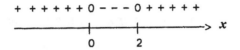

diagram, we see that the critical point $x = 0$ gives a relative maximum, whereas, $x = 2$ gives a relative minimum. The values are $g(0) = 4$ and $g(2) = 8 - 12 + 4 = 0$.

57. $f(x) = \frac{1}{2}x^4 - x^2$. $f'(x) = 2x^3 - 2x = 2x(x^2 - 1) = 2x(x+1)(x-1)$ is continuous everywhere and has zeros as $x = -1$, $x = 0$, and $x = 1$, the critical points of f. Using the First Derivative Test and the following sign diagram of f'

$$- - - - \; 0+ \; + + \; 0 - - \; \; 0 + + +$$
$$\xrightarrow{\hspace{4cm}} x$$
$$\quad -1 \quad\quad 0 \quad\quad 1$$

we see that $f(-1) = -1/2$ and $f(1) = -1/2$ are relative minima of f and $f(0) = 0$ is a relative maximum of f.

59. $F(x) = \frac{1}{3}x^3 - x^2 - 3x + 4$. Setting $F'(x) = x^2 - 2x - 3 = (x - 3)(x + 1) = 0$ gives $x = -1$ and $x = 3$ as critical points. From the sign diagram

$$+ \; + + + \; 0 - - - - - - -0 + + + + +$$
$$\xrightarrow{\hspace{4cm}} x$$
$$\quad -1 \quad 0 \quad\quad\quad 3$$

we see that $x = -1$ gives a relative maximum and $x = 3$ gives a relative minimum. The values are
$$F(-1) = -\frac{1}{3} - 1 + 3 + 4 = \frac{17}{3} \quad \text{and} \quad F(3) = 9 - 9 - 9 + 4 = -5,$$
respectively.

61. $g(x) = x^4 - 4x^3 + 8$. Setting $g'(x) = 4x^3 - 12x^2 = 4x^2(x - 3) = 0$ gives $x = 0$ and $x = 3$ as critical points. From the sign diagram

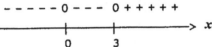

we see that $x = 3$ gives a relative minimum. Its value is $g(3) = 3^4 - 4(3)^3 + 8 = -19$.

63. $g'(x) = \dfrac{d}{dx}\left(1 + \dfrac{1}{x}\right) = -\dfrac{1}{x^2}$. Observe that g' is never zero for all values of x.

Furthermore, g' is undefined at $x = 0$, but $x = 0$ is not in the domain of g.
Therefore g has no critical points and so g has no relative extrema.

65. $f(x) = x + \dfrac{9}{x} + 2$. Setting $f'(x) = 1 - \dfrac{9}{x^2} = \dfrac{x^2 - 9}{x^2} = \dfrac{(x+3)(x-3)}{x^2} = 0$

gives $x = -3$ and $x = 3$ as critical points. From the sign diagram

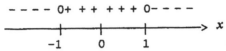

we see that $(-3,-4)$ is a relative maximum and $(3,8)$ is a relative minimum.

67. $f(x) = \dfrac{x}{1+x^2}$. $f'(x) = \dfrac{(1+x^2)(1) - x(2x)}{(1+x^2)^2} = \dfrac{1-x^2}{(1+x^2)^2} = \dfrac{(1-x)(1+x)}{(1+x^2)^2} = 0$ if $x = \pm 1$,

and these are critical points of f. From the sign diagram of f'

we see that f has a relative minimum at $(-1,-\tfrac{1}{2})$ and a relative maximum at $(1,\tfrac{1}{2})$.

69. $f(x) = \dfrac{x^2}{x^2 - 4}$. $f'(x) = \dfrac{(x^2-4)(2x) - x^2(2x)}{(x^2-4)^2} = -\dfrac{8x}{(x^2-4)^2}$ is continuous

everywhere except at $x \pm 2$ and has a zero at $x = 0$. Therefore, $x = 0$ is the only
critical point of f (the points $x = \pm 2$ do not lie in the domain of f).Using the
following sign diagram of f'

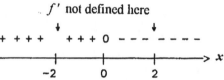

f' not defined here

+ + + +   + + + 0 - - - - - - - -

⟶ x

−2   0   2

and the First Derivative Test, we conclude that $f(0) = 0$ is a relative maximum of f.

71. $f(x) = (x - 1)^{2/3}$. $f'(x) = \dfrac{2}{3}(x-1)^{-1/3} = \dfrac{2}{3(x-1)^{1/3}}$.

$f'(x)$ is discontinuous at $x = 1$. The sign diagram for f' is

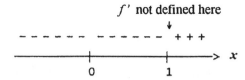

f' not defined here

− − − − − −   − − − − − −   + + +

⟶ x

0   1

We conclude that $f(1) = 0$ is a relative minimum.

73. $h(t) = -16t^2 + 64t + 80$. $h'(t) = -32t + 64 = -32(t - 2)$ and has sign diagram

+ + + + 0 − − − −

⟶ t

0   2

This tells us that the stone is rising on the time interval $(0,2)$ and falling when $t > 2$. It hits the ground when $h(t) = -16t^2 + 64t + 80 = 0$
or $t^2 - 4t - 5 = (t - 5)(t + 1) = 0$ or $t = 5$ (we reject the root $t = -1$.)

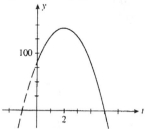

75. $P'(x) = \dfrac{d}{dx}(0.0726x^2 + 0.7902x + 4.9623) = 0.1452x + 0.7902$.

Since $P'(x) > 0$ on $(0, 25)$, we see that P is increasing on the interval in question. Our result tells us that the percent of the population afflicted with Alzheimer's disease increases with age for those that are 65 and over.

77. $I(t) = \frac{1}{3}t^3 - \frac{5}{2}t^2 + 80$; $I'(t) = t^2 - 5t = t(t - 5) = 0$ if $t = 0$ or 5. From the sign

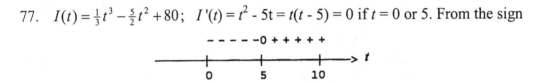

diagram, we see that I is decreasing on $(0,5)$ and increasing on $(5,10)$. After declining from 1984 through 1989, the index begins to increase after 1989.

79. $\overline{C}(x) = -0.0001x + 2 + \dfrac{2000}{x}$. $\overline{C}'(x) = -0.0001 - \dfrac{2000}{x^2} < 0$ for all values of x and

so $\overline{C}$ is always decreasing.

81. $A(t) = -96.6t^4 + 403.6t^3 + 660.9t^2 + 250$
$A'(t) = -386.4t^3 + 1210.8t^2 + 1321.8t = t(386.4t^2 + 1210.8t + 1321.8)$.
Solving $A'(t) = 0$, we find $t = 0$ and
$$t = \frac{-1210.8 \pm \sqrt{(1210.8)^2 - 4(-386.4)(1321.8)}}{-2(386.4)} = \frac{-1210.8 \pm 1873.2}{-2(386.4)} \approx 4.$$
Since t lies in the interval $[0,5]$, we see that the continuous function A' has zeros at $t = 0$ and $t = 4$. From the sign diagram

we see that f is increasing on $(0,4)$ and decreasing on $(4,5)$. We conclude that the cash in the Central Provident Trust Funds will be increasing from 1995 to 2035 and decreasing from 2035 to 2045.

83. a. $f'(t) = \dfrac{d}{dt}(-0.05t^3 + 0.56t^2 + 5.47t + 7.5) = -0.15t^2 + 1.12t + 5.47$.

Setting $f'(t) = 0$ gives $-0.15t^2 + 1.12t + 5.47 = 0$. Using the quadratic formula, we find
$$t = \frac{-1.12 \pm \sqrt{(1.12)^2 - 4(-0.15)(5.47)}}{-0.3}$$
that is, $t = -3.37$, or 10.83. Since f' is continuous, the only critical points of f are $t = -3.4$ and $t = 10.8$, both of which lie outside the interval of interest. Nevertheless this result can be used to tell us that f' does not change sign in the

interval (-3.4, 10.8). Using $t = 0$ as the test point, we see that $f'(0) = 5.47 > 0$ and so we see that f is increasing on (-3.4, 10.8), and , in particular, in the interval (0, 6). Thus, we conclude that f is increasing on (0, 6).

b. The result of part (a) tells us that sales in the Web-hosting industry will be increasing throughout the years from 1999 through 2005.

85. a. $N'(t) = \dfrac{d}{dt}(0.09444t^3 - 1.44167t^2 + 10.65695t + 52)$

$\qquad = (0.28332t^2 - 2.88334t + 10.65695)$

Observe that N' is continuous everywhere. Setting $N'(t) = 0$ gives

$$t = \dfrac{2.88334 \pm \sqrt{2.88334^2 - 4(0.28332)(10.65695)}}{2(0.28332)}$$

Since the expression under the radical is $-3.76 < 0$, we see that there is no solution. Now, $N'(0) = 10.65695 > 0$, and we conclude that N is always increasing on (0, 6).

b. The result of part (a) shows that the number of subscribers is always increasing over the period in question.

87. $C(t) = \dfrac{t^2}{2t^3 + 1}$; $C'(t) = \dfrac{(2t^3 + 1)(2t) - t^2(6t^2)}{(2t^3 + 1)^2} = \dfrac{2t - 2t^4}{(2t^3 + 1)^2} = \dfrac{2t(1 - t^3)}{(2t^3 + 1)^2}$.

From the sign diagram of C' on $(0, \infty)$,

```
                + + 0 - - - - - - -
        ────────┼──┼──────────┼──> t
                0  1          4
```

We see that the drug concentration is increasing on $(0,1)$ and decreasing on $(1,4)$.

89. $A(t) = \dfrac{136}{1 + 0.25(t - 4.5)^2} + 28$.

$A'(t) = 136\dfrac{d}{dt}[1 + 0.25(t - 4.5)^2]^{-1} = -136[1 + 0.25(t - 4.5)^2]^{-2}2(0.25)(t - 4.5)$

$\qquad = -\dfrac{68(t - 4.5)}{[1 + 0.25(t - 4.5)^2]^2}$.

Observe that $A'(t) > 0$ if $t < 4.5$ and $A'(t) < 0$ if $t > 4.5$, so the pollution is increasing from 7 A.M. to 11:30 A.M. and decreasing from 11:30 A.M. to 6 P.M.

91. We compute $f'(x) = m$. If $m > 0$, then $f'(x) > 0$ for all x and f is increasing; if $m < 0$, then $f'(x) < 0$ for all x and f is decreasing; if $m = 0$, then $f'(x) = 0$ for all x and f is a constant function.

93. False. The function $f(x) = \begin{cases} -x+1 & x < 0 \\ -\dfrac{1}{2}x+1 & x \geq 0 \end{cases}$

is decreasing on (-1,1), but $f'(0)$ does not exist.

95. False. Let $f(x) = -x$ and $g(x) = -2x$. then both f and g are decreasing on $(-\infty, \infty)$, but $f(x) - g(x) = x - (-2x) = x$ is increasing on $(-\infty, \infty)$.

97. False. Let $f(x) = x^3$. then $f'(0) = 3x^2\big|_{x=0} = 0$. But f does not have a relative extremum at $x = 0$.

99. $f'(x) = 3x^2 + 1$ is continuous on $(-\infty, \infty)$ and is always greater than or equal to 1. So f has no critical points in $(-\infty, \infty)$. Therefore f has no relative extrema on $(-\infty, \infty)$.

101. a. $f'(x) = -2x$ if $x \neq 0$. $f'(-1) = 2$ and $f'(1) = -2$ so $f'(x)$ changes sign from positive to negative as we move across $x = 0$.
b. f does not have a relative maximum at $x = 0$ because $f(0) = 2$ but a neighborhood of $x = 0$, for example $(-\frac{1}{2}, \frac{1}{2})$, contains points with values larger than 2. This does not contradict the First Derivative Test because f is not continuous at $x = 0$.

103. $f(x) = ax^2 + bx + c$. Setting $f'(x) = 2ax + b = 2a\left(x + \frac{b}{2a}\right) = 0$ gives $x = -\frac{b}{2a}$ as the only critical point of f. If $a < 0$, we have the sign diagram

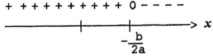

from which we see that $x = -b/2a$ gives a relative maximum. Similarly, you can show that if $a > 0$, then $x = -b/2a$ gives a relative minimum.

105. a. $f'(x) = 3x^2 + 1$ and so $f'(x) > 1$ on the interval $(0,1)$. Therefore, f is increasing on $(0,1)$.

b. $f(0) = -1$ and $f(1) = 1 + 1 - 1 = 1$. So the Intermediate Value Theorem guarantees that there is at least one root of $f(x) = 0$ in $(0,1)$. Since f is increasing on $(0,1)$, the graph of f can cross the x-axis at only one point in $(0,1)$. So $f(x) = 0$ has exactly one root.

USING TECHNOLOGY EXERCISES 12.1, page 745

1. a. f is decreasing on $(-\infty,-0.2934)$ and increasing on $(-0.2934,\infty)$.
   b. Relative minimum: $f(-0.2934) = -2.5435$

3. a. f is increasing on $(-\infty,-1.6144) \cup (0.2390,\infty)$ and decreasing on $(-1.6144, 0.2390)$
   b. Relative maximum: $f(-1.6144) = 26.7991$; relative minimum: $f(0.2390) = 1.6733$

5. a. f is decreasing on $(-\infty,-1) \cup (0.33,\infty)$ and increasing on $(-1,0.33)$
   b. Relative maximum: $f(0.33) = 1.11$; relative minimum: $f(-1) = -0.63$

7. a. f is decreasing on $(-1,-0.71)$ and increasing on $(-0.71,1)$.
   b. f has a relative minimum at $(-0.71,-1.41)$.

9. a.

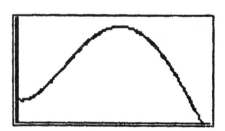

b. f is decreasing on $(0,0.2398) \cup (6.8758,12)$ and increasing on $(0.2398,6.8758)$
   c. $(6.8758, 200.14)$; The rate at which the number of banks were failing reached a peak of 200/yr during the latter part of 1988 ($t = 6.8758$).

11. a.

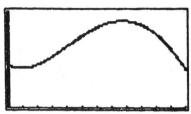

b. f is decreasing on (0, 0.8343) $\cup$ (7.6726, 12) and increasing on
(0.8343, 7.6726).   The rate at which single-family homes in the greater Boston
area were selling was decreasing during most of 1984, but started increasing in
late 1984 and continued increasing until mid 1991 when it started decreasing again
until 1996.

13. f is decreasing on the interval (0,1) and increasing on (1,4). The relative minimum
occurs at the point (1,32). These results indicate that the speed of traffic flow drops
between 6 A.M. and 7 A.M. reaching a low of 32 mph. Thereafter, it increases till
10 A.M.

EXERCISES 12.2, page 758

1. f is concave downward on (-∞,0) and concave upward on (0,∞). f has an inflection
point at (0,0).

3. f is concave downward on (-∞,0) $\cup$ (0,∞).

5. f is concave upward on (-∞,0) $\cup$ (1,∞) and concave downward on (0,1).
(0,0) and (1,-1) are inflection points of f.

7. f is concave downward on (-∞,-2) $\cup$ (-2,2) $\cup$ (2,∞).

9. a   11. b

13. a. $D_1'(t) > 0$, $D_2'(t) > 0$, $D_1''(t) > 0$, and $D_2''(t) < 0$ on (0,12).
b. With or without the proposed promotional campaign, the deposits will increase,
but with the promotion, the deposits will increase at an increasing rate whereas
without the promotion, the deposits will increase at a decreasing rate.

15. The significance of the inflection point Q is that the restoration process is working

at its peak at the time t_0 corresponding to its t-coordinate.

17. $f(x) = 4x^2 - 12x + 7$. $f'(x) = 8x - 12$ and $f''(x) = 8$. So, $f''(x) > 0$ everywhere and therefore f is concave upward everywhere.

19. $f(x) = \dfrac{1}{x^4} = x^{-4}$; $f'(x) = -\dfrac{4}{x^5}$ and $f''(x) = \dfrac{20}{x^6} > 0$ for all values of x in $(-\infty, 0) \cup (0, \infty)$ and so f is concave upward everywhere.

21. $g(x) = -\sqrt{4 - x^2}$. $g'(x) = \dfrac{d}{dx}\left[-(4 - x^2)^{1/2}\right] = -\tfrac{1}{2}(4 - x^2)^{-1/2}(-2x) = x(4 - x^2)^{-1/2}$.

$g''(x) = (4 - x^2)^{-1/2} + x(-\tfrac{1}{2})(4 - x^2)^{-3/2}(-2x)$

$= (4 - x^2)^{-3/2}[(4 - x^2) + x^2] = \dfrac{4}{(4 - x^2)^{3/2}} > 0$,

whenever it is defined and so g is concave upward wherever it is defined.

23. $f(x) = 2x^2 - 3x + 4$; $f'(x) = 4x - 3$ and $f''(x) = 4 > 0$ for all values of x. So f is concave upward on $(-\infty, \infty)$.

25. $f(x) = x^3 - 1$. $f'(x) = 3x^2$ and $f''(x) = 6x$. The sign diagram of f'' follows.

```
- - - - - - 0 + + + + + + +
——————————+————————————> x
          0
```

We see that f is concave downward on $(-\infty, 0)$ and concave upward on $(0, \infty)$.

27. $f(x) = x^4 - 6x^3 + 2x + 8$; $f'(x) = 4x^3 - 18x^2 + 2$ and $f''(x) = 12x^2 - 36x = 12x(x - 3)$. The sign diagram of f''

```
+ + + + + + 0 - - - - -0 + + +
——————————+———————+————> x
          0       3
```

shows that f is concave upward on $(-\infty, 0) \cup (3, \infty)$ and concave downward on $(0, 3)$.

29. $f(x) = x^{4/7}$. $f'(x) = \dfrac{4}{7}x^{-3/7}$ and $f''(x) = -\dfrac{12}{49}x^{-10/7} = -\dfrac{12}{49x^{10/7}}$.

Observe that $f''(x) < 0$ for all x different from zero. So f is concave downward on $(-\infty, 0) \cup (0, \infty)$.

31. $f(x) = (4-x)^{1/2}$. $f'(x) = \dfrac{1}{2}(4-x)^{-1/2}(-1) = -\dfrac{1}{2}(4-x)^{-1/2}$;

$f''(x) = \dfrac{1}{4}(4-x)^{-3/2}(-1) = -\dfrac{1}{4(4-x)^{3/2}} < 0.$

whenever it is defined. So f is concave downward on $(-\infty,4)$.

33. $f'(x) = \dfrac{d}{dx}(x-2)^{-1} = -(x-2)^{-2}$ and $f''(x) = 2(x-2)^{-3} = \dfrac{2}{(x-2)^3}$.

The sign diagram of f'' shows that f is concave downward on $(-\infty,2)$ and concave upward on $(2,\infty)$.

f'' is not defined here

$- - - - - - - - + + + + +$

$\longrightarrow x$

$0 \qquad 2$

35. $f'(x) = \dfrac{d}{dx}(2+x^2)^{-1} = -(2+x^2)^{-2}(2x) = -2x(2+x^2)^{-2}$ and

$f''(x) = -2(2+x^2)^{-2} - 2x(-2)(2+x^2)^{-3}(2x)$

$= 2(2+x^2)^{-3}[-(2+x^2)+4x^2] = \dfrac{2(3x^2-2)}{(2+x^2)^3} = 0$ if $x = \pm\sqrt{2/3}$.

From the sign diagram of f''

$+ + + + \; 0 \quad - - - - - - \quad 0 + + + +$

$\longrightarrow x$

$-\sqrt{2/3} \qquad 0 \qquad \sqrt{2/3}$

we see that f is concave upward on $(-\infty, -\sqrt{2/3}) \cup (\sqrt{2/3}, \infty)$ and concave downward on $(-\sqrt{2/3}, \sqrt{2/3})$.

37. $h(t) = \dfrac{t^2}{t-1}$; $h'(t) = \dfrac{(t-1)(2t) - t^2(1)}{(t-1)^2} = \dfrac{t^2-2t}{(t-1)^2}$;

$h''(t) = \dfrac{(t-1)^2(2t-2) - (t^2-2t)2(t-1)}{(t-1)^4}$

$= \dfrac{(t-1)(2t^2 - 4t + 2 - 2t^2 + 4t)}{(t-1)^4} = \dfrac{2}{(t-1)^3}.$

The sign diagram of h'' is

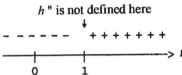

h" is not defined here
$\downarrow$
$-\ -\ -\ -\ -\ -$ $+ + + + + + +$

$\xrightarrow{\quad\quad\quad\quad\quad\quad\quad\quad} t$

0   1

and tells us that h is concave downward on $(-\infty,1)$ and concave upward on $(1,\infty)$.

39. $g(x) = x + \dfrac{1}{x^2}$. $g'(x) = 1 - 2x^{-3}$ and $g''(x) = 6x^{-4} = \dfrac{6}{x^4} > 0$ whenever $x \neq 0$.

Therefore, g is concave upward on $(-\infty,0) \cup (0,\infty)$.

41. $g(t) = (2t - 4)^{1/3}$. $g'(t) = = \dfrac{1}{3}(2t - 4)^{-2/3}(2) = \dfrac{2}{3}(2t - 4)^{-2/3}$.

$g''(t) = -\dfrac{4}{9}(2t - 4)^{-5/3} = -\dfrac{4}{9(2t - 4)^{5/3}}$.   The sign diagram of g''

g" is not defined here
$\downarrow$
$+ + + + + + + + + +\ \ -\ -\ -\ -$

$\xrightarrow{\quad\quad\quad\quad\quad\quad\quad} x$

0   2

tells us that g is concave upward on $(-\infty,2)$ and concave downward on $(2,\infty)$.

43. $f(x) = \dfrac{x^2}{x^2 - 1}$. $f'(x) = \dfrac{(x^2 - 1)(2x) - x^2(2x)}{(x^2 - 1)^2} = -\dfrac{2x}{(x^2 - 1)^2}$.

$f''(x) = -\dfrac{(x^2 - 1)^2(2) - (-2x)(2)(x^2 - 1)(2x)}{(x^2 - 1)^4} = -\dfrac{2(x^2 - 1)[x^2 - 1) + 4x^2]}{(x^2 - 1)^4}$

$= \dfrac{2(3x^2 + 1)}{(x^2 - 1)^3}$.

The sign diagram of f'' is

f" is not defined here
$\downarrow$ $\downarrow$
$+ + + +$ $-\ -\ -\ -\ -\ -$ $+ + + + +$

$\xrightarrow{\quad\quad\quad\quad\quad\quad\quad\quad} x$

-1   0   1

and we see that f is concave upward on $(-\infty,-1) \cup (1,\infty)$ and concave downward on

(-1,1).

45. $f(x) = x^3 - 2. f'(x) = 3x^2$ and $f''(x) = 6x. f''(x)$ is continuous everywhere and has a
zero at $x = 0$. From the sign diagram of f''

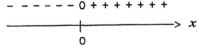

we conclude that $(0,-2)$ is an inflection point of f.

47. $f(x) = 6x^3 - 18x^2 + 12x - 15; f'(x) = 18x^2 - 36x + 12$ and
$f''(x) = 36x - 36 = 36(x - 1) = 0$ if $x = 1$. The sign diagram of f''

$$- - - - - - - - -0 + + + + + $$
$$\xrightarrow{\quad\quad\;\; | \quad\quad | \quad\quad\quad\quad} x$$
$$\quad\quad\quad\quad 0 \quad\quad 1$$

tells us that f has an inflection point at $(1,-15)$.

49. $f(x) = 3x^4 - 4x^3 + 1. f'(x) = 12x^3 - 12x^2$ and $f''(x) = 36x^2 - 24x = 12x(3x - 2) = 0$ if
$x = 0$ or $2/3$. These are candidates for inflection points. The sign diagram of f''

$$+ + + + + + 0 - - -0 + + + + +$$
$$\xrightarrow{\quad\quad\;\; | \quad\quad | \quad\quad\quad\quad} x$$
$$\quad\quad\quad\quad 0 \quad\quad \frac{2}{3}$$

shows that $(0,1)$ and $(\frac{2}{3},\frac{11}{27})$ are inflection points of f.

51. $g(t) = t^{1/3}, g'(t) = \frac{1}{3}t^{-2/3}$ and $g''(t) = -\frac{2}{9}t^{-5/3} = -\dfrac{2}{9t^{5/3}}$. Observe that $t = 0$ is in the
domain of g. Next, since $g''(t) > 0$ if $t < 0$ and $g''(t) < 0$, if $t > 0$, we see that $(0,0)$ is
an inflection point of g.

53. $f(x) = (x - 1)^3 + 2. f'(x) = 3(x - 1)^2$ and $f''(x) = 6(x - 1)$. Observe that $f''(x) < 0$
if $x < 1$ and $f''(x) > 0$ if $x > 1$ and so $(1,2)$ is an inflection point of f.

55. $f(x) = \dfrac{2}{1+x^2} = 2(1+x^2)^{-1}. f'(x) = -2(1+x^2)^{-2}(2x) = -4x(1+x^2)^{-2}.$

$f''(x) = -4(1+x^2)^{-2} - 4x(-2)(1+x^2)^{-3}(2x)$

$= 4(1+x^2)^{-3}[-(1+x^2)+4x^2] = \dfrac{4(3x^2-1)}{(1+x^2)^3},$

is continuous everywhere and has zeros at $x = \pm\frac{\sqrt{3}}{3}$. From the sign diagram of f''

we conclude that $(-\frac{\sqrt{3}}{3},\frac{3}{2})$ and$(\frac{\sqrt{3}}{3},\frac{3}{2})$ are inflection points of f.

57. $f(x) = -x^2 + 2x + 4$ and $f'(x) = -2x + 2$. The critical point of f is $x = 1$. Since $f''(x) = -2$ and $f''(1) = -2 < 0$, we conclude that $f(1) = 5$ is a relative maximum of f.

59. $f(x) = 2x^3 + 1; f'(x) = 6x^2 = 0$ if $x = 0$ and this is a critical point of f. Next, $f''(x) = 12x$ and so $f''(0) = 0$. Thus, the Second Derivative Test fails. But the First Derivative Test shows that $(0,0)$ is not a relative extremum.

61. $f(x) = \frac{1}{3}x^3 - 2x^2 - 5x - 10.$ $f'(x) = x^2 - 4x - 5 = (x - 5)(x + 1)$ and this gives $x = -1$ and $x = 5$ as critical points of f. Next, $f''(x) = 2x - 4$. Since $f''(-1) = -6 < 0$, we see that $(-1,-\frac{22}{3})$ is a relative maximum. Next, $f''(5) = 6 > 0$ and this shows that $(5,-\frac{130}{3})$ is a relative minimum.

63. $g(t) = t + \dfrac{9}{t}$. $g'(t) = 1 - \dfrac{9}{t^2} = \dfrac{t^2 - 9}{t^2} = \dfrac{(t+3)(t-3)}{t^2}$ and this shows that $t = \pm 3$ are critical points of g. Now, $g''(t) = 18t^{-3} = \dfrac{18}{t^3}$. Since $g''(-3) = -\dfrac{18}{27} < 0$ the Second Derivative Test implies that g has a relative maximum at $(-3,-6)$. Also, $g''(3) = \dfrac{18}{27} > 0$ and so g has a relative minimum at $(3,6)$.

65. $f(x) = \dfrac{x}{1-x}$. $f'(x) = \dfrac{(1-x)(1) - x(-1)}{(1-x)^2} = \dfrac{1}{(1-x)^2}$ is never zero. So there are no critical points and f has no relative extrema.

67. $f(t) = t^2 - \dfrac{16}{t}$. $f'(t) = 2t + \dfrac{16}{t^2} = \dfrac{2t^3 + 16}{t^2} = \dfrac{2(t^3 + 8)}{t^2}$. Setting $f'(t) = 0$ gives $t = -2$ as a critical point. Next, we compute $f''(t) = \dfrac{d}{dt}(2t + 16t^{-2}) = 2 - 32t^{-3} = 2 - \dfrac{32}{t^3}$. Since $f''(-2) = 2 - \dfrac{32}{(-8)} = 6 > 0$, we see that $(-2,12)$ is a relative minimum.

69. $g(s) = \dfrac{s}{1+s^2}$; $g'(s) = \dfrac{(1+s^2)(1) - s(2s)}{(1+s^2)^2} = \dfrac{1-s^2}{(1+s^2)^2} = 0$ gives $s = -1$ and $s = 1$

as critical points of g. Next, we compute

$$g''(s) = \frac{(1+s^2)^2(-2s) - (1-s^2)2(1+s^2)(2s)}{(1+s^2)^4}$$

$$= \frac{2s(1+s^2)(-1-s^2-2+2s^2)}{(1+s^2)^4} = \frac{2s(s^2-3)}{(1+s^2)^3}.$$

Now, $g''(-1) = \frac{1}{2} > 0$ and so $g(-1) = -\frac{1}{2}$ is a relative minimum of g. Next, $g''(1) = -\frac{1}{2} < 0$ and so $g(1) = \frac{1}{2}$ is a relative maximum of g.

71. $f(x) = \dfrac{x^4}{x-1}$.

$$f'(x) = \frac{(x-1)(4x^3) - x^4(1)}{(x-1)^2} = \frac{4x^4 - 4x^3 - x^4}{(x-1)^2} = \frac{3x^4 - 4x^3}{(x-1)^2} = \frac{x^3(3x-4)}{(x-1)^2}$$

and so $x = 0$ and $x = 4/3$ are critical points of f. Next,

$$f''(x) = \frac{(x-1)^2(12x^3 - 12x^2) - (3x^4 - 4x^3)(2)(x-1)}{(x-1)^4}$$

$$= \frac{(x-1)(12x^4 - 12x^3 - 12x^3 + 12x^2 - 6x^4 + 8x^3)}{(x-1)^4}$$

$$= \frac{6x^4 - 16x^3 + 12x^2}{(x-1)^3} = \frac{2x^2(3x^2 - 8x + 6)}{(x-1)^3}.$$

Since $f''(\frac{4}{3}) > 0$, we see that $f(\frac{4}{3}) = \frac{256}{27}$ is a relative minimum. Since $f''(0) = 0$, the Second Derivative Test fails. Using the sign diagram for f',

f' is not defined here

$$+ \; + \; + \; + \quad 0 \; - \; - \; - \quad \downarrow \; - \; 0+ \; + \; + \; + \; +$$

```
———————|————————|—|——————————> x
        0        1  4
                    3
```

and the First Derivative Test, we see that $f(0) = 0$ is a relative maximum.

73. $g(x) = \dfrac{2-x}{(x+2)^3}$.

$$g'(x) = \frac{(x+2)^3(-1)-(2-x)(3)(x+2)^2}{(x+2)^6}$$

$$= \frac{(x+2)^2[-(x+2)-3(2-x)]}{(x+2)^6} = \frac{2(x-4)}{(x+2)^4}.$$

We see that $x = 4$ is a critical point of g.

$$g''(x) = \frac{(x+2)^4(2)-2(x-4)(4)(x+2)^3}{(x+2)^8}$$

$$= \frac{2(x+2)^3[(x+2)-4(x-4)]}{(x+2)^8} = -\frac{2(3x-18)}{(x+2)^5}.$$

Since $g''(4) = \frac{12}{(6)^5} > 0,$ we see that g has a relative minimum at $(4, -\frac{1}{108})$.

75.   77.

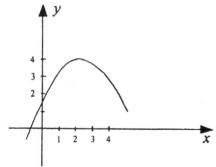   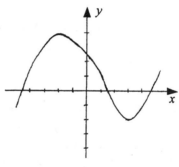

79. a. $N'(t)$ is positive because N is increasing on $(0,12)$.

b. $N''(t) < 0$ on $(0,6)$ and $N''(t) > 0$ on $(6,12)$.

c. The rate of growth of the number of help-wanted advertisements was decreasing over the first six months of the year and increasing over the last six months.

81. $f(t)$ increases at an increasing rate until the water level reaches the middle of the vase at which time (and this corresponds to the inflection point of f), $f(t)$ is increasing at the fastest rate. Though $f(t)$ still increases until the vase is filled, it does so at a decreasing rate.

83. a. $f'(t) = \frac{d}{dt}(0.0117t^3 + 0.0037t^2 + 0.7563t + 4.1)$

$= 0.0351t^2 + 0.0074t + 0.7563 \geq 0.7563$

for all t in the interval $[0, 9]$. This shows that f is increasing on $(0, 9)$. It tells us that the projected amount of AMT will keep on increasing over the years in question.

b. $f''(t) = \dfrac{d}{dt}(0.03511t^2 + 0.0074t + 0.7563) = 0.0702t + 0.0074 \geq 0.0074$.

This shows that f' is increasing on $(0, 9)$. Out result tells us that not only is the amount of AMT paid increasing over the period in question, but it is actually accelerating!

85. $S(x) = -0.002x^3 + 0.6x^2 + x + 500$; $S'(x) = -0.006x^2 + 1.2x + 1$; $S''(x) = -0.012x + 1.2$. $x = 100$ is a candidate for an inflection point of S. The sign diagram for S'' is

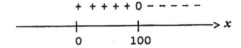

We see that $(100, 4600)$ is an inflection point of S.

87. We wish to find the inflection point of the function $N(t) = -t^3 + 6t^2 + 15t$. Now, $N'(t) = -3t^2 + 12t + 15$ and $N''(t) = -6t + 12 = -6(t - 2)$ giving $t = 2$ as the only candidate for an inflection point of N. From the sign diagram

for N'', we conclude that $t = 2$ gives an inflection point of N. Therefore, the average worker is performing at peak efficiency at 10 A.M.

89. $s = f(t) = -t^3 + 54t^2 + 480t + 6$. The velocity of the rocket is
$$v = f'(t) = -3t^2 + 108t + 480$$
and its acceleration is $a = f''(t) = -6t + 108 = -6(t - 18)$. From the sign diagram

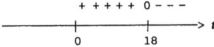

we see that $(18, 20,310)$ is an inflection point of f. Our computations reveal that the maximum velocity of the rocket is attained when $t = 18$. The maximum velocity is
$$f'(18) = -3(18)^2 + 108(18) + 480 = 1452, \text{ or } 1452 \text{ ft/sec}$$

91. $A(t) = 1.0974t^3 - 0.0915t^4$. $A'(t) = 3.2922t^2 - 0.366t^3$ and $A''(t) = 6.5844t - 1.098t^2$.
Setting $A'(t) = 0$, we obtain $t^2(3.2922 - 0.366t) = 0$, and this gives $t = 0$ or
$t \approx 8.995 \approx 9$. Using the Second Derivative Test, we find
$A''(9) = 6.5844(9) - 1.098(81) = -29.6784 < 0$, and this tells us that $t \approx 9$ gives rise
to a relative maximum of A. Our analysis tells us that on that May day, the level of
ozone peaked at approximately 4 P.M. in the afternoon.

93. True. If f' is increasing on (a,b), then $-f'$ is decreasing on (a,b), and so if the
graph of f is concave upward on (a,b), the graph of $-f$ must be concave
downward on (a,b).

95. True. The given conditions imply that $f''(0) < 0$ and the Second Derivative Test
gives the desired conclusion.

97. $f(x) = ax^2 + bx + c$. $f'(x) = 2ax + b$ and $f''(x) = 2a$. So $f''(x) > 0$ if $a > 0$, and the
parabola opens upward. If $a < 0$, then $f''(x) < 0$ and the parabola opens downward.

99. $f(x) = ax^3 + bx^2 + c$. $f'(x) = 2ax + b$ and $f''(x) = 6ax + 2b = 6a\left(x + \dfrac{b}{a}\right)$.

If $a > 0$, the sign diagram of f'' is

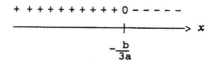

and if a $a < 0$ the sign diagram of f'' is

$$
\begin{array}{c}
+\ +\ +\ +\ +\ +\ +\ +\ +\ +\ 0\ -\ -\ -\ -\ - \\
\rule{6cm}{0.4pt} \longrightarrow x \\
-\dfrac{b}{3a}
\end{array}
$$

So $\left(-\dfrac{b}{3a}, f\left(-\dfrac{b}{3a}\right)\right)$ is an inflection point of f.

USING TECHNOLOGY EXERCISES 12.2, page 765

1. a. f is concave upward on $(-\infty, 0) \cup (1.1667, \infty)$ and concave downward on

(0, 1.1667).
b. (1.1667, 1.1153); (0,2)

3. a. f is concave downward on $(-\infty,0)$ and concave upward on $(0, \infty)$.
b. (0,2)

5. a. f is concave downward on $(-\infty,0)$ and concave upward on $(0, \infty)$.
b. (0,0)

7. a. f is concave downward on $(-\infty,-2.4495) \cup (0, 2.4495)$; f is concave upward on $(-2.4495, 0) \cup (2.4495, \infty)$.   b. (-2.4495, -0.3402); (2.4495, 0.3402)

9. a.

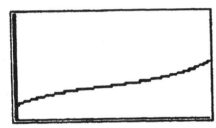

b. (5.5318, 35.9483)
c. $t = 5.5318$

11. a.

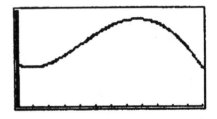

b. (3.9024, 77.0919);
sales of houses were increasing
at the fastest rate in late 1988.

13. a.

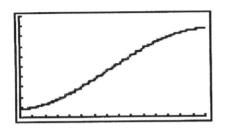

b. April 1993 ($t = 7.36$)

EXERCISES 12.3, page 774

1. $y = 0$ is a horizontal asymptote.

3. $y = 0$ is a horizontal asymptote and $x = 0$ is a vertical asymptote.

5. $y = 0$ is a horizontal asymptote and $x = -1$ and $x = 1$ are vertical asymptotes.

7. $y = 3$ is a horizontal asymptote and $x = 0$ is a vertical asymptote.

9. $y = 1$ and $y = -1$ are horizontal asymptotes.

11. $\lim\limits_{x \to \infty} \dfrac{1}{x} = 0$ and so $y = 0$ is a horizontal asymptote. Next, since the numerator of the

    rational expression is not equal to zero and the denominator is zero at $x = 0$, we see
    that $x = 0$ is a vertical asymptote.

13. $f(x) = -\dfrac{2}{x^2}$. $\lim\limits_{x \to \infty} -\dfrac{2}{x^2} = 0$, so $y = 0$ is a horizontal asymptote. Next, the

    denominator of $f(x)$ is equal to zero at $x = 0$. Since the numerator of $f(x)$ is not

    equal to zero at $x = 0$, we see that $x = 0$ is a vertical asymptote.

15. $\lim\limits_{x \to \infty} \dfrac{x-1}{x+1} = \lim\limits_{x \to \infty} \dfrac{1-\frac{1}{x}}{1+\frac{1}{x}} = 1$, and so $y = 1$ is a horizontal asymptote. Next, the

    denominator is equal to zero at $x = -1$ and the numerator is not equal to zero at this
    point, so $x = -1$ is a vertical asymptote.

17. $h(x) = x^3 - 3x^2 + x + 1$. $h(x)$ is a polynomial function and, therefore, it does not
    have any horizontal or vertical asymptotes.

19. $\lim\limits_{t \to \infty} \dfrac{t^2}{t^2 - 9} = \lim\limits_{t \to \infty} \dfrac{1}{1 - \frac{9}{t^2}} = 1$, and so $y = 1$ is a horizontal asymptote. Next, observe

    that the denominator of the rational expression $t^2 - 9 = (t + 3)(t - 3) = 0$ if $t = -3$
    and $t = 3$. But the numerator is not equal to zero at these points. Therefore, $t = -3$
    and $t = 3$ are vertical asymptotes.

21. $\lim\limits_{x \to \infty} \dfrac{3x}{x^2 - x - 6} = \lim\limits_{x \to \infty} \dfrac{\frac{3}{x}}{1 - \frac{1}{x} - \frac{6}{x^2}} = 0$ and so $y = 0$ is a horizontal asymptote. Next,

    observe that the denominator $x^2 - x - 6 = (x - 3)(x + 2) = 0$ if $x = -2$ or $x = 3$. But the numerator $3x$ is not equal to zero at these points. Therefore, $x = -2$ and $x = 3$ are vertical asymptotes.

23. $\displaystyle\lim_{t \to \infty} \left[2 + \frac{5}{(t-2)^2} \right] = 2$, and so $y = 2$ is a horizontal asymptote. Next observe that

$$\lim_{t \to 2^+} g(t) = \lim_{t \to 2^-} \left[2 + \frac{5}{(t-2)^2} \right] = \infty, \text{ and so } t = 2 \text{ is a vertical asymptote.}$$

25. $\displaystyle\lim_{x \to \infty} \frac{x^2 - 2}{x^2 - 4} = \lim_{x \to \infty} \frac{1 - \frac{2}{x^2}}{1 - \frac{4}{x^2}} = 1$ and so $y = 1$ is a horizontal asymptote. Next, observe

that the denominator $x^2 - 4 = (x + 2)(x - 2) = 0$ if $x = -2$ or 2. Since the numerator

$x^2 - 2$ is not equal to zero at these points, the lines $x = -2$ and $x = 2$ are vertical asymptotes.

27. $g(x) = \dfrac{x^3 - x}{x(x+1)}$; Rewrite $g(x)$ as $g(x) = \dfrac{x^2 - 1}{x + 1}$ $(x \neq 0)$ and note that

$$\lim_{x \to -\infty} g(x) = \lim_{x \to -\infty} \frac{x - \frac{1}{x}}{1 + \frac{1}{x}} = -\infty \text{ and } \lim_{x \to \infty} g(x) = \infty. \text{ Therefore, there are no horizontal}$$

asymptotes. Next, note that the denominator of $g(x)$ is equal to zero at $x = 0$ and $x = -1$. However, since the numerator of $g(x)$ is also equal to zero when $x = 0$, we see that $x = 0$ is not a vertical asymptote. Also, the numerator of $g(x)$ is equal to zero when $x = -1$, so $x = -1$ is not a vertical asymptote.

29. f is the derivative function of the function g. Observe that at a relative maximum (relative minimum) of g, $f(x) = 0$.

31.

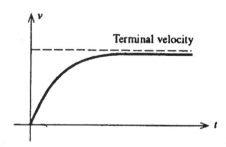

Terminal velocity

33.

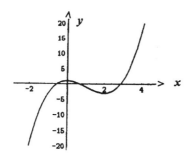

35.

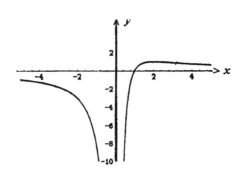

37. $g(x) = 4 - 3x - 2x^3$.

We first gather the following information on the graph of f.

1. The domain of f is $(-\infty, \infty)$.

2. Setting $x = 0$ gives $y = 4$ as the y-intercept. Setting $y = g(x) = 0$ gives a cubic equation which is not easily solved and we will not attempt to find the x-intercepts.

3. $\lim\limits_{x \to -\infty} g(x) = \infty$ and $\lim\limits_{x \to \infty} g(x) = -\infty$. 4. There are no asymptotes of g.

5. $g'(x) = -3 - 6x^2 = -3(2x^2 + 1) < 0$ for all values of x and so g is decreasing on $(-\infty, \infty)$.

6. The results of 5 show that g has no critical points and hence has no relative extrema.

7. $g''(x) = -12x$. Since $g''(x) > 0$ for $x < 0$ and $g''(x) < 0$ for $x > 0$, we see that g is concave upward on $(-\infty, 0)$ and concave downward on $(0, \infty)$.

8. From the results of (7), we see that $(0,4)$ is an inflection point of g. The graph of g follows.

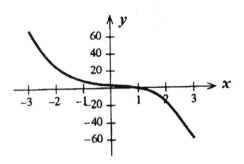

39. $h(x) = x^3 - 3x + 1$

We first gather the following information on the graph of h.

1. The domain of h is $(-\infty, \infty)$.

2. Setting $x = 0$ gives 1 as the y-intercept. We will not find the x-intercept.

3. $\lim\limits_{x \to -\infty} (x^3 - 3x + 1) = -\infty$ and $\lim\limits_{x \to \infty} (x^3 - 3x + 1) = \infty$

4. There are no asymptotes since $h(x)$ is a polynomial.

5. $h'(x) = 3x^2 - 3 = 3(x + 1)(x - 1)$, and we see that $x = -1$ and $x = 1$ are critical points. From the sign diagram

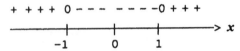

we see that h is increasing on $(-\infty, -1) \cup (1, \infty)$ and decreasing on $(-1, 1)$.

6. The results of (5) shows that $(-1, 3)$ is a relative maximum and $(1, -1)$ is a relative minimum.

7. $h''(x) = 6x$ and $h''(x) < 0$ if $x < 0$ and $h''(x) > 0$ if $x > 0$. So the graph of h is concave downward on $(-\infty, 0)$ and concave upward on $(0, \infty)$.

8. The results of (7) show that $(0, 1)$ is an inflection point of h.

The graph of h follows.

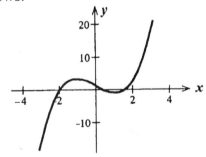

41. $f(x) = -2x^3 + 3x^2 + 12x + 2$

We first gather the following information on the graph of f.

1. The domain of f is $(-\infty, \infty)$.

2. Setting $x = 0$ gives 2 as the y-intercept.

3. $\lim\limits_{x \to -\infty} (-2x^3 + 3x^2 + 12x + 2) = \infty$ and $\lim\limits_{x \to \infty}(-2x^3 + 3x^2 + 12x + 2) = -\infty$

4. There are no asymptotes because $f(x)$ is a polynomial function.

5. $f'(x) = -6x^2 + 6x + 12 = -6(x^2 - x - 2) = -6(x - 2)(x + 1) = 0$ if $x = -1$ or $x = 2$, the critical points of f. From the sign diagram

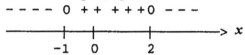

we see that f is decreasing on $(-\infty, -1) \cup (2, \infty)$ and increasing on $(-1, 2)$.

6. The results of (5) show that $(-1, -5)$ is a relative minimum and $(2, 22)$ is a relative maximum.

7. $f''(x) = -12x + 6 = 0$ if $x = 1/2$. The sign diagram of f''

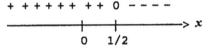

shows that the graph of f is concave upward on $(-\infty, 1/2)$ and concave downward on $(1/2, \infty)$.

8. The results of (7) show that $(\frac{1}{2}, \frac{17}{2})$ is an inflection point.

The graph of f follows.

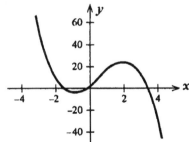

43. $h(x) = \frac{3}{2}x^4 - 2x^3 - 6x^2 + 8$

We first gather the following information on the graph of h.

1. The domain of h is $(-\infty, \infty)$.

2. Setting $x = 0$ gives 8 as the y-intercept.

3. $\lim\limits_{x \to -\infty} h(x) = \lim\limits_{x \to \infty} h(x) = \infty$

4. There are no asymptotes.

5. $h'(x) = 6x^3 - 6x^2 - 12x = 6x(x^2 - x - 2) = 6x(x-2)(x+1) = 0$ if $x = -1, 0,$ or $2,$

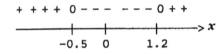

and these are the critical points of h. The sign diagram of h' is
and this tells us that h is increasing on $(-1, 0) \cup (2, \infty)$ and decreasing on
$(-\infty, -1) \cup (0,2)$.

6. The results of (5) show that $(-1, \frac{11}{2})$ and $(2,-8)$ are relative minima of h and $(0,8)$
is a relative maximum of h.

7. $h''(x) = 18x^2 - 12x - 12 = 6(3x^2 - 2x - 2)$. The zeros of h'' are

$$x = \frac{2 \pm \sqrt{4 + 24}}{6} \approx -0.5 \text{ or } 1.2.$$

The sign diagram of h'' is

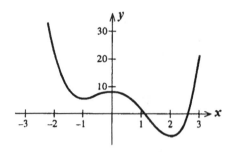

and tells us that the graph of h is concave upward on $(-\infty, -0.5) \cup (1.2, \infty)$ and is
concave downward on $(0.5, 1.2)$.

8. The results of (7) also show that $(-0.5, 6.8)$ and $(1.2, -1)$ are inflection points.
The graph of h follows.

45. $f(t) = \sqrt{t^2 - 4}.$

We first gather the following information on f.

1. The domain of f is found by solving $t^2 - 4 \geq 0$ giving it as $(-\infty, -2] \cup [2, \infty)$.

2. Since $t \neq 0$, there is no y-intercept. Next, setting $y = f(t) = 0$ gives the

t-intercepts as -2 and 2.

3. $\lim\limits_{t \to -\infty} f(t) = \lim\limits_{t \to \infty} f(t) = \infty$ 　　　　4. There are no asymptotes.

5. $f'(t) = \dfrac{1}{2}(t^2 - 4)^{-1/2}(2t) = t(t^2 - 4)^{-1/2} = \dfrac{t}{\sqrt{t^2 - 4}}$.

Setting $f'(t) = 0$ gives $t = 0$. But $t = 0$ is not in the domain of f and so there are no critical points. The sign diagram for f' is

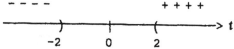

We see that f is increasing on $(2,\infty)$ and decreasing on $(-\infty,-2)$.

6. From the results of (5) we see that there are no relative extrema.

7. $f''(t) = (t^2 - 4)^{-1/2} + t(-\tfrac{1}{2})(t^2 - 4)^{-3/2}(2t) = (t^2 - 4)^{-3/2}(t^2 - 4 - t^2)$

$$= -\dfrac{4}{(t^2 - 4)^{3/2}}.$$

8. Since $f''(t) < 0$ for all t in the domain of f, we see that f is concave downward everywhere. From the results of (7), we see that there are no inflection points. The graph of f follows.

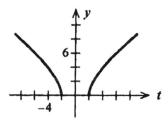

47. $g(x) = \tfrac{1}{2}x - \sqrt{x}$.

We first gather the following information on g.

1. The domain of g is $[0,\infty)$.

2. The y-intercept is 0. To find the x-intercept, set $y = 0$, giving

$$\tfrac{1}{2}x - \sqrt{x} = 0$$

$$x = 2\sqrt{x}$$

$$x^2 = 4x$$

$$x(x - 4) = 0, \text{ and } x = 0 \text{ or } x = 4$$

3. $\lim\limits_{x \to \infty} (\tfrac{1}{2}x - \sqrt{x}) = \lim\limits_{x \to \infty} \tfrac{1}{2}x(1 - \tfrac{2}{\sqrt{x}}) = \infty$.

4. There are no asymptotes.

5. $g'(x) = \frac{1}{2} - \frac{1}{2}x^{-1/2} = \frac{1}{2}x^{-1/2}(x^{1/2} - 1) = \dfrac{\sqrt{x}-1}{2\sqrt{x}}$

which is zero when $x = 1$. From the sign diagram for g'

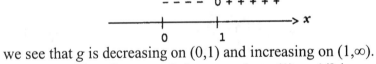

we see that g is decreasing on $(0,1)$ and increasing on $(1,\infty)$.

6. From the sign diagram of g', we see that $g(1) = -1/2$ is a relative minimum.

7. $g''(x) = (-\frac{1}{2})(-\frac{1}{2})x^{-3/2} = \dfrac{1}{4x^{3/2}} > 0$ for $x > 0$, and so g is concave upward on

$(0,\infty)$.

8. There are no inflection points.

The graph of g follows.

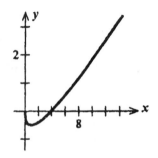

49. $g(x) = \dfrac{2}{x-1}$.   We first gather the following information on g.

1. The domain of g is $(-\infty,1) \cup (1,\infty)$.

2. Setting $x = 0$ gives -2 as the y-intercept. There are no x-intercepts since

$\dfrac{2}{x-1} \neq 0$ for all values of x.

3. $\lim\limits_{x \to -\infty} \dfrac{2}{x-1} = 0$ and $\lim\limits_{x \to \infty} \dfrac{2}{x-1} = 0$.

4. The results of (3) show that $y = 0$ is a horizontal asymptote. Furthermore, the denominator of $g(x)$ is equal to zero at $x = 1$ but the numerator is not equal to zero there. Therefore, $x = 1$ is a vertical asymptote.

5. $g'(x) = -2(x-1)^{-2} = -\dfrac{2}{(x-1)^2} < 0$ for all $x \neq 1$ and so g is decreasing on

$(-\infty,1)$ and $(1,\infty)$.

6. Since g has no critical points, there are no relative extrema.

7. $g''(x) = \dfrac{4}{(x-1)^3}$ and so $g''(x) < 0$ if $x < 1$ and $g''(x) > 0$ if $x > 1$. Therefore, the graph of g is concave downward on $(-\infty,1)$ and concave upward on $(1,\infty)$.

8. Since $g''(x) \neq 0$, there are no inflection points.

The graph of g follows.

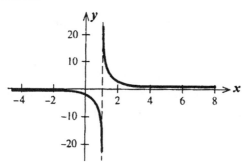

51. $h(x) = \dfrac{x+2}{x-2}$.

We first gather the following information on the graph of h.

1. The domain of h is $(-\infty,2) \cup (2,\infty)$.

2. Setting $x = 0$ gives $y = -1$ as the y-intercept. Next, setting $y = 0$ gives $x = -2$ as the x-intercept.

3. $\displaystyle\lim_{x\to\infty} h(x) = \lim_{x\to-\infty} \dfrac{1+\dfrac{2}{x}}{1-\dfrac{2}{x}} = \lim_{x\to-\infty} h(x) = 1.$

4. Setting $x - 2 = 0$ gives $x = 2$. Furthermore,

$$\lim_{x\to 2^+} \dfrac{x+2}{x-2} = \infty \quad \text{and} \quad \lim_{x\to 2^+} \dfrac{x+2}{x-2} = -\infty$$

So $x = 2$ is a vertical asymptote of h. Also, from the resultsof (3), we see that $y = 1$ is a horizontal asymptote of h.

5. $h'(x) = \dfrac{(x-2)(1)-(x+2)(1)}{(x-2)^2} = -\dfrac{4}{(x-2)^2}.$

We see that there are no critical points of h. (Note $x = 2$ does not belong to the domain of h.) The sign diagram of h' follows.

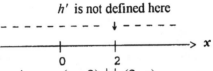

h' is not defined here

We see that *h* is decreasing on $(-\infty,2) \cup (2,\infty)$.

6. From the results of (5), we see that there is no relative extremum.

7. $h''(x) = \dfrac{8}{(x-2)^3}$. Note that $x = 2$ is not a candidate for an inflection point

because *h*(2) is not defined. Since $h''(x) < 0$ for $x < 2$ and $h''(x) > 0$ for $x > 2$, we see that *h* is concave downward on $(-\infty,2)$ and concave upward on $(2,\infty)$.

8. From the results of (7), we see that there are no inflection points. The graph of *h* follows.

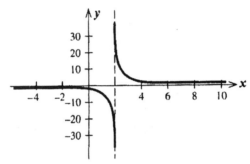

53. $f(t) = \dfrac{t^2}{1+t^2}$.

We first gather the following information on the graph of *f*.

1. The domain of *f* is $(-\infty, \infty)$.

2. Setting $t = 0$ gives the y-intercept as 0. Similarly, setting $y = 0$ gives the t-intercept as 0.

3. $\lim\limits_{t \to -\infty} \dfrac{t^2}{1+t^2} = \lim\limits_{t \to \infty} \dfrac{t^2}{1+t^2} = 1.$

4. The results of (3) show that $y = 1$ is a horizontal asymptote. There are no vertical asymptotes since the denominator is not equal to zero.

5. $f'(t) = \dfrac{(1+t^2)(2t) - t^2(2t)}{(1+t^2)^2} = \dfrac{2t}{(1+t^2)^2} = 0$, if $t = 0$, the only critical point of *f*.

Since $f'(t) < 0$ if $t < 0$ and $f'(t) > 0$ if $t > 0$, we see that *f* is decreasing on $(-\infty,0)$

and increasing on $(0,\infty)$.

6. The results of (5) show that $(0,0)$ is a relative minimum.

7. $f''(t) = \dfrac{(1+t^2)^2(2) - 2t(2)(1+t^2)(2t)}{(1+t^2)^4} = \dfrac{2(1+t^2)[(1+t^2) - 4t^2]}{(1+t^2)^4}$

$= \dfrac{2(1-3t^2)}{(1+t^2)^3} = 0$ if $t = \pm\dfrac{\sqrt{3}}{3}$.

The sign diagram of f'' is

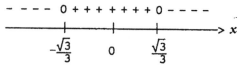

and shows that f is concave downward on $(-\infty, -\frac{\sqrt{3}}{3}) \cup (\frac{\sqrt{3}}{3}, \infty)$ and concave

upward on $(-\frac{\sqrt{3}}{3}, \frac{\sqrt{3}}{3})$.

8. The results of (7) show that $(-\frac{\sqrt{3}}{3}, \frac{1}{4})$ and $(\frac{\sqrt{3}}{3}, \frac{1}{4})$ are inflection points.
The graph of f follows.

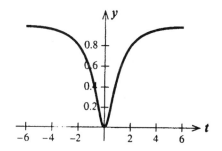

55. $g(t) = -\dfrac{t^2 - 2}{t - 1}$.

First we obtain the following information on g.
1. The domain of g is $(-\infty, 1) \cup (1, \infty)$.
2. Setting $t = 0$ gives -2 as the y-intercept.
3. $\lim\limits_{t \to -\infty} -\dfrac{t^2 - 2}{t - 1} = \infty$ and $\lim\limits_{t \to \infty} -\dfrac{t^2 - 2}{t - 1} = -\infty$.

4. There are no horizontal asymptotes. The denominator is equal to zero at $t = 1$ at

which point the numerator is not equal to zero. Therefore $t = 1$ is a vertical asymptote.

5. $g'(t) = -\dfrac{(t-1)(2t)-(t^2-2)(1)}{(t-1)^2} = -\dfrac{t^2-2t+2}{(t-1)^2} \neq 0$

for all values of t. The sign diagram of g'

g' is not defined here

$\downarrow$

$- - - - - - \quad - - - -$

$\xrightarrow{\qquad \; + \qquad + \qquad\qquad\qquad} t$

$0 \qquad 1$

shows that g is decreasing on $(-\infty,1) \cup (1, \infty)$.

6. Since there are no critical points, g has no relative extrema.

7. $g''(t) = -\dfrac{(t-1)^2(2t-2)-(t^2-2t+2)(2)(t-1)}{(t-1)^4}$

$= \dfrac{-2(t-1)(t^2-2t+1-t^2+2t-2)}{(t-1)^4} = \dfrac{2}{(t-1)^3}.$

The sign diagram of g''

g" is not defined here

$\downarrow$

$- - - - - - \quad + + + + + + +$

$\xrightarrow{\qquad \; + \quad + \qquad\qquad\qquad} t$

$0 \qquad 1$

shows that the graph of g is concave upward on $(1,\infty)$ and concave downward on $(-\infty,1)$.

8. There are no inflection points since $g''(x) \neq 0$ for all x.
The graph of g follows.

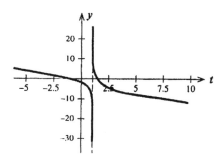

57. $g(t) = \dfrac{t^2}{t^2 - 1}$.

We first gather some information on the graph of g.

1. Since $t^2 - 1 = 0$ if $t = \pm 1$, we see that the domain of g is $(-\infty, -1) \cup (-1, 1) \cup (1, \infty)$.

2. Setting $t = 0$ gives 0 as the y-intercept. Setting $y = 0$ gives 0 as the t-intercept.

3. $\lim_{t \to -\infty} g(t) = \lim_{t \to \infty} g(t) = 1$.

4. The results of (3) show that $y = 1$ is a horizontal asymptote. Since the denominator (but not the numerator) is zero at $t = \pm 1$, we see that $t = \pm 1$ are vertical asymptotes.

5. $g'(t) = \dfrac{(t^2 - 1)(2t) - (t^2)(2t)}{(t^2 - 1)^2} = -\dfrac{2t}{(t^2 - 1)^2} = 0$, if $t = 0$.

The sign diagram of g' is

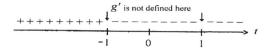

We see that g is increasing on $(-\infty, -1) \cup (-1, 0)$ and decreasing on $(0, 1) \cup (1, \infty)$.

6. From the results of (5), we see that g has a relative maximum at $t = 0$.

7. $g''(t) = \dfrac{(t^2 - 1)^2(-2) - (-2t)(2)(t^2 - 1)(2t)}{(t^2 - 1)^4}$

$= \dfrac{2(t^2 - 1)^2[-(t^2 - 1) + 4t^2]}{(t^2 - 1)^3}$

$$= \frac{2(-t^2 + 1 + 4t^2)}{(t^2 - 1)^3} = \frac{2(3t^2 + 1)}{(t^2 - 1)^3}$$

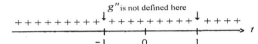

g'' is not defined here

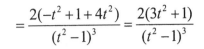

From the sign diagram we see that the graph of g is concave up on $(-\infty, -1) \cup (-1, 1) \cup (1, \infty)$.

8. From (7), we see that the graph of g has no inflection points.
    The graph of g follows.

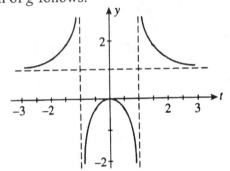

59. $h(x) = (x - 1)^{2/3} + 1$.

    We begin by obtaining the following information on h.
    1. The domain of h is $(-\infty, \infty)$.
    2. Setting $x = 0$ gives 2 as the y-intercept; since $h(x) \neq 0$ there is no x-intercept.
    3. $\lim_{x \to \infty} [(x - 1)^{2/3} + 1] = \infty$. Similarly, $\lim_{x \to -\infty} [(x - 1)^{2/3} + 1] = \infty$.
    4. There are no asymptotes.
    5. $h'(x) = \frac{2}{3}(x - 1)^{-1/3}$ and is positive if $x > 1$ and negative if $x < 1$. So h is
    increasing on $(1, \infty)$, and decreasing on $(-\infty, 1)$.
    6. From (5), we see that h has a relative minimum at $(1, 1)$.
    7. $h''(x) = \frac{2}{3}(-\frac{1}{3})(x - 1)^{-4/3} = -\frac{2}{9}(x - 1)^{-4/3} = -\frac{2}{(x - 1)^{4/3}}$. Since $h''(x) < 0$ on

    $(-\infty, 1) \cup (1, \infty)$, we see that h is concave downward on $(-\infty, 1) \cup (1, \infty)$. Note that
    $h''(x)$ is not defined at $x = 1$.
    8. From the results of (7), we see h has no inflection points.
    The graph of h follows.

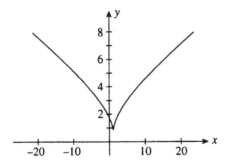

61. a. The denominator of $C(x)$ is equal to zero if $x = 100$. Also,

$$\lim_{x \to 100^-} \frac{0.5x}{100 - x} = \infty \quad \text{and} \quad \lim_{x \to 100^+} \frac{0.5x}{100 - x} = -\infty$$

Therefore, $x = 100$ is a vertical asymptote of C.

b. No, because the denominator will be equal to zero in that case.

63. a. Since $\displaystyle \lim_{t \to \infty} C(t) = \lim_{t \to \infty} \frac{0.2t}{t^2 + 1} = \lim_{t \to \infty} \left[\frac{0.2}{t + \frac{1}{t^2}} \right] = 0$, $y = 0$ is a horizontal asymptote.

b. Our results reveal that as time passes, the concentration of the drug decreases and approaches zero.

65. $G(t) = -0.2t^3 + 2.4t^2 + 60$.

We first gather the following information on the graph of G.
1. The domain of G is $(0,\infty)$.
2. Setting $t = 0$ gives 60 as the y-intercept.
Note that Step 3 is not necessary in this case because of the restricted domain.
4. There are no asymptotes since G is a polynomial function.
5. $G'(t) = -0.6t^2 + 4.8t = -0.6t(t - 8) = 0$, if $t = 0$ or $t = 8$. But these points do not lie in the interval $(0,8)$, so they are not critical points. The sign diagram of G'

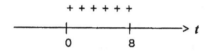

shows that G is increasing on $(0,8)$.
6. The results of (5) tell us that there are no relative extrema.
7. $G''(t) = -1.2t + 4.8 = -1.2(t - 4)$. The sign diagram of G'' is

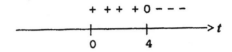

and shows that G is concave upward on $(0,4)$ and concave downward on $(4,8)$.
6. The results of (7) shows that $(4,85.6)$ is an inflection point.

The graph of G follows.

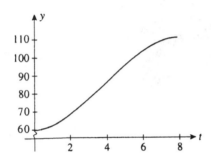

67. $C(t) = \dfrac{0.2t}{t^2 + 1}$.

We first gather the following information on the function C.
1. The domain of C is $[0,\infty)$.
2. If $t = 0$, then $y = 0$. Also, if $y = 0$, then $t = 0$.
3. $\lim\limits_{t \to \infty} \dfrac{0.2t}{t^2 + 1} = 0$.
4. The results of (3) imply that $y = 0$ is a horizontal asymptote.
5. $C'(t) = \dfrac{(t^2 + 1)(0.2) - 0.2t(2t)}{(t^2 + 1)^2} = \dfrac{0.2(t^2 + 1 - 2t^2)}{(t^2 + 1)^2} = \dfrac{0.2(1 - t^2)}{(t^2 + 1)^2}$

and this is equal to zero at $t = \pm 1$, so $t = 1$ is a critical point of C. The sign diagram of C' is

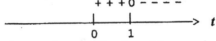

and tells us that C is decreasing on $(1,\infty)$ and increasing on $(0,1)$.
6. The results of (5) tell us that $(1,0.1)$ is a relative maximum.

7. $C''(t) = 0.2 \left[\dfrac{(t^2+1)^2(-2t) - (1-t^2)2(t^2+1)(2t)}{(t^2+1)^4} \right]$

$= \dfrac{0.2(t^2+1)(2t)(-t^2-1-2+2t^2)}{(t^2+1)^4} = \dfrac{0.4t(t^2-3)}{(t^2+1)^3}.$

The sign diagram of C'' is

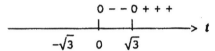

and so the graph of C is concave downward on $(0, \sqrt{3})$ and concave upward on $(\sqrt{3}, \infty)$.

8. The results of (7) show that $(\sqrt{3}, 0.05\sqrt{3})$ is an inflection point.
The graph of C follows.

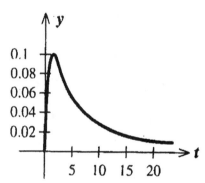

69. $T(x) = \dfrac{120x^2}{x^2+4}$.

We first gather the following information on the function T.
1. The domain of T is $[0,\infty)$.
2. Setting $x = 0$ gives 0 as the y-intercept.
3. $\displaystyle\lim_{x \to \infty} \dfrac{120x^2}{x^2+4} = 120.$
4. The results of (3) show that $y = 120$ is a horizontal asymptote.
5. $T'(x) = 120\left[\dfrac{(x^2+4)2x - x^2(2x)}{(x^2+4)^2} \right] = \dfrac{960x}{(x^2+4)^2}.$ Since $T'(x) > 0$

if $x > 0$, we see that T is increasing on $(0,\infty)$.
6. There are no relative extrema in $(0,\infty)$.

7. $T''(x) = 960 \left[\dfrac{(x^2+4)^2 - x(2)(x^2+4)(2x)}{(x^2+4)^4} \right]$

$= \dfrac{960(x^2+4)[(x^2+4)-4x^2]}{(x^2+4)^4} = \dfrac{960(4-3x^2)}{(x^2+4)^3}.$

The sign diagram for T'' is

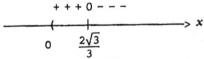

$$+ + + 0 - - -$$

$0 \qquad \dfrac{2\sqrt{3}}{3}$

We see that T is concave downward on $(\dfrac{2\sqrt{3}}{3}, \infty)$ and concave upward on $(0, \dfrac{2\sqrt{3}}{3})$.

8. We see from the results of (7) that $(\dfrac{2\sqrt{3}}{3}, 30)$ is an inflection point.

The graph of T follows.

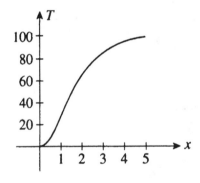

USING TECHNOLOGY EXERCISES 12.3, page 779

1.

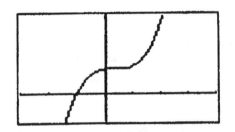

3.

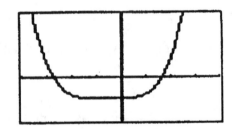

5. -0.9733; 2.3165, 4.6569   7. -1.1301; 2.9267   9. 1.5142

EXERCISES 12.4, page 789

1. f has no absolute extrema.

3. f has an absolute minimum at $(0,0)$.

5. f has an absolute minimum at $(0,-2)$ and an absolute maximum at $(1,3)$.

7. f has an absolute minimum at $(\frac{3}{2},-\frac{27}{16})$ and an absolute maximum at $(-1,3)$.

9. The graph of $f(x) = 2x^2 + 3x - 4$ is a parabola that opens upward. Therefore, the vertex of the parabola is the absolute minimum of f. To find the vertex, we solve the equation $f'(x) = 4x + 3 = 0$ giving $x = -3/4$. We conclude that the absolute minimum value is $f(-\frac{3}{4}) = -\frac{41}{8}$.

11. Since $\lim\limits_{x \to -\infty} x^{1/3} = -\infty$ and $\lim\limits_{x \to \infty} x^{1/3} = \infty$, we see that h is unbounded. Therefore it has no absolute extrema.

13. $f(x) = \dfrac{1}{1+x^2}$.

Using the techniques of graphing, we sketch the graph of f (see Fig. 12.40, page 754, in the text). The absolute maximum of f is $f(0) = 1$. Alternatively, observe that $1 + x^2 \geq 1$ for all real values of x. Therefore, $f(x) \leq 1$ for all x, and we see that the absolute maximum is attained when $x = 0$.

15. $f(x) = x^2 - 2x - 3$ and $f'(x) = 2x - 2 = 0$, so $x = 1$ is a critical point. From the table,

x	-2	1	3
$f(x)$	5	-4	0

we conclude that the absolute maximum value is $f(-2) = 5$ and the absolute minimum value is $f(1) = -4$.

17. $f(x) = -x^2 + 4x + 6$; The function f is continuous and defined on the closed interval $[0,5]$. $f'(x) = -2x + 4$ and $x = 2$ is a critical point. From the table

x	0	2	5
$f(x)$	6	10	1

we conclude that $f(2) = 10$ is the absolute maximum value and $f(5) = 1$ is the absolute minimum value.

19. The function $f(x) = x^3 + 3x^2 - 1$ is continuous and defined on the closed interval $[-3,2]$ and differentiable in $(-3,2)$. The critical points of f are found by solving
$$f'(x) = 3x^2 + 6x = 3x(x + 2)$$
giving $x = -2$ and $x = 0$. Next, we compute the values of f given in the following table.

x	-3	-2	0	2
$f(x)$	-1	3	-1	19

From the table, we see that the absolute maximum value of f is $f(2) = 19$ and the absolute minimum value is $f(-3) = -1$ and $f(0) = -1$.

21. The function $g(x) = 3x^4 + 4x^3$ is continuous and differentiable on the closed interval $[-2,1]$ and differentiable in $(-2,1)$. The critical points of g are found by solving

$$g'(x) = 12x^3 + 12x^2 = 12x^2(x + 1)$$

giving $x = 0$ and $x = -1$. We next compute the values of g shown in the following table.

x	-2	-1	0	1
$g(x)$	16	-1	0	7

From the table we see that $g(-2) = 16$ is the absolute maximum value of g and

$g(-1) = -1$ is the absolute minimum value of g.

23. $f(x) = \dfrac{x+1}{x-1}$ on [2,4]. Next, we compute,
$$f'(x) = \frac{(x-1)(1) - (x+1)(1)}{(x-1)^2} = -\frac{2}{(x-1)^2}.$$
Since there are no critical points, ($x = 1$ is not in the domain of f), we need only test the endpoints. From the table

x	2	4
$g(x)$	3	5/3

we conclude that $f(4) = 5/3$ is the absolute minimum value and $f(2) = 3$ is the absolute maximum value.

25. $f(x) = 4x + \dfrac{1}{x}$ is continuous on [1,3] and differentiable in (1,3). To find the critical points of f, we solve $f'(x) = 4 - \frac{1}{x^2} = 0$, obtaining $x = \pm\frac{1}{2}$. Since these critical points lie outside the interval [1,3], they are not candidates for the absolute extrema of f. Evaluating f at the endpoints of the interval [1,3], we find that the absolute maximum value of f is $f(3) = \frac{37}{3}$, and the absolute minimum value of f is $f(1) = 5$.

27. $f(x) = \frac{1}{2}x^2 - 2\sqrt{x} = \frac{1}{2}x^2 - 2x^{1/2}$. To find the critical points of f, we solve
$$f'(x) = x - x^{-1/2} = 0, \quad \text{or} \quad x^{3/2} - 1 = 0,$$
obtaining $x = 1$. From the table

x	0	1	3
$f(x)$	0	$-\frac{3}{2}$	$\frac{9}{2} - 2\sqrt{3} \approx 1.04$

we conclude that $f(3) \approx 1.04$ is the absolute maximum value and $f(1) = -3/2$ is the absolute minimum value.

29. The graph of $f(x) = 1/x$ over the interval $(0,\infty)$ follows.

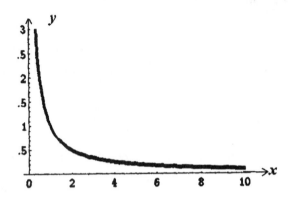

From the graph of f, we conclude that f has no absolute extrema.

31. $f(x) = 3x^{2/3} - 2x$. The function f is continuous on $[0,3]$ and differentiable on $(0,3)$. To find the critical points of f, we solve
$$f'(x) = 2x^{-1/3} - 2 = 0$$
obtaining $x = 1$ as the critical point. From the table,

x	0	1	3
$f(x)$	0	1	$3^{5/3} - 6 \approx 0.24$

we conclude that the absolute maximum value is $f(1) = 1$ and the absolute minimum value is $f(0) = 0$.

33. $f(x) = x^{2/3}(x^2 - 4)$. $f'(x) = x^{2/3}(2x) + \frac{2}{3}x^{-1/3}(x^2 - 4) = \frac{2}{3}x^{-1/3}[3x^2 + (x^2 - 4)]$
$$= \frac{8(x^2 - 1)}{3x^{1/3}} = 0.$$

Observe that f' is not defined at $x = 0$. Furthermore, $f'(x) = 0$ at $x \pm 1$. So the critical points of f are -1, 0, 1. From the following table,

x	-1	0	1	2
$f(x)$	-3	0	-3	0

we see that f has an absolute minimum at $(-1,-3)$ and $(1,-3)$ and absolute maxima at $(0,0)$ and $(2,0)$.

35. $f(x) = \dfrac{x}{x^2+2}$. To find the critical points of f, we solve

$$f'(x) = \frac{(x^2+2) - x(2x)}{(x^2+2)^2} = \frac{2-x^2}{(x^2+2)^2} = 0$$

obtaining $x = \pm\sqrt{2}$. Since $x = -\sqrt{2}$ lies outside $[-1,2]$, $x = \sqrt{2}$ is the only critical point in the given interval. From the table

x	-1	$\sqrt{2}$	2
$f(x)$	$-\frac{1}{3}$	$\sqrt{2}/4 \approx 0.35$	$\frac{1}{3}$

we conclude that $f(\sqrt{2})) = \sqrt{2}/4 \approx 0.35$ is the absolute maximum value and $f(-1) = -1/3$ is the absolute minimum value.

37. The function $f(x) = \dfrac{x}{\sqrt{x^2+1}} = \dfrac{x}{(x^2+1)^{1/2}}$ is continuous and defined on the closed interval $[-1,1]$ and differentiable on $(-1,1)$. To find the critical points of f, we first compute

$$f'(x) = \frac{(x^2+1)^{1/2}(1) - x(\frac{1}{2})(x^2+1)^{-1/2}(2x)}{[(x^2+1)^{1/2}]^2}$$

$$= \frac{(x^2+1)^{-1/2}[x^2+1-x^2]}{x^2+1} = \frac{1}{(x^2+1)^{3/2}}$$

which is never equal to zero. Next, we compute the values of f shown in the following table.

x	-1	1
$f(x)$	$-\sqrt{2}/2$	$\sqrt{2}/2$

We conclude that $f(-1) = -\sqrt{2}/2$ is the absolute minimum value and $f(1) = \sqrt{2}/2$ is the absolute maximum value.

39. $h(t) = -16t^2 + 64t + 80$. To find the maximum value of h, we solve
$$h'(t) = -32t + 64 = -32(t - 2) = 0$$
giving $t = 2$ as the critical point of h. Furthermore, this value of t gives rise to the absolute maximum value of h since the graph of h is parabola that opens downward. The maximum height is given by
$$h(2) = -16(4) + 64(2) + 80 = 144, \text{ or } 144 \text{ feet.}$$

41. $P(x) = -0.04x^2 + 240x - 10,000$. We compute $P'(x) = -0.08x + 240$. Setting $P'(x) = 0$ gives $x = 3000$. The graph of P is a parabola that opens downward and so $x = 3000$ gives rise to the absolute maximum of P. Thus, to maximize profits, the company should produce 3000 cameras per month.

43. $N(t) = 0.81t - 1.14\sqrt{t} + 1.53$. $N'(t) = 0.81 - 1.14(\frac{1}{2}t^{-1/2}) = 0.81 - \dfrac{0.57}{t^{1/2}}$. Setting $N'(t) = 0$ gives $t^{1/2} = \dfrac{0.57}{0.81}$, or $t = 0.4952$ as a critical point of N. Evaluating $N(t)$ at the endpoints $t = 0$ and $t = 6$ as well as at the critical point, we have

t	0	0.4952	6
$N(t)$	1.53	1.13	3.60

From the table, we see that the absolute maximum of N occurs at $t = 6$ and the absolute minimum occurs at $t \approx 0.5$. Our results tell us that the number of nonfarm full-time self-employed women over the time interval from 1963 to 1993 was the highest in 1993 and stood at approximately 3.6 million.

45. $P(x) = -0.000002x^3 + 6x - 400$. $P'(x) = -0.000006x^2 + 6 = 0$ if $x = \pm 1000$. We

reject the negative root. Next, we compute $P''(x) = -0.000012x$. Since $P''(1000) = -0.012 < 0$, the Second Derivative Test shows that $x = 1000$ affords a relative maximum of f. From physical considerations, or from a sketch of the graph of f, we see that the maximum profit is realized if 1000 cases are produced per day. The profit is $P(1000) = -0.000002(1000)^3 + 6(1000) - 400$, or \$3600/day.

47. The revenue is $R(x) = px = -0.0004x^2 + 10x$, and the profit is
$$P(x) = R(x) - C(x) = -0.0004x^2 + 10x - (400 + 4x + 0.0001x^2)$$
$$= -0.0005x^2 + 6x - 400.$$
$$P'(x) = -0.001x + 6 = 0$$
if $x = 6000$, a critical point. Since $P''(x) = -0.001 < 0$ for all x, we see that the graph of P is a parabola that opens downward. Therefore, a level of production of 6000 rackets/day will yield a maximum profit.

49. The total cost function is given by
$$C(x) = V(x) + 20,000$$
$$= 0.000001x^3 - 0.01x^2 + 50x + 20,000$$
The profit function is
$$P(x) = R(x) - C(x)$$
$$= -0.02x^2 + 150x - 0.000001x^3 + 0.01x^2 - 50x + 20,000$$
$$= -0.000001x^3 - 0.01x^2 + 100x - 20,000$$
We want to maximize P on [0, 7000].
$$P'(x) = -0.000003x^2 - 0.02x + 100$$
Setting $P'(x) = 0$ gives $3x^2 + 20,000x - 100,000,000 = 0$
or $x = \dfrac{-20,000 \pm \sqrt{20,000^2 + 1,200,000,000}}{6} = -10,000$ or $3,333.3$.
So $x = 3,333.30$ is a critical point in the interval [0, 7000].

x	0	3,333	7,000
$P(x)$	-20,000	165,185	-519,700

From the table, we see that a level of production of 3,333 pagers per week will yield a maximum profit of \$165,185 per week.

12 Applications of the Derivative

51. a. $\overline{C}(x) = \dfrac{C(x)}{x} = 0.0025x + 80 + \dfrac{10,000}{x}$.

b. $\overline{C}'(x) = 0.0025 - \dfrac{10,000}{x^2} = 0$ if $0.0025x^2 = 10,000$, or $x = 2000$.

Since $\overline{C}''(x) = \dfrac{20,000}{x^3}$, we see that $\overline{C}''(x) > 0$ for $x > 0$ and so $\overline{C}$ is concave upward on $(0, \infty)$. Therefore, $x = 2000$ yields a minimum.

c. We solve $\overline{C}(x) = C'(x)$. $0.0025x + 80 + \dfrac{10,000}{x} = 0.005x + 80$,

$0.0025x^2 = 10,000$, or $x = 2000$.

d. It appears that we can solve the problem in two ways.
REMARK This can be proved.

53. The demand equation is $p = \sqrt{800 - x} = (800 - x)^{1/2}$. The revenue function is $R(x) = xp = x(800 - x)^{1/2}$. To find the maximum of R, we compute

$$
\begin{aligned}
R'(x) &= \tfrac{1}{2}(800 - x)^{-1/2}(-1)(x) + (800 - x)^{1/2} \\
&= \tfrac{1}{2}(800 - x)^{-1/2}[-x + 2(800 - x)] \\
&= \tfrac{1}{2}(800 - x)^{-1/2}(1600 - 3x).
\end{aligned}
$$

Next, $R'(x) = 0$ implies $x = 800$ or $x = 1600/3$ are critical points of R. Next, we compute the values of R given in the following table.

x	0	800	1600/3
$R(x)$	0	0	8709

We conclude that $R(\tfrac{1600}{3}) = 8709$ is the absolute maximum value. Therefore, the revenue is maximized by producing $1600/3 \approx 533$ dresses.

55. $f(t) = 100\left[\dfrac{t^2 - 4t + 4}{t^2 + 4}\right]$.

a. $f'(t) = 100\left[\dfrac{(t^2 + 4)(2t - 4) - (t^2 - 4t + 4)(2t)}{(t^2 + 4)^2}\right] = \dfrac{400(t^2 - 4)}{(t^2 + 4)^2}$

$$= \frac{400(t-2)(t+2)}{(t^2+4)^2}.$$

From the sign diagram for f'

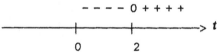

we see that $t = 2$ gives a relative minimum, and we conclude that the oxygen content is the lowest 2 days after the organic waste has been dumped into the pond.

b.

$$f''(t) = 400\left[\frac{(t^2+4)^2(2t)-(t^2-4)2(t^2+4)(2t)}{(t+4)^4}\right] = 400\left[\frac{(2t)(t^2+4)(t^2+4-2t^2+8)}{(t^2+4)^4}\right]$$

$$= -\frac{800t(t^2-12)}{(t^2+4)^3}$$

and $f''(t) = 0$ when $t = 0$ and $t = \pm 2\sqrt{3}$. We reject $t = 0$ and $t = -2\sqrt{3}$. From the sign diagram for f'',

we see that $f'(2\sqrt{3})$ gives an inflection point of f and we conclude that this is an absolute maximum. Therefore, the rate of oxygen regeneration is greatest 3.5 days after the organic waste has been dumped into the pond.

57. We compute $\bar{R}'(x) = \frac{xR'(x)-R(x)}{x^2}$. Setting $\bar{R}'(x) = 0$ gives $xR'(x)-R(x)=0$

or $R'(x) = \frac{R(x)}{x} = \bar{R}(x)$, so a critical point of $\bar{R}$ occurs when $\bar{R}(x) = R'(x)$.

Next, we compute

$$\bar{R}''(x) = \frac{x^2[R'(x)+xR''(x)-R'(x)]-[xR'(x)-R(x)](2x)}{x^4} = \frac{R''(x)}{x} < 0.$$

So, by the Second Derivative Test, the critical point does give a maximum revenue.

59. The growth rate is $G'(t) = -0.6t^2 + 4.8t$. To find the maximum growth rate, we compute $G''(t) = -1.2t + 4.8$. Setting $G''(t) = 0$ gives $t = 4$ as a critical point.

t	0	4	8
$G'(t)$	0	9.6	0

From the table, we see that G is maximal at $t = 4$; that is, the growth rate is greatest in 1997.

61. $f(t) = -0.0129t^4 + 0.3087t^3 + 2.1760t^2 + 62.8466t + 506.2955$.
    To find the maximum of $f(t)$, we first compute
    $$f'(t) = -0.0516t^3 + 0.9261t^2 + 4.352t + 62.8466.$$

Then
$$f'(23.6811) = -0.0516(23.6811)^3 + 0.9261(23.6811)^2 + 4.352(23.6811) + 62.8466$$
$$\approx 0$$

Next, we compute $f'(23) \approx 25.03$ and $f'(24) = -15.18$. Since f is a polynomial function it is continuous. We conclude that $f(t)$ is maximized when $t = 23.6811$

since f' changes sign from positive to negative as we move across the critical point $t = 23.6811$. These results may be confirmed by graphing the derivative function f' on your graphing calculator.

63. $R = D^2 \left(\dfrac{k}{2} - \dfrac{D}{3} \right) = \dfrac{kD^2}{2} - \dfrac{D^3}{3}$. $\dfrac{dR}{dD} = \dfrac{2kD}{2} - \dfrac{3D^2}{3} = kD - D^2 = D(k - D)$

Setting $\dfrac{dR}{dD} = 0$, we have $D = 0$ or $k = D$. We only consider $k = D$

(since $D > 0$). If $k > 0$, $\dfrac{dR}{dD} > 0$ and if $k < 0$, $\dfrac{dR}{dD} < 0$. Therefore $k = D$ provides a relative maximum. The nature of the problem suggests that $k = D$ gives the absolute maximum of R. We can also verify this by graphing R.

65. False. Let $f(x) = \begin{cases} |x| & \text{if } x \neq 0 \\ 1 & \text{if } x = 0 \end{cases}$ on $[-1, 1]$.

67. False. Let $f(x) = \begin{cases} -x & \text{if } -1 \le x < 0 \\ \dfrac{1}{2} & \text{if } 0 \le x < 1 \end{cases}$. Then f is discontinuous at $x = 0$. But f has an absolute maximum value of 1 attained at $x = -1$.

69. Since $f(x) = c$ for all x, the function f satisfies $f(x) \le c$ for all x and so f has an absolute maximum at all points of x. Similarly, f has an absolute minimum at all points of x.

71. a. f is not continuous at $x = 0$ because $\lim\limits_{x \to 0} f(x)$ does not exist.

   b. $\lim\limits_{x \to 0} f(x) = \lim\limits_{x \to 0^-} \dfrac{1}{x} = -\infty$ and $\lim\limits_{x \to 0^+} f(x) = \lim\limits_{x \to 0^+} \dfrac{1}{x} = \infty$

   c.

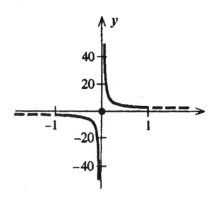

USING TECHNOLOGY EXERCISES 12.4, page 792

1. Absolute maximum value: 145.8985; absolute minimum value: -4.3834

3. Absolute maximum value: 16; absolute minimum value: -0.1257

5. Absolute maximum value: 2.8889; absolute minimum value: 0

7. a.

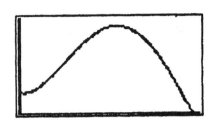

b. 200.1410 banks/yr

9. a.

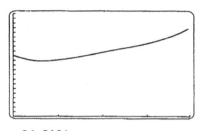

b. 21.51%

11. a.

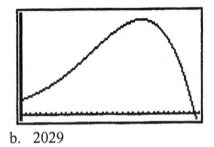

b. 2029

EXERCISES 12.5, page 803

1. Refer to the following figure.

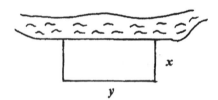

We have $2x + y = 3000$ and we want to maximize the function
$$A = f(x) = xy = x(3000 - 2x) = 3000x - 2x^2$$
on the interval $[0,1500]$. The critical point of A is obtained by solving
$f'(x) = 3000 - 4x = 0$, giving $x = 750$. From the table of values

x	0	750	1500
$f(x)$	0	1,125,000	0

we conclude that $x = 750$ yields the absolute maximum value of A. Thus, the
required dimensions are 750×1500 yards. The maximum area is 1,125,000 sq yd.

3. Let x denote the length of the side made of wood and y the length of the side made of steel. The cost of construction will be $C = 6(2x) + 3y$. But $xy = 800$. So $y = 800/x$ and therefore $C = f(x) = 12x + 3\left(\dfrac{800}{x}\right) = 12x + \dfrac{2400}{x}$. To minimize C, we compute

$$f'(x) = 12 - \frac{2400}{x^2} = \frac{12x^2 - 2400}{x^2} = \frac{12(x^2 - 200)}{x^2}.$$

Setting $f'(x) = 0$ gives $x = \pm\sqrt{200}$ as critical points of f. The sign diagram of f'

```
        - - - -  0 + + + + + +
    ────────────┼──────────────────>  x
                0    √200
```

shows that $x = \pm\sqrt{200}$ gives a relative minimum of f. $f''(x) = \dfrac{4800}{x^3} > 0$

if $x > 0$ and so f is concave upward for $x > 0$. Therefore $x = \sqrt{200} = 10\sqrt{2}$ actually yields the absolute minimum. So the dimensions of the enclosure should be

$$10\sqrt{2} \text{ ft} \times \frac{800}{10\sqrt{2}} \text{ ft, or } 14.1 \text{ ft x } 56.6 \text{ ft.}$$

5. Let the dimensions of each square that is cut out be $x'' \times x''$. Refer to the following diagram.

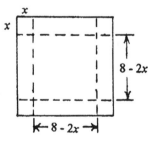 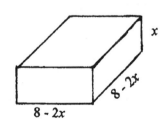

Then the dimensions of the box will be $(8 - 2x)''$ by $(8 - 2x)''$ by x''. Its volume will be $V = f(x) = x(8 - 2x)^2$. We want to maximize f on $[0,4]$.

$$f'(x) = (8 - 2x)^2 + x(2)(8 - 2x)(-2) \qquad \text{[Using the Product Rule.]}$$
$$= (8 - 2x)[(8 - 2x) - 4x] = (8 - 2x)(8 - 6x) = 0$$

if $x = 4$ or $4/3$. The latter is a critical point in $(0,4)$.

x	0	4/3	4
$f(x)$	0	1024/27	0

We see that $x = 4/3$ yields an absolute maximum for f. So the dimensions of the box should be $\frac{16}{3}" \times \frac{16}{3}" \times \frac{4}{3}"$.

7. Let x denote the length of the sides of the box and y denote its height. Referring to the following figure, we see that the volume of the box is given by $x^2y = 128$. The

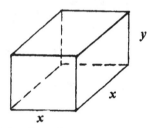

amount of material used is given by

$$S = f(x) = 2x^2 + 4xy$$

$$= 2x^2 + 4x\left(\frac{128}{x^2}\right)$$

$$= 2x^2 + \frac{512}{x} \text{ square inches.}$$

We want to minimize f subject to the condition that $x > 0$. Now

$$f'(x) = 4x - \frac{512}{x^2} = \frac{4x^3 - 512}{x^2} = \frac{4(x^3 - 128)}{x^2}.$$

Setting $f'(x) = 0$ yields $x = 5.04$, a critical point of f. Next,

$$f''(x) = 4 + \frac{1024}{x^3} > 0$$

for all $x > 0$. Thus, the graph of f is concave upward and so $x = 5.04$ yields an absolute minimum of f. Thus, the required dimensions are 5.04" × 5.04" × 5.04".

9. The length plus the girth of the box is $4x + h = 108$ and $h = 108 - 4x$. Then
$$V = x^2h = x^2(108 - 4x) = 108x^2 - 4x^3$$
and $V' = 216x - 12x^2$. We want to maximize V on the interval $[0,27]$. Setting $V'(x) = 0$ and solving for x, we obtain $x = 18$ and $x = 0$. Evaluating $V(x)$ at $x = 0$,

$x = 18$, and $x = 27$, we obtain
$$V(0) = 0, \; V(18) = 11{,}664, \text{ and } V(27) = 0$$
Thus, the dimensions of the box are 18" $\times$ 18" $\times$ 36" and its maximum volume is approximately 11,664 cu in.

11. We take $2\pi r + \ell = 108$. We want to maximize
$$V = \pi r^2 \ell = \pi r^2 (-2\pi r + 108) = -2\pi^2 r^3 + 108\pi r^2$$
subject to the condition that $0 \le r \le \frac{54}{\pi}$. Now
$$V'(r) = -6\pi^2 r^2 + 216\pi r = -6\pi r(\pi r - 36).$$
Since $V' = 0$, we find $r = 0$ or $r = 36/\pi$, the critical points of V. From the table

r	0	36/π	54/π
V	0	46,656/π	0

we conclude that the maximum volume occurs when $r = 36/\pi \approx 11.5$ inches and $\ell = 108 - 2\pi\left(\frac{36}{\pi}\right) = 36$ inches and its volume is $46{,}656/\pi$ cu in .

13. Let y denote the height and x the width of the cabinet. Then $y = (3/2)x$. Since the volume is to be 2.4 cu ft, we have $xyd = 2.4$, where d is the depth of the cabinet.

We have $\quad x\left(\dfrac{3}{2}x\right)d = 2.4 \; \text{ or } \; d = \dfrac{2.4(2)}{3x^2} = \dfrac{1.6}{x^2}.$

The cost for constructing the cabinet is
$$C = 40(2xd + 2yd) + 20(2xy) = 80\left[\frac{1.6}{x} + \left(\frac{3}{2}x\right)\left(\frac{1.6}{x^2}\right)\right] + 40x\left(\frac{3}{2}x\right)$$

$$= \frac{320}{x} + 60x^2.$$

$$C'(x) = -\frac{320}{x^2} + 120x = \frac{120x^3 - 320}{x^2} = 0 \; \text{ if } x = \sqrt[3]{\frac{8}{3}} = \frac{2}{\sqrt[3]{3}} = \frac{2}{3}\sqrt[3]{9}$$

Therefore, $x = \frac{2}{3}\sqrt[3]{9}$ is a critical point of C. The sign diagram

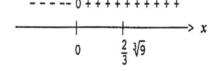

shows that $x = \frac{2}{3}\sqrt[3]{9}$ gives a relative minimum. Next, $C''(x) = \dfrac{640}{x^3} + 120 > 0$

for all $x > 0$ tells us that the graph of C is concave upward. So $x = \frac{2}{3}\sqrt[3]{9}$ yields an

absolute minimum. The required dimensions are $\frac{2}{3}\sqrt[3]{9}\,' \times \sqrt[3]{9}\,' \times \frac{2}{3}\sqrt[3]{9}\,'$.

15. We want to maximize the function
$$R(x) = (200 + x)(300 - x) = -x^2 + 100x + 60000.$$
$$R'(x) = -2x + 100 = 0$$
gives $x = 50$ and this is a critical point of R. Since $R''(x) = -2 < 0$, we see that $x = 50$ gives an absolute maximum of R. Therefore, the number of passengers should be 250. The fare will then be \$250/passenger and the revenue will be \$62,500.

17. Let x denote the number of people beyond 20 who sign up for the cruise. Then the revenue is $R(x) = (20 + x)(600 - 4x) = -4x^2 + 520x + 12,000$. We want to maximize R on the closed bounded interval $[0, 70]$.
$$R'(x) = -8x + 520 = 0 \text{ implies } x = 65,$$
a critical point of R. Evaluating R at this critical point and the endpoints, we have

x	0	65	70
$R(x)$	12,000	28,900	28,800

From this table, we see that R is maximized if $x = 65$. Therefore, 85 passengers will result in a maximum revenue of \$28,900. The fare would be \$340/passenger.

19. We want to maximize $S = kh^2w$. But $h^2 + w^2 = 24^2$ or $h^2 = 576 - w^2$. So $S = f(w) = kw(576 - w^2) = k(576w - w^3)$. Now, setting
$$f'(w) = k(576 - 3w^2) = 0$$
gives $w = \pm\sqrt{192} \approx \pm 13.86$. Only the positive root is a critical point of interest. Next, we find $f''(w) = -6kw$, and in particular,
$$f''(\sqrt{192}) = -6\sqrt{192}\,k < 0,$$
so that $w = \pm\sqrt{192} \approx \pm 13.86$ gives a relative maximum of f. Since $f''(w) < 0$ for $w > 0$, we see that the graph of f is concave downward on $(0, \infty)$ and so, $w = \sqrt{192}$ gives an absolute maximum of f. We find $h^2 = 576 - 192 = 384$ or $h \approx 19.60$. So the width and height of the log should be approximately 13.86 inches and 19.60 inches, respectively.

21. We want to minimize $C(x) = 1.50(10,000 - x) + 2.50\sqrt{3000^2 + x^2}$ subject to $0 \le x \le 10,000$. Now

$$C'(x) = -1.50 + 2.5(\tfrac{1}{2})(9,000,000 + x^2)^{-1/2}(2x) = -1.50 + \frac{2.50x}{\sqrt{9,000,000 + x^2}}$$

$$C'(x) = 0 \Rightarrow 2.5x = 1.50\sqrt{9,000,000 + x^2}$$

$6.25x^2 = 2.25(9,000,000 + x^2)$ or $4x^2 = 20250000$, $x = 2250$.

x	0	2250	10000
$f(x)$	22500	21000	26101

From the table, we see that $x = 2250$ gives the absolute minimum.

23. The time taken for the flight is

$$T = f(x) = \frac{12 - x}{6} + \frac{\sqrt{x^2 + 9}}{4}.$$

$$f'(x) = -\frac{1}{6} + \frac{1}{4}\left(\frac{1}{2}\right)(x^2 + 9)^{-1/2}(2x) = -\frac{1}{6} + \frac{x}{4\sqrt{x^2 + 9}}$$

$$= \frac{3x - 2\sqrt{x^2 + 9}}{12\sqrt{x^2 + 9}}.$$

Setting $f'(x) = 0$ gives $3x = 2\sqrt{x^2 + 9}$, $9x^2 = 4(x^2 + 9)$ or $5x^2 = 36$. Therefore, $x = \pm 6/\sqrt{5} = \pm 6\sqrt{5}/5$. Only the critical point $x = 6\sqrt{5}/5$ is of interest. The nature of the problem suggests $x \approx 2.68$ gives an absolute minimum for T.

25. Let x denote the number of motorcycle tires in each order. We want to minimize

$$C(x) = 400\left(\frac{40,000}{x}\right) + x = \frac{16,000,000}{x} + x.$$

We compute $C'(x) = -\dfrac{16,000,000}{x^2} + 1 = \dfrac{x^2 - 16,000,000}{x^2}.$

Setting $C'(x) = 0$ gives $x = 4000$, a critical point of C. Since

$$C''(x) = \frac{32,000,000}{x^3} > 0 \text{ for all } x > 0,$$

we see that the graph of C is concave upward and so $x = 4000$ gives an absolute minimum of C. So there should be 10 orders per year, each order of 4000 tires.

27. We want to minimize the function $C(x) = \dfrac{500,000,000}{x} + 0.2x + 500,000$ on the interval (0, 1,000,000). Differentiating C(x), we have $C'(x) = -\dfrac{500,000,000}{x^2} + 0.2$.

Setting $C'(x) = 0$ and solving the resulting equation, we find $0.2x^2 = 500,000,000$ and $x = \sqrt{2,500,000,000}$ or $x = 50,000$. Next, we find

$C''(x) = \dfrac{1,000,000,000}{x^3} > 0$ for all x and so the graph of C is concave upward on

(0,∞). Thus, $x = 50,000$ gives rise to the absolute minimum of C. So, the company should produce 50,000 containers of cookies per production run.

CHAPTER 12 REVIEW, page 808

1. a. $f(x) = \frac{1}{3}x^3 - x^2 + x - 6.$ $f'(x) = x^2 - 2x + 1 = (x - 1)^2.$ $f'(x) = 0$ gives $x = 1$, the critical point of f. Now, $f'(x) > 0$ for all $x \neq 1$. Thus, f is increasing on (-∞,1) $\cup$ (1,∞).
   b. Since $f'(x)$ does not change sign as we move across the critical point $x = 1$, the First Derivative Test implies that $x = 1$ does not give rise to a relative extremum of f.
   c. $f''(x) = 2(x - 1)$. Setting $f''(x) = 0$ gives $x = 1$ as a candidate for an inflection point of f. Since $f''(x) < 0$ for $x < 1$, and $f''(x) > 0$ for $x > 1$, we see that f is concave downward on (-∞,1) and concave upward on (1,∞).
   d. The results of (c) imply that $(1, -\frac{17}{3})$ is an inflection point.

3. a. $f(x) = x^4 - 2x^2.$ $f'(x) = 4x^3 - 4x = 4x(x^2 - 1) = 4x(x + 1)(x - 1)$. The sign diagram of f' shows that f is decreasing on (-∞,-1) $\cup$ (0,1) and increasing on (-1,0) $\cup$ (1,∞).

```
    - - -0 + + 0 - - 0 + + +
   ─────┼───┼───┼──────> x
       -1   0   1
```

   b. The results of (a) and the First Derivative Test show that (-1,-1) and (1,-1) are relative minima and (0,0) is a relative maximum.
   c. $f''(x) = 12x^2 - 4 = 4(3x^2 - 1) = 0$ if $x = \pm\sqrt{3}/3$. The sign diagram

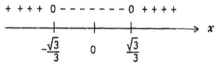

shows that f is concave upward on $(-\infty, -\sqrt{3}/3) \cup (\sqrt{3}/3, \infty)$ and concave downward on $(-\sqrt{3}/3, \sqrt{3}/3)$.

d. The results of (c) show that $(-\sqrt{3}/3, -5/9)$ and $(\sqrt{3}/3, -5/9)$ are inflection points.

5. a. $f(x) = \dfrac{x^2}{x-1}$. $f'(x) = \dfrac{(x-1)(2x) - x^2(1)}{(x-1)^2} = \dfrac{x^2 - 2x}{(x-1)^2} = \dfrac{x(x-2)}{(x-1)^2}$.

The sign diagram of f'

f' is not defined here

$+ \ + \ + \ 0 \ - \ - \ \ - \ - \ 0 \ + \ + \ + \ + \ + \quad \to x$

$\qquad\qquad 0 \qquad 1 \qquad 2$

shows that f is increasing on $(-\infty, 0) \cup (2, \infty)$ and decreasing on $(0,1) \cup (1,2)$.

b. The results of (a) show that $(0,0)$ is a relative maximum and $(2,4)$ is a relative minimum.

c. $f''(x) = \dfrac{(x-1)^2(2x-2) - x(x-2)2(x-1)}{(x-1)^4} = \dfrac{2(x-1)[(x-1)^2 - x(x-2)]}{(x-1)^4}$

$\qquad = \dfrac{2}{(x-1)^3}$.

Since $f''(x) < 0$ if $x < 1$ and $f''(x) > 0$ if $x > 1$, we see that f is concave downward on $(-\infty, 1)$ and concave upward on $(1, \infty)$.

d. Since $x = 1$ is not in the domain of f, there are no inflection points.

7. $f(x) = (1-x)^{1/3}$. $f'(x) = -\dfrac{1}{3}(1-x)^{-2/3} = -\dfrac{1}{3(1-x)^{2/3}}$.

The sign diagram for f' is

f' not defined here

$- \ - \ - \ - \ - \ - \ - \ - \ - \ \downarrow \ - \ - \ - \ - \ -$

$\qquad\qquad 0 \qquad 1 \qquad\qquad \to x$

a. f is decreasing on $(-\infty, 1) \cup (1, \infty)$.

b. There are no relative extrema.

c. Next, we compute $f''(x) = -\dfrac{2}{9}(1-x)^{-5/3} = -\dfrac{2}{9(1-x)^{5/3}}$.

The sign diagram for f'' is

f'' not defined here

$$-\; -\; -\; -\; -\; -\; -\; -\; -\; \downarrow +\; +\; +\; +\; +\; +$$

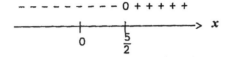

$$\begin{array}{ccc} & 0 & 1 \end{array} \longrightarrow x$$

We find f is concave downward on $(-\infty,1)$ and concave upward on $(1,\infty)$.

d. $x = 1$ is a candidate for an inflection point of f. Referring to the sign diagram for f'', we see that $(1,0)$ is an inflection point.

9. a. $f(x) = \dfrac{2x}{x+1}$. $f'(x) = \dfrac{(x+1)(2) - 2x(1)}{(x+1)^2} = \dfrac{2}{(x+1)^2} > 0$ if $x \neq -1$.

Therefore f is increasing on $(-\infty,-1) \cup (-1,\infty)$.

b. Since there are no critical points, f has no relative extrema.

c. $f''(x) = -4(x+1)^{-3} = -\dfrac{4}{(x+1)^3}$. Since $f''(x) > 0$ if $x < -1$ and $f''(x) < 0$ if $x > -1$,

we see that f is concave upward on $(-\infty,-1)$ and concave downward on $(-1,\infty)$.

d. There are no inflection points since $f''(x) \neq 0$ for all x in the domain of f.

11. $f(x) = x^2 - 5x + 5$

1. The domain of f is $(-\infty, \infty)$.

2. Setting $x = 0$ gives 5 as the y-intercept.

3. $\lim\limits_{x \to -\infty} (x^2 - 5x + 5) = \lim\limits_{x \to \infty} (x^2 - 5x + 5) = \infty$.

4. There are no asymptotes because f is a quadratic function.

5. $f'(x) = 2x - 5 = 0$ if $x = 5/2$. The sign diagram

$$-\; -\; -\; -\; -\; -\; -\; -\; -\; 0\; +\; +\; +\; +\; +$$

$$\begin{array}{ccc} & 0 & \frac{5}{2} \end{array} \longrightarrow x$$

shows that f is increasing on $(\frac{5}{2},\infty)$ and decreasing on $(-\infty,\frac{5}{2})$.

6. The First Derivative Test implies that $(\frac{5}{2},-\frac{5}{4})$ is a relative minimum.

7. $f''(x) = 2 > 0$ and so f is concave upward on $(-\infty, \infty)$.

8. There are no inflection points.

The graph of f follows.

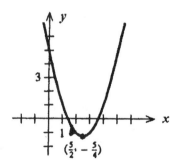

$(\frac{5}{2}, -\frac{5}{4})$

13. $g(x) = 2x^3 - 6x^2 + 6x + 1$.
   1. The domain of g is $(-\infty, \infty)$.
   2. Setting $x = 0$ gives 1 as the y-intercept.
   3. $\lim\limits_{x \to -\infty} g(x) = -\infty$, $\lim\limits_{x \to \infty} g(x) = \infty$.
   4. There are no vertical or horizontal asymptotes.
   5. $g'(x) = 6x^2 - 12x + 6 = 6(x^2 - 2x + 1) = 6(x - 1)^2$. Since $g'(x) > 0$ for all $x \neq 1$, we see that g is increasing on $(-\infty,1) \cup (1,\infty)$.
   6. $g'(x)$ does not change sign as we move across the critical point $x = 1$, so there is no extremum.
   7. $g''(x) = 12x - 12 = 12(x - 1)$. Since $g''(x) < 0$ if $x < 1$ and $g''(x) > 0$ if $x > 1$, we see that g is concave upward on $(1,\infty)$ and concave downward on $(-\infty,1)$.
   8. The point $x = 1$ gives rise to the inflection point $(1,3)$.
   9. The graph of g follows.

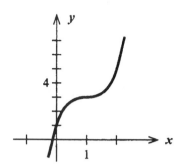

15. $h(x) = x\sqrt{x - 2}$.
   1. The domain of h is $[2,\infty)$.
   2. There are no y-intercepts. Next, setting $y = 0$ gives 2 as the x-intercept.
   3. $\lim\limits_{x \to \infty} x\sqrt{x - 2} = \infty$.

4. There are no asymptotes.

5. $h'(x) = (x-2)^{1/2} + x(\frac{1}{2})(x-2)^{-1/2} = \frac{1}{2}(x-2)^{-1/2}[2(x-2)+x]$

$$= \frac{3x-4}{2\sqrt{x-2}} > 0 \quad \text{on } [2,\infty)$$

and so h is increasing on $[2,\infty)$.

6. Since h has no critical points in $(2,\infty)$, there are no relative extrema.

7. $h''(x) = \frac{1}{2}\left[\dfrac{(x-2)^{1/2}(3) - (3x-4)\frac{1}{2}(x-2)^{-1/2}}{x-2} \right]$

$$= \frac{(x-2)^{-1/2}[6(x-2)-(3x-4)]}{4(x-2)} = \frac{3x-8}{4(x-2)^{3/2}}.$$

The sign diagram for h''

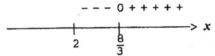

$$--\ -\ 0\ +\ +\ +\ +\ +$$

shows that h is concave downward on $(2,\frac{8}{3})$ and concave upward on $(\frac{8}{3},\infty)$.

8. The results of (7) tell us that $(\frac{8}{3}, \frac{8\sqrt{6}}{9})$ is an inflection point.

The graph of h follows.

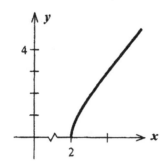

17. $f(x) = \dfrac{x-2}{x+2}$.

1. The domain of f is $(-\infty,-2) \cup (-2,\infty)$.

2. Setting $x = 0$ gives -1 as the y-intercept. Setting $y = 0$ gives 2 as the x-intercept.

3. $\displaystyle\lim_{x\to-\infty} \frac{x-2}{x+2} = \lim_{x\to\infty}\frac{x-2}{x+2} = 1.$

4. The results of (3) tell us that $y = 1$ is a horizontal asymptote. Next, observe that the denominator of $f(x)$ is equal to zero at $x = -2$, but its numerator is not equal to zero there. Therefore, $x = -2$ is a vertical asymptote.

5. $$f'(x) = \frac{(x+2)(1)-(x-2)(1)}{(x+2)^2} = \frac{4}{(x+2)^2}.$$

The sign diagram of f'

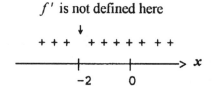

tells us that f is increasing on $(-\infty,-2) \cup (-2,\infty)$.

6. The results of (5) tells us that there are no relative extrema.

7. $f''(x) = -\dfrac{8}{(x+2)^3}$. The sign diagram of f'' follows

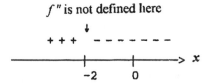

and it shows that f is concave upward on $(-\infty,-2)$ and concave downward on $(-2,\infty)$.

8. There are no inflection points.
The graph of f follows.

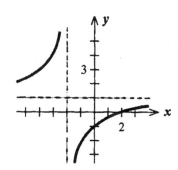

19. $\lim\limits_{x\to-\infty}\dfrac{1}{2x+3}=\lim\limits_{x\to\infty}\dfrac{1}{2x+3}=0$ and so $y=0$ is a horizontal asymptote. Since the denominator is equal to zero at $x=-3/2$, but the numerator is not equal to zero there, we see that $x=-3/2$ is a vertical asymptote.

21. $\lim\limits_{x\to-\infty}\dfrac{5x}{x^2-2x-8}=\lim\limits_{x\to\infty}\dfrac{5x}{x^2-2x-8}=0$ and so $y=0$ is a horizontal asymptote. Next, note that the denominator is zero if $x^2 - 2x - 8 = (x-4)(x+2)=0$, or $x=-2$ or $x=4$. Since the numerator is not equal to zero at these points, we see that $x=-2$ and $x=4$ are vertical asymptotes.

23. $f(x)=2x^2+3x-2$; $f'(x)=4x+3$. Setting $f'(x)=0$ gives $x=-3/4$ as a critical point of f. Next, $f''(x)=4>0$ for all x, so f is concave upward on $(-\infty,\infty)$. Therefore, $f(-\frac{3}{4})=-\frac{25}{8}$ is an absolute minimum of f. There is no absolute maximum.

25. $g(t)=\sqrt{25-t^2}=(25-t^2)^{1/2}$. Differentiating $g(t)$, we have
$$g'(t)=\tfrac{1}{2}(25-t^2)^{-1/2}(-2t)=-\dfrac{t}{\sqrt{25-t^2}}.$$
Setting $g'(t)=0$ gives $t=0$ as a critical point of g. The domain of g is given by solving the inequality $25 - t^2 \ge 0$ or $(5-t)(5+t)\ge 0$ which implies that $t\in[-5,5]$. From the table

t	-5	0	5
$g(t)$	0	5	0

we conclude that $g(0)=5$ is the absolute maximum of g and $g(-5)=0$ and $g(5)=0$ is the absolute minimum value of g.

27. $h(t)=t^3-6t^2$. $h'(t)=3t^2-12t=3t(t-4)=0$ if $t=0$ or $t=4$, critical points of h. But only $t=4$ lies in $(2,5)$.

t	2	4	5
$h(t)$	-16	-32	-25

From the table, we see that there is an absolute minimum at (4,-32) and an absolute maximum at (2,-16).

29. $f(x) = x - \dfrac{1}{x}$ on [1,3]. $f'(x) = 1 + \dfrac{1}{x^2}$. Since $f'(x)$ is never zero, f has no critical point.

x	1	3
$f(x)$	0	$\frac{8}{3}$

We see that $f(1) = 0$ is the absolute minimum value and $f(3) = 8/3$ is the absolute maximum value.

31. $f(s) = s\sqrt{1-s^2}$ on [-1,1]. The function f is continuous on [-1,1] and differentiable on (-1,1). Next,

$$f'(s) = (1-s^2)^{1/2} + s(\tfrac{1}{2})(1-s^2)^{-1/2}(-2s) = \frac{1-2s^2}{\sqrt{1-s^2}}.$$

Setting $f'(s) = 0$, we have $s = \pm\sqrt{2}/2$, giving the critical points of f. From the table

x	-1	$-\sqrt{2}/2$	$\sqrt{2}/2$	1
$f(x)$	0	-1/2	1/2	0

we see that $f(-\sqrt{2}/2) = -1/2$ is the absolute minimum value and $f(\sqrt{2}/2) = 1/2$ is the absolute maximum value of f.

33. We want to maximize $P(x) = -x^2 + 8x + 20$. Now, $P'(x) = -2x + 8 = 0$ if $x = 4$, a critical point of P. Since $P''(x) = -2 < 0$, the graph of P is concave downward. Therefore, the critical point $x = 4$ yields an absolute maximum. So, to maximize

profit, the company should spend $4000 on advertising per month.

35. a. $I(t) = \dfrac{50t^2 + 600}{t^2 + 10}$.

$I'(t) = \dfrac{(t^2 + 10)(100t) - (50t^2 + 600)(2t)}{(t^2 + 10)^2} = -\dfrac{200t}{(t^2 + 10)^2} < 0$ on (0,10) and so I is

decreasing on (0,10).

b. $I''(t) = -200\left[\dfrac{(t^2 + 10)^2(1) - t(2)(t^2 + 10)(2t)}{(t^2 + 10)^4}\right]$

$= \dfrac{-200(t^2 + 10)[(t^2 + 10) - 4t^2]}{(t^2 + 10)^4} = -\dfrac{200(10 - 3t^2)}{(t^2 + 10)^3}$.

The sign diagram of I'' (for $t > 0$)

```
              - - - -  0 + + + + + + +
        ─────────────┼───────┼──────────────> t
              0      √10/3 ≈ 1.8
```

shows that I is concave downward on $(0, \sqrt{10/3})$ and concave upward on
$(\sqrt{10/3}, \infty)$.

c.

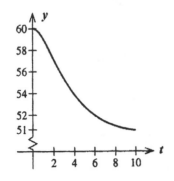

d. The rate of decline in the environmental quality of the wildlife was increasing
the first 1.8 years. After that time the rate of decline decreased.

37. a. $C(x) = 0.001x^2 + 100x + 4000$.

$\overline{C}(x) = \dfrac{C(x)}{x} = \dfrac{0.001x^2 + 100x + 4000}{x} = 0.001x + 100 + \dfrac{4000}{x}$.

b. $\overline{C}'(x) = 0.001 - \dfrac{4000}{x^2} = \dfrac{0.001x^2 - 4000}{x^2} = \dfrac{0.001(x^2 - 4,000,000)}{x^2}$.

Setting $\overline{C}'(x) = 0$ gives $x = \pm 2000$. We reject the negative root.

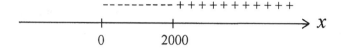

The sign diagram of $\overline{C}'$ shows that $x = 2000$ gives rise to a relative minimum of $\overline{C}$.

Since $\overline{C}''(x) = \dfrac{8000}{x^3} > 0$ if $x > 0$, we see that $\overline{C}$ is concave upward on $(0,\infty)$. So

$x = 2000$ yields an absolute minimum. So the required production level is 2000 units.

39. $R'(x) = k\dfrac{d}{dx}x(M - x) = k[(M - x) + x(-1)] = k(M - 2x)$

Setting $R'(x) = 0$ gives $M - 2x = 0$, or $x = \frac{M}{2}$, a critical point of R. Since $R''(x) = -2k < 0$, we see that $x = M/2$ affords a maximum; that is R is greatest when half the population is infected.

41. Suppose the radius is r and the height is h. Then the capacity is $\pi r^2 h$ and we want

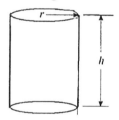

it to be 32π cu ft; that is, $\pi r^2 h = 32\pi$. Let the cost for the side by \$c/sq ft. Then the cost of construction is $C = 2\pi rhc + 2(\pi r^2)(2c) = 2\pi crh + 4\pi cr^2$. But

$h = \dfrac{32\pi}{\pi r^2} = \dfrac{32}{r^2}$. Therefore,

$$C = f(r) = -\dfrac{64\pi c}{r^2} + 8\pi cr = \dfrac{-64\pi c + 8\pi cr^3}{r^2} = \dfrac{8\pi c(-8 + r^3)}{r^2}$$

Setting $f'(r) = 0$ gives $r^3 = 8$ or $r = 2$. Next, $f''(r) = \dfrac{128\pi c}{r^3} + 8\pi c$ and so

$f''(2) > 0$. Therefore, $r = 2$ minimizes f. The required dimensions are $r = 2$ and

$h = \dfrac{32}{4} = 8$. That is, its radius is 2 ft and its height is 8 ft.

43. Let x denote the number of cases in each order. Then the average number of cases of beer in storage during the year is $x/2$. The storage cost is $2(x/2)$, or x dollars. Next, we see that the number of orders required is $800,000/x$, and so the ordering cost is

$$\frac{500(800,000)}{x} = \frac{400,000,000}{x}$$

dollars. Thus, the total cost incurred by the company per year is given by

$$C(x) = x + \frac{400,000,000}{x}.$$

We want to minimize C in the interval $(0, \infty)$. Now

$$C'(x) = 1 - \frac{400,000,000}{x^2}.$$

Setting $C'(x) = 0$ gives $x^2 = 400,000,000$, or $x = 20,000$ (we reject $x = -20,000$).

Next, $C''(x) = \dfrac{800,000,000}{x^3} > 0$ for all x, so C is concave upward. Thus,

$x = 20,000$ gives rise to the absolute minimum of C. Thus, the company should order 20,000 cases of beer per order.

CHAPTER 13

EXERCISES 13.1, page 818

1. a. $4^{-3} \times 4^5 = 4^{-3+5} = 4^2 = 16$
    b. $3^{-3} \times 3^6 = 3^{6-3} = 3^3 = 27.$

3. a. $9(9)^{-1/2} = \dfrac{9}{9^{1/2}} = \dfrac{9}{3} = 3.$
    b. $5(5)^{-1/2} = 5^{1/2} = \sqrt{5}.$

5. a. $\dfrac{(-3)^4(-3)^5}{(-3)^8} = (-3)^{4+5-8} = (-3)^1 = -3.$
    b. $\dfrac{(2^{-4})(2^6)}{2^{-1}} = 2^{-4+6+1} = 2^3 = 8.$

7. a. $\dfrac{5^{3.3} \cdot 5^{-1.6}}{5^{-0.3}} = \dfrac{5^{3.3-1.6}}{5^{-0.3}} = 5^{1.7+(0.3)} = 5^2 = 25.$

    b. $\dfrac{4^{2.7} \cdot 4^{-1.3}}{4^{-0.4}} = 4^{2.7-1.3+0.4} = 4^{1.8} \approx 12.1257.$

9. a. $(64x^9)^{1/3} = 64^{1/3}(x^{9/3}) = 4x^3.$
    b. $(25x^3y^4)^{1/2} = 25^{1/2}(x^{3/2})(y^{4/2}) = 5x^{3/2}y^2 = 5xy^2\sqrt{x}.$

11. a. $\dfrac{6a^{-5}}{3a^{-3}} = 2a^{-5+3} = 2a^{-2} = \dfrac{2}{a^2}.$
    b. $\dfrac{4b^{-4}}{12b^{-6}} = \dfrac{1}{3}b^{-4+6} = \dfrac{1}{3}b^2.$

13. a. $(2x^3y^2)^3 = 2^3 \times x^{3(3)} \times y^{2(3)} = 8x^9y^6.$
    b. $(4x^2y^2z^3)^2 = 4^2 \times x^{2(2)} \times y^{2(2)} \times z^{3(2)} = 16x^4y^4z^6.$

15. a. $\dfrac{5^0}{(2^{-3}x^{-3}y^2)^2} = \dfrac{1}{2^{-3(2)}x^{-3(2)}y^{2(2)}} = \dfrac{2^6x^6}{y^4} = \dfrac{64x^6}{y^4}.$
    b. $\dfrac{(x+y)(x-y)}{(x-y)^0} = (x+y)(x-y).$

17. $6^{2x} = 6^4$ if and only if $2x = 4$ or $x = 2.$

19. $3^{3x-4} = 3^5$ if and only if $3x - 4 = 5, \ 3x = 9, \text{ or } x = 3.$

21. $(2.1)^{x+2} = (2.1)^5$ if and only if $x + 2 = 5$, or $x = 3$.

23. $8^x = (\frac{1}{32})^{x-2}$, $(2^3)^x = (32)^{2-x} = (2^5)^{2-x}$, so $2^{3x} = 2^{5(2-x)}$, $3x = 10 - 5x$, $8x = 10$, or $x = 5/4$.

25. Let $y = 3^x$, then the given equation is equivalent to
$$y^2 - 12y + 27 = 0$$
$$(y-9)(y-3) = 0$$
giving $y = 3$ or 9. So $3^x = 3$ or $3^x = 9$, and therefore, $x = 1$ or $x = 2$.

27. $y = 2^x$, $y = 3^x$, and $y = 4^x$

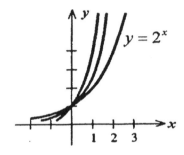

29. $y = 2^{-x}$, $y = 3^{-x}$, and $y = 4^{-x}$

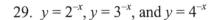
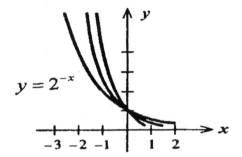

31. $y = 4^{0.5x}$, $y = 4x$, and $y = 4^{2x}$

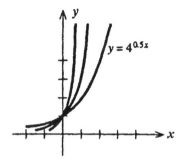

33. $y = e^{0.5x}$, $y = e^x$, $y = e^{1.5x}$

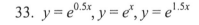
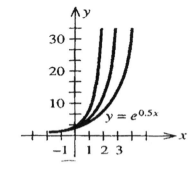

35. $y = 0.5e^{-x}$, $y = e^{-x}$, and $y = 2e^{-x}$

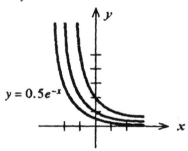

$y = 0.5e^{-x}$

37. $A = 5000e^{0.08(4)} = 6885.64$, or $6,885.64.

39. $P = Ae^{-rt} = 59673e^{-(0.08)5} \approx 40,000$, or $40,000.

41. a. If inflation over the next 15 years is 6 percent, then Carmen's first year's pension will be worth $P = 40,000e^{-0.9} = 16,262.79$, or $16,262.79.
b. If inflation over the next 15 years is 8 percent, then Carmen's first year's pension will be worth $P = 40,000e^{-1.2} = 12,047.77$, or $12,047.77.
c. If inflation over the next 15 years is 12 percent, then Ms. Carmen's first year's pension will be worth $P = 40,000e^{-1.8} = 6611.96$, or $6,611.96.

43. a.

Year	0	1	2	3	4	5
Number (billions)	0.45	0.80	1.41	2.49	4.39	7.76

b.

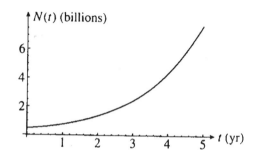

45. a. The initial concentration is given by $C(0) = 0.3(0) - 18(1 - e^{-(0)/60})$, or 0 g/cm^3

    b. The concentration after 10 seconds is given by
    $$C(10) = 0.3(10) - 18(1 - e^{-(10)/60}) = 0.23667, \text{ or } 0.2367 \text{ g/cm}^3.$$

    c. The concentration after 30 seconds is given by
    $$C(30) = 18e^{-(30)/60} - 12e^{-(30-20)/60} = 0.75977, \text{ or } 0.7598 \text{ g/cm}^3.$$

    d. The concentration of the drug in the long run is given by
    $$\lim_{t \to \infty} C(t) = \lim_{t \to \infty}[18e^{-t/60} - 12e^{-\frac{t-20}{60}}] = 0$$

47. False. $(x^2 + 1)^3 = x^6 + 3x^4 + 3x^2 + 1$.

49. True. $f(x) = e^x$ is an increasing function and so if $x < y$, then $f(x) < f(y)$, or $e^x < e^y$.

51. $r_{eff} = \lim_{m \to \infty}\left(1 + \dfrac{r}{m}\right)^m - 1 = e^r - 1$.

53. The effective rate of interest at Bank A is given by $R = \left(1 + \dfrac{0.07}{4}\right)^4 - 1 = 0.07186$,

    or 7.186 percent. The effective rate of interest at Bank B is given by
    $R = e^e - 1 = e^{0.07125} - 1 = 0.07385$, or 7.385 percent.
    We conclude that Bank B has the higher effective rate of interest.

USING TECHNOLOGY EXERCISES 13.1, page 820

1.

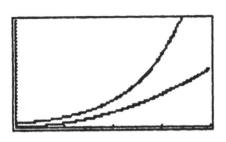

3.

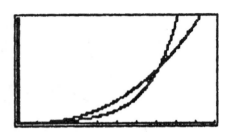

5..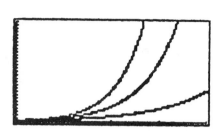

7.

9.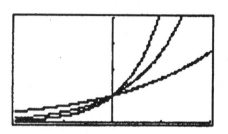

11. a.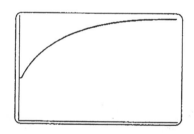

13.a.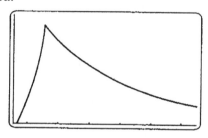

b. 0.08 g/cm^3 c. 0.12 g/cm^3
d. 0.2 g/cm^3

b. 20 sec.   c. 35.1 sec

EXERCISES 13.2 , page 831

1. $\log_2 64 = 6$ 3. $\log_3 \dfrac{1}{9} = -2$ 5. $\log_{1/3} \dfrac{1}{3} = 1$ 7. $\log_{32} 8 = \dfrac{3}{5}$

9. $\log_{10} 0.001 = -3$

11. $\log 12 = \log 4 \times 3 = \log 4 + \log 3 = 0.6021 + 0.4771 = 1.0792.$

13. $\log 16 = \log 4^2 = 2 \log 4 = 2(0.6021) = 1.2042$.

15. $\log 48 = \log 3 \times 4^2 = \log 3 + 2 \log 4 = 0.4771 + 2(0.6021) = 1.6813$.

17. $\log x(x+1)^4 = \log x + \log (x+1)^4 = \log x + 4 \log (x+1)$.

19. $\log \dfrac{\sqrt{x+1}}{x^2+1} = \log (x+1)^{1/2} - \log(x^2+1) = \frac{1}{2} \log (x+1) - \log (x^2+1)$

21. $\ln xe^{-x^2} = \ln x - x^2$.

23. $\ln \left(\dfrac{x^{1/2}}{x^2 \sqrt{1+x^2}} \right) = \ln x^{1/2} - \ln x^2 - \ln (1+x^2)^{1/2}$

$\qquad = \frac{1}{2} \ln x - 2 \ln x - \frac{1}{2} \ln (1+x^2) = -\frac{3}{2} \ln x - \frac{1}{2} \ln (1+x^2)$.

25. $\ln x^x = x \ln x$.

27. $y = \log_3 x$

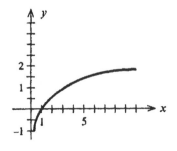

29. $y = \ln 2x$

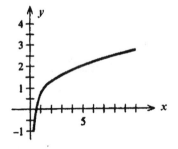

31. $y = 2^x$ and $y = \log_2 x$

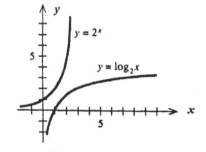

33. $e^{0.4t} = 8$, $0.4t \ln e = \ln 8$, and $0.4t = \ln 8$ ($\ln e = 1$.) So, $t = \dfrac{\ln 8}{0.4} = 5.1986$.

35. $5e^{-2t} = 6$, $e^{-2t} = \frac{6}{5} = 1.2$. Taking the logarithm, we have

$$-2t \ln e = \ln 1.2, \text{ or } t = -\dfrac{\ln 1.2}{2} \approx -0.0912.$$

37. $2e^{-0.2t} - 4 = 6$, $2e^{-0.2t} = 10$. Taking the logarithm on both sides of this last equation, we have $\ln e^{-0.2t} = \ln 5$; $-0.2t \ln e = \ln 5$; $-0.2t = \ln 5$;

and $\quad t = -\dfrac{\ln 5}{0.2} \approx -8.0472.$

39. $\dfrac{50}{1+4e^{0.2t}} = 20$, $1 + 4e^{0.2t} = \dfrac{50}{20} = 2.5$, $4e^{0.2t} = 1.5$,

$e^{0.2t} = \dfrac{1.5}{4} = 0.375$, $\ln e^{0.2t} = \ln 0.375$, $0.2t = \ln 0.375$. So $t = \dfrac{\ln 0.375}{0.2} \approx -4.9041.$

41. Taking the logarithm on both sides, we obtain

$\ln A = \ln Be^{-t/2}$, $\ln A = \ln B + \ln e^{-t/2}$, $\ln A - \ln B = -t/2 \ln e$,

$\ln \dfrac{A}{B} = -\dfrac{t}{2}$ or $t = -2 \ln \dfrac{A}{B} = 2 \ln \dfrac{B}{A}$

43. We use the compound interest formula with $A = 7500$, $P = 5000$, $m = 12$, and $t = 3$. Thus

$$7500 = 5000\left(1+\tfrac{r}{12}\right)^{36};$$
$$\left(1+\tfrac{r}{12}\right)^{36} = \tfrac{7500}{5000} = \tfrac{3}{2}, \ \ln(1+\tfrac{r}{12})^{36} = \ln 1.5;$$
$$36\left(1+\tfrac{r}{12}\right) = \ln 1.5$$
$$\left(1+\tfrac{r}{12}\right) = \tfrac{\ln 1.5}{36} = 0.0112629$$
$$1+\tfrac{r}{12} = e^{0.0112629} = 1.011327; \tfrac{r}{12} = 0.011327;$$
$$r = 0.13592$$

So the interest rate is 13.59% per year.

45. We use the compound interest formula) with $A = 8000$, $P = 4000$, $m = 2$, and $t = 4$.

Thus

$$8000 = 4000\left(1+\tfrac{r}{2}\right)^8$$

$$\left(1+\tfrac{r}{2}\right)^8 = \tfrac{8000}{5000} = 1.6, \ \ln\left(1+\tfrac{r}{2}\right)^8 = \ln 1.6;$$

$$8\ln\left(1+\tfrac{r}{2}\right) = \ln 1.6$$

$$\ln\left(1+\tfrac{r}{2}\right) = \tfrac{\ln 1.6}{8} = 0.05875$$

$$1+\tfrac{r}{2} = e^{0.05875} = 1.06051; \ \tfrac{r}{2} = 0.06051$$

$$r = 0.1210$$

So the required interest rate is 12.1% per year.

47. We use the compound interest formula with $A = 4000$, $P = 2000$, $m = 1$, and $t = 5$. Thus

$$4000 = 2000(1+r)^5; \quad (1+r)^5 = 2\,; 5\ln(1+r) = \ln 2; \ \ln(1+r) = \tfrac{\ln 2}{5} = 0.138629$$

$$1+r = e^{0.138629} = 1.148698; \ r = 0.1487$$

So the required interst rate is 14.87% per year.

49. We use the compound interest formula with $A = 6500$, $P = 5000$, $m = 12$, and $r = 0.12$. Thus

$$6500 = 5000\left(1+\frac{0.12}{12}\right)^{12t}; \quad (1.01)^{12t} = \frac{6500}{5000} = 1.3; \ 12t\ln(1.01) = \ln 1.3$$

$$t = \frac{\ln 1.3}{12\ln 1.01} \approx 2.197$$

So, it will take approximately 2.2 years.

51. We use the compound interest formula with $A = 4000$, $P = 2000$, $m = 12$, and $r = 0.09$. Thus,

$$4000 = 2000\left(1+\tfrac{0.09}{12}\right)^{12t}$$

$$\left(1+\tfrac{0.09}{12}\right)^{12t} = 2$$

$$12t\ln\left(1+\tfrac{0.09}{12}\right) = \ln 2 \ \text{ and } \ t = \frac{\ln 2}{12\ln\left(1+\tfrac{0.09}{12}\right)} \approx 7.73.$$

So it will take approximately 7.7 years.

53. We use the continuous compound interest formula with $A = 6000$, $P = 5000$, and $t = 3$. Thus,

$$6000 = 5000e^{3r}$$

$$e^{3r} = \frac{6000}{5000} = 1.2$$

$$3r = \ln 1.2$$

$$r = \frac{\ln 1.2}{3} \approx 0.6077$$

So the interest rate is 6.08% per year.

55. We use the continuous compound interest formula with $A = 7000, P = 6000,$ and $r = 0.075$. Thus

$$7000 = 6000e^{0.075t}$$

$$e^{0.075t} = \tfrac{7000}{6000} = \tfrac{7}{6}$$

$$0.075t \ln e = \ln \tfrac{7}{6}$$

and $\quad t = \dfrac{\ln \tfrac{7}{6}}{0.075} \approx 2.055.$

So, it will take 2.06 years.

57. $p(x) = 19.4 \ln x + 18$. For a child weighing 92 lb, we find
$p(92) = 19.4 \ln 92 + 18 = 105.72$ millimeters of mercury.

59. a. $30 = 10 \log \dfrac{I}{I_0}$; $\quad 3 = \log \dfrac{I}{I_0}$; $\quad \dfrac{I}{I_0} = 10^3 = 1000.$ $\quad$ So $I = 1000\, I_0.$

b. When $D = 80, I = 10^8 I_0$ and when $D = 30, I = 10^3 I_0$. Therefore, an 80–decibel sound is $10^8/10^3$ or $10^5 = 100,000$ times louder than a 30–decibel sound.

c. It is $10^{15}/10^8 = 10^7$, or 10,000,000, times louder.

61. We solve the following equation for t. Thus,

$$\frac{160}{1+240e^{-0.2t}} = 80$$

$$1+240e^{-0.2t} = \frac{160}{80}$$

$$240e^{-0.2t} = 2-1 = 1; \quad e^{-0.2t} = \frac{1}{240}; \quad -0.2t = \ln \frac{1}{240}$$

$$t = -\frac{1}{0.2} \ln \frac{1}{240} \approx 27.40 \text{, or approximately 27.4 years old.}$$

63. We solve the following equation for t:
$$200(1 - 0.956e^{-0.18t}) = 140$$
$$1 - 0.956e^{-0.18t} = \frac{140}{200} = 0.7$$
$$-0.956e^{-0.18t} = 0.7 - 1 = -0.3$$
$$e^{-0.18t} = \frac{0.3}{0.956}$$
$$-0.18t = \ln\left(\frac{0.3}{0.956}\right)$$
$$t = -\frac{\ln\left(\frac{0.3}{0.956}\right)}{0.18} \approx 6.43875.$$

So, its approximate age is 6.44 years.

65. a. We solve the equation $0.08 + 0.12e^{-0.02t} = 0.18$.
$$0.12e^{-0.02t} = 0.1;\ e^{-0.02t} = \frac{0.1}{0.12} = \frac{1}{1.2}$$
$$\ln e^{-0.02t} = \ln\frac{1}{1.2} = \ln 1 - \ln 1.2 = -\ln 1.2$$
$$-0.02t = -\ln 1.2$$
$$t = \frac{\ln 1.2}{0.02} \approx 9.116, \quad \text{or } 9.12 \text{ sec.}$$

b. We solve the equation $0.08 + 0.12e^{-0.02t} = 0.16$.
$$0.12e^{-0.02t} = 0.08;\ e^{-0.02t} = \frac{0.08}{0.12} = \frac{2}{3};\ -0.02t = \ln\frac{2}{3}$$
$$t = -\frac{\ln\left(\frac{2}{3}\right)}{0.02} \approx 20.2733, \quad \text{or } 20.27 \text{ sec.}$$

67. False. Take $x = e$. Then $(\ln e)^3 = 1^3 = 1 \neq 3 \ln e = 3$.

69. True. $g(x) = \ln x$ is continuous and greater than zero on $(1, \infty)$. Therefore,
$$f(x) = \frac{1}{\ln x} \text{ is continuous on } (1, \infty).$$

71. a. Taking the logarithm on both sides gives $\ln 2^x = \ln e^{kx}$, $x \ln 2 = kx(\ln e) = kx$.
   So, $x(\ln 2 - k) = 0$ for all x and this implies that $k = \ln 2$.
   b. Tracing the same steps as done in (a), we find that $k = \ln b$.

73. Let $\log_b m = p$, then $m = b^p$. Therefore, $m^n = (b^p)^n = b^{np}$. Therefore,
$$\log_b m^n = \log_b b^{np} = np \log_b b = np \qquad \text{(Since } \log_b b = 1.\text{)}$$
$$= n \log_b m, \qquad \text{as was to be shown.}$$

EXERCISES 13.3 , page 839

1. $f(x) = e^{3x};\ f'(x) = 3e^{3x}$

3. $g(t) = e^{-t};\ g'(t) = -e^{-t}$

5. $f(x) = e^x + x;\ f'(x) = e^x + 1$

7. $f(x) = x^3 e^x,\ f'(x) = x^3 e^x + e^x(3x^2) = x^2 e^x(x + 3).$

9. $f(x) = \dfrac{2e^x}{x},\quad f'(x) = \dfrac{x(2e^x) - 2e^x(1)}{x^2} = \dfrac{2e^x(x - 1)}{x^2}.$

11. $f(x) = 3(e^x + e^{-x});\ f'(x) = 3(e^x - e^{-x}).$

13. $f(w) = \dfrac{e^w + 1}{e^w} = 1 + \dfrac{1}{e^w} = 1 + e^{-w}.\ f'(w) = -e^{-w} = -\dfrac{1}{e^w}.$

15. $f(x) = 2e^{3x-1},\ f'(x) = 2e^{3x-1}(3) = 6e^{3x-1}.$

17. $h(x) = e^{-x^2};\ h'(x) = e^{-x^2}(-2x) = -2xe^{-x^2}.$

19. $f(x) = 3e^{-1/x};\ f'(x) = 3e^{-1/x} \cdot \dfrac{d}{dx}\left(-\dfrac{1}{x}\right) = 3e^{-1/x}\left(\dfrac{1}{x^2}\right) = \dfrac{3e^{-1/x}}{x^2}.$

21. $f(x) = (e^x + 1)^{25},\ f'(x) = 25(e^x + 1)^{24} e^x = 25e^x(e^x + 1)^{24}.$

23. $f(x) = e^{\sqrt{x}};\ f'(x) = e^{\sqrt{x}} \dfrac{d}{dx} x^{1/2} = e^{\sqrt{x}} \dfrac{1}{2} x^{-1/2} = \dfrac{e^{\sqrt{x}}}{2\sqrt{x}}.$

25. $f(x) = (x - 1)e^{3x+2};\ f'(x) = (x - 1)(3)e^{3x+2} + e^{3x+2} = e^{3x+2}(3x - 3 + 1) = e^{3x+2}(3x - 2).$

27. $f(x) = \dfrac{e^x - 1}{e^x + 1}$; $f'(x) = \dfrac{(e^x + 1)(e^x) - (e^x - 1)(e^x)}{(e^x + 1)^2} = \dfrac{e^x(e^x + 1 - e^x + 1)}{(e^x + 1)^2} = \dfrac{2e^x}{(e^x + 1)^2}$.

29. $f(x) = e^{-4x} + 2e^{3x}$; $f'(x) = -4e^{-4x} + 6e^{3x}$ and
$f''(x) = 16e^{-4x} + 18e^{3x} = 2(8e^{-4x} + 9e^{3x})$.

31. $f(x) = 2xe^{3x}$; $f'(x) = 2e^{3x} + 2xe^{3x}(3) = 2(3x + 1)e^{3x}$.
$f''(x) \quad = 6e^{3x} + 2(3x + 1)e^{3x}(3) = 6(3x + 2)e^{3x}$.

33. $y = f(x) = e^{2x-3}$. $f'(x) = 2e^{2x-3}$. To find the slope of the tangent line to the graph
of f at $x = 3/2$, we compute $f'(\frac{3}{2}) = 2e^{3-3} = 2$. Next, using the point–slope form of
the equation of a line, we find that
$$y - 1 = 2(x - \tfrac{3}{2})$$
$$= 2x - 3, \qquad \text{or} \qquad y = 2x - 2.$$

35. $f(x) = e^{-x^2/2}$, $f'(x) = e^{-x^2/2}(-x) = -xe^{-x^2/2}$. Setting $f'(x) = 0$, gives $x = 0$ as the only
critical point of f. From the sign diagram,

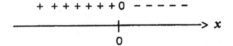

we conclude that f is increasing on $(-\infty, 0)$ and decreasing on $(0, \infty)$.

37. $f(x) = \frac{1}{2}e^x - \frac{1}{2}e^{-x}$, $f'(x) = \frac{1}{2}(e^x + e^{-x})$, $f''(x) = \frac{1}{2}(e^x - e^{-x})$. Setting $f''(x) = 0$,
gives $e^x = e^{-x}$ or $e^{2x} = 1$, and $x = 0$. From the sign diagram for f'',

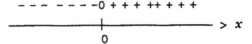

we conclude that f is concave upward on $(0, \infty)$ and concave downward on $(-\infty, 0)$.

39. $f(x) = xe^{-2x}$. $f'(x) = e^{-2x} + xe^{-2x}(-2) = (1 - 2x)e^{-2x}$.
$f''(x) = -2e^{-2x} + (1 - 2x)e^{-2x}(-2) = 4(x - 1)e^{-2x}$.
Observe that $f''(x) = 0$ if $x = 1$. The sign diagram of f''

$$-\ -\ -\ -\ -\ -\ -\ -\ 0\ +\ +\ +\ ++\ +\ +\ +$$
$$\xrightarrow{\hspace{4cm}} x$$
$$1$$

shows that $(1, e^{-2})$ is an inflection point.

41. $f(x) = e^{-x^2}$. $f'(x) = -2xe^{-x^2} = 0$ if $x = 0$, the only critical point of f.

x	-1	0	1
$f(x)$	e^{-1}	1	e^{-1}

From the table, we see that f has an absolute minimum value of e^{-1} attained at $x = -1$ and $x = 1$. It has an absolute maximum at $(0,1)$.

43. $g(x) = (2x - 1)e^{-x}$; $g'(x) = 2e^{-x} + (2x - 1)e^{-x}(-1) = (3 - 2x)e^{-x} = 0$, if $x = 3/2$. The graph of g shows that $(\frac{3}{2}, 2e^{-3/2})$ is an absolute maximum, and $(0,-1)$ is an absolute minimum.

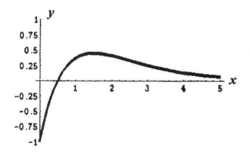

45. $f(t) = e^t - t$;
   We first gather the following information on f.
   1. The domain of f is $(-\infty,\infty)$.
   2. Setting $t = 0$ gives 1 as the y–intercept.
   3. $\lim_{t \to -\infty} (e^t - t) = \infty$ and $\lim_{t \to \infty} (e^t - t) = \infty$.
   4. There are no asymptotes.
   5. $f'(t) = e^t - 1$ if $t = 0$, a critical point of f. From the sign diagram for f'

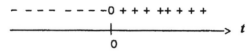

   we see that f is decreasing on $(-\infty,0)$ and increasing on $(0,\infty)$.
   6. From the results of (5), we see that $(0,1)$ is a relative minimum of f.
   7. $f''(t) = e^t > 0$ for all t in $(-\infty,\infty)$. So the graph of f is concave upward on $(-\infty,\infty)$.
   8. There are no inflection points.
   The graph of f follows.

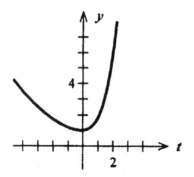

47. $f(x) = 2 - e^{-x}$.

We first gather the following information on f.

1. The domain of f is $(-\infty,\infty)$.

2. Setting $x = 0$ gives 1 as the y-intercept.

3. $\lim\limits_{x \to -\infty} (2 - e^{-x}) = -\infty$ and $\lim\limits_{x \to \infty} (2 - e^{-x}) = 2$,

4. From the results of (3), we see that $y = 2$ is a horizontal asymptote of f.

5. $f'(x) = e^{-x}$. Observe that $f'(x) > 0$ for all x in $(-\infty,\infty)$ and so f is increasing on $(-\infty,\infty)$.

6. Since there are no critical points, f has no relative extrema.

7. $f''(x) = -e^{-x} < 0$ for all x in $(-\infty,\infty)$ and so the graph of f is concave downward on $(-\infty,\infty)$.

8. There are no inflection points

The graph of f follows.

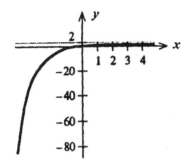

49. $P'(t) = 20.6(-0.009)e^{-0.009t} = -0.1854e^{-0.009t}$

$P'(10) = -0.1694$, $P'(20) = -0.1549$, and $P'(30) = -0.1415$,

and this tells us that the percentage of the total population relocating was decreasing at the rate of 0.17% in 1970, 0.15% in 1980, and 0.14% in 1990.

51. a. The number of air passengers in 2000 is $N(0) = 666$, or 666 million. The number in 2005 is $N(5) = 818,759$, or 819 million.

b. $f'(t) = 666(0.0413)e^{0.0413t} = 27.5058e^{0.0413t}$. The required rate is
$f'(5) = 33.8147$, or 33.8 million/yr.

53. $S(t) = 20,000(1 + e^{-0.5t})$

$S'(t) = 20,000(-0.5e^{-0.5t}) = -10,000e^{-0.5t}$;

$S'(1) = -10,000e^{-0.5} = -6065$, or $-\$6065$/day.

$S'(2) = -10,000e^{-1} = -3679$, or $-\$3679$/day.

$S'(3) = -10,000(e^{-1.5}) = -2231$, or $-\$2231$/day.

$S'(4) = -10,000e^{-2} = -1353$, or $-\$1353$/day.

55. $N(t) = 5.3e^{0.095t^2 - 0.85t}$.

a. $N'(t) = 5.3e^{0.095t^2 - 0.85t}(0.19t - 0.85)$. Since $N'(t)$ is negative for $(0 \le t \le 4)$, we see that $N(t)$ is decreasing over that interval.

b. To find the rate at which the number of polio cases was decreasing at the beginning of 1959, we compute
$$N'(0) = 5.3e^{0.095(0^2) - 0.85(0)}(0.85) = 5.3(-0.85) = -4.505$$
(t is measured in thousands), or 4,505 cases per year. To find the rate at which the number of polio cases was decreasing at the beginning of 1962, we compute
$$N'(3) = 5.3e^{0.095(9) - 0.85(3)}(0.57 - 0.85)$$
$$= (-0.28)(0.9731) \approx -0.273, \text{ or } 273 \text{ cases per year.}$$

57. $R(x) = px = 100xe^{-0.0001x}$.

$R'(x) = 100e^{-0.0001x} + 100xe^{-0.0001x} \cdot (-0.0001)$

$= 100(1 - 0.0001x)e^{-0.0001x}$.

Setting $R'(x) = 0$ gives $x = 10,000$, a critical point of R. From the graph of R

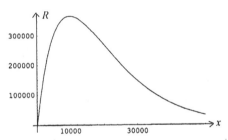

we see that the revenue is maximized when $x = 10,000$. So 10,000 pairs must be sold, yielding a maximum revenue of $R(10,000) = 367,879.44$, or \$367,879.

59. $p = 240\left(1 - \dfrac{3}{3 + e^{-0.0005x}}\right) = 240[1 - 3(3 + e^{-0.0005x})^{-1}].$

$p' = 720(3 + e^{-0.0005x})^{-2}(-0.0005e^{-0.0005x})$

$p'(1000) = 720(3 + e^{-0.0005(1000)})^{-2}(-0.0005e^{-0.0005(1000)})$

$ = -\dfrac{0.36(0.606531)}{(3 + 0.606531)^2} \approx -0.0168, \quad$ or -1.68 cents per case.

$p(1000) = 240(1 - \dfrac{3}{3.606531}) \approx 40.36, \quad$ or \$40.36/case.

61. a. $W = 2.4e^{1.84h}$; $W = 2.4e^{1.84(16)} \approx 45.58$, or approximately 45.6 kg.

b. $\Delta W \approx dW = (2.4)(1.84)e^{1.84h}dh$. With $h = 1.6$ and $dh = \Delta h = 1.65 - 1.6 = 0.05$, we find

$$\Delta W \approx (2.4)(1.84)e^{1.84(1.6)} \cdot (0.05) \approx 4.19,$$ or approximately 4.2 kg.

63. $P(t) = 80,000\, e^{\sqrt{t}/2 - 0.09t} = 80,000\, e^{\frac{1}{2}t^{1/2} - 0.09t}$.

$P'(t) = 80,000(\tfrac{1}{4}t^{-1/2} - 0.09)e^{\frac{1}{2}t^{1/2} - 0.09t}$.

Setting $P'(t) = 0$, we have

$$\tfrac{1}{4}t^{-1/2} = 0.09,\ t^{-1/2} = 0.36,$$

$$\dfrac{1}{\sqrt{t}} = 0.36,\ t = \left(\dfrac{1}{0.36}\right)^2 \approx 7.72.$$

Evaluating $P(t)$ at each of its endpoints and at the point $t = 7.72$, we find

t	$P(t)$
0	80,000

7.72	160,207.69
8	160,170.71

We conclude that P is optimized at $t = 7.72$. The optimal price is $160,207.69.

65. $f(t) = 1.5 + 1.8te^{-1.2t}$

$$f'(t) = 1.8\frac{d}{dt}(te^{-1.2t}) = 1.8[e^{-1.2t} + te^{-1.2t}(-1.2)] = 1.8e^{-1.2t}(1 - 1.2t).$$

$f'(0) = 1.8$, $f'(1) = -0.11$, $f'(2) = -0.23$, and $f'(3) = -0.13$,
and this tells us that the rate of change of the amount of oil used is 1.8 barrels per
$1000 of output per decade in 1965; it is decreasing at the rate of 0.11 barrels per
$1000 of output per decade in 1966, and so on.

67. a. The price at $t = 0$ is $8 + 4$, or 12, dollars per unit.

  b. $\dfrac{dp}{dt} = -8e^{-2t} + e^{-2t} - 2te^{-2t}$.

$$\frac{dp}{dt}\bigg|_{t=0} = -8e^{-2t} + e^{-2t} - 2te^{-2t}\bigg|_{t=0} = -8 + 1 = -7.$$

  That is, the price is decreasing at the rate of $7/week.

  c. The equilibrium price is $\lim_{t \to \infty}(8 + 4e^{-2t} + te^{-2t}) = 8 + 0 + 0$, or $8 per unit.

69. We are given that
$$c(1 - e^{-at/V}) < m$$

$$1 - e^{-at/V} < \frac{m}{c}$$

$$-e^{-at/V} < \frac{m}{c} - 1 \quad \text{and} \quad e^{-at/V} > 1 - \frac{m}{c}.$$

Taking the log of both sides of the inequality, we have
$$-\frac{at}{V}\ln e > \ln\frac{c - m}{c}$$

$$-\frac{at}{V} > \ln\frac{c - m}{c}$$

$$-t > \frac{V}{a}\ln\frac{c - m}{c} \quad \text{or} \quad t < \frac{V}{a}\left(-\ln\frac{c - m}{c}\right) = \frac{V}{a}\ln\left(\frac{c}{c - m}\right).$$

Therefore the liquid must not be allowed to enter the organ for a time longer than

$$t = \frac{V}{a}\ln\left(\frac{c}{c-m}\right) \text{ minutes.}$$

71. $C'(t) = \begin{cases} 0.3 + 18e^{-t/60}(-\frac{1}{60}) & 0 \le t \le 20 \\ -\frac{18}{60}e^{-t/60} + \frac{12}{60}e^{-(t-20)/60} & t > 20 \end{cases} = \begin{cases} 0.3(1-e^{-t/60}) & 0 \le t \le 20 \\ -0.3e^{-t/60} + 0.2e^{-(t-20)/60} & t > 20 \end{cases}$

a. $C'(10) = 0.3(1 - e^{-10/60}) \approx 0.05$ or 0.05 g/cm^3/sec.

b. $C'(30) = -0.3e^{-30/60} + 0.2e^{-10/60} \approx -0.01$, or decreasing at the rate of 0.01 g/ cm^3/sec.

c. On the interval (0, 20), $C'(t) = 0$ implies $1 - e^{-t/60} = 0$, or $t = 0$.
   Therefore, C attains its absolute maximum value at an endpoint. In this case, at $t = 20$. On the interval $[20, \infty)$, $C'(t) = 0$ implies

   $$-0.3e^{-t/60} = -0.2e^{-(t-20)/60}$$

   $$\frac{e^{-\left(\frac{t-20}{60}\right)}}{e^{-t/60}} = \frac{3}{2}; \text{ or } e^{1/3} = \frac{3}{2},$$

   which is not possible. Therefore $C'(t) \ne 0$ on $[20, \infty)$. Since $C(t) \to 0$ as $t \to \infty$, the absolute maximum of c occurs at $t = 20$. Thus, the concentration of the drug reaches a maximum at $t = 20$.

d. The maximum concentration is $C(20) = 0.90$ g/cm^3.

73. False. $f(x) = 3^x = e^{x\ln 3}$ and so $f'(x) = e^{x\ln 3} \cdot \frac{d}{dx}(x\ln 3) = (\ln 3)e^{x\ln 3} = (\ln 3)3^x$.

75. False. $f'(x) = (\ln \pi)\pi^x$.

USING TECHNOLOGY EXERCISES 13.3, page 842

1. 5.4366   3. 12.3929   5. 0.1861

7. a. The initial population of crocodiles is $P(0) = \frac{300}{6} = 50$.

   b. $\lim_{t \to 0} P(t) = \lim_{t \to 0} \frac{300e^{-0.024t}}{5e^{-0.024t} + 1} = \frac{0}{0+1} = 0$.

c.

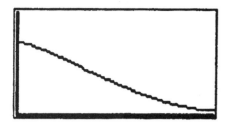

9. a.

b. 4.2720 billion/half century

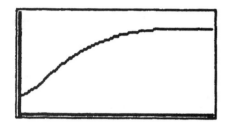

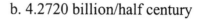

11. a. Using the function evaluation capabilities of a graphing utility, we find
$$f(11) = 153.024 \text{ and } g(11) = 235.180977624$$
and this tells us that the number of violent-crime arrests will be 153,024 at the beginning of the year 2000, but if trends like inner-city drug use and wider availability of guns continue, then the number of arrests will be 235,181.
b. Using the differentiation capability of a graphing utility, we find
$$f'(11) = -0.634 \text{ and } g'(11) = 18.4005596893$$
and this tells us that the number of violent-crime arrests will be decreasing at the rate of 634 per year at the beginning of the year 2000. But if the trends like inner-city drug use and wider availability of guns continues, then the number of arrests will be increasing at the rate of 18,401 per year at the beginning of the year 2000.

13. a. $P(10) = \dfrac{74}{1 + 2.6e^{-0.166(10)+0.04536(10)^2-0.0066(10)^3}} \approx 69.63$ percent.

b. $P'(10) = 5.09361$, or 5.09361%/decade

1. $f(x) = 5 \ln x; f'(x) = 5\left(\dfrac{1}{x}\right) = \dfrac{5}{x}.$

3. $f(x) = \ln(x + 1); f'(x) = \dfrac{1}{x+1}.$

5. $f(x) = \ln x^8; f'(x) = \dfrac{8x^7}{x^8} = \dfrac{8}{x}.$

7. $f(x) = \ln x^{1/2}; \quad f'(x) = \dfrac{\frac{1}{2}x^{-1/2}}{x^{1/2}} = \dfrac{1}{2x}.$

9. $f(x) = \ln\left(\dfrac{1}{x^2}\right) = \ln x^{-2} = -2 \ln x; \quad f'(x) = -\dfrac{2}{x}.$

11. $f(x) = \ln(4x^2 - 6x + 3); \quad f'(x) = \dfrac{8x - 6}{4x^2 - 6x + 3} = \dfrac{2(4x - 3)}{4x^2 - 6x + 3}.$

13. $f(x) = \ln\left(\dfrac{2x}{x+1}\right) = \ln 2x - \ln(x + 1).$

$f'(x) = \dfrac{2}{2x} - \dfrac{1}{x+1} = \dfrac{2(x+1) - 2x}{2x(x+1)} = \dfrac{2x + 2 - 2x}{2x(x+1)} = \dfrac{2}{2x(x+1)} = \dfrac{1}{x(x+1)}.$

15. $f(x) = x^2 \ln x; \; f'(x) = x^2\left(\frac{1}{x}\right) + (\ln x)(2x) = x + 2x \ln x = x(1 + 2 \ln x)$

17. $f(x) = \dfrac{2 \ln x}{x}. \quad f'(x) = \dfrac{x\left(\frac{2}{x}\right) - 2 \ln x}{x^2} = \dfrac{2(1 - \ln x)}{x^2}.$

19. $f(u) = \ln(u - 2)^3; f'(u) = \dfrac{3(u-2)^2}{(u-2)^3} = \dfrac{3}{u-2}.$

21. $f(x) = (\ln x)^{1/2}$ and $f'(x) = \dfrac{1}{2}(\ln x)^{-1/2}\left(\dfrac{1}{x}\right) = \dfrac{1}{2x\sqrt{\ln x}}.$

23. $f(x) = (\ln x)^3; \; f'(x) = 3(\ln x)^2\left(\dfrac{1}{x}\right) = \dfrac{3(\ln x)^2}{x}.$

25. $f(x) = \ln(x^3 + 1); \; f'(x) = \dfrac{3x^2}{x^3 + 1}.$

27. $f(x) = e^x \ln x. \; f'(x) = e^x \ln x + e^x\left(\dfrac{1}{x}\right) = \dfrac{e^x(x \ln x + 1)}{x}.$

29. $f(t) = e^{2t} \ln (t + 1)$

$$f'(t) = e^{2t} \left(\frac{1}{t+1} \right) + \ln(t+1) \cdot (2e^{2t}) = \frac{[2(t+1)\ln(t+1)+1]e^{2t}}{t+1}.$$

31. $f(x) \dfrac{\ln x}{x}$. $f'(x) = \dfrac{x(\frac{1}{x}) - \ln x}{x^2} = \dfrac{1 - \ln x}{x^2}.$

33. $f(x) = \ln 2 + \ln x$; So $f'(x) = \dfrac{1}{x}$ and $f''(x) = -\dfrac{1}{x^2}.$

35. $f(x) = \ln (x^2 + 2)$; $f'(x) = \dfrac{2x}{(x^2 + 2)}$ and

$$f''(x) = \frac{(x^2+2)(2) - 2x(2x)}{(x^2+2)^2} = \frac{2(2-x^2)}{(x^2+2)^2}.$$

37. $y = (x + 1)^2(x + 2)^3$

$\ln y = \ln (x + 1)^2(x + 2)^3 = \ln (x + 1)^2 + \ln (x + 2)^3$

$\quad = 2 \ln (x + 1) + 3 \ln (x + 2).$

$$\frac{y'}{y} = \frac{2}{x+1} + \frac{3}{x+2} = \frac{2(x+2)+3(x+1)}{(x+1)(x+2)} = \frac{5x+7}{(x+1)(x+2)}$$

$$y' = \frac{(5x+7)(x+1)^2(x+2)^3}{(x+1)(x+2)} = (5x+7)(x+1)(x+2)^2.$$

39. $y = (x - 1)^2(x + 1)^3(x + 3)^4$

$\ln y = 2 \ln (x - 1) + 3 \ln (x + 1) + 4 \ln (x + 3)$

$$\frac{y'}{y} = \frac{2}{x-1} + \frac{3}{x+1} + \frac{4}{x+3}$$

$$= \frac{2(x+1)(x+3) + 3(x-1)(x+3) + 4(x-1)(x+1)}{(x-1)(x+1)(x+3)}$$

$$= \frac{2x^2 + 8x + 6 + 3x^2 + 6x - 9 + 4x^2 - 4}{(x-1)(x+1)(x+3)} = \frac{9x^2 + 14x - 7}{(x-1)(x+1)(x+3)}.$$

Therefore,

$$y' = \frac{9x^2 + 14x - 7}{(x-1)(x+1)(x+3)} \cdot y$$

$$= \frac{(9x^2 + 14x - 7)(x-1)^2(x+1)^3(x+3)^4}{(x-1)(x+1)(x+3)}$$

$$= (9x^2 + 14x - 7)(x-1)(x+1)^2(x+3)^3.$$

41. $y = \dfrac{(2x^2 - 1)^5}{\sqrt{x+1}}.$

$$\ln y = \ln \frac{(2x^2 - 1)^5}{(x+1)^{1/2}} = 5\ \ln(2x^2 - 1) - \frac{1}{2}\ln(x+1)$$

So $\dfrac{y'}{y} = \dfrac{20x}{2x^2 - 1} - \dfrac{1}{2(x+1)} = \dfrac{40x(x+1) - (2x^2 - 1)}{2(2x^2 - 1)(x+1)}$

$$= \frac{38x^2 + 40x + 1}{2(2x^2 - 1)(x+1)}.$$

$$y' = \frac{38x^2 + 40x + 1}{2(2x^2 - 1)(x+1)} \cdot \frac{(2x^2 - 1)^5}{\sqrt{x+1}} = \frac{(38x^2 + 40x + 1)(2x^2 - 1)^4}{2(x+1)^{3/2}}.$$

43. $y = 3^x;$ $\quad \ln y = x \ln 3;$ $\quad \dfrac{1}{y} \cdot \dfrac{dy}{dx} = \ln 3\ ;$ $\quad \dfrac{dy}{dx} = y \ln 3 = 3^x \ln 3.$

45. $y = (x^2 + 1)^x;\ \ln y = \ln (x^2 + 1)^x = x \ln (x^2 + 1).$ So

$$\frac{y'}{y} = \ln(x^2 + 1) + x\left(\frac{2x}{x^2 + 1}\right) = \frac{(x^2 + 1)\ln(x^2 + 1) + 2x^2}{x^2 + 1}.$$

$$y' = \frac{[(x^2 + 1)\ln(x^2 + 1) + 2x^2](x^2 + 1)^x}{x^2 + 1}$$

47. $y = x \ln x.$ The slope of the tangent line at any point is

$$y' = \ln x + x\left(\tfrac{1}{x}\right) = \ln x + 1.$$

In particular, the slope of the tangent line at $(1,0)$ where $x = 1$ is $m = \ln 1 + 1 = 1.$

So, an equation of the tangent line is $y - 0 = 1(x - 1)$ or $y = x - 1.$

49. $f(x) = \ln x^2 = 2 \ln x$ and so $f'(x) = 2/x$. Since $f'(x) < 0$ if $x < 0$, and $f'(x) > 0$ if $x > 0$, we see that f is decreasing on $(-\infty, 0)$ and increasing on $(0, \infty)$.

51. $f(x) = x^2 + \ln x^2$; $f'(x) = 2x + \dfrac{2x}{x^2} = 2x + \dfrac{2}{x}$; $f''(x) = 2 - \dfrac{2}{x^2}$.

To find the intervals of concavity for f, we first set $f''(x) = 0$ giving

$$2 - \frac{2}{x^2} = 0, \quad 2 = \frac{2}{x^2}, \quad 2x^2 = 2$$

or $\qquad x^2 = 1$ and $x = \pm 1$.

Next, we construct the sign diagram for f''

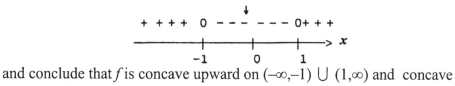

and conclude that f is concave upward on $(-\infty, -1) \cup (1, \infty)$ and concave downward on $(-1, 0) \cup (0, 1)$.

53. $f(x) = \ln(x^2 + 1)$. $f'(x) = \dfrac{2x}{x^2 + 1}$; $f''(x) = \dfrac{(x^2 + 1)(2) - (2x)(2x)}{(x^2 + 1)^2} = -\dfrac{2(x^2 - 1)}{(x^2 + 1)^2}$.

Setting $f''(x) = 0$ gives $x = \pm 1$ as candidates for inflection points of f.
From the sign diagram for f''

$$- \; - \; - \; - \; 0 + + + 0 + + + 0 - - -$$

we see that $(-1, \ln 2)$ and $(1, \ln 2)$ are inflection points of f.

55. $f(x) = x - \ln x$; $f'(x) = 1 - \dfrac{1}{x} = \dfrac{x-1}{x} = 0$ if $x = 1$, a critical point of f.

x	1/2	1	3
$f(x)$	1/2 + ln 2 ln 3	1	3 −

From the table, we see that f has an absolute minimum at $(1,1)$ and an absolute maximum at $(3, 3 - \ln 3)$.

57. $f(x) = 7.2956 \ln(0.0645012 x^{0.95} + 1)$

$$f'(x) = 7.2956 \cdot \frac{\frac{d}{dx}(0.0645012 x^{0.95} + 1)}{0.0645012 x^{0.95} + 1} = \frac{7.2956(0.0645012)(0.95 x^{-0.05})}{0.0645012 x^{0.95} + 1}$$

$$= \frac{0.4470462}{x^{0.05}(0.0645012 x^{0.95} + 1)}$$

So $\quad f'(100) = 0.05799$, or approximately 0.0580 percent/kg and

$\quad\quad f'(500) = 0.01329$, or approximately 0.0133 percent/kg.

59. a. If $0 < r < 100$, then $c = 1 - \frac{r}{100}$ satisfies $0 < c < 1$. It suffices to show that

$A_1(n) = -(1 - \frac{r}{100})^n$ is increasing, (why?), or equivalently $A_2(n) = -A_1(n) = \left(1 - \frac{r}{100}\right)^n$

is decreasing. Let $y = \left(1 - \frac{r}{100}\right)^n$. Then $\ln y = \ln\left(1 - \frac{r}{100}\right)^n = \ln c^n = n \ln c$.

Differentiating both sides with respect to n, we find

$$\frac{y'}{y} = \ln c \text{ and so } y' = (\ln c)\left(1 - \frac{r}{100}\right)^n < 0$$

since $\ln c < 0$ and $\left(1 - \frac{r}{m}\right)^n > 0$ for $0 < r < 100$. Therefore, A is an increasing

function of n on $(0, \infty)$.

b.

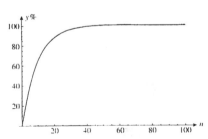

c. $\displaystyle\lim_{n \to \infty} A(n) = \lim_{n \to \infty} 100\left[1 - \left(1 - \frac{r}{100}\right)^n\right] = 100$

61. a. $R = \log \dfrac{10^6 I_0}{I_0} = \log 10^6 = 6$.

b. $I = I_0 10^R$ by definition. Taking the natural logarithm on both sides, we find

$\ln I = \ln I_0 10^R = \ln I_0 + \ln 10^R = \ln I_0 + R \ln 10$.

Differentiating implicitly with respect to R, we obtain

$$\frac{I'}{I} = \ln 10 \text{ or } \frac{dI}{dR} = (\ln 10)I .$$

Therefore, $\Delta I \approx dI = \dfrac{dI}{dR} \Delta R = (\ln 10) I \Delta R$. With $|\Delta R| \le (0.02)(6) = 0.12$ and

$I = 1,000,000 I_0$, (see part a), we have

$$|\Delta I| \le (\ln 10)(1,000,000 I_0)(0.12) = 276310.21 I_o$$

So the error is at most 276,310 times the standard reference intensity.

63. $f(x) = \ln (x - 1)$.

   1. The domain of f is obtained by requiring that $x - 1 > 0$. We find the domain to be $(1, \infty)$.

   2. Since $x \ne 0$, there are no y-intercepts. Next, setting $y = 0$ gives $x - 1 = 1$ or $x = 2$ as the x-intercept.

   3. $\lim\limits_{x \to 1^+} \ln (x - 1) = -\infty$.

   4. There are no horizontal asymptotes. Observe that $\lim\limits_{x \to 1^+} \ln (x - 1) = -\infty$ so $x = 1$ is a vertical asymptote.

   5. $f'(x) = \dfrac{1}{x - 1}$.

     The sign diagram for f' is

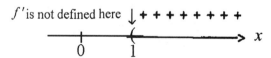

     We conclude that f is increasing on $(1, \infty)$.

   6. The results of (5) show that f is increasing on $(1, \infty)$.

   7. $f''(x) = -\dfrac{1}{(x - 1)^2}$. Since $f''(x) < 0$ for $x > 1$, we see that f is concave downward on $(1, \infty)$.

   8. From the results of (7), we see that f has no inflection points. The graph of f follows.

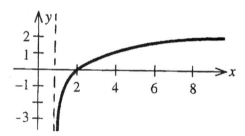

65. False. ln 5 is a constant function and $f'(x) = 0$.

67. If $x \leq 0$, then $|x| = -x$. Therefore, $\ln |x| = \ln(-x)$. Writing $f(x) = \ln |x|$ we have $|x| = -x = e^{f(x)}$. Differentiating both sides with respect to x and using the Chain Rule, we have $\qquad -1 = e^{f(x)} \cdot f'(x) \quad$ or $\quad f'(x) = -\dfrac{1}{e^{f(x)}} = -\dfrac{1}{-x} = \dfrac{1}{x}$.

EXERCISES 13.5, page 859

1. a. The growth constant is $k = 0.05$. b. Initially, the quantity present is 400 units.
   c.

t	0	10	20	100	1000
Q	400	660	1087	59365	2.07×10^{24}

3. a. $Q(t) = Q_0 e^{kt}$. Here $Q_0 = 100$ and so $Q(t) = 100 e^{kt}$. Since the number of cells doubles in 20 minutes, we have
   $$Q(20) = 100 e^{20k} = 200, \ e^{20k} = 2, \ 20k = \ln 2, \ \text{or} \ k = \tfrac{1}{20} \ln 2 \approx 0.03466.$$
   $$Q(t) = 100 e^{0.03466t}$$
   b. We solve the equation $100 e^{0.03466t} = 1,000,000$. We obtain
   $$e^{0.03466t} = 10000 \ \text{or} \ 0.03466t = \ln 10000,$$
   $$t = \frac{\ln 10,000}{0.03466} \approx 266, \qquad \text{or 266 minutes.}$$

   c. $Q(t) = 1000 e^{0.03466t}$.

5. a. We solve the equation
   $$5.3 e^{0.0198t} = 3(5.3) \ \text{or} \ e^{0.0198t} = 3,$$
   or $\qquad 0.0198t = \ln 3$ and $t = \dfrac{\ln 3}{0.0198} \approx 55.5$.
   So the world population will triple in approximately 55.5 years.

   b. If the growth rate is 1.8 percent, then proceeding as before, we find
   $1.018(5.3) = 5.3 e^{k}$, and $\quad k = \ln 1.018 \approx 0.0178$.
   So $N(t) = 5.3 e^{0.0178t}$. If $t = 55.5$, the population would be
   $\qquad N(55.5) = 5.3 e^{0.0178(55.5)} \approx 14.23$, or approximately 14.23 billion.

7. $P(h) = p_0 e^{-kh}$, $P(0) = 15$, therefore, $p_0 = 15$.

$$P(4000) = 15e^{-4000k} = 12.5; \quad e^{-4000k} = \frac{12.5}{15},$$

$$-4000k = \ln\left(\frac{12.5}{15}\right) \quad \text{and } k = 0.00004558.$$

Therefore, $P(12,000) = 15e^{-0.00004558(12,000)} = 8.68$, or 8.7 lb/sq in.

The rate of change of the atmospheric pressure with respect to altitude is given by

$$P'(h) = \frac{d}{dh}(15e^{-0.00004558h}) = -0.0006837e^{-0.00004558h}.$$

So, the rate of change of the atmospheric pressure with respect to altitude when the altitude is 12,000 feet is $P'(12,000) = -0.0006837e^{-0.00004558(12,000)} \approx -0.00039566$.
That is, it is dropping at the rate of approximately 0.0004 lbs per square inch/foot.

9. Suppose the amount of phosphorus 32 at time t is given by
$$Q(t) = Q_0 e^{-kt}$$
where Q_0 is the amount present initially and k is the decay constant. Since this element has a half–life of 14.2 days, we have

$$\tfrac{1}{2}Q_0 = Q_0 e^{-14.2k}, \quad e^{-14.2k} = \tfrac{1}{2}, \; -14.2k = \ln\tfrac{1}{2}, \; k = -\frac{\ln\tfrac{1}{2}}{14.2} \approx 0.0488.$$

Therefore, the amount of phosphorus 32 present at any time t is given by
$$Q(t) = 100e^{-0.0488t}$$
The amount left after 7.1 days is given by
$$Q(7.1) = 100e^{-0.0488(7.1)} = 100e^{-0.3465}$$
$$= 70.71, \quad \text{or } 70.71 \text{ grams.}$$
The rate at which the phosphorus 32 is decaying when $t = 7.1$ is given by

$$Q'(t) = \frac{d}{dt}[100e^{-0.0488t}] = 100(-0.0488)e^{-0.0488t} = -4.88e^{-0.0488t}.$$

Therefore, $Q'(7.1) = -4.88e^{-0.0488(7.1)} \approx -3.451$; that is, it is changing at the rate of 3.451 gms/day.

11. We solve the equation $0.2Q_0 = Q_0 e^{-0.00012t}$

obtaining $\qquad t = \dfrac{\ln 0.2}{-0.00012} \approx 13{,}412$, or approximately 13,412 years.

13. The graph of $Q(t)$ follows.

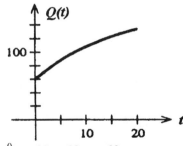

a. $Q(0) = 120(1 - e^0) + 60 = 60$, or 60 w.p.m.
b. $Q(10) = 120(1 - e^{-0.5}) + 60 = 107.22$, or approximately 107 w.p.m.
c. $Q(20) = 120(1 - e^{-1}) + 60 = 135.65$, or approximately 136 w.p.m.

15. The graph of $D(t)$ follows.

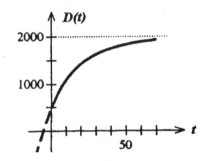

a. After one month, the demand is $D(1) = 2000 - 1500e^{-0.05} \approx 573$.
After twelve months, the demand is $D(12) = 2000 - 1500e^{-0.6} \approx 1177$.
After twenty-four months the demand is $D(24) = 2000 - 1500e^{-1.2} \approx 1548$.
After sixty months, the demand is $D(60) = 2000 - 1500e^{-3} \approx 1925$.
b. $\lim_{t \to \infty} D(t) = \lim_{t \to \infty} 2000 - 1500e^{-0.05t} = 2000$

and we conclude that the demand is expected to stabilize at 2000 computers per
month.
c. $D'(t) = -1500e^{-0.05t}(-0.05) = 75e^{-0.05t}$. Therefore, the rate of growth after ten
months is given by $D'(10) = 75e^{-0.5} \approx 45.49$, or approximately 46 computers per
month.

17. a. The length is given by $f(5) = 200(1 - 0.956e^{-0.18(5)}) \approx 122.26$, or approximately
122.3 cm.
b. $f'(t) = 200(-0.956)e^{-0.18t}(-0.18) = 34.416e^{-0.18t}$. So, a 5-yr old is growing at the

rate of $f'(5) = 34.416e^{-0.18(5)} \approx 13.9925$, or approximately 14 cm/yr.

c. The maximum length is given by $\lim_{t \to \infty} 200(1 - 0.956e^{-0.18t}) = 200$, or 200 cm.

19. a. The percent of lay teachers is $f(3) = \dfrac{98}{1 + 2.77e^{-3}} \approx 86.1228$, or 86.12%.

b. $f'(t) = \dfrac{d}{dt}[98(1 + 2.77e^{-t})^{-1}] = 98(-1)(1 + 2.77e^{-t})^{-2}(2.77e^{-t})(-1)$

$= \dfrac{271.46e^{-t}}{(1 + 2.77e^{-t})^2}$

$f'(3) = \dfrac{271.46e^{-3}}{(1 + 2.77e^{-3})^2} \approx 10.4377.$

So it is increasing at the rate of 10.44%/yr.

c. $f''(t) = 271.46 \left[\dfrac{(1 + 2.77e^{-t})^2(-e^{-t}) - e^{-t} \cdot 2(1 + 2.77e^{-t})(-2.77e^{-t})}{(1 + 2.77e^{-t})^4} \right]$

$= \dfrac{271.46[-(1 + 2.77e^{-t} + 5.54e^{-t}]}{e^t(1 + 2.77e^{-t})^3} = \dfrac{271.46(2.77e^{-t} - 1)}{e^t(1 + 2.77e^{-t})^3}.$

Setting $f''(t) = 0$ gives $2.77e^{-t} = 1$

$$e^{-t} = \dfrac{1}{2.77}; \quad -t = \ln\left(\tfrac{1}{2.77}\right), \quad \text{and} \quad t = 1.0188.$$

The sign diagram of f'' shows that $t = 1.02$ gives an inflection point of P. So, the

percent of lay teachers was increasing most rapidly in 1970.

21. $P(t) = \dfrac{68}{1 + 21.67e^{-0.62t}}.$

The percentage of households that owned VCRs at the beginning of 1985 is given

by $P(0) = \dfrac{68}{1 + 21.67e^{-0.62(0)}} = \dfrac{68}{22.67} \approx 3$, or approximately 3 percent.

The percentage of households that owned VCRs at the beginning of 1995 is given

by $P(10) = \dfrac{68}{1 + 21.67e^{-0.62(10)}} \approx 65.14$, or approximately 65.14 percent.

23. The first of the given conditions implies that $f(0) = 300$, that is,

$$300 = \frac{3000}{1 + Be^0} = \frac{3000}{1 + B}.$$

So $1 + B = 10$, or $B = 9$. Therefore, $f(t) = \dfrac{3000}{1 + 9e^{-kt}}$. Next, the condition

$f(2) = 600$ gives the equation

$$600 = \frac{3000}{1 + 9e^{-2k}}, \quad 1 + 9e^{-2k} = 5, \; e^{-2k} = \frac{4}{9}, \; \text{or } k = -\frac{1}{2}\ln\left(\frac{4}{9}\right).$$

Therefore, $f(t) = \dfrac{3000}{1 + 9e^{(1/2)t \cdot \ln(4/9)}} = \dfrac{3000}{1 + 9(\frac{4}{9})^{t/2}}$.

The number of students who had heard about the policy four hours later is given by

$$f(4) = \frac{3000}{1 + 9(\frac{4}{9})^2} = 1080, \quad \text{or } 1080 \text{ students.}$$

To find the rate at which the rumor was spreading at any time time, we compute

$$f'(t) = \frac{d}{dt}\left[3000(1 + 9e^{-0.405465t})^{-1}\right]$$

$$= (3000)(-1)(1 + 9e^{-0.405465})^{-2}\frac{d}{dt}(9e^{-0.405465t})$$

$$= -3000(9)(-0.405465)e^{-0.405465t}(1 + 9e^{-0.405465t})^{-2} = \frac{10947.555\,e^{-0.405465t}}{(1 + 9e^{-0.405465t})^2}$$

In particular, the rate at which the rumor was spreading 4 hours after the ceremony

is given by $f'(4) = \dfrac{10947.555e^{-0.405465(4)}}{(1 + 9e^{-0.405465(4)})^2} \approx 280.25737.$

So , the rumor is spreading at the rate of 280 students per hour.

25. $x(t) = \dfrac{15\left(1 - (\frac{2}{3})^{3t}\right)}{1 - \frac{1}{4}(\frac{2}{3})^{3t}}$; $\lim\limits_{t \to \infty} x(t) = \lim\limits_{t \to \infty}\dfrac{15\left(1 - (\frac{2}{3})^{3t}\right)}{1 - \frac{1}{4}(\frac{2}{3})^{3t}} = \dfrac{15(1 - 0)}{1 - 0} = 15$

or 15 lbs.

27. a. $C(t) = \dfrac{k}{b - a}\left(e^{-at} - e^{-bt}\right)$;

$$C'(t) = \frac{k}{b-a}(-ae^{-at} + be^{-bt}) = \frac{kb}{b-a}\left[e^{-bt} - \left(\frac{a}{b}\right)e^{-at}\right]$$

$$= \frac{kb}{b-a}e^{-bt}\left[1 - \frac{a}{b}e^{(b-a)t}\right]$$

$C'(t) = 0$ implies that $1 = \frac{a}{b}e^{(b-a)t}$ or $t = \frac{\ln\left(\frac{b}{a}\right)}{b-a}$.

The sign diagram of C'

$$0 \; + + + \; + + \; + \quad - \; - - \; -$$
$$\xrightarrow{\hspace{5cm}} t$$
$$\quad\quad 0 \quad\quad\quad \frac{\ln\left(\frac{b}{a}\right)}{(b-a)}$$

shows that this value of t gives a minimum.

b. $\displaystyle\lim_{t\to\infty} C(t) = \frac{k}{b-a}$.

29. a. We solve $Q_0 e^{-kt} = \frac{1}{2}Q_0$ for t. Proceeding, we have

$$e^{-kt} = \frac{1}{2}, \; \ln e^{-kt} = \ln\frac{1}{2} = \ln 1 = \ln 2 = -\ln 2;$$

$$-kt = -\ln 2;$$

So $\quad \bar{t} = \dfrac{\ln 2}{k}$

b. $\quad \bar{t} = \dfrac{\ln 2}{0.0001238} \approx 5598.927,$ or approximately 5599 years.

USING TECHNOLOGY EXERCISES 13.5, page 863

1. a.

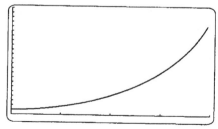

b. 12.146%/yrc. 9.474%/yr

3. a.

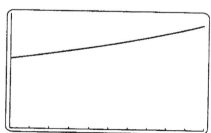

b. 666 million, 818.8 million
c. 33.8 million/yr

5. a.

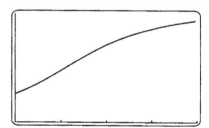

b. 86.12%/yr c. 10.44%/yr
d. 1970

7. a.

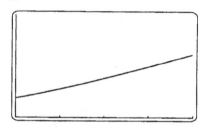

b. 325 million
c. 76.84 million/decade

9. a.

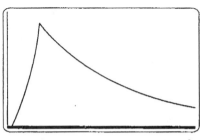

b. 0   c. 0.237 g/cm^3 d. 0.760 g/cm^3 e. 0

CHAPTER 13 REVIEW EXERCISES, page 866

1. a-b

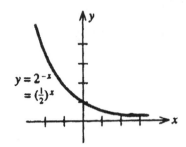

Since $y = \left(\dfrac{1}{2}\right)^x = \dfrac{1}{2^x} = 2^{-x}$, it has the same graph as that of $y = 2^{-x}$.

3. $16^{-3/4} = 0.125$ is equivalent to $-\dfrac{3}{4} = \log_{16} 0.125$.

5. $\quad\quad\quad \ln(x-1) + \ln 4 = \ln(2x+4) - \ln 2$

$\ln(x-1) - \ln(2x+4) = -\ln 2 - \ln 4 = -(\ln 2 + \ln 4)$

$$\ln\left(\frac{x-1}{2x+4}\right) = -\ln 8 = \ln \tfrac{1}{8}..$$

$$\left(\frac{x-1}{2x+4}\right) = \frac{1}{8}$$

$$8x - 8 = 2x + 4$$
$$6x = 12, \text{ or } x = 2.$$

CHECK: l.h.s. $\ln(2-1) + \ln 4 = \ln 4$

r.h.s $\ln(4+4) - \ln 2 = \ln 8 - \ln 2 = \ln \tfrac{8}{2} = \ln 4$.

7. $\ln 3.6 = \ln \tfrac{36}{10} = \ln 36 - \ln 10 = \ln 6^2 - \ln 2 \cdot 5 = 2 \ln 6 - \ln 2 - \ln 5$

$$= 2(\ln 2 + \ln 3) - \ln 2 - \ln 5 = 2(x+y) - x - z = x + 2y - z.$$

9.   We first sketch the graph of $y = 2^{x-3}$. Then we take the reflection of this graph with respect to the line $y = x$.

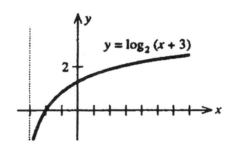

11. $f(x) = xe^{2x}; f'(x) = e^{2x} + xe^{2x}(2) = (1 + 2x)e^{2x}$.

13. $g(t) = \sqrt{t}e^{-2t}; g'(t) = \tfrac{1}{2}t^{-1/2}e^{-2t} + \sqrt{t}e^{-2t}(-2) = \dfrac{1-4t}{2\sqrt{t}e^{2t}}$.

15. $y = \dfrac{e^{2x}}{1+e^{-2x}}; y' = \dfrac{(1+e^{-2x})e^{2x}(2) - e^{2x} \cdot e^{-2x}(-2)}{(1+e^{-2x})^2} = \dfrac{2(e^{2x}+2)}{(1+e^{-2x})^2}$.

17. $f(x) = xe^{-x^2}; f'(x) = e^{-x^2} + xe^{-x^2}(-2x) = (1 - 2x^2)e^{-x^2}$.

19. $f(x) = x^2 e^x + e^x$;
$f'(x) = 2xe^x + x^2 e^x + e^x = (x^2 + 2x + 1)e^x = (x + 1)^2 e^x$.

21. $f(x) = \ln(e^{x^2} + 1)$; $f'(x) = \dfrac{e^{x^2}(2x)}{e^{x^2} + 1} = \dfrac{2xe^{x^2}}{e^{x^2} + 1}$.

23. $f(x) = \dfrac{\ln x}{x + 1}$. $\quad f'(x) = \dfrac{(x+1)\left(\dfrac{1}{x}\right) - \ln x}{(x+1)^2} = \dfrac{1 + \dfrac{1}{x} - \ln x}{(x+1)^2} = \dfrac{x - x\,\ln x + 1}{x(x+1)^2}$.

25. $y = \ln(e^{4x} + 3)$; $y' = \dfrac{e^{4x}(4)}{e^{4x} + 3} = \dfrac{4e^{4x}}{e^{4x} + 3}$.

27. $f(x) = \dfrac{\ln x}{1 + e^x}$;

$$f'(x) = \dfrac{(1 + e^x)\dfrac{d}{dx}\ln x - \ln x \dfrac{d}{dx}(1 + e^x)}{(1 + e^x)^2} = \dfrac{(1 + e^x)\left(\dfrac{1}{x}\right) - (\ln x)e^x}{(1 + e^x)^2}$$

$$= \dfrac{1 + e^x - xe^x \ln x}{x(1 + e^x)^2} = \dfrac{1 + e^x(1 - x\ln x)}{x(1 + e^x)^2}.$$

29. $y = \ln(3x + 1)$; $y' = \dfrac{3}{3x + 1}$;

$$y'' = 3\dfrac{d}{dx}(3x + 1)^{-1} = -3(3x + 1)^{-2}(3) = -\dfrac{9}{(3x + 1)^2}.$$

31. $h'(x) = g'(f(x))f'(x)$. But $g'(x) = 1 - \dfrac{1}{x^2}$ and $f'(x) = e^x$.

So $f(0) = e^0 = 1$ and $f'(0) = e^0 = 1$. Therefore,

$$h'(0) = g'(f(0))f'(0) = g'(1)f'(0) = 0 \cdot 1 = 0.$$

33. $y = (2x^3 + 1)(x^2 + 2)^3$. $\quad \ln y = \ln(2x^3 + 1) + 3\ln(x^2 + 2)$.

$$\dfrac{y'}{y} = \dfrac{6x^2}{2x^3 + 1} + \dfrac{3(2x)}{x^2 + 2} = \dfrac{6x^2(x^2 + 2) + 6x(2x^3 + 1)}{(2x^3 + 1)(x^2 + 2)}$$

$$= \frac{6x^4 + 12x^2 + 12x^4 + 6x}{(2x^3 + 1)(x^2 + 2)} = \frac{18x^4 + 12x^2 + 6x}{(2x^3 + 1)(x^2 + 2)}.$$

Therefore, $y' = 6x(3x^3 + 2x + 1)(x^2 + 2)^2$.

35. $y = e^{-2x}$. $y' = -2e^{-2x}$ and this gives the slope of the tangent line to the graph of $y = e^{-2x}$ at any point (x, y). In particular, the slope of the tangent line at $(1, e^{-2})$ is

$y'(1) = -2e^{-2}$. The required equation is $y - e^{-2} = -2e^{-2}(x - 1)$ or $y = \dfrac{1}{e^2}(-2x + 3)$.

37. $f(x) = xe^{-2x}$.

We first gather the following information on f.
1. The domain of f is $(-\infty, \infty)$.
2. Setting $x = 0$ gives 0 as the y-intercept.
3. $\lim\limits_{x \to -\infty} xe^{-2x} = -\infty$ and $\lim\limits_{x \to \infty} xe^{-2x} = 0$.
4. The results of (3) show that $y = 0$ is a horizontal asymptote.
5. $f'(x) = e^{-2x} + xe^{-2x}(-2) = (1 - 2x)e^{-2x}$. Observe that $f'(x) = 0$ if $x = 1/2$, a critical point of f. The sign diagram of f'

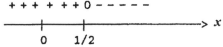

shows that f is increasing on $(-\infty, \frac{1}{2})$ and decreasing on $(\frac{1}{2}, \infty)$.

6. The results of (5) show that $(\frac{1}{2}, \frac{1}{2}e^{-1})$ is a relative maximum.

7. $f''(x) = -2e^{-2x} + (1 - 2x)e^{-2x}(-2) = 4(x - 1)e^{-2x}$ and is equal to zero if $x = 1$. The sign diagram of f''

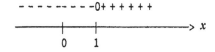

shows that the graph of f is concave downward on $(-\infty, 1)$ and concave upward on $(1, \infty)$.
The graph of f follows.

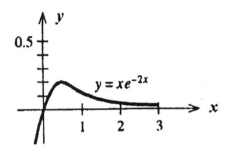

39. $f(t) = te^{-t}$. $f'(t) = e^{-t} + t(-e^{-t}) = e^{-t}(1 - t)$. Setting $f'(t) = 0$ gives $t = 1$ as the only critical point of f. From the sign diagram of f'

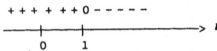

we see that $f(1) = e^{-1} = 1/e$ is the absolute maximum value of f.

41. We want to find r where r satisfies the equation $8.2 = 4.5\, e^{r(5)}$. We have

$$e^{5r} = \frac{8.2}{4.5} \quad \text{or} \quad r = \frac{1}{5}\ln\left(\frac{8.2}{4.5}\right) \approx 0.12$$

and so the annual rate of return is 12 percent per year.

43. a. $Q(t) = 2000e^{kt}$. Now $Q(120) = 18{,}000$ gives $2000e^{120k} = 18{,}000$, $e^{120k} = 9$, or $120k = \ln 9$. So $k = \frac{1}{120}\ln 9 \approx 0.01831$ and $Q(t) = 2000e^{0.01831t}$.
    b. $Q(4) = 2000e^{0.01831(240)} \approx 161{,}992$, or approximately 162,000.

45.

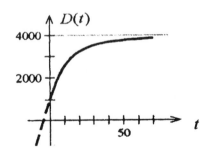

   a. $D(1) = 4000 - 3000\,e^{-0.06} = 1175$, $D(12) = 4000 - 3000\,e^{-0.72} = 2540$, and
      $D(24) = 4000 - 3000\,e^{-1.44} = 3289$.
   b. $\lim_{t\to\infty} D(t) = \lim_{t\to\infty} (4000 - 3000e^{-0.06t}) = 4000$.

CHAPTER 14

1. $F(x) = \frac{1}{3}x^3 + 2x^2 - x + 2$; $F'(x) = x^2 + 4x - 1 = f(x)$.

3. $F(x) = (2x^2 - 1)^{1/2}$; $F'(x) = \frac{1}{2}(2x^2 - 1)^{-1/2}(4x) = 2x(2x^2 - 1)^{-1/2} = f(x)$.

5. a. $G'(x) = \dfrac{d}{dx}(2x) = 2 = f(x)$ b. $F(x) = G(x) + C = 2x + C$

    c.

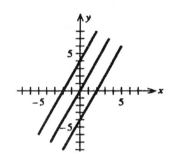

7. a. $G'(x) = \dfrac{d}{dx}(\frac{1}{3}x^3) = x^2 = f(x)$ b. $F(x) = G(x) + C = \frac{1}{3}x^3 + C$

    c.

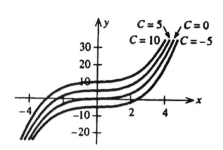

9-39 odd

9. $\displaystyle\int 6\,dx = 6x + C.$ 11. $\displaystyle\int x^3\,dx = \frac{1}{4}x^4 + C$

13. $\displaystyle\int x^{-4}\,dx = -\frac{1}{3}x^{-3} + C$ 15. $\displaystyle\int x^{2/3}\,dx = \frac{3}{5}x^{5/3} + C$

17. $\int x^{-5/4}dx = -4x^{-1/4} + C$

19. $\int \dfrac{2}{x^2}\, dx = 2\int x^{-2}dx = 2(-1x^{-1}) + C = -\dfrac{2}{x} + C$

21. $\int \pi\sqrt{t}\, dt = \pi\int t^{1/2}dt = \pi(\tfrac{2}{3}t^{3/2}) + C = \dfrac{2\pi}{3}t^{3/2} + C$

23. $\int (3-2x)\, dx = \int 3\, dx - 2\int x\, dx = 3x - x^2 + C$

25. $\int (x^2 + x + x^{-3})\, dx = \int x^2\, dx + \int x\, dx + \int x^{-3}\, dx = \tfrac{1}{3}x^3 + \tfrac{1}{2}x^2 - \tfrac{1}{2}x^{-2} + C$

27. $\int 4e^x\, dx = 4e^x + C$

29. $\int (1+x+e^x)\, dx = x + \tfrac{1}{2}x^2 + e^x + C$

31. $\int (4x^3 - \dfrac{2}{x^2} - 1)\, dx = \int (4x^3 - 2x^{-2} - 1)\, dx = x^4 + 2x^{-1} - x + C = x^4 + \dfrac{2}{x} - x + C$

33. $\int (x^{5/2} + 2x^{3/2} - x)\, dx = \tfrac{2}{7}x^{7/2} + \tfrac{4}{5}x^{5/2} - \tfrac{1}{2}x^2 + C$

35. $\int (x^{1/2} + 3x^{-1/2})\, dx = \tfrac{2}{3}x^{3/2} + 6x^{1/2} + C$

37. $\int \left(\dfrac{u^3 + 2u^2 - u}{3u}\right) du = \dfrac{1}{3}\int (u^2 + 2u - 1)\, du = \dfrac{1}{9}u^3 + \dfrac{1}{3}u^2 - \dfrac{1}{3}u + C$

39. $\int (2t+1)(t-2)\, dt = \int (2t^2 - 3t - 2)\, dt = \tfrac{2}{3}t^3 - \tfrac{3}{2}t^2 - 2t + C$

41. $\displaystyle \int \frac{1}{x^2}(x^4 - 2x^2 + 1)\,dx = \int (x^2 - 2 + x^{-2})\,dx = \frac{1}{3}x^3 - 2x - x^{-1} + C$

$$= \frac{1}{3}x^3 - 2x - \frac{1}{x} + C$$

43. $\displaystyle \int \frac{ds}{(s+1)^{-2}} = \int (s+1)^2\,ds = \int (s^2 + 2s + 1)\,ds = \tfrac{1}{3}s^3 + s^2 + s + C$

45. $\displaystyle \int (e^t + t^e)\,dt = e^t + \frac{1}{e+1}t^{e+1} + C$

47. $\displaystyle \int \left(\frac{x^3 + x^2 - x + 1}{x^2} \right)dx = \int \left(x + 1 - \frac{1}{x} + \frac{1}{x^2} \right)dx = \frac{1}{2}x^2 + x - \ln|x| - x^{-1} + C$

49. $\displaystyle \int \left(\frac{(x^{1/2} - 1)^2}{x^2} \right)dx = \int \left(\frac{x - 2x^{1/2} + 1}{x^2} \right)dx = \int (x^{-1} - 2x^{-3/2} + x^{-2})\,dx$

$$= \ln|x| + 4x^{-1/2} - x^{-1} + C = \ln|x| + \frac{4}{\sqrt{x}} - \frac{1}{x} + C$$

51. $\displaystyle \int f'(x)\,dx = \int (2x + 1)\,dx = x^2 + x + C.$ The condition $f(1) = 3$ gives
   $f(1) = 1 + 1 + C = 3$, or $C = 1$. Therefore, $f(x) = x^2 + x + 1.$

53. $f'(x) = 3x^2 + 4x - 1;\ f(x) = x^3 + 2x^2 - x + C.$ Using the given initial condition,
   we have $f(2) = 8 + 2(4) - 2 + C = 9$, so $16 - 2 + C = 9$, or $C = -5.$ Therefore,
   $f(x) = x^3 + 2x^2 - x - 5.$

55. $\displaystyle f(x) = \int f'(x)\,dx = \int \left(1 + \frac{1}{x^2} \right)dx = \int (1 + x^{-2})\,dx = x - \frac{1}{x} + C.$

   Using the given initial condition, we have $f(1) = 1 - 1 + C = 2$, or $C = 2.$

   Therefore, $\displaystyle f(x) = x - \frac{1}{x} + 2.$

57. $\displaystyle f(x) = \int \frac{x+1}{x}\,dx = \int \left(1 + \frac{1}{x} \right)dx = x + \ln|x| + C.$ Using the initial condition, we

have $f(1) = 1 + \ln 1 + C = 1 + C = 1$, or $C = 0$. So $f(x) = x + \ln|x|$.

59. $f(x) = \int f'(x)\,dx = \int \frac{1}{2}x^{-1/2}\,dx = \frac{1}{2}(2x^{1/2}) + C = x^{1/2} + C;\ f(2) = \sqrt{2} + C = \sqrt{2}$

implies $C = 0$. So $f(x) = \sqrt{x}$.

61. $f'(x) = e^x + x;\ f(x) = e^x + \frac{1}{2}x^2 + C;\ f(0) = e^0 + \frac{1}{2}(0) + C = 1 + C$

So $3 = 1 + C$ or $2 = C$. Therefore, $f(x) = e^x + \frac{1}{2}x^2 + 2$.

63. The position of the car is

$s(t) = \int f(t)\,dt = \int 2\sqrt{t}\,dt = \int 2t^{1/2}\,dt = 2(\frac{2}{3}t^{3/2}) + C = \frac{4}{3}t^{3/2} + C.$

$s(0) = 0$ implies $s(0) = C = 0$. So $s(t) = \frac{4}{3}t^{3/2}.$

Marginal cost

65. $C(x) = \int C'(x)\,dx = \int (0.000009x^2 - 0.009x + 8)\,dx$

$= 0.000003x^3 - 0.0045x^2 + 8x + k.$

FIX COST

$C(0) = k = 120$ and so $C(x) = 0.000003x^3 - 0.0045x^2 + 8x + 120.$

$C(500) = 0.000003(500)^3 - 0.0045(500)^2 + 8(500) + 120,$ or $\$3370.$

Marginal profit

$\$16,000/m = $ fixed Cost of producing & selling

67. $P'(x) = -0.004x + 20,\ P(x) = -0.002x^2 + 20x + C$. Since $C = -16,000$, we find
that $P(x) = -0.002x^2 + 20x - 16,000$. The company realizes a maximum profit
when $P'(x) = 0$, that is, when $x = 5000$ units. Next,

$P(5000) = -0.002(5000)^2 + 20(5000) - 16,000 = 34,000.$

Thus, a maximum profit of $\$34,000$ is realized at a production level of 5000 units.

69. a. $N(t) = \int N'(t)\,dt = \int (-3t^2 + 12t + 45)\,dt = -t^3 + 6t^2 + 45t + C$. But $N(0) = C = 0$

and so $N(t) = -t^3 + 6t^2 + 45t.$
b. The number is $N(4) = -4^3 + 6(4)^2 + 45(4) = 212.$

71. a. We have the initial-value problem:

$C'(t) = 12.288t^2 - 150.5594t + 695.23$

$C(0) = 3142$

Integrating, we find

$$C(t) = \int C'(t)\,dt = \int (12.288t^2 - 150.5594t + 695.23)\,dt$$
$$= 4.096t^3 - 75.2797t^2 + 695.23t + k$$

Using the initial condition, we find

$C(0) = 0 + k = 3142$, and so $k = 3142$.

Therefore, $C(t) = 4.096t^3 - 75.2797t^2 + 695.23t + 3142$.

The projected average out-of-pocket costs for beneficiaries is 2010 is
$$C(2) = 4.096(8) - 75.2797(4) + 695.23(2) + 3142 = 4264.1092$$
or $4264.11.

73. The number of new subscribers at any time is
$$N(t) = \int (100 + 210t^{3/4})\,dt = 100t + 120t^{7/4} + C.$$

The given condition implies that $N(0) = 5000$. Using this condition, we find $C = 5000$. Therefore, $N(t) = 100t + 120t^{7/4} + 5000$. The number of subscribers 16 months from now is
$$N(16) = 100(16) + 120(16)^{7/4} + 5000, \text{ or } 21{,}960.$$

75. a. We have the initial-value problem *grow at the rate of* *percent/yr* *yr 2000 $(t=0) = 2.9\%$*
$$S'(t) = R(t) = -0.033t^2 + 0.3428t + 0.07 \quad \text{and } S(0) = 2.9$$
where $S(t)$ denotes the share of online advertisement, worldwide, as a precent of the total ad market. Integrating, we find
$$S(t) = \int S'(t)\,dt = \int (-0.033t^2 + 0.3428t + 0.07)\,dt$$
$$= -0.011t^3 + 0.1714t^2 + 0.07t + C$$

Using the initial condition, we find $S(0) = 0 + C = 2.9$ and so $C = 2.9$.

Therefore, $S(t) = -0.011t^3 + 0.1714t^2 + 0.07t + 2.9$.

b. The projected online ad market share at the beginning of 2005 is
$$S(5) = -0.011(125) + 0.1714(25) + 0.07(5) + 2.9 = 6.16, \text{ or } 6.16\%.$$

77. $A(t) = \int A'(t)\,dt = \int (3.2922t^2 - 0.366t^3)\,dt = 1.0974t^3 - 0.0915t^4 + C$

Now, $A(0) = C = 0$. So, $A(t) = 1.0974t^3 - 0.0915t^4$.

79. $h(t) = \int h'(t)\,dt = \int (-3t^2 + 192t + 120)\,dt = -t^3 + 96t^2 + 120t + C$
$$= -t^3 + 96t^2 + 120t + C.$$

$h(0) = C = 0$ implies $h(t) = -t^3 + 96t^2 + 120t$.
The altitude 30 seconds after lift-off is
$h(30) = -30^3 + 96(30)^2 + 120(30) = 63{,}000$ ft.

81. $v(r) = \int v'(r)\,dr = \int -kr\,dr = -\tfrac{1}{2}kr^2 + C.$

But $v(R) = 0$ and so $v(R) = -\tfrac{1}{2}kR^2 + C = 0,$ or $C = \tfrac{1}{2}kR^2.$ Therefore,

$v(R) = -\tfrac{1}{2}kr^2 + \tfrac{1}{2}kR^2 = \tfrac{1}{2}k(R^2 - r^2).$

83. Denote the constant deceleration by k (ft/sec^2). Then $f''(t) = -k$, so
$f'(t) = v(t) = -kt + C_1$. Next, the given condition implies that $v(0) = 88$. This gives
$C_1 = 88$, or $f'(t) = -kt + 88$.
$$s = f(t) = \int f'(t)\,dt = \int (-kt + 88)\,dt = -\tfrac{1}{2}kt^2 + 88t + C_2.$$

Also, $f(0) = 0$ gives $s = f(t) = -\tfrac{1}{2}kt^2 + 88t$. Since the car is brought to rest in 9
seconds, we have $v(9) = -9k + 88 = 0$, or $k = \tfrac{88}{9}$, or $9\tfrac{7}{9}$. So the deceleration is
$9\tfrac{7}{9}$ ft/sec^2. The distance covered is
$$s = f(9) = -\tfrac{1}{2}\left(\tfrac{88}{9}\right)(81) + 88(9) = 396.$$
So the stopping distance is 396 ft.

85. a. We have the initial-value problem $R'(t) = \dfrac{8}{(t+4)^2}$ and $R(0) = 0$

Integrating, we find $R(t) = \displaystyle\int \dfrac{8}{(t+4)^2}\,dt = 8\int (t+4)^{-2}\,dt = -\dfrac{8}{t+4} + C$

$R(0) = 0$ implies $-\dfrac{8}{4} + C = 0$ or $C = 2.$

Therefore, $R(t) = -\dfrac{8}{t+4} + 2 = \dfrac{-8 + 2t + 8}{t+4} = \dfrac{2t}{t+4}.$

b. After 1 hr, $R(1) = \dfrac{2}{5} = 0.4$, or 0.4" had fallen. After 2 hr, $R(2) = \dfrac{4}{6} = \dfrac{2}{3}$, or

$\dfrac{2}{3}$" had fallen.

87. The net amount on deposit in Branch A is given by the area under the graph of f
from $t = 0$ to $t = 180$. On the other hand, the net amount on deposit in Branch B is

given by the area under the graph of g over the same interval. Evidently the first area is longer than the second. Therefore, we see that Branch A has the larger net deposit.

89. True. See proof in Section 14.1 in the text.

91. True. Use the Sum Rule followed by the Constant Multiple Rule.

EXERCISES 14.2, page 890

1 – 13 odd

1. Put $u = 4x + 3$ so that $du = 4\,dx$, or $dx = \frac{1}{4}\,du$. Then

$$\int 4(4x+3)^4\,dx = \int u^4\,du = \frac{1}{5}u^5 + C = \frac{1}{5}(4x+3)^5 + C.$$

3. Let $u = x^3 - 2x$ so that $du = (3x^2 - 2)\,dx$. Then

$$\int (x^3 - 2x)^2(3x^2 - 2)\,dx = \int u^2\,du = \frac{1}{3}u^3 + C = \frac{1}{3}(x^3 - 2x)^3 + C.$$

5. Let $u = 2x^2 + 3$ so that $du = 4x\,dx$. Then

$$\int \frac{4x}{(2x^2+3)^3}\,dx = \int \frac{1}{u^3}\,du = \int u^{-3}\,du = -\frac{1}{2}u^{-2} + C = -\frac{1}{2(2x^2+3)^2} + C.$$

7. Put $u = t^3 + 2$ so that $du = 3t^2\,dt$ or $t^2\,dt = \frac{1}{3}\,du$. Then

$$\int 3t^2\sqrt{t^3+2}\,dt = \int u^{1/2}\,du = \frac{2}{3}u^{3/2} + C = \frac{2}{3}(t^3+2)^{3/2} + C$$

9. Let $u = x^2 - 1$ so that $du = 2x\,dx$ and $x\,dx = \frac{1}{2}\,du$. Then,

$$\int (x^2 - 1)^9 x\,dx = \int \frac{1}{2}u^9\,du = \frac{1}{20}u^{10} + C = \frac{1}{20}(x^2 - 1)^{10} + C.$$

11. Let $u = 1 - x^5$ so that $du = -5x^4\,dx$ or $x^4\,dx = -\frac{1}{5}\,du$. Then

$$\int \frac{x^4}{1-x^5}\,dx = -\frac{1}{5}\int \frac{du}{u} = -\frac{1}{5}\ln|u| + C = -\frac{1}{5}\ln\left|1-x^5\right| + C.$$

13. Let $u = x - 2$ so that $du = dx$. Then

$$\int \frac{2}{x-2} dx = 2\int \frac{du}{u} = 2\ln|u| + C = \ln u^2 + C = \ln(x-2)^2 + C$$

15. Let $u = 0.3x^2 - 0.4x + 2$. Then $du = (0.6x - 0.4) dx = 2(0.3x - 0.2) dx$.

$$\int \frac{0.3x - 0.2}{0.3x^2 - 0.4x + 2} dx = \int \frac{1}{2u} du = \frac{1}{2}\ln|u| + C = \frac{1}{2}\ln(0.3x^2 - 0.4x + 2) + C.$$

17-27 odd

17. Let $u = 3x^2 - 1$ so that $du = 6x\, dx$, or $x\, dx = \frac{1}{6} du$. Then

$$\int \frac{x}{3x^2 - 1} dx = \frac{1}{6}\int \frac{du}{u} = \frac{1}{6}\ln|u| + C = \frac{1}{6}\ln\left|3x^2 - 1\right| + C.$$

19. Let $u = -2x$ so that $du = -2\, dx$ or $dx = -\frac{1}{2} du$. Then

$$\int e^{-2x} dx = -\frac{1}{2}\int e^u\, du = -\frac{1}{2}e^u + C = -\frac{1}{2}e^{-2x} + C.$$

21. Let $u = 2 - x$ so that $du = -dx$ or $dx = -du$. Then

$$\int e^{2-x} dx = -\int e^u\, du = -e^u + C = -e^{2-x} + C.$$

23. Let $u = -x^2$, then $du = -2x\, dx$ or $x\, dx = -\frac{1}{2} du$.

$$\int xe^{-x^2} dx = \int -\frac{1}{2}e^u\, du = -\frac{1}{2}e^u + C = -\frac{1}{2}e^{-x^2} + C.$$

25. $\displaystyle\int (e^x - e^{-x}) dx = \int e^x\, dx - \int e^{-x}\, dx = e^x - \int e^{-x}\, dx.$

To evaluate the second integral on the right, let $u = -x$ so that $du = -dx$ or $dx = -du$.   Therefore,

$$\int (e^x - e^{-x}) dx = e^x + \int e^u\, du = e^x + e^u + C = e^x + e^{-x} + C.$$

27. Let $u = 1 + e^x$ so that $du = e^x\, dx$. Then

$$\int \frac{e^x}{1+e^x} dx = \int \frac{du}{u} = \ln|u| + C = \ln(1 + e^x) + C.$$

29. Let $u = \sqrt{x} = x^{1/2}$. Then $du = \frac{1}{2} x^{-1/2}\, dx$ or $2\, du = x^{-1/2}\, dx$.

$$\int \frac{e^{\sqrt{x}}}{\sqrt{x}}\, dx = \int 2e^u\, du = 2e^u + C = 2e^{\sqrt{x}} + C.$$

31. Let $u = e^{3x} + x^3$ so that $du = (3e^{3x} + 3x^2)\, dx = 3(e^{3x} + x^2)\, dx$ or $(e^{3x} + x^2)\, dx = \frac{1}{3} du$.
    Then

$$\int \frac{e^{3x} + x^2}{(e^{3x} + x^3)^3}\, dx = \frac{1}{3} \int \frac{du}{u^3} = \frac{1}{3} \int u^{-3}\, du = -\frac{1}{6} u^{-2} + C = -\frac{1}{6(e^{3x} + x^3)^2} + C.$$

33. Let $u = e^{2x} + 1$, so that $du = 2e^{2x}\, dx$, or $\frac{1}{2} du = e^{2x}\, dx$.

$$\int e^{2x}(e^{2x} + 1)^3\, dx = \int \frac{1}{2} u^3\, du = \frac{1}{8} u^4 + C = \frac{1}{8}(e^{2x} + 1)^4 + C.$$

35. Let $u = \ln 5x$ so that $du = \dfrac{1}{x}\, dx$. Then

$$\int \frac{\ln 5x}{x}\, dx = \int u\, du = \frac{1}{2} u^2 + C = \frac{1}{2}(\ln 5x)^2 + C.$$

37. Let $u = \ln x$ so that $du = \frac{1}{x} dx$. Then

$$\int \frac{1}{x \ln x}\, dx = \int \frac{du}{u} = \ln |u| + C = \ln |\ln x| + C.$$

39. Let $u = \ln x$ so that $du = \frac{1}{x} dx$. Then

$$\int \frac{\sqrt{\ln x}}{x}\, dx = \int \sqrt{u}\, du = \frac{2}{3} u^{3/2} + C = \frac{2}{3}(\ln x)^{3/2} + C$$

41. $$\int \left(xe^{x^2} - \frac{x}{x^2 + 2} \right) dx = \int xe^{x^2} - \int \frac{x}{x^2 + 2}\, dx.$$

To evaluate the first integral, let $u = x^2$ so that $du = 2x\, dx$, or $x\, dx = \frac{1}{2} du$. Then

$$\int xe^{x^2}\, dx = \frac{1}{2} \int e^u\, du + C_1 = \frac{1}{2} e^u + C_1 = \frac{1}{2} e^{x^2} + C_1.$$

To evaluate the second integral, let $u = x^2 + 2$ so that $du = 2x\,dx$, or $x\,dx = \frac{1}{2}\,du$. Then

$$\int \frac{x}{x^2+2}\,dx = \frac{1}{2}\int \frac{du}{u} = \frac{1}{2}\ln|u| + C_2 = \frac{1}{2}\ln(x^2+2) + C_2.$$

Therefore, $\displaystyle\int \left(xe^{x^2} - \frac{x}{x^2+2} \right) dx = \frac{1}{2}e^{x^2} - \frac{1}{2}\ln(x^2+2) + C.$

43. Let $u = \sqrt{x} - 1$ so that $du = \frac{1}{2}x^{-1/2}\,dx = \frac{1}{2\sqrt{x}}\,dx$ or $dx = 2\sqrt{x}\,du$.

Also, we have $\sqrt{x} = u + 1$, so that $x = (u+1)^2 = u^2 + 2u + 1$ and $dx = 2(u+1)\,du$. So

$$\int \frac{x+1}{\sqrt{x}-1}\,dx = \int \frac{u^2 + 2u + 2}{u} \cdot 2(u+1)\,du = 2\int \frac{(u^3 + 3u^2 + 4u + 2)}{u}\,du$$

$$= 2\int \left(u^2 + 3u + 4 + \frac{2}{u} \right) du = 2\left(\frac{1}{3}u^3 + \frac{3}{2}u^2 + 4u + 2\ln|u| \right) + C$$

$$= 2\left[\frac{1}{3}(\sqrt{x}-1)^3 + \frac{3}{2}(\sqrt{x}-1)^2 + 4(\sqrt{x}-1) + 2\ln|\sqrt{x}-1| \right] + C.$$

45. Let $u = x - 1$ so that $du = dx$. Also, $x = u + 1$ and so

$$\int x(x-1)^5\,dx = \int (u+1)u^5\,du = \int (u^6 + u^5)\,du$$

$$= \frac{1}{7}u^7 + \frac{1}{6}u^6 + C = \frac{1}{7}(x-1)^7 + \frac{1}{6}(x-1)^6 + C$$

$$= \frac{(6x+1)(x-1)^6}{42} + C.$$

47. Let $u = 1 + \sqrt{x}$ so that $du = \frac{1}{2}x^{-1/2}\,dx$ and $dx = 2\sqrt{x} = 2(u-1)\,du$

$$\int \frac{1-\sqrt{x}}{1+\sqrt{x}}\,dx = \int \left(\frac{1-(u-1)}{u} \right) \cdot 2(u-1)\,du = 2\int \frac{(2-u)(u-1)}{u}\,du$$

$$= 2\int \frac{-u^2 + 3u - 2}{u}\,du = 2\int \left(-u + 3 - \frac{2}{u} \right) du = -u^2 + 6u - 4\ln|u| + C$$

$$= -(1+\sqrt{x})^2 + 6(1+\sqrt{x}) - 4\ln(1+\sqrt{x}) + C$$
$$= -1 - 2\sqrt{x} - x + 6 + 6\sqrt{x} - 4\ln(1+\sqrt{x}) + C$$
$$= -x + 4\sqrt{x} + 5 - 4\ln(1+\sqrt{x}) + C.$$

49. $I = \int v^2(1-v)^6\, dv$. Let $u = 1 - v$, then $du = -dv$. Also, $1 - u = v$, and $(1-u)^2 = v^2$. Therefore,

$$I = \int -(1-2u+u^2)u^6\, du = \int -(u^6 - 2u^7 + u^8)\, du = -\left(\frac{u^7}{7} - \frac{2u^8}{8} + \frac{u^9}{9}\right) + C$$

$$= -u^7\left(\frac{1}{7} - \frac{1}{4}u + \frac{1}{9}u^2\right) + C = -\frac{1}{252}(1-v)^7[36 - 63(1-v) + 28(1 - 2v + v^2)]$$

$$= -\frac{1}{252}(1-v)^7[36 - 63 + 63v + 28 - 56v + 28v^2]$$

$$= -\frac{1}{252}(1-v)^7(28v^2 + 7v + 1) + C.$$

51. $f(x) = \int f'(x)\, dx = 5\int (2x-1)^4\, dx$. Let $u = 2x - 1$ so that $du = 2x-1$ so that $du = 2\, dx$, or $dx = \frac{1}{2}\, du$. Then

$$f(x) = \frac{5}{2}\int u^4\, du = \frac{1}{2}u^5 + C = \frac{1}{2}(2x-1)^5 + C.$$

Next, $f(1) = 3$ implies $\frac{1}{2} + C = 3$ or $C = \frac{5}{2}$. Therefore,

$$f(x) = \frac{1}{2}(2x-1)^5 + \frac{5}{2}.$$

53. $f(x) = \int -2xe^{-x^2+1}\, dx$. Let $u = -x^2 + 1$ so that $du = -2x\, dx$. Then

$$f(x) = \int e^u\, du = e^u + C = e^{-x^2+1} + C.$$ The condition $f(1) = 0$ implies

$f(1) = 1 + C = 0$, or $C = -1$. Therefore, $f(x) = e^{-x^2+1} - 1$.

55. $N'(t) = 2000(1 + 0.2t)^{-3/2}$. Let $u = 1 + 0.2t$. Then $du = 0.2\, dt$ and $5\, du = dt$. Therefore, $N(t) = (5)(2000)$

$$\int u^{-3/2}\, du = -20{,}000u^{-1/2} + C = -20{,}000(1+0.2t)^{-1/2} + C.$$

Next, $N(0) = -20{,}000(1)^{-1/2} + C = 1000$. Therefore, $C = 21{,}000$ and

$$N(t) = -\frac{20{,}000}{\sqrt{1+0.2t}} + 21{,}000. \quad \text{In particular, } N(5) = -\frac{20{,}000}{\sqrt{2}} + 21{,}000 \approx 6{,}858.$$

57. $p(x) = \displaystyle\int -\frac{250x}{(16+x^2)^{3/2}}\, dx = -250\int \frac{x}{(16+x^2)^{3/2}}\, dx.$

Let $u = 16 + x^2$ so that $du = 2x\, dx$ and $x\, dx = \frac{1}{2}\, du$.

Then $p(x) = -\frac{250}{2}\displaystyle\int u^{-3/2}\, du = (-125)(-2)u^{-1/2} + C = \frac{250}{\sqrt{16+x^2}} + C.$

$p(3) = \dfrac{250}{\sqrt{16+9}} + C = 50$ implies $C = 0$ and $p(x) = \dfrac{250}{\sqrt{16+x^2}}.$

59. Let $u = 2t + 4$, so that $du = 2\, dt$. Then

$$r(t) = \int \frac{30}{\sqrt{2t+4}}\, dt = 30\int \frac{1}{2}u^{-1/2}\, du = 30u^{1/2} + C = 30\sqrt{2t+4} + C.$$

$r(0) = 60 + C = 0$, and $C = -60$. Therefore, $r(t) = 30\left(\sqrt{2t+4} - 2\right)$. Then

$r(16) = 30\left(\sqrt{36} - 2\right) = 120\,\text{ft}$. Therefore, the polluted area is

$$\pi r^2 = \pi(120)^2 = 14{,}400\pi, \qquad \text{or } 14{,}400\pi \text{ sq ft.}$$

61. The population t years from now will be

$$P(t) = \int r(t)\, dt = \int 400\left(1 + \frac{2t}{24+t^2}\right) dt = \int 400\, dt + 800 \int \frac{t}{24+t^2}\, dt$$

In order to evaluate the second integral on the right, let
$$u = 24 + t^2, \quad du = 2t\, dt, \quad \text{or } t\, dt = \tfrac{1}{2}\, du$$

We obtain $P(t) = 400t + 800\displaystyle\int \frac{\frac{1}{2}\, du}{u} = 400t + 400\left|\ln u\right| + C$

$$= 400[t + \ln(24+t^2)] + C$$

To find C, use the condition $P(0) = 60{,}000$ giving
$$400[0 + \ln 24] + C = 60{,}000 \quad \text{or} \quad C = 58728.78$$

14 Integration

504

So $P(t) = 400[t + \ln(24 + t^2)] + 58728.78$. Therefore, the population 5 years from now will be

$400[5 + \ln(24 + 25)] + 58728.78 \approx 62,285.51$, or approximately, 62,286.

63. $N(t) = \int N'(t)\,dt = 6\int e^{-0.05t}\,dt = \dfrac{6}{-0.05}e^{-0.05t}$ (Let u = -0.05t)

$$= -120e^{-0.05t} + C.$$

$N(0) = 60$ implies $-120 + C = 60$ or $C = 180$. Therefore, $N(t) = -120e^{-0.05t} + 180$.

65. $A(t) = \int A'(t)\,dt = r\int e^{-at}\,dt$. Let $u = -at$ so that $du = -a\,dt$, or $dt = -\frac{1}{a}\,du$.

$$A(t) = r\left(-\tfrac{1}{a}\right)\int e^u\,du = -\tfrac{r}{a}e^u + C = -\tfrac{r}{a}e^{-at} + C$$

$A(0)$ implies $-\dfrac{r}{a} + C = 0$, or $C = \dfrac{r}{a}$. So, $A(t) = -\dfrac{r}{a}e^{-at} + \dfrac{r}{a} = \dfrac{r}{a}(1 - e^{-at})$.

EXERCISES 14.3, page 901

1. $\frac{1}{3}(1.9 + 1.5 + 1.8 + 2.4 + 2.7 + 2.5) = \frac{12.8}{3} \approx 4.27$.

3.   a. $A = \frac{1}{2}(2)(6) = 6$ sq units.

     b. $\Delta x = \frac{2}{4} = \frac{1}{2}$; $x_1 = 0$, $x_2 = \frac{1}{2}$, $x_3 = 1$, $x_4 = \frac{3}{2}$.

$$A \approx \tfrac{1}{2}[3(0) + 3(\tfrac{1}{2}) + 3(1) + 3(\tfrac{3}{2})] = \tfrac{9}{2}$$
$$= 4.5 \text{ sq units.}$$

     c. $\Delta x = \frac{2}{8} = \frac{1}{4}$. $x_1 = 0, \ldots, x_8 = \frac{7}{4}$.

$$A \approx \tfrac{1}{4}\left[3(0) + 3(\tfrac{1}{4}) + 3(\tfrac{1}{2}) + 3(\tfrac{3}{4}) + 3(1) + 3(\tfrac{5}{4}) + 3(\tfrac{3}{2}) + 3(\tfrac{7}{4})\right]$$
$$= \tfrac{21}{4} = 5.25 \text{ sq units.}$$

     d. Yes.

5. a. $A = 4$

b. $\Delta = \frac{2}{5} = 0.4$; $x_1 = 0, x_2 = 0.4, x_3 = 0.8, x_4 = 1.2$

$x_5 = 1.6$,
$A \approx 0.4\{[4 - 2(0)] + [4 - 2(0.4)] + [4 - 2(0.8)]$
$\quad + [4 - 2(1.2)] + [4 - 2(1.6)]\}$
$\quad = 4.8$

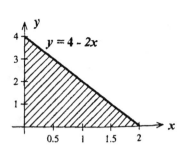

c. $\Delta x = \frac{2}{10} = 0.2$, $x_1 = 0, x_2 = 0.2, x_3 = 0.4, ..., x_{10} = 1.8$.
$A \approx 0.2\{[4 - 2(0)] + [4 - 2(0.2)] + [4 - 2(0.4)]$
$\quad + \cdots + [4 - 2(1.8)]\} = 4.4$

d. Yes.

7. a. $\Delta x = \dfrac{4-2}{2} = 1$; $x_1 = 2.5, x_2 = 3.5$;   The Riemann sum is $[(2.5)^2 + (3.5)^2] = 18.5$.

b. $\Delta x = \dfrac{4-2}{5} = 0.4$; $x_1 = 2.2, x_2 = 2.6, x_3 = 3.0, x_4 = 3.4, x_5 = 3.8$
The Riemann sum is $0.4[2.2^2 + 2.6^2 + 3.0^2 + 3.4^2 + 3.8^2] = 18.64$.

c. $\Delta x = \dfrac{4-2}{10} = 0.2$; $x_1 = 2.1, x_2 = 2.3, x_2 = 2.5, ..., x_{10} = 3.9$
The Riemann sum is $0.2[2.1^2 + 2.3^2 + 2.5^2 + \cdots + 3.9^2] = 18.66$.
The area seems to be $18\frac{2}{3}$ sq units.

9. a. $\Delta x = \dfrac{4-2}{2} = 1$; $x_1 = 3, x_2 = 4$. The Riemann sum is $(1)[3^2 + 4^2] = 25$.

b. $\Delta x = \dfrac{4-2}{5} = 0.4$; $x_1 = 2.4, x_2 = 2.8, x_3 = 3.2, x_4 = 3.6, x_5 = 4$.
The Riemann sum is $0.4[2.4^2 + 2.8^2 + \cdots + 4^2] = 21.12$.

c. $\Delta x = \dfrac{4-2}{10} = 0.2$; $x_1 = 2.2, x_2 = 2.4, x_3 = 2.6, ..., x_{10} = 4$.
The Riemann sum is $0.2[2.2^2 + 2.4^2 + 2.6^2 + \cdots + 4^2] = 19.88$.
d. 19.9 sq units.

11. a. $\Delta x = \dfrac{1}{2}$, $x_1 = 0, x_2 = \dfrac{1}{2}$. The Riemann sum is

$$f(x_1)\Delta x + f(x_2)\Delta x = \left[(0)^3 + (\tfrac{1}{2})^3\right]\tfrac{1}{2} = \tfrac{1}{16} = 0.0625.$$

b. $\Delta x = \dfrac{1}{5}$, $x_1 = 0$, $x_2 = \dfrac{1}{5}$, $x_3 = \dfrac{2}{5}$, $x_4 = \dfrac{3}{5}$, $x_5 = \dfrac{4}{5}$. The Riemann sum

is $f(x_1)\Delta x + f(x_2)\Delta x + \cdots f(x_5)\Delta x = \left[(\tfrac{1}{5})^3 + (\tfrac{2}{5})^3 + \cdots + (\tfrac{4}{5})^3\right]\tfrac{1}{5} = \tfrac{100}{625} = 0.16.$

c. $\Delta x = \dfrac{1}{10}$; $x_1 = 0$, $x_2 = \dfrac{1}{10}$, $x_3 = \dfrac{2}{10}$, $\cdots$, $x_{10} = \dfrac{9}{10}$.

The Riemann sum is

$$f(x_1)\Delta x + f(x_2)\Delta x + \cdots + f(x_{10})\Delta x = \left[(\tfrac{1}{10})^3 + (\tfrac{2}{10})^3 + \cdots + (\tfrac{9}{10})^3\right]\tfrac{1}{10}$$
$$= \tfrac{2025}{10,000} = 0.2025 \approx 0.2 \text{ sq units.}$$

The Riemann sum seems to approach 0.2.

13. $\Delta x = \dfrac{2-0}{5} = \dfrac{2}{5}$; $x_1 = \dfrac{1}{5}$, $x_2 = \dfrac{3}{5}$, $x_3 = \dfrac{5}{5}$, $x_4 = \dfrac{7}{5}$, $x_5 = \dfrac{9}{5}$.

$$A \approx \left\{\left[(\tfrac{1}{5})^2 + 1\right] + \left[(\tfrac{3}{5})^2 + 1\right] + \left[(\tfrac{5}{5})^2 + 1\right] + \left[(\tfrac{7}{5})^2 + 1\right] + \left[(\tfrac{9}{5})^2 + 1\right](\tfrac{2}{5})\right\}$$
$$= \tfrac{580}{125} = 4.64 \text{ sq units.}$$

15. $\Delta x = \dfrac{3-1}{4} = \dfrac{1}{2}$; $x_1 = \dfrac{3}{2}$, $x_2 = \dfrac{4}{2}$, $x_3 = \dfrac{5}{2}$, $x_4 = 3$.

$$A \approx \left[\tfrac{1}{\tfrac{3}{2}} + \tfrac{1}{\tfrac{4}{2}} + \tfrac{1}{\tfrac{5}{2}} + \tfrac{1}{3}\right]\tfrac{1}{2} \approx 0.95 \text{ sq units.}$$

17. $A = 20[f(10) + f(30) + f(50) + f(70) + f(90)]$
$= 20(80 + 100 + 110 + 100 + 80) = 9400 \text{ sq ft.}$

EXERCISES 14.4, page 910

1. $A = \displaystyle\int_1^4 2\,dx = 2x\Big|_1^4 = 2(4-1) = 6$, or 6 square

units. The region is a rectangle whose area is

$3 \cdot 2$, or 6, square units.

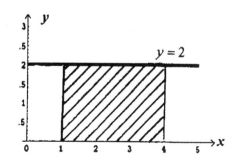

3. $A = \int_{1}^{3} 2x\,dx = x^2\Big|_{1}^{3} = 9 - 1 = 8$, or 8 sq units.

The region is a parallelogram of area
$(1/2)(3 - 1)(2 + 6) = 8$ sq units.

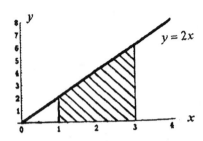

5. $A = \int_{-1}^{2} (2x + 3)\,dx = x^2 + 3x\Big|_{-1}^{2} = (4 + 6) - (1 - 3) = 12$, or 12 sq units.

7. $A = \int_{-1}^{2} (-x^2 + 4)\,dx = -\dfrac{1}{3}x^3 + 4x\Big|_{-1}^{2}$

$\quad = \left(-\dfrac{8}{3} + 8\right) - \left(\dfrac{1}{3} - 4\right) = 9$, or 9 sq units.

9. $A = \int_{1}^{2} \dfrac{1}{x}\,dx = \ln|x|\Big|_{1}^{2} = \ln 2 - \ln 1 = \ln 2$, or $\ln 2$ sq units.

11. $A = \int_{1}^{9} \sqrt{x}\,dx = \dfrac{2}{3}x^{3/2}\Big|_{1}^{9} = \dfrac{2}{3}(27 - 1) = \dfrac{52}{3}$, or $17\tfrac{1}{3}$ sq units.

13. $A = \int_{-8}^{-1} (1 - x^{1/3})\,dx = x - \tfrac{3}{4}x^{4/3}\Big|_{-8}^{-1} = (-1 - \tfrac{3}{4}) - (-8 - 12) = 18\tfrac{1}{4}$, or $18\tfrac{1}{4}$ sq units.

15. $A = \int_{0}^{2} e^x\,dx = e^x\Big|_{0}^{2} = (e^2 - 1)$, or approximately 6.39 sq units.

17. $\int_{2}^{4} 3\,dx = 3x\Big|_{2}^{4} = 3(4 - 2) = 6.$

19. $\int_{1}^{3} (2x + 3)\,dx = x^2 + 3x\Big|_{1}^{3} = (9 + 9) - (1 + 3) = 14.$

21. $\int_{-1}^{3} 2x^2\,dx = \tfrac{2}{3}x^3\Big|_{-1}^{3} = \tfrac{2}{3}(27) - \tfrac{2}{3}(-1) = \tfrac{56}{3}.$

23. $\int_{-2}^{2} (x^2 - 1)\,dx = \tfrac{1}{3}x^3 - x\Big|_{-2}^{2} = \left(\tfrac{8}{3} - 2\right) - \left(-\tfrac{8}{3} + 2\right) = \tfrac{4}{3}.$

25. $\int_1^8 4x^{1/3} \, dx = (4)(\frac{3}{4})x^{4/3}\Big|_1^8 = 3(16-1) = 45.$

27–37 odd

27. $\int_0^1 (x^3 - 2x^2 + 1) \, dx = \frac{1}{4}x^4 - \frac{2}{3}x^3 + x\Big|_0^1 = \frac{1}{4} - \frac{2}{3} + 1 = \frac{7}{12}$

29. $\int_2^4 \frac{1}{x} \, dx = \ln|x|\Big|_2^4 = \ln 4 - \ln 2 = \ln(\frac{4}{2}) = \ln 2.$

31. $\int_0^4 x(x^2 - 1) \, dx = \int_0^4 (x^3 - x) \, dx = \frac{1}{4}x^4 - \frac{1}{2}x^2\Big|_0^4 = 64 - 8 = 56.$

33. $\int_1^3 (t^2 - t)^2 \, dt = \int_1^3 t^4 - 2t^3 + t^2) \, dt = \frac{1}{5}t^5 - \frac{1}{2}t^4 + \frac{1}{3}t^3\Big|_1^3$

$$= \left(\frac{243}{5} - \frac{81}{2} + \frac{27}{3}\right) - \left(\frac{1}{5} - \frac{1}{2} + \frac{1}{3}\right) = \frac{512}{30} = \frac{256}{15}.$$

35. $\int_{-3}^{-1} x^{-2} \, dx = -\frac{1}{x}\Big|_{-3}^{-1} = 1 - \frac{1}{3} = \frac{2}{3}.$

37. $\int_1^4 \left(\sqrt{x} - \frac{1}{\sqrt{x}}\right) dx = \int_1^4 (x^{1/2} - x^{-1/2}) \, dx = \frac{2}{3}x^{3/2} - 2x^{1/2}\Big|_1^4$

$$= \left(\frac{16}{3} - 4\right) - \left(\frac{2}{3} - 2\right) = \frac{8}{3}.$$

39. $\int_1^4 \frac{3x^3 - 2x^2 + 4}{x^2} \, dx = \int_1^4 (3x - 2 + 4x^{-2}) \, dx = \frac{3}{2}x^2 - 2x - \frac{4}{x}\Big|_1^4$

$$= (24 - 8 - 1) - (\frac{3}{2} - 2 - 40) = \frac{39}{2}.$$

41. a. $C(300) - C(0) = \int_0^{300} (0.0003x^2 - 0.12x + 20) \, dx = 0.0001x^3 - 0.06x^2 + 20x\Big|_0^{300}$

$$= 0.0001(300)^3 - 0.06(300)^2 + 20(300) = 3300.$$

b. $\int_{200}^{300} C'(x)\,dx = (0.0001x^3 - 0.06x^2 + 20x)\big|_{200}^{300}$

$$= [0.0001(300)^3 - 0.06(300)^2 + 20(300)]$$
$$-[0.0001(200)^3 - 0.06(200)^2 + 20(200)]$$
$$= 900 \text{ or } \$900.$$

43. a. The profit is $\int_0^{200} (-0.0003x^2 + 0.02x + 20)\,dx + P(0)$

$$= -0.0001x^3 + 0.01x^2 + 20x\big|_0^{200} + P(0)$$
$$= 3600 + P(0) = 3600 - 800, \text{ or } \$2800.$$

b. $\int_{200}^{220} P'(x)\,dx = P(220) - P(200) = -0.0001x^3 + 0.01x^2 + 20x\big|_{200}^{220}$

$$= 219.20, \text{ or } \$219.20.$$

45. The distance is

$$\int_0^{20} v(t)\,dt = \int_0^{20} (-t^2 + 20t + 440)\,dt = -\tfrac{1}{3}t^3 + 10t^2 + 440t\big|_0^{20} \approx 10{,}133\tfrac{1}{3}\text{ ft.}$$

47. The number will be

$$\int (0.00933t^3 + 0.019t^2 - 0.10833t + 1.3467)\,dt$$

$$= 0.0023325t^4 + 0.0063333t^3 - 0.054165t^2 + 1.3467\,t\big|_0^{10} = 37.7,$$

or approximately 37.7 million Americans.

49. The average population over the period in question is

$$A = \tfrac{1}{3}\int \frac{85}{1 + 1.859e^{-0.66t}}\,dt$$

Multiplying the integrand by $e^{0.66t}/e^{0.66t}$ gives

$$A = \frac{85}{3}\int_0^3 \frac{e^{0.66t}}{e^{0.66t} + 1.859}\,dt$$

Let $u = 1.859 + e^{0.66t}$, $du = 0.66e^{0.66t}\,dt$, or $e^{0.66t}\,dt = \dfrac{du}{0.66}$

If $t = 0$, then $u = 2.859$. If $t = 3$, then $u = 9.1017$

Substituting

$$A = \frac{85}{3}\int_{2.859}^{9.1017} \frac{du}{(0.66)u} = \frac{85}{3(0.66)}\ln u\,\Big|_{2.859}^{9.1017}$$

$$= \frac{85}{3(0.66)}(\ln 9.1017 - \ln 2.859) = \frac{85}{3(0.66)}\ln\frac{9.1017}{2.859} \approx 49.712,$$

or approximately 49.7 million people.

51. False. The integrand $f(x) = \frac{1}{x^3}$ is discontinuous at $x = 0$.

53. False. $f(x)$ is not nonnegative on $[0, 2]$.

USING TECHNOLOGY EXERCISES 14.4, page 912

1. 6.1787

3. 0.7873

5. −0.5888

7. 2.7044

9. 3.9973

11. 37.7 million

13. 333, 209

15. 903,213

EXERCISES 14.5, page 920

1. Let $u = x^2 - 1$ so that $du = 2x\, dx$ or $x\, dx = \frac{1}{2}\, du$. Also, if $x = 0$,
    then $u = -1$ and if $x = 2$, then $u = 3$. So
    $$\int_0^2 x(x^2 - 1)^3\, dx = \frac{1}{2}\int_{-1}^3 u^3\, du = \frac{1}{8}u^4\Big|_{-1}^3 = \frac{1}{8}(81) - \frac{1}{8}(1) = 10.$$

3. Let $u = 5x^2 + 4$ so that $du = 10x\, dx$ or $x\, dx = \frac{1}{10}\, du$. Also, if
    $x = 0$, then $u = 4$, and if $x = 1$, then $u = 9$. So
    $$\int_0^1 x\sqrt{5x^2 + 4}\, dx = \frac{1}{10}\int_4^9 u^{1/2}\, du = \frac{1}{15}u^{3/2}\Big|_4^9 = \frac{1}{15}(27) - \frac{1}{15}(8) = \frac{19}{15}.$$

5. Let $u = x^3 + 1$ so that $du = 3x^2\, dx$ or $x^2\, dx = \frac{1}{3}\, du$. Also, if $x = 0$,
    then $u = 1$, and if $x = 2$, then $u = 9$. So,
    $$\int_0^2 x^2(x^3 + 1)^{3/2}\, dx = \frac{1}{3}\int_1^9 u^{3/2}\, du = \frac{2}{15}u^{5/2}\Big|_1^9 = \frac{2}{15}(243) - \frac{2}{15}(1) = \frac{484}{15}.$$

7. Let $u = 2x + 1$ so that $du = 2\, dx$ or $dx = \frac{1}{2}\, du$. Also, if $x = 0$,

511

then $u = 1$ and if $x = 1$ then $u = 3$. So

$$\int_0^1 \frac{1}{\sqrt{2x+1}}\, dx = \frac{1}{2}\int_1^3 \frac{1}{\sqrt{u}}\, du = \frac{1}{2}\int_1^3 u^{-1/2}\, du = u^{1/2}\Big|_1^3 = \sqrt{3} - 1.$$

9. $\int_1^2 (2x-1)^4\, dx$. Put $u = 2x - 1$ so that $du = 2\, dx$ or $dx = \frac{1}{2}\, du$.

Then $\int_1^2 (2x-1)^4\, dx = \frac{1}{2}\int_1^3 u^4\, du = \frac{1}{10} u^5 \Big|_1^3 = \frac{1}{10}(243 - 1) = \frac{121}{5} = 24\frac{1}{5}.$

11-25 odd

11. Let $u = x^3 + 1$ so that $du = 3x^2\, dx$ or $x^2\, dx = \frac{1}{3}\, du$. Also, if $x = -1$,

then $u = 0$ and if $x = 1$, then $u = 2$. So

$$\int_{-1}^1 x^2(x^3+1)^4\, dx = \frac{1}{3}\int_0^2 u^4\, du = \frac{1}{15} u^5 \Big|_0^2 = \frac{32}{15}.$$

13. Let $u = x - 1$ so that $du = dx$. Then if $x = 1$, $u = 0$, and if $x = 5$, then $u = 4$.

$$\int_1^5 x\sqrt{x-1}\, dx = \int_0^4 (u+1)u^{1/2}\, du = \int_0^4 (u^{3/2} + u^{1/2})\, du$$

$$= \frac{2}{5} u^{5/2} + \frac{2}{3} u^{3/2} \Big|_0^4 = \frac{2}{5}(32) + \frac{2}{3}(8) = 18\frac{2}{15}.$$

15. Let $u = x^2$ so that $du = 2x\, dx$ or $x\, dx = \frac{1}{2}\, du$. If $x = 0$, $u = 0$ and if

$x = 2$, $u = 4$. So

$$\int_0^2 xe^{x^2}\, dx = \frac{1}{2}\int_0^4 e^u\, du = \frac{1}{2} e^u \Big|_0^4 = \frac{1}{2}(e^4 - 1).$$

17. $\int_0^1 (e^{2x} + x^2 + 1)\, dx = \frac{1}{2} e^{2x} + \frac{1}{3} x^3 + x \Big|_0^1 = (\frac{1}{2} e^2 + \frac{1}{3} + 1) - \frac{1}{2}$

$$= \frac{1}{2} e^2 + \frac{5}{6}.$$

19. Put $u = x^2 + 1$ so that $du = 2x\, dx$ or $x\, dx = \frac{1}{2}\, du$. Then

$$\int_{-1}^1 xe^{x^2+1}\, dx = \frac{1}{2}\int_2^2 e^u\, du = \frac{1}{2} e^u \Big|_2^2 = 0$$

(Since the upper and lower limits are equal.)

21. Let $u = x - 2$ so that $du = dx$. If $x = 3$, $u = 1$ and if $x = 6$, $u = 4$. So

$$\int_3^6 \frac{2}{x-2}\,dx = 2\int_1^4 \frac{du}{u} = 2\ln|u|\Big|_1^4 = 2\ln 4.$$

23. Let $u = x^3 + 3x^2 - 1$ so that $du = (3x^2 + 6x)dx = 3(x^2 + 2x)dx$. If $x = 1$, $u = 3$, and if $x = 2$, $u = 19$. So

$$\int_1^2 \frac{x^2 + 2x}{x^3 + 3x^2 - 1}\,dx = \frac{1}{3}\int_3^{19} \frac{du}{u} = \frac{1}{3}\ln u\Big|_3^{19} = \frac{1}{3}(\ln 19 - \ln 3).$$

25. $\displaystyle\int_1^2 \left(4e^{2u} - \frac{1}{u}\right) du = 2e^{2u} - \ln u\Big|_1^2 = (2e^4 - \ln 2) - (2e^2 - 0) = 2e^4 - 2e^2 - \ln 2.$

27. $\displaystyle\int_1^2 (2e^{-4x} - x^{-2})\,dx = -\frac{1}{2}e^{-4x} + \frac{1}{x}\Big|_1^2 = (-\frac{1}{2}e^{-8} + \frac{1}{2}) - (-\frac{1}{2}e^{-4} + 1)$

$$= -\frac{1}{2}e^{-8} + \frac{1}{2}e^{-4} - \frac{1}{2} = \frac{1}{2}(e^{-4} - e^{-8} - 1).$$

29. $\displaystyle AV = \frac{1}{2}\int_0^2 (2x+3)\,dx = \frac{1}{2}(x^2 + 3x)\Big|_0^2 = \frac{1}{2}(10) = 5.$

31. $\displaystyle AV = \frac{1}{2}\int_1^3 (2x^2 - 3)\,dx = \frac{1}{2}(\frac{2}{3}x^3 - 3x)\Big|_1^3 = \frac{1}{2}(9 + \frac{7}{3}) = \frac{17}{3}.$

33. $\displaystyle AV = \frac{1}{3}\int_{-1}^2 (x^2 + 2x - 3)\,dx = \frac{1}{3}(\frac{1}{3}x^3 + x^2 - 3x)\Big|_{-1}^2$

$$= \frac{1}{3}[(\frac{8}{3} + 4 - 6) - (-\frac{1}{3} + 1 + 3)] = \frac{1}{3}(\frac{8}{3} - 2 + \frac{1}{3} - 4) = -1.$$

35. $\displaystyle AV = \frac{1}{4}\int_0^4 (2x+1)^{1/2}\,dx = (\frac{1}{4})(\frac{1}{2})(\frac{2}{3})(2x+1)^{3/2}\Big|_0^4 = \frac{1}{12}(27 - 1) = \frac{13}{6}$

37. $\displaystyle AV = \frac{1}{2}\int_0^2 xe^{x^2}\,dx = \frac{1}{4}e^{x^2}\Big|_0^2 = \frac{1}{4}(e^4 - 1).$

39. The amount produced was

$$\int_0^{20} 3.5e^{0.05t} dt = \frac{3.5}{0.05} e^u \Big|_0^{20} \qquad \text{(Use the substitution } u = 0.05t.)$$

$$= 70(e-1) \approx 120.3, \quad \text{or } 120.3 \text{ billion metric tons.}$$

41. The amount is $\int_1^2 t(\frac{1}{2}t^2 + 1)^{1/2} dt$. Let $u = \frac{1}{2}t^2 + 1$, so that $du = t \, dt$. Therefore,

$$\int_1^2 t(\frac{1}{2}t^2 + 1)^{1/2} dt = \int_{3/2}^3 u^{1/2} du = \frac{2}{3} u^{3/2} \Big|_{3/2}^3 = \frac{2}{3}[(3)^{3/2} - (\frac{3}{2})^{3/2}]$$

$$\approx 2.24 \text{ million dollars.}$$

43. The tractor will depreciate

$$\int_0^5 13388.61e^{-0.22314t} dt = \frac{13388.61}{-0.22314} e^{-0.22314t} \Big|_0^5$$

$$= -60,000.94e^{-0.22314t} \Big|_0^5 = -60,000.94(-0.672314)$$

$$= 40,339.47, \quad \text{or } \$40,339.$$

45. $\quad \overline{A} = \frac{1}{5} \int (\frac{1}{12}t^2 + 2t + 44) dt = \frac{1}{5} \left[\frac{1}{36}t^3 + t^2 + 44t \Big|_0^5 \right]$

$$= \frac{1}{5} \left[\frac{125}{36} + 25 + 220 \right] = \frac{125 + 900 + 7920}{5(36)} \approx 49.69, \text{ or } 49.7 \text{ ft/sec.}$$

47. The average whale population will be

$$\frac{1}{10} \int_0^{10} (3t^3 + 2t^2 - 10t + 600) dt = \frac{1}{10}(\frac{3}{4}t^4 + \frac{2}{3}t^3 - 5t^2 + 600t \Big|_0^{10}$$

$$\approx \frac{1}{10}(7500 + 666.67 - 500 + 6000) \approx 1367 \text{ whales.}$$

49. The average yearly sales of the company over its first 5 years of operation is given

by $\quad \frac{1}{5-0} \int_0^5 t(0.2t^2 + 4)^{1/2} dt = \frac{1}{5}[(\frac{5}{2})(\frac{2}{3})(0.2t^2 + 4)^{3/2}] \Big|_0^5 \qquad$ [Let $u = -0.2t^2 + 4$.]

$$= \frac{1}{5}[\frac{5}{3}(5+4)^{3/2} - \frac{5}{3}(4)^{3/2}] = \frac{1}{3}(27-8) = \frac{19}{3}, \text{ or } 6\frac{1}{3} \text{ million dollars.}$$

51. The average velocity is $\frac{1}{4} \int_0^4 3t\sqrt{16-t^2} \, dt = \frac{1}{4}(64) = 16$, or 16 ft/sec.

(Using the results of Exercise 44.)

53. $\int_0^5 p\,dt = \frac{1}{5}\left[18t + \frac{3}{2}e^{-2t} + 18e^{-t/3}\right]_0^5$

$\qquad = \frac{1}{5}\left[18(5) + \frac{3}{2}e^{-10} + 18e^{-5/3} - \frac{3}{2} - 18\right] = 14.78,$ or $\$14.78.$

55. $\int_a^b f(x)\,dx = F(x)\Big|_a^b = F(b) - F(a) = -[F(a) - F(b)]$

$\qquad = -F(x)\Big|_b^a = -\int_b^a f(x)\,dx$

57. $\int_a^b cf(x)\,dx = xF(x)\Big|_a^b = c[F(b) - F(a)] = c\int_a^b f(x)\,dx.$

59. $\int_0^1 (1 + x - e^x)\,dx = x + \frac{1}{2}x^2 - e^x\Big|_0^1 = (1 + \frac{1}{2} - e) + 1 = \frac{5}{2} - e.$

$\qquad \int_0^1 dx + \int_0^1 x\,dx - \int_0^1 e^x\,dx = x\Big|_0^1 + \frac{1}{2}x^2\Big|_0^1 - e^x\Big|_0^1 = (1 - 0) + (\frac{1}{2} - 0) - (e - 1) = \frac{5}{2} - e.$

61. $\int_0^3 (1 + x^3)\,dx = x + \frac{1}{4}x^4\Big|_0^3 = 3 + \frac{81}{4} = \frac{93}{4}.$

$\qquad \int_0^1 (1 + x^3)\,dx + \int_1^2 (1 + x^3)\,dx + \int_2^3 (1 + x^3)\,dx$

$\qquad\qquad = (x + \frac{1}{4}x^4)\Big|_0^1 + (x + \frac{1}{4}x^4)\Big|_1^2 + (x + \frac{1}{4}x^4)\Big|_2^3$

$\qquad\qquad = (1 + \frac{1}{4}) + (2 + 4) - (1 + \frac{1}{4}) + (3 + \frac{81}{4}) - (2 + 4) = \frac{93}{4}.$

63. $\int_3^0 f(x)\,dx = -\int_0^3 f(x)\,dx = -4.$ (Property 2)

65. a. $\int_{-1}^2 [2f(x) + g(x)]\,dx = 2\int_{-1}^2 f(x)\,dx + \int_{-1}^2 g(x)\,dx = 2(-2) + 3 = -1.$

b. $\int_{-1}^2 [g(x) - f(x)]\,dx = \int_{-1}^2 g(x)\,dx - \int_{-1}^2 f(x)\,dx = 3 - (-2) = 5.$

c. $\int_{-1}^2 [2f(x) - 3g(x)]\,dx = 2\int_{-1}^2 f(x)\,dx - 3\int_{-1}^2 g(x)\,dx = 2(-2) - 3(3) = -13.$

67. True. This follows from Property 1 of the definite integral.

69. False. Only a constant can be "moved out" of the integral sign.

14   Integration

71. True. This follows from Properties 3 and 4 of the definite integral.

USING TECHNOLOGY EXERCISES 14.5, page 922

1. 7.71667   3. 17.5649   5. 10,140   7. 60.5/day

EXERCISES 14.6, page 932

1. $-\int_0^6 (x^3 - 6x^2)\,dx = -\frac{1}{4}x^4 + 2x^3\Big|_0^6 = -\frac{1}{4}(6^4) + 2(6^3) = 108$ sq units.

3. $A = -\int_{-1}^0 x\sqrt{1-x^2}\,dx + \int_0^1 x\sqrt{1-x^2}\,dx = 2\int_0^1 x(1-x^2)^{1/2}\,dx$ (by symmetry). Let

   $u = 1 - x^2$ so that $du = -2x\,dx$ or $x\,dx = -\frac{1}{2}\,du$. Also, if $x = 0$, then $u = 1$ and

   if $x = 1$, $u = 0$. So $A = (2)(-\frac{1}{2})\int_0^1 u^{1/2}\,du = -\frac{2}{3}u^{3/2}\Big|_1^0 = \frac{2}{3}$, or $\frac{2}{3}$ sq units.

   5- 25 odd

5. $A = -\int_0^4 (x - 2\sqrt{x})\,dx = \int_0^4 (-x + 2x^{1/2})\,dx = -\frac{1}{2}x^2 + \frac{4}{3}x^{3/2}\Big|_0^4$

   $= 8 + \frac{32}{3} = \frac{8}{3}$ sq units.

7. The required area is given by

   $\int_{-1}^0 (x^2 - x^{1/3})\,dx + \int_0^1 (x^{1/3} - x^2)\,dx = \frac{1}{3}x^3 - \frac{3}{4}x^{4/3}\Big|_{-1}^0 + \frac{3}{4}x^{4/3} - \frac{1}{3}x^3\Big|_0^1$

   $= -(-\frac{1}{3} - \frac{3}{4}) + (\frac{3}{4} - \frac{1}{3}) = 1\frac{1}{2}$ sq units.

9. The required area is given by

   $-\int_{-1}^2 -x^2\,dx = \frac{1}{3}x^3\Big|_{-1}^2 = \frac{8}{3} + \frac{1}{3} = 3$ sq units.

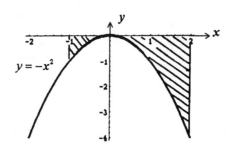

$y = -x^2$

11. $y = x^2 - 5x + 4 = (x - 4)(x - 1) = 0$
    if $x = 1$ or 4. These give the x-intercepts.

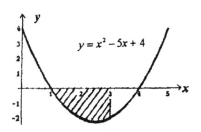

$$A = -\int_1^3 (x^2 - 5x + 4)\, dx = -\frac{1}{3}x^3 + \frac{5}{2}x^2 - 4x\Big|_1^3$$

$$= (-9 + \frac{45}{2} - 12) - (-\frac{1}{3} + \frac{5}{2} - 4) = \frac{10}{3} = 3\frac{1}{3}.$$

13. The required area is given by

$$-\int_0^9 -(1 + \sqrt{x})\, dx = x + \frac{2}{3}x^{3/2}\Big|_0^9 = 9 + 18 = 27.$$

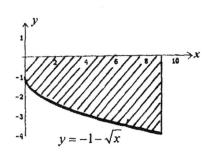

15. $-\int_{-2}^4 -e^{(1/2)x}\, dx = 2e^{(1/2)x}\Big|_{-2}^4$

$$= 2(e^2 - e^{-1})\, \text{sq units.}$$

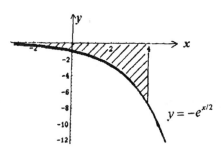

17. $A = \int_1^3 [(x^2 + 3) - 1]\, dx$

$$= \int_1^3 (x^2 + 2)\, dx = \frac{1}{3}x^3 + 2x\Big|_1^3$$

$$= (9 + 6) - (\frac{1}{3} + 2) = \frac{38}{3}.$$

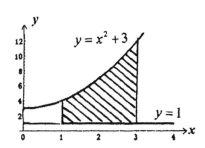

19. $A = \int_0^2 (-x^2 + 2x + 3 + x - 3)\, dx$

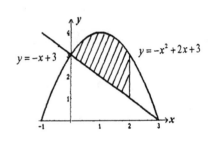

$= \int_0^2 (-x^2 + 3x)\, dx$

$= -\frac{1}{3}x^3 + \frac{3}{2}x^2 \Big|_0^2 = -\frac{1}{3}(8) + \frac{3}{2}(4)$

$= 6 - \frac{8}{3} = \frac{10}{3}$ sq units

21. $A = \int_{-1}^2 [(x^2 + 1) - \frac{1}{3}x^3]\, dx$

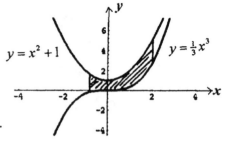

$= \int_{-1}^2 (-\frac{1}{3}x^3 + x^2 + 1)\, dx$

$= -\frac{1}{12}x^4 + \frac{1}{3}x^3 + x \Big|_{-1}^2$

$= (-\frac{4}{3} + \frac{8}{3} + 2) - (-\frac{1}{12} - \frac{1}{3} - 1) = 4\frac{3}{4}$ sq units.

23. $A = \int_1^4 \left[(2x - 1) - \frac{1}{x} \right] dx = \int_1^4 \left(2x - 1 - \frac{1}{x} \right) dx$

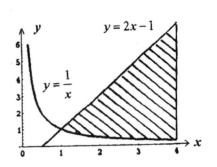

$= (x^2 - x - \ln x) \Big|_1^4$

$= (16 - 4 - \ln 4) - (1 - 1 - \ln 1)$

$= 12 - \ln 4 \approx 10.6$ sq units.

25. $A = \int_1^2 \left(e^x - \frac{1}{x} \right) dx = e^x - \ln x \Big|_1^2$

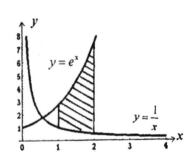

$= (e^2 - \ln 2) - e = (e^2 - e - \ln 2)$ sq units.

27.

$$A = -\int_{-1}^{0} x\, dx + \int_{0}^{2} x\, dx$$

$$= -\tfrac{1}{2}x^2\Big|_{-1}^{0} + \tfrac{1}{2}x^2\Big|_{0}^{2}$$

$$= \tfrac{1}{2} + 2 = 2\tfrac{1}{2} \text{ sq units.}$$

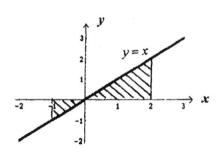

29. The x–intercepts are found by solving

$x^2 - 4x + 3 = (x - 3)(x - 1) = 0$ giving $x = 1$
or 3. The region is shown in the figure.

$$A = -\int_{-1}^{1} [(-x^2 + 4x - 3)\, dx + \int_{1}^{2} (-x^2 + 4x - 3)\, dx$$

$$= \tfrac{1}{3}x^3 - 2x^2 + 3x\Big|_{-1}^{1} + (-\tfrac{1}{3}x^3 + 2x^2 - 3x)\Big|_{1}^{2}$$

$$= (\tfrac{1}{3} - 2 + 3) - (-\tfrac{1}{3} - 2 - 3)$$

$$+ (-\tfrac{8}{3} + 8 - 6) - (-\tfrac{1}{3} + 2 - 3) = \tfrac{22}{3} \text{ sq units.}$$

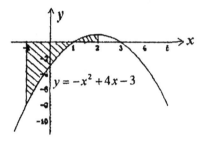

31. The region is shown in the figure at the right.

$$A = \int_{0}^{1} (x^3 - 4x^2 + 3x)\, dx - \int_{1}^{2} (x^3 - 4x^2 + 3x)\, dx$$

$$= (\tfrac{1}{4}x^4 - \tfrac{4}{3}x^3 + \tfrac{3}{2}x^2)\Big|_{0}^{1}$$

$$-(\tfrac{1}{4}x^4 - \tfrac{4}{3}x^3 + \tfrac{3}{2}x^2)\Big|_{1}^{2} = \tfrac{3}{2} \text{ sq units.}$$

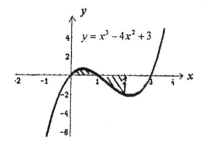

33. The region is shown in the figure at the right.

$$A = -\int_{-1}^{0}(e^x - 1)\,dx + \int_{0}^{3}(e^x - 1)\,dx$$

$$= (-e^x + x)\Big|_{-1}^{0} + (e^x - x)\Big|_{0}^{3}$$

$$= -1 - (-e^{-1} - 1) + (e^3 - 3) - 1$$

$$= e^3 - 4 + \tfrac{1}{e} \approx 16.5 \quad \text{sq units.}$$

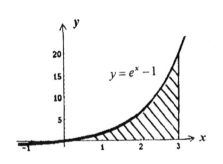

35. To find the points of intersection of the two curves, we solve the equation
$$x^2 - 4 = x + 2$$
$$x^2 - x - 6 = (x - 3)(x + 2) = 0, \text{ obtaining}$$
$x = -2$ or $x = 3$. The region is shown in the figure at the right.

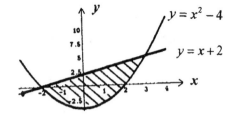

$$A = \int_{-2}^{3}[(x+2) - (x^2 - 4)]\,dx = \int_{-2}^{3}(-x^2 + x + 6)\,dx = (-\tfrac{1}{3}x^3 + \tfrac{1}{2}x^2 + 6x)\Big|_{-2}^{3}$$
$$= (-9 + \tfrac{9}{2} + 18) - (\tfrac{8}{3} + 2 - 12) = \tfrac{125}{6}\text{ sq units.}$$

37. To find the points of intersection of the two curves, we solve the equation $x^3 = x^2$ or $x^3 - x^2 = x^2(x - 1) = 0$ giving $x = 0$ or 1. The region is shown in the figure.

$$A = -\int_{0}^{1}(x^2 - x^3)\,dx$$

$$= (\tfrac{1}{3}x^3 - \tfrac{1}{4}x^4)\Big|_{0}^{1} = \tfrac{1}{3} - \tfrac{1}{4} = \tfrac{1}{12}\text{ sq units}$$

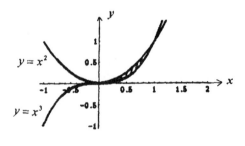

39. To find the points of intersection of the two curves,

we solve the equation
$$x^3 - 6x^2 + 9x = x^2 - 3x,$$
or $x^3 - 7x^2 + 12x = x(x - 4)(x - 3) = 0$
obtaining $x = 0$, 3, or 4.

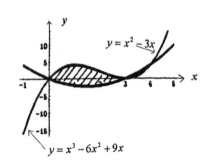

$$A = \int_0^3 [(x^3 - 6x^2 + 9x) - (x^2 + 3x)]\,dx$$

$$+ \int_3^4 [(x^2 - 3x) - (x^3 - 6x^2 + 9x)]\,dx$$

$$= \int_0^3 (x^3 - 7x^2 + 12x)\,dx - \int_3^4 (x^3 - 7x^2 + 12x)\,dx$$

$$= (\tfrac{1}{4}x^4 - \tfrac{7}{3}x^3 + 6x^2)\Big|_0^3 - (\tfrac{1}{4}x^4 - \tfrac{7}{3}x^3 + 6x^2)\Big|_3^4$$

$$= (\tfrac{81}{4} - 63 + 54) - (64 - \tfrac{448}{3} + 96) + (\tfrac{81}{4} - 63 + 54) = \tfrac{71}{6}.$$

41. By symmetry, $A = 2\int_0^3 x(9 - x^2)^{1/2}\,dx$. We integrate by substitution with

$u = 9 - x^2$, $du = -2x\,dx$. If $x = 0, u = 9$, and if $x = 3, u = 0$. So

$$A = 2\int_9^0 -\tfrac{1}{2}u^{1/2}\,du = -\int_9^0 u^{1/2}\,du = -\tfrac{2}{3}u^{3/2}\Big|_9^0 = \tfrac{2}{3}(9)^{3/2} = 18 \text{ sq units.}$$

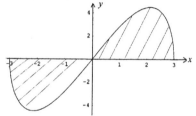

43. S gives the additional revenue that the company would realize if it used a different advertising agency. $S = \int_0^b [g(x) - f(x)]\,dx.$

45. Shortfall $= \int_{2010}^{2050} [f(t) - g(t)]\,dt$

47. a. $\int_{T_1}^{T} [g(t) - f(t)]\,dt - \int_0^{T_1} [f(t) - g(t)]\,dt = A_2 - A_1.$

    b. The number $A_2 - A_1$ gives the distance car 2 is ahead of car 1 after t seconds.

49. The turbo-charged model is moving at

$$A = \int_0^{10} [(4 + 1.2t + 0.03t^2) - (4 + 0.8t)]\,dt$$

$$= \int_0^{10} (0.4t + 0.03t^2)\,dt = (0.2t^2 + 0.1t^3)\Big|_0^{10}$$

$= 20 + 10,$ or 30 ft/sec faster than the standard model.

51. The additional number of cars will be given by

$$\int_0^5 (5e^{0.3t} - 5 - 0.5t^{3/2})\,dt = \frac{5}{0.3}e^{0.3t} - 5t - 0.2t^{5/2}\Big|_0^5$$

$$= \frac{5}{0.3}e^{1.5} - 25 - 0.2(5)^{5/2} - \frac{5}{0.3} = 74.695 - 25 - 0.2(5)^{5/2} - \frac{50}{3}$$

$\approx 21.85,$ or 21,850 cars. (Remember t is measured in thousands.)

53. True. If $f(x) \ge g(x)$ on $[a, b]$, then the area of the said region is

$$\int_a^b [f(x) - g(x)]\,dx = \int_a^b |f(x) - g(x)|\,dx$$

If $f(x) \le g(x)$ on $[a, b]$, then the area of the region is

$$\int_a^b [g(x) - f(x)]\,dx = \int_a^b -[f(x) - g(x)]\,dx = \int_a^b |f(x) - g(x)|\,dx$$

55. The area of R' is

$$A = \int_a^b \{[f(x) + C] - [g(x) + C]\}\,dx = \int_a^b [f(x) + C - g(x) - C]\,dx$$

$$= \int_a^b [f(x) - g(x)]\,dx$$

USING TECHNOLOGY EXERCISES 14.6, page 936

1. a.

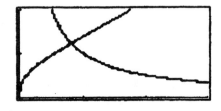

    b. 1074.2857

3. a.

    b. 0.9961 sq units

5. a.

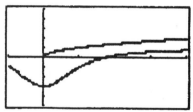

    b. 5.4603 sq units

7. a.

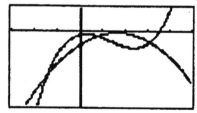

    b. 25.8549 sq units

9. a.

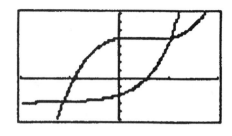

    b. 10.5144 sq units

11. a.

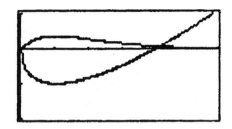

    b. 3.5799 sq units

13. 207.43 sq units

EXERCISES 14.7, page 949   1-17 odd

1. When $p = 4$, $-0.01x^2 - 0.1x + 6 = 4$ or $x^2 + 10x - 200 = 0$, $(x - 10)(x + 20) = 0$
    and $x = 10$ or -20. We reject the root $x = -20$. The consumers' surplus is
    $$CS = \int_0^{10}(-0.01x^2 - 0.1x + 6)\,dx - (4)(10)$$
    $$= -\frac{0.01}{3}x^3 - 0.05x^2 + 6x\Big|_0^{10} - 40 \approx 11.667 \text{, or } \$11,667.$$

3. Setting $p = 10$, we have $\sqrt{225 - 5x} = 10$, $225 - 5x = 100$, or $x = 25$.
    Then $CS = \int_0^{25}\sqrt{225 - 5x}\,dx - (10)(25) = \int_0^{25}(225 - 5x)^{1/2}\,dx - 250$.
    To evaluate the integral, let $u = 225 - 5x$ so that $du = -5\,dx$ or
    $dx = -\frac{1}{5}\,du$. If $x = 0$, $u = 225$ and if $x = 25$, $u = 100$. So

$$CS = -\frac{1}{5}\int_{225}^{100} u^{1/2}\, du - 250 = -\frac{2}{15}u^{3/2}\Big|_{225}^{100} - 250$$
$$= -\frac{2}{15}(1000 - 3375) - 250 = 66.667, \text{ or } \$6,667.$$

5. To find the equilibrium point, we solve
$$0.01x^2 + 0.1x + 3 = -0.01x^2 - 0.2x + 8, \quad \text{or } 0.02x^2 + 0.3x - 5 = 0,$$
$$2x^2 + 30x - 500 = (2x - 20)(x + 25) = 0$$
obtaining $x = -25$ or 10. So the equilibrium point is $(10,5)$. Then
$$PS = (5)(10) - \int_0^{10} (0.01x^2 + 0.1x + 3)\, dx$$
$$= 50 - (\frac{0.01}{3}x^3 + 0.05x^2 + 3x)\Big|_0^{10} = 50 - \frac{10}{3} - 5 - 30 = \frac{35}{3},$$
or approximately $\$11,667$.

7. To find the market equilibrium, we solve
$$-0.2x^2 + 80 = 0.1x^2 + x + 40, \quad 0.3x^2 + x - 40 = 0,$$
$$3x^2 + 10x - 400 = 0, \ (3x + 40)(x - 10) = 0$$
giving $x = -\frac{40}{3}$ or $x = 10$. We reject the negative root. The corresponding
equilibrium price is $\$60$. The consumers' surplus is
$$CS = \int_0^{10} (-0.2x^2 + 80)dx - (60)(10) = -\frac{0.2}{3}x^3 + 80x\Big|_0^{10} - 600 = 133\tfrac{1}{3},$$
or $\$13,333$. The producers' surplus is
$$PS = 600 - \int_0^{10} (0.1x^2 + x + 40)dx = 600 - [\frac{0.1}{3}x^3 + \frac{1}{2}x^2 + 40x]\Big|_0^{10}$$
$$= 116\tfrac{2}{3}, \text{ or } \$11,667.$$

9. Here $P = 200,000$, $r = 0.08$, and $T = 5$. So
$$PV = \int_0^5 200,000e^{-0.08t}\, dt = -\frac{200,000}{0.08}e^{-0.08t}\Big|_0^5 = -2,500,000(e^{-0.4} - 1)$$
$$\approx 824,199.85, \text{ or } \$824,200.$$

11. Here $P = 250$, $m = 12$, $T = 20$, and $r = 0.08$. So
$$A = \frac{mP}{r}(e^{rT} - 1) = \frac{12(250)}{0.08}(e^{1.6} - 1) \approx 148,238.70$$
or approximately $\$148,239$.

13. Here $P = 150$, $m = 12$, $T = 15$, and $r = 0.08$. So
$$A = \frac{12(150)}{0.08}(e^{1.2} - 1) \approx 52{,}202.60, \text{ or approximately } \$52{,}203.$$

15. Here $P = 2000$, $m = 1$, $T = 15.75$, and $r = 0.1$. So
$$A = \frac{1(2000)}{0.1}(e^{1.575} - 1) \approx 76{,}615, \text{ or approximately } \$76{,}615.$$

17. Here $P = 1200$, $m = 12$, $T = 15$, and $r = 0.1$. So
$$PV = \frac{12(1200)}{0.1}(1 - e^{-1.5}) \approx 111{,}869, \text{ or approximately } \$111{,}869.$$

19. We want the present value of an annuity with $P = 300$, $m = 12$, $T = 10$, and $r = 0.12$. So
$$PV = \frac{12(300)}{0.12}(1 - e^{-1.2}) \approx 20{,}964, \text{ or approximately } \$20{,}964.$$

21. a.

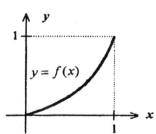

b. $f(0.4) = \frac{15}{16}(0.16) + \frac{1}{16}(0.4) \approx 0.175$; $f(0.9) = \frac{15}{16}(0.81) + \frac{1}{16}(0.9) \approx 0.816$.

So, the lowest 40 percent of the people receive 17.5 percent of the total income and the lowest 90 percent of the people receive 81.6 percent of the income.

23. a.

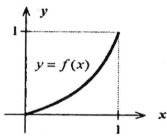

b. $f(0.3) = \frac{14}{15}(0.09) + \frac{1}{15}(0.3) = 0.104$

$$f(0.7) = \tfrac{14}{15}(0.49) + \tfrac{1}{15}(0.7) \approx 0.504.$$

USING TECHNOLOGY EXERCISES 14.7, page 952

1. Consumer's surplus: $18,000,000; producer's surplus: $11,700,000.

3. Consumer's surplus: $33,120; producer's surplus: $2,880.

5. Investment A

CHAPTER 14 REVIEW EXERCISES, page 954

1. $\displaystyle\int (x^3 + 2x^2 - x)\,dx = \tfrac{1}{4}x^4 + \tfrac{2}{3}x^3 - \tfrac{1}{2}x^2 + C.$

3. $\displaystyle\int \left(x^4 - 2x^3 + \frac{1}{x^2}\right) dx = \frac{x^5}{5} - \frac{1}{2}x^4 - \frac{1}{x} + C$

5. $\displaystyle\int x(2x^2 + x^{1/2})\,dx = \int (2x^3 + x^{3/2})\,dx = \tfrac{1}{2}x^4 + \tfrac{2}{5}x^{5/2} + C.$

7. $\displaystyle\int (x^2 - x + \tfrac{2}{x} + 5)\,dx = \int x^2\,dx - \int x\,dx + 2\int \tfrac{dx}{x} + 5\int dx$
$$= \tfrac{1}{3}x^3 - \tfrac{1}{2}x^2 + 2\ln|x| + 5x + C.$$

9. Let $u = 3x^2 - 2x + 1$ so that $du = (6x - 2)\,dx = 2(3x - 1)\,dx$ or $(3x - 1)\,dx = \tfrac{1}{2}\,du.$
So $\displaystyle\int (3x - 1)(3x^2 - 2x + 1)^{1/3}\,dx = \tfrac{1}{2}\int u^{1/3}\,du = \tfrac{3}{8}u^{4/3} + C = \tfrac{3}{8}(3x^2 - 2x + 1)^{4/3} + C.$

11. Let $u = x^2 - 2x + 5$ so that $du = 2(x - 1)\,dx$ or $(x - 1)\,dx = \tfrac{1}{2}\,du.$
$$\int \frac{x - 1}{x^2 - 2x + 5}\,dx = \frac{1}{2}\int \frac{du}{u} = \frac{1}{2}\ln|u| + C = \frac{1}{2}\ln(x^2 - 2x + 5) + C.$$

13. Put $u = x^2 + x + 1$ so that $du = (2x + 1)\,dx = 2(x + \tfrac{1}{2})\,dx$ and $(x + \tfrac{1}{2})\,dx = \tfrac{1}{2}\,du.$
$$\int (x + \tfrac{1}{2})e^{x^2 + x + 1}\,dx = \tfrac{1}{2}\int e^u\,du = \tfrac{1}{2}e^u + C = \tfrac{1}{2}e^{x^2 + x + 1} + C.$$

15. Let $u = \ln x$ so that $du = \frac{1}{x}\,dx$. Then

$$\int \frac{(\ln x)^5}{x}\,dx = \int u^5\,du = \frac{1}{6}u^6 + C = \frac{1}{6}(\ln x)^6 + C.$$

17. Let $u = x^2 + 1$ so that $du = 2x\,dx$ or $x\,dx = \frac{1}{2}\,du$. Then

$$\int x^3(x^2+1)^{10}\,dx = \frac{1}{2}\int (u-1)u^{10}\,du \qquad (x^2 = u - 1)$$

$$= \frac{1}{2}\int (u^{11} - u^{10})\,du = \frac{1}{2}(\frac{1}{12}u^{12} - \frac{1}{11}u^{11}) + C$$

$$= \frac{1}{264}u^{11}(11u - 12) + C = \frac{1}{264}(x^2+1)^{11}(11x^2 - 1) + C.$$

19. Put $u = x - 2$ so that $du = dx$. Then $x = u + 2$ and

$$\int \frac{x}{\sqrt{x-2}}\,dx = \int \frac{u+2}{\sqrt{u}}\,du = \int (u^{1/2} + 2u^{-1/2})\,du = \int u^{1/2}\,du + 2\int u^{-1/2}\,du$$

$$= \frac{2}{3}u^{3/2} + 4u^{1/2} + C = \frac{2}{3}u^{1/2}(u+6) + C = \frac{2}{3}\sqrt{x-2}(x-2+6) + C$$

$$= \frac{2}{3}(x+4)\sqrt{x-2} + C.$$

21. $\displaystyle\int_0^1 (2x^3 - 3x^2 + 1)\,dx = \frac{1}{2}x^4 - x^3 + x\Big|_0^1 = \frac{1}{2} - 1 + 1 = \frac{1}{2}.$

23. $\displaystyle\int_1^4 (x^{1/2} + x^{-3/2})\,dx = \frac{2}{3}x^{3/2} - 2x^{-1/2}\Big|_1^4 = \frac{2}{3}x^{3/2} - \frac{2}{\sqrt{x}}\Big|_1^4 = (\frac{16}{3} - 1) - (\frac{2}{3} - 2) = \frac{17}{3}.$

25. Put $u = x^3 - 3x^2 + 1$ so that $du = (3x^2 - 6x)\,dx = 3(x^2 - 2x)\,dx$ or $(x^2 - 2x)\,dx = \frac{1}{3}\,du$. Then if $x = -1$, $u = -3$, and if $x = 0$, $u = 1$,

$$\int_{-1}^0 12(x^2 - 2x)(x^3 - 3x^2 + 1)^3\,dx = (12)(\frac{1}{3})\int_{-3}^1 u^3\,du = 4(\frac{1}{4})u^4\Big|_{-3}^1$$

$$= 1 - 81 = -80.$$

27. Let $u = x^2 + 1$ so that $du = 2x\,dx$ or $x\,dx = \frac{1}{2}\,du$. Then, if $x = 0$, $u = 1$, and if $x = 2$, $u = 5$, so

$$\int_0^2 \frac{x}{x^2+1}\,dx = \frac{1}{2}\int_1^5 \frac{du}{u} = \frac{1}{2}\ln u\Big|_1^5 = \frac{1}{2}\ln 5.$$

527

29. Let $u = 1 + 2x^2$ so that $du = 4x\,dx$ or $x\,dx = \frac{1}{4}\,du$. If $x = 0$, then $u = 1$ and if $x = 2$, then $u = 9$.

$$\int_0^2 \frac{4x}{\sqrt{1+2x^2}}\,dx = \int_1^9 \frac{du}{u^{1/2}} = 2u^{1/2}\Big|_1^9 = 2(3-1) = 4.$$

31. Let $u = 1 + e^{-x}$ so that $du = -e^{-x}\,dx$ and $e^{-x}\,dx = -\,du$. Then

$$\int_{-1}^0 \frac{e^{-x}}{(1+e^{-x})^2}\,dx = -\int_{1+e}^2 \frac{du}{u^2} = \frac{1}{u}\Big|_{1+e}^2 = \frac{1}{2} - \frac{1}{1+e} = \frac{e-1}{2(1+e)}.$$

33. $f(x) = \int f'(x)\,dx = \int (3x^2 - 4x + 1)\,dx = 3\int x^2\,dx - 4\int x\,dx + \int dx$

$$= x^3 - 2x^2 + x + C.$$

The given condition implies that $f(1) = 1$ or $1 - 2 + 1 + C = 1$, and $C = 1$.
Therefore, the required function is $f(x) = x^3 - 2x^2 + x + 1$.

35. $f(x) = \int f'(x)\,dx = \int (1 - e^{-x})\,dx = x + e^{-x} + C$, $f(0) = 2$ implies $0 + 1 + C = 2$ or $C = 1$. So $f(x) = x + e^{-x} + 1$.

37. $\Delta x = \frac{2-1}{5} = \frac{1}{5}$; $x_1 = \frac{6}{5}$, $x_2 = \frac{7}{5}$, $x_3 = \frac{8}{5}$, $x_4 = \frac{9}{5}$, $x_5 = \frac{10}{5}$. The Riemann sum is

$$f(x_1)\Delta x + \cdots + f(x_5)\Delta x = \left\{\left[-2\left(\tfrac{6}{5}\right)^2 + 1\right] + \left[-2\left(\tfrac{7}{5}\right)^2 + 1\right] + \cdots + \left[-2\left(\tfrac{10}{5}\right)^2 + 1\right]\right\}\left(\tfrac{1}{5}\right)$$
$$= \tfrac{1}{5}(-1.88 - 2.92 - 4.12 - 5.48 - 7) = -4.28.$$

39. a. $R(x) = \int R'(x)\,dx = \int (-0.03x + 60)\,dx = -0.015x^2 + 60x + C$.

  $R(0) = 0$ implies that $C = 0$. So, $R(x) = -0.015x^2 + 60x$.

  b. From $R(x) = px$, we have $-0.015x^2 + 60x = px$ or $p = -0.015x + 60$.

41. The total number of systems that Vista may expect to sell t months from the time they are put on the market is given by $f(t) = 3000t - 50,000(1 - e^{-0.04t})$.

  The number is $\displaystyle\int_0^{12} (3000 - 2000e^{-0.04t})\,dt = \left(3000t - \frac{2000}{-0.04}e^{-0.04t}\right)\Big|_0^{12}$

  $$= 3000(12) + 50,000e^{-0.48} - 50,000 = 16,939.$$

43. $C(x) = \int C'(x)\,dx = \int (0.00003x^2 - 0.03x + 10)\,dx$

$\qquad = 0.00001x^3 - 0.015x^2 + 10x + k.$
But $C(0) = 600$ and this implies that $k = 600$. Therefore,
$\qquad C(x) = 0.00001x^3 - 0.015x^2 + 10x + 600.$
The total cost incurred in producing the first 500 corn poppers is
$\qquad C(500) = 0.00001(500)^3 - 0.015(500)^2 + 10(500) + 600$
$\qquad\qquad = 3{,}100$, or $\$3{,}100$.

45. $A = \int_{-1}^{2}(3x^2 + 2x + 1)\,dx = x^3 + x^2 + x\big|_{-1}^{2} = [2^3 + 2^2 + 2] - [(-1)^3 + 1 - 1]$
$\qquad = 14 - (-1) = 15$ sq units.

47. $A = \int_{1}^{3}\dfrac{1}{x^2}\,dx = \int_{1}^{3} x^{-2}\,dx = -\dfrac{1}{x}\bigg|_{1}^{3} = -\dfrac{1}{3} + 1 = \dfrac{2}{3}$ sq units

49.

$A = \int_{a}^{b}[f(x) - g(x)]\,dx$

$\quad = \int_{0}^{2}(e^x - x)\,dx$

$\quad = \left(e^x - \dfrac{1}{2}x^2\right)\bigg|_{0}^{2}$

$\quad = (e^2 - 2) - (1 - 0) = e^2 - 3$ sq units

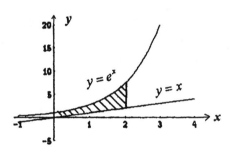

51. $A = \int_{0}^{1}(x^3 - 3x^2 + 2x)\,dx - \int_{1}^{2}(x^3 - 3x^2 + 2x)\,dx$

$\quad = \dfrac{x^4}{4} - x^3 + x^2\Big|_{0}^{1} - \left(\dfrac{x^4}{4} - x^3 + x^2\right)\Big|_{1}^{2}$

$\quad = \tfrac{1}{4} - 1 + 1 - [(4 - 8 + 4) - (\tfrac{1}{4} - 1 + 1)]$

$\quad = \tfrac{1}{4} + \tfrac{1}{4} = \tfrac{1}{2}$ sq units

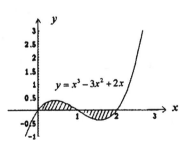

53.

$A = \dfrac{1}{3}\int_{0}^{3}\dfrac{x}{\sqrt{x^2 + 16}}\,dx = \dfrac{1}{3}\cdot\dfrac{1}{2}\cdot 2(x^2 + 16)^{1/2}\Big|_{0}^{3}$

$$= \frac{1}{3}(x^2 + 16)^{1/2}\Big|_0^3 = \frac{1}{3}(5 - 4) = \frac{1}{3} \text{ sq units.}$$

55. To find the equilibrium point, we solve $0.1x^2 + 2x + 20 = -0.1x^2 - x + 40$
$0.2x^2 + 3x - 20 = 0$, $x^2 + 15x - 100 = 0$, $(x + 20)(x - 5) = 0$, or $x = 5$.
Therefore, $p = -0.1(25) - 5 + 40 = 32.5$.
$$CS = \int_0^5 (-0.1x^2 - x + 40)\,dx - (5)(32.5) = -\frac{0.1}{3}x^3 - \frac{1}{2}x^2 + 40x\Big|_0^5 - 162.5$$
$$= 20.833, \text{ or } \$2083.$$

$$PS = (5)(32.5) - \int_0^5 (0.1x^2 + 2x + 20)\,dx = 162.5 - \tfrac{0.1}{3}x^3 + x^2 + 20x)\Big|_0^5$$
$$= 33.333, \text{ or } \$3,333.$$

57. Use Equation (18) with $P = 925$, $m = 12$, $T = 30$, and $r = 0.12$, obtaining
$$PV = \frac{mP}{r}(1 - e^{-rT}) = \frac{(12)(925)}{(0.12)}(1 - e^{-0.12(30)}) = 89972.56,$$
and we conclude that the present value of the purchase price of the house is
$89,972.56 + \$9000$, or $98,972.56.

59. a.

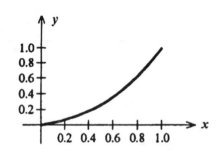

b. $f(0.3) = \tfrac{17}{18}(0.3)^2 + \tfrac{1}{18}(0.3) \approx 0.1$ so that 30 percent of the people receive 10
percent of the total income. $f(0.6) = \tfrac{17}{18}(0.6)^2 + \tfrac{1}{18}(0.6) \approx 0.37$ so that 60 percent of
the people receive 37 percent of the total revenue.
c. The coefficient of inequality for this curve is
$$L = 2\int_0^1 [x - \tfrac{17}{18}x^2 - \tfrac{1}{18}x]\,dx = \tfrac{17}{9}\int_0^1 (x - x^2)\,dx = \tfrac{17}{9}\left(\tfrac{1}{2}x^2 - \tfrac{1}{3}x^3\right)\Big|_0^1$$
$$= \tfrac{17}{54} \approx 0.315.$$

CHAPTER 15

EXERCISES 15.1, page 963

1. $I = \int xe^{2x}\, dx$. Let $u = x$ and $dv = e^{2x}\, dx$. Then $du = dx$ and $v = \frac{1}{2}e^{2x}$. Therefore,

$I = uv - \int v\, du = \frac{1}{2}xe^{2x} - \int \frac{1}{2}e^{2x}\, dx = \frac{1}{2}xe^{2x} - \frac{1}{4}e^{2x} = \frac{1}{4}e^{2x}(2x - 1) + C.$

3. $I = \int xe^{x/4}\, dx$. Let $u = x$ and $dv = e^{x/4}\, dx$. Then $du = dx$ and $v = 4e^{x/4}$.

$\int xe^{x/4}\, dx = uv - \int v\, du = 4xe^{x/4} - 4\int e^{x/4}\, dx = 4xe^{x/4} - 16e^{x/4} + C$

$= 4(x - 4)e^{x/4} + C.$

5. $\int (e^x - x)^2\, dx = \int (e^{2x} - 2xe^x + x^2)\, dx = \int e^{2x}\, dx - 2\int xe^x\, dx + \int x^2\, dx.$

Using the result $\int xe^x\, dx = (x - 1)e^x + k$, from Example 1, we see that

$\int (e^x - x)^2\, dx = \frac{1}{2}e^{2x} - 2(x - 1)e^x + \frac{1}{3}x^3 + C.$

7. $I = \int (x + 1)e^x\, dx$. Let $u = x + 1$, $dv = e^x\, dx$. Then $du = dx$ and $v = e^x$. Therefore,

$I = (x + 1)e^x - \int e^x\, dx = (x + 1)e^x - e^x + C = xe^x + C.$

9. Let $u = x$ and $dv = (x + 1)^{-3/2}\, dx$. Then $du = dx$ and $v = -2(x + 1)^{-1/2}$.

$\int x(x + 1)^{-3/2}\, dx = uv - \int v\, du = -2x(x + 1)^{-1/2} + 2\int (x + 1)^{-1/2}\, dx$

$= -2x(x + 1)^{-1/2} + 4(x + 1)^{1/2} + C$

$= 2(x + 1)^{-1/2}[-x + 2(x + 1)] + C = \dfrac{2(x + 2)}{\sqrt{x + 1}} + C.$

11. $I = \int x(x - 5)^{1/2}\, dx$. Let $u = x$ and $dv = (x - 5)^{1/2}\, dx$. Then $du = dx$ and

$v = \frac{2}{3}(x - 5)^{3/2}$. Therefore,

$$I = \tfrac{2}{3}x(x-5)^{3/2} - \int \tfrac{2}{3}(x-5)^{3/2}\,dx = \tfrac{2}{3}x(x-5)^{3/2} - \tfrac{2}{3}\cdot\tfrac{2}{5}(x-5)^{5/2} + C$$
$$= \tfrac{2}{3}(x-5)^{3/2}[x - \tfrac{2}{5}(x-5)] + C = \tfrac{2}{15}(x-5)^{3/2}(5x - 2x + 10) + C$$
$$= \tfrac{2}{15}(x-5)^{3/2}(3x + 10) + C.$$

13. $I = \int x \ln 2x\, dx$. Let $u = \ln 2x$ and $dv = x\, dx$. Then $du = \tfrac{1}{x}\, dx$ and $v = \tfrac{1}{2}x^2$.

Therefore, $I = \tfrac{1}{2}x^2 \ln 2x - \int \tfrac{1}{2}x\, dx = \tfrac{1}{2}x^2 \ln 2x - \tfrac{1}{4}x^2 + C = \tfrac{1}{4}x^2(2\ln 2x - 1) + C.$

15. Let $u = \ln x$ and $dv = x^3\, dx$, then $du = \tfrac{1}{x}\, dx$, and $v = \tfrac{1}{4}x^4$.

$$\int x^3 \ln x\, dx = \tfrac{1}{4}x^4 \ln x - \tfrac{1}{4}\int x^3\, dx = \tfrac{1}{4}x^4 \ln x - \tfrac{1}{16}x^4 + C$$
$$= \tfrac{1}{16}x^4(4\ln x - 1) + C.$$

17. Let $u = \ln x^{1/2}$ and $dv = x^{1/2}\, dx$. Then $du = \tfrac{1}{2x}\, dx$ and $v = \tfrac{2}{3}x^{3/2}$,

and $\int \sqrt{x} \ln \sqrt{x}\, dx = uv - \int v\, du = \tfrac{2}{3}x^{3/2} \ln x^{1/2} - \tfrac{1}{3}\int x^{1/2}\, dx$
$$= \tfrac{2}{3}x^{3/2} \ln x^{1/2} - \tfrac{2}{9}x^{3/2} + C = \tfrac{2}{9}x\sqrt{x}(3\ln\sqrt{x} - 1) + C.$$

19. Let $u = \ln x$ and $dv = x^{-2}\, dx$. Then $du = \tfrac{1}{x}\, dx$ and $v = -x^{-1}$,

$$\int \frac{\ln x}{x^2}\, dx = uv - \int v\, du = -\frac{\ln x}{x} + \int x^{-2}\, dx = -\frac{\ln x}{x} - \frac{1}{x} + C$$
$$= -\frac{1}{x}(\ln x + 1) + C.$$

21. Let $u = \ln x$ and $dv = dx$. Then $du = \tfrac{1}{x}\, dx$ and $v = x$ and

$$\int \ln x\, dx = uv - \int v\, du = x\ln x - \int dx = x\ln x - x + C = x(\ln x - 1) + C.$$

23. Let $u = x^2$ and $dv = e^{-x}\, dx$. Then $du = 2x\, dx$ and $v = -e^{-x}$, and

$$\int x^2 e^{-x}\, dx = uv - \int v\, du = -x^2 e^{-x} + 2\int xe^{-x}\, dx.$$

We can integrate by parts again, or, using the result of Problem 2, we find

$$\int x^2 e^{-x}\, dx = -x^2 e^{-x} + 2[-(x+1)e^{-x}] + C = -x^2 e^{-x} - 2(x+1)e^{-x} + C$$
$$= -(x^2 + 2x + 2)e^{-x} + C.$$

25. $I = \int x(\ln x)^2\, dx$. Let $u = (\ln x)^2$ and $dv = x\, dx$, so that

$du = 2(\ln x)\left(\dfrac{1}{x}\right) = \dfrac{2\ln x}{x}$ and $v = \frac{1}{2}x^2$. Then $I = \frac{1}{2}x^2(\ln x)^2 - \int x\ln x\, dx$.

Next, we evaluate $\int x\ln x\, dx$, by letting $u = \ln x$ and $dv = x\, dx$, so that $du = \frac{1}{x}dx$

and $v = \frac{1}{2}x^2$. Then $\int x\ln x\, dx = \frac{1}{2}x^2(\ln x) - \frac{1}{2}\int x\, dx = \frac{1}{2}x^2\ln x - \frac{1}{4}x^2 + C.$

Therefore, $\int x(\ln x)^2\, dx = \frac{1}{2}x^2(\ln x)^2 - \frac{1}{2}x^2\ln x + \frac{1}{4}x^2 + C$
$$= \frac{1}{4}x^2[2(\ln x)^2 - 2\ln x + 1] + C.$$

27. $\displaystyle\int_0^{\ln 2} xe^x\, dx = (x-1)e^x\Big|_0^{\ln 2}$ (Using the results of Example 1.)

$= (\ln 2 - 1)e^{\ln 2} - (-e^0) = 2(\ln 2 - 1) + 1$ (Recall $e^{\ln 2} = 2$.) $= 2\ln 2 - 1$.

29. We first integrate $I = \int \ln x\, dx$. Integrating by parts with $u = \ln x$ and $dv = dx$ so

that $du = \frac{1}{x}\, dx$ and $v = x$, we find
$$I = x\ln x - \int dx = x\ln x - x + C = x(\ln x - 1) + C.$$

Therefore, $\displaystyle\int_1^4 \ln x\, dx = x(\ln x - 1)\Big|_1^4 = 4(\ln 4 - 1) - 1(\ln 1 - 1) = 4\ln 4 - 3.$

31. Let $u = x$ and $dv = e^{2x}\, dx$. Then $u = dx$ and $v = \frac{1}{2}e^{2x}$ and
$$\int_0^2 xe^{2x}\, dx = \frac{1}{2}xe^{2x}\Big|_0^2 - \frac{1}{2}\int_0^2 e^{2x}\, dx = e^4 - \frac{1}{4}e^{2x}\Big|_0^2$$
$$= e^4 - \frac{1}{4}e^4 + \frac{1}{4} = \frac{1}{4}(3e^4 + 1).$$

33. Let $u = x$ and $dv = e^{-2x}\, dx$, so that $du = dx$ and $v = -\frac{1}{2}e^{-2x}$.

$f(x) = \int xe^{-2x} \, dx = -\frac{1}{2}xe^{-2x} - \frac{1}{4}e^{-2x} + C$; $f(0) = -\frac{1}{4} + C = 3$ and $C = \frac{13}{4}$.

Therefore, $y = -\frac{1}{2}xe^{-2x} - \frac{1}{4}e^{-2x} + \frac{13}{4}$.

35. The required area is given by $\int_1^5 \ln x \, dx$. We first find $\int \ln x \, dx$. Using the

technique of integration by parts with $u = \ln x$ and $dv = dx$ so that $du = \frac{1}{x} dx$ and $v = x$, we have

$$\int \ln x \, dx = x \ln x - \int dx = x \ln x - x = x(\ln x - 1) + C.$$

Therefore, $\int_1^5 \ln x \, dx = x(\ln x - 1)\Big|_1^5 = 5(\ln 5 - 1) - 1(\ln 1 - 1) = 5 \ln 5 - 4$

and the required area is $(5 \ln 5 - 4)$ sq units.

37. The distance covered is given by $\int_0^{10} 100te^{-0.2t} \, dt = 100\int_0^{10} te^{-0.2t} \, dt$.

We integrate by parts, letting $u = t$ and $dv = e^{-0.2t} \, dt$ so that $du = dt$ and

$v = -\dfrac{1}{0.2}e^{-0.2t} = -5e^{-0.2t}$. Therefore,

$$100\int_0^{10} te^{-0.2t} \, dt = 100\left[-5te^{-0.2t}\Big|_0^{10}\right] + 5\int_0^{10} e^{-0.2t} \, dt$$

$$= 100[-5te^{-0.2t} - 25e^{-0.2t}]\Big|_0^{10} = -500e^{-0.2t}(t+5)\Big|_0^{10}$$

$$= -500e^{-2}(15) + 500(5) = 1485, \text{ or } 1485 \text{ feet.}$$

39. The average concentration is $C = \dfrac{1}{12}\int_0^{12} 3te^{-t/3} \, dt = \dfrac{1}{4}\int_0^{12} te^{-t/3} \, dt$.

Let $u = t$ and $dv = e^{-t/3} \, dt$. So $du = dt$ and $v = -3e^{-t/3}$. Then

$$C = \frac{1}{4}\left[-3te^{-t/3}\Big|_0^{12} + 3\int_0^{12} e^{-t/3} \, dt\right] = \frac{1}{4}\left\{-36e^{-4} - \left[9e^{-t/3}\Big|_0^{12}\right]\right\}$$

$$= \frac{1}{4}(-36e^{-4} - 9e^{-4} + 9) \approx 2.04 \text{ mg/ml.}$$

41. $N = 2\int te^{-0.1t} \, dt$. Let $u = t$ and $dv = e^{-0.1t}$, so that $du = dt$ and $v = -10e^{-0.1t}$. Then

$v = -10e^{-0.1t}$. Then

$$N(t) = 2[-10te^{-0.1t} + 10\int e - 0.1t \, dt] = 2(-10te^{-0.1t} - 100e^{-0.1t}) + C$$

$$= -20e^{-0.1t}(t+10) + 200. \qquad\qquad [N(0) = 0]$$

43. $$PV = \int_0^5 (30{,}000 + 800t)e^{-0.08t}\,dt = 30{,}000\int_0^5 e^{-0.08t}\,dt + 800\int_0^5 te^{-0.08t}\,dt\,.$$

Let $I = \int te^{-0.08t}\,dt$. To evaluate I by parts, let $u = t$, $dv = e^{-0.08t}\,dt$

and $du = dt$, $v = -\dfrac{1}{0.08}e^{-0.08t} = -12.5e^{-0.08t}$.

Therefore, $I = -12.5te^{-0.08t} + 12.5\int e^{-0.08t}\,dt = -12.5te^{-0.08t} - 156.25e^{-0.08t} + C.$

$$PV = \left[-\frac{30{,}000}{0.08}e^{-0.08t} - 800(12.5)te^{-0.08t} - 800(156.25)e^{-0.08t} \right]_0^5$$

$$= -375{,}000\,e^{-0.4} + 375{,}000 - 50{,}000e^{-0.4} - 125{,}000e^{-0.4} + 125{,}000$$

$$= 500{,}000 - 550{,}000e^{-0.4} = 131{,}323.97, \text{ or approximately } \$131{,}324.$$

45. The membership will be
$$N(5) = N(0) + \int_0^5 9\sqrt{t+1}\ln\sqrt{t+1}\,dt = 50 + 9\int_0^5 \sqrt{t+1}\ \ln\sqrt{t+1}\,dt$$
To evaluate the integral, let $u = t+1$ so that $du = dt$. Also, if $t = 0$, then $u = 1$ and
if $t = 5$, then $u = 6$. So $9\int_0^5 \sqrt{t+1}\ \ln\sqrt{t+1}\,dt = 9\int_1^6 \sqrt{u}\ln\sqrt{u}\,du.$

Using the results of Problem 17, we find $\quad 9\int_1^6 \sqrt{u}\ln\sqrt{u}\,du = 2u\sqrt{u}(3\ln\sqrt{u} - 1)\Big|_1^6.$
Therefore, $N = 50 + 51.606 \approx 101.606$ or $101{,}606$ people.

47. True. This is just the integration by parts formula.

EXERCISES 15.2, page 976

1. $\Delta x = \frac{2}{6} = \frac{1}{3}, x_0 = 0, x_1 = \frac{1}{3}, x_2 = \frac{2}{3}, x_3 = 1, x_4 = \frac{4}{3}, x_5 = \frac{5}{3}, x_6 = 2.$
   Trapezoidal Rule:
   $$\int_0^2 x^2\,dx \approx \frac{1}{6}\Big[0 + 2(\tfrac{1}{3})^2 + 2(\tfrac{2}{3})^2 + 2(1)^2 + 2(\tfrac{4}{3})^2 + 2(\tfrac{5}{3})^2 + 2^2\Big]$$
   $$\approx \frac{1}{6}\,(0.22222 + 0.88889 + 2 + 3.55556 + 5.55556 + 4) \approx 2.7037.$$
   Simpson's Rule:
   $$\int x^2\,dx = \frac{1}{9}[0 + 4(\tfrac{1}{3})^2 + 2(\tfrac{2}{3})^2 + 4(1)^2 + 2(\tfrac{4}{3})^2 + 4(\tfrac{5}{3})^2 + 2^2]$$

$$\approx \tfrac{1}{9}\,(0.44444 + 0.88889 + 4 + 3.55556 + 11.11111 + 4) \approx 2.6667.$$

Exact Value: $\displaystyle\int_0^2 x^2\,dx = \tfrac{1}{3}x^3\Big|_0^2 = \tfrac{8}{3} = 2\tfrac{2}{3}.$

3. $\Delta x = \dfrac{b-a}{n} = \dfrac{1-0}{4} = \tfrac{1}{4}; x_0 = 0, x_1 = \tfrac{1}{4}, x_2 = \tfrac{1}{2}, x_3 = \tfrac{3}{4}, x_4 = 1.$

Trapezoidal Rule:

$$\int_0^1 x^3\,dx \approx \tfrac{1}{8}\Big[0 + 2(\tfrac{1}{4})^3 + 2(\tfrac{1}{2})^3 + 2(\tfrac{3}{4})^3 + 1^3\Big] \approx \tfrac{1}{8}(0 + 0.3125 + 0.25 + 0.8)$$
$$\approx 0.265625.$$

Simpson's Rule:

$$\int_0^1 x^3\,dx \approx \tfrac{1}{12}\Big[0 + 4(\tfrac{1}{4})^3 + 2(\tfrac{1}{2})^3 + 4(\tfrac{3}{4})^3 + 1\Big] \approx \tfrac{1}{12}[0 + 0.625 + 0.25 + 1.6875 + 1]$$
$$\approx 0.25.$$

Exact Value: $\displaystyle\int_0^1 x^3\,dx = \tfrac{1}{4}x^4\Big|_0^1 = \tfrac{1}{4} - 0 = \tfrac{1}{4}.$

5. a. Here $a = 1$, $b = 2$, and $n = 4$; so $\Delta x = \tfrac{2-1}{4} = \tfrac{1}{4} = 0.25$, and $x_0 = 1$, $x_1 = 1.25$, $x_2 = 1.5$, $x_3 = 1.75$, $x_4 = 2$.

Trapezoidal Rule:

$$\int_1^2 \frac{1}{x}\,dx \approx \frac{0.25}{2}\left[1 + 2\left(\frac{1}{1.25}\right) + 2\left(\frac{1}{1.5}\right) + 2\left(\frac{1}{1.75}\right) + \frac{1}{2}\right] \approx 0.697.$$

Simpson's Rule:

$$\int_1^2 \frac{1}{x}\,dx \approx \frac{0.25}{3}\left[1 + 4\left(\frac{1}{1.25}\right) + 2\left(\frac{1}{1.5}\right) + 4\left(\frac{1}{1.75}\right) + \frac{1}{2}\right] \approx 0.6933.$$

$$\int_1^2 \frac{1}{x}\,dx = \ln x\Big|_1^2 = \ln 2 - \ln 1 \approx 0.6931.$$

7. $\Delta x = \tfrac{1}{4}$, $x_0 = 1$, $x_1 = \tfrac{5}{4}$, $x_2 = \tfrac{3}{2}$, $x_3 = \tfrac{7}{4}$, $x_4 = 2$.

Trapezoidal Rule:

$$\int_1^2 \frac{1}{x^2}\,dx \approx \tfrac{1}{8}\Big[1 + 2(\tfrac{4}{5})^2 + 2(\tfrac{2}{3})^2 + 2(\tfrac{4}{7})^2 + (\tfrac{1}{2})^2\Big] \approx 0.5090.$$

Simpson's Rule:

$$\int_1^2 \frac{1}{x^2}\,dx \approx \tfrac{1}{12}\Big[1 + 4(\tfrac{4}{5})^2 + 2(\tfrac{2}{3})^2 + 4(\tfrac{4}{7})^2 + (\tfrac{1}{2})^2\Big] \approx 0.5004.$$

Exact Value: $\displaystyle\int_1^2 \frac{1}{x^2}\,dx = -\frac{1}{x}\Big|_1^2 = -\frac{1}{2}+1 = \frac{1}{2}.$

9. $\Delta x = \frac{b-a}{n} = \frac{4-0}{8} = \frac{1}{2}; x_0 = 0, x_1 = \frac{1}{2}, x_2 = \frac{2}{2}, x_3 = \frac{3}{2}, \dots, x_8 = \frac{8}{2}.$

Trapezoidal Rule:

$$\int_0^4 \sqrt{x}\,dx \approx \frac{\tfrac{1}{2}}{2}\Big[0+2\sqrt{0.5}+2\sqrt{1}+2\sqrt{1.5}+\cdots+2\sqrt{3.5}+\sqrt{4}\Big] \approx 5.26504.$$

Simpson's Rule:

$$\int_0^4 \sqrt{x}\,dx \approx \frac{\tfrac{1}{2}}{3}\Big[0+4\sqrt{0.5}+2\sqrt{1}+4\sqrt{1.5}+\cdots+4\sqrt{3.5}+\sqrt{4}\Big] \approx 5.30463.$$

The actual value is $\displaystyle\int_0^4 \sqrt{x}\,dx \approx \frac{2}{3}x^{3/2}\Big|_0^4 = \frac{2}{3}(8) = \frac{16}{3} \approx 5.333333.$

11. $\Delta x = \frac{1-0}{6} = \frac{1}{6}; x_0 = 0, x_1 = \frac{1}{6}, x_2 = \frac{2}{6}, \dots, x_6 = \frac{6}{6}.$

Trapezoidal Rule:

$$\int_0^1 e^{-x}\,dx \approx \frac{\tfrac{1}{6}}{2}[1+2e^{-1/6}+2e^{-2/6}+\cdots+2e^{-5/6}+e^{-1}] \approx 0.633583.$$

Simpson's Rule:

$$\int_0^1 e^{-x}\,dx \approx \frac{\tfrac{1}{6}}{3}[1+4e^{-1/6}+2e^{-2/6}+\cdots+4e^{-5/6}+e^{-1}] \approx 0.632123.$$

The actual value is $\displaystyle\int_0^1 e^{-x}\,dx = -e^{-x}\Big|_0^1 = -e^{-1}+1 \approx 0.632121.$

13. $\Delta x = \frac{1}{4}; x_0 = 0, x_1 = \frac{5}{4}, x_2 = \frac{3}{2}, x_3 = \frac{7}{4}, x_4 = 2.$

Trapezoidal Rule:

$$\int_1^2 \ln x\,dx \approx \frac{1}{8}[\ln 1 + 2\ln\tfrac{5}{4}+2\ln\tfrac{3}{2}+2\ln\tfrac{7}{4}+\ln 2] \approx 0.38370.$$

Simpson's Rule:

$$\int_1^2 \ln x\,dx \approx \frac{1}{12}[\ln 1 + 4\ln\tfrac{5}{4}+2\ln\tfrac{3}{2}+4\ln\tfrac{7}{4}+\ln 2] \approx 0.38626.$$

Exact Value: $\displaystyle\int_1^2 \ln x\,dx \approx x(\ln x - 1)\Big|_1^2 = 2(\ln 2 - 1)+1 = 2\ln 2 - 1.$

15. $\Delta x = \frac{1-0}{4} = \frac{1}{4}; x_0 = 0, x_1 = \frac{1}{4}, x_2 = \frac{2}{4}, x_3 = \frac{3}{4}, x_4 = \frac{4}{4}.$

Trapezoidal Rule:

$$\int_0^1 \sqrt{1+x^3}\, dx \approx \tfrac{1}{2}\left[\sqrt{1}+2\sqrt{1+(\tfrac{1}{4})^3}+\cdots+2\sqrt{1+(\tfrac{3}{4})^3}+\sqrt{2}\right]\approx 1.1170.$$

Simpson's Rule:

$$\int_0^1 \sqrt{1+x^3}\, dx \approx \tfrac{1}{3}\left[\sqrt{1}+4\sqrt{1+(\tfrac{1}{4})^3}+2\sqrt{1+(\tfrac{2}{4})^3}\cdots+4\sqrt{1+(\tfrac{3}{4})^3}+\sqrt{2}\right]\approx 1.1114.$$

17. $\Delta x = \tfrac{2-0}{4}=\tfrac{1}{2}; x_0=0, x_1=\tfrac{1}{2}, x_2=\tfrac{2}{2}, x_3=\tfrac{3}{2}, x_4=\tfrac{4}{2}.$

Trapezoidal Rule:

$$\int_0^2 \frac{1}{\sqrt{x^3+1}}\, dx = \frac{\tfrac{1}{2}}{2}\left[1+\frac{2}{\sqrt{(\tfrac{1}{2})^3+1}}+\frac{2}{\sqrt{(1)^3+1}}+\frac{2}{\sqrt{(\tfrac{3}{2})^3+1}}+\frac{1}{\sqrt{(2)^3+1}}\right]$$

$$\approx 1.3973$$

Simpson's Rule:

$$\int_0^2 \frac{1}{\sqrt{x^3+1}}\, dx = \frac{\tfrac{1}{2}}{3}\left[1+\frac{4}{\sqrt{(\tfrac{1}{2})^3+1}}+\frac{2}{\sqrt{(1)^3+1}}+\frac{4}{\sqrt{(\tfrac{3}{2})^3+1}}+\frac{1}{\sqrt{(2)^3+1}}\right]$$

$$\approx 1.4052$$

19. $\Delta x = \tfrac{2}{4}=\tfrac{1}{2}; x_0=0, x_1=\tfrac{1}{2}, x_2=1, x_3=\tfrac{3}{2}, x_4=2.$

Trapezoidal Rule:

$$\int_0^2 e^{-x^2}\, dx = \tfrac{1}{4}[e^{-0}+2e^{-(1/2)^2}+2e^{-1}+2e^{-(3/2)^2}+e^{-4}]\approx 0.8806.$$

Simpson's Rule:

$$\int_0^2 e^{-x^2}\, dx = \tfrac{1}{6}[e^{-0}+4e^{-(1/2)^2}+2e^{-1}+4e^{-(3/2)^2}+e^{-4}]\approx 0.8818.$$

21. $\Delta x = \tfrac{2-1}{4}=\tfrac{1}{4}; x_0=1, x_1=\tfrac{5}{4}, x_2=\tfrac{6}{4}, x_3=\tfrac{7}{4}, x_4=\tfrac{8}{4}.$

Trapezoidal Rule:

$$\int_1^2 x^{-1/2}e^x\, dx = \frac{\tfrac{1}{4}}{2}\left[e+\frac{2e^{5/4}}{\sqrt{\tfrac{5}{4}}}+\cdots+\frac{2e^{7/4}}{\sqrt{\tfrac{7}{4}}}+\frac{e^2}{\sqrt{2}}\right]\approx 3.7757.$$

Simpson's Rule:

$$\int_1^2 x^{-1/2}e^x\, dx = \frac{\tfrac{1}{4}}{3}\left[e+\frac{4e^{5/4}}{\sqrt{\tfrac{5}{4}}}+\cdots+\frac{4e^{7/4}}{\sqrt{\tfrac{7}{4}}}+\frac{e^2}{\sqrt{2}}\right]\approx 3.7625.$$

23. a. Here $a = -1$, $b = 2$, $n = 10$, and $f(x) = x^5$. $f'(x) = 5x^4$ and $f''(x) = 20x^3$.
Because $f'''(x) = 60x^2 > 0$ on $(-1,0) \cup (0,2)$, we see that $f''(x)$ is increasing on
$(-1,0) \cup (0,2)$. So, we take $M = f''(2) = 20(2^3) = 160$.
Using (7), we see that the maximum error incurred is
$$\frac{M(b-a)^3}{12n^2} = \frac{160[2-(-1)]^3}{12(100)} = 3.6.$$
b. We compute $f''' = 60x^2$ and $f^{(iv)}(x) = 120x$. $f^{(iv)}(x)$ is clearly increasing on
$(-1,2)$, so we can take $M = f^{(iv)}(2) = 240$. Therefore, using (8), we see that an
error bound is $\dfrac{M(b-a)^3}{180n^4} = \dfrac{240(3)^5}{180(10^4)} \approx 0.0324$.

25. a. Here $a = 1$, $b = 3$, $n = 10$, and $f(x) = \dfrac{1}{x}$. We find $f'(x) = -\dfrac{1}{x^2}$, $f'''(x) = \dfrac{2}{x^3}$.

Since $f'''(x) = -\dfrac{6}{x^4} < 0$ on $(1,3)$, we see that $f''(x)$ is decreasing there. We
may take $M = f''(1) = 2$. Using (7), we find an error bound is
$$\frac{M(b-a)^3}{12n^2} = \frac{2(3-1)^3}{12(100)} \approx 0.013.$$
b. $f'''(x) = -\dfrac{6}{x^4}$ and $f^{(iv)}(x) = \dfrac{24}{x^5}$. $f^{(iv)}(x)$ is decreasing on $(1,3)$, so we can

take $M = f^{(iv)}(1) = 24$. Using (8), we find an error bound is $\dfrac{24(3-1)^5}{180(10^4)} \approx 0.00043$.

27. a. Here $a = 0$, $b = 2$, $n = 8$, and $f(x) = (1+x)^{-1/2}$. We find
$$f'(x) = -\tfrac{1}{2}(1+x)^{-3/2}, \; f''(x) = \tfrac{3}{4}(1+x)^{-5/2}.$$
Since f'' is positive and decreasing on $(0,2)$, we see that $|f''(x)| \le \tfrac{3}{4}$.
So the maximum error is $\dfrac{\tfrac{3}{4}(2-0)^3}{12(8)^2} = 0.0078125$.
b. $f''' = -\tfrac{15x}{8}(1+x)^{-7/2}$ and $f^{(4)}(x) = \dfrac{105}{16}(1+x)^{-9/2}$. Since $f^{(4)}$ is positive
and decreasing on $(0,2)$, we find $\left| f^{(4)}(x) \right| \le \tfrac{105}{16}$.

Therefore, the maximum error is $\dfrac{\frac{105}{16}(2-0)^5}{180(8)^4} = 0.000285.$

29. The distance covered is given by

$$d = \int_0^2 V(t)\,dt = \tfrac{\frac14}{2}\big[V(0) + 2V(\tfrac14) + \cdots + 2V(\tfrac74) + V(2)\big]$$
$$= \tfrac18[19.5 + 2(24.3) + 2(34.2) + 2(40.5) + 2(38.4) + 2(26.2)$$
$$+ 2(18) + 2(16) + 8] \approx 52.84, \text{ or } 52.84 \text{ miles.}$$

31. $\dfrac{1}{13}\displaystyle\int_0^{13} f(t)\,dt = (\tfrac{1}{13})(\tfrac12)\{13.2 + 2[14.8 + 16.6 + 17.2 + 18.7 + 19.3 + 22.6 + 24.2 + 25$
$$+ 24.6 + 25.6 + 26.4 + 26.6] + 26.6\} \approx 21.65.$$

33. Think of the "upper" curve as the graph of f and the lower curve as the graph of g. Then the required area is given by

$$A = \int_0^{150} [f(x) - g(x)]\,dx = \int_0^{150} h(x)\,dx$$

where $h = f - g$. Using Simpson's rule,
$$A \approx \tfrac{15}{3}[h(0) + 4h(1) + 2f(2) + 4f(3) + 2f(4) + \cdots 4f(9) + f(10)]$$
$$= 5[0 + 4(25) + 2(40) + 4(70) + 2(80) + 4(90) +$$
$$2(65) + 4(50) + 2(60) + 4(35) + 0]$$
$$= 7850 \text{ or } 7850 \text{ sq ft.}$$

35. We solve the equation $8 = \sqrt{0.01x^2 + 0.11x + 38}$.
$64 = 0.01x^2 + 0.11x + 38$, $0.01x^2 + 0.11x - 26 = 0$, $x^2 + 11x - 2600 = 0$,

and $x = \dfrac{-11 \pm \sqrt{121 + 10{,}400}}{2} \approx 45.786.$ Therefore

$$PS = (8)(45.786) - \int_0^{45.786} \sqrt{0.01x^2 + 0.11x + 38}\,dx.$$

a. $\Delta x = \dfrac{45.786}{8} = 5.72;\ x_0 = 0,\ x_1 = 5.72,\ x_2 = 11.44,\ \dots,\ x_8 = 45.79$

$$PS = 366.288 - \dfrac{5.72}{2}\Big[\sqrt{38} + 2\sqrt{0.01(5.72)^2 + 0.11(5.72) + 38} + \cdots$$
$$+ \sqrt{0.01(45.79)^2 + 0.11(45.79) + 38}\Big] \approx 51{,}558, \text{ or } \$51{,}558.$$

$$PS = 366.288 - \frac{5.72}{2}\left[\sqrt{38} + 4\sqrt{0.01(5.72)^2 + 0.11(5.72) + 38} + \cdots\right.$$

$$\left. +\sqrt{0.01(45.79)^2 + 0.11(45.79) + 38}\right] \approx 51,708, \text{ or } \$51,708.$$

37. The percentage of the nonfarm work force in a certain country, will continue to grow at the rate of $A = 30 + \int_0^1 5e^{1/(t+1)}\,dt$ percent, t decades from now.

$\Delta t = \frac{1}{10} = 0.1, \quad t_0 = 0, \; t_1 = 0.1, \; ..., \; t_{10} = 1.$

Using Simpson's Rule we have

$A = 30 + \frac{1}{3}(5e^1 + 4 \cdot 5e^{1/1.1} + 2 \cdot 5e^{1/1.2} + 4 \cdot 5e^{1/1.3} + \cdots + 4 \cdot 5e^{1/1.9} + 5e^{1/2})$

$\quad = 40.1004, \quad$ or approximately 40.1 percent.

39. $\Delta x = \frac{40,000 - 30,000}{10} = 1000; \; x_0 = 30,000, \; x_1 = 31,000, \; x_2, \; ..., \; x_{10} = 40,000.$

$$P = \frac{100}{2000\sqrt{2\pi}} \int_{30,000}^{40,000} e^{-0.5[x - 40,000]/[2000]^2}\,dx$$

$$P = \frac{100(1000)}{2000\sqrt{2\pi}}\left[e^{-0.5[30,000 - 40,000)/[2000]^2} + 4e^{-0.5[(31,000 - 40,000)/2000]^2} + \cdots + 1]\right]$$

$\quad \approx 0.50, \text{ or } 50 \text{ percent.}$

41. False. The number n can be odd or even.

43. True.

45. Taking the limit and recalling the definition of the Riemann sum, we find

$$\lim_{\Delta t \to 0}[c(t_1)R\Delta t + c(t_2)R\Delta t + \cdots + c(t_n)R\Delta t]/60 = D$$

$$\frac{R}{60}\lim_{\Delta t \to 0}[c(t_1)\Delta t + c(t_2)\Delta t + \cdots + c(t_n)\Delta t] = D$$

$$\frac{R}{60}\int_0^T c(t)\,dt = D, \text{ or } R = \frac{60D}{\int_0^T c(t)\,dt}.$$

EXERCISES 15.3, page 988

1. The required area is given by

$$\int_3^\infty \frac{2}{x^2}\,dx = \lim_{b\to\infty}\int_3^b \frac{2}{x^2}\,dx = \lim_{b\to\infty}\left(-\frac{2}{x}\right)\Big|_3^b = \lim_{b\to\infty}\left(-\frac{2}{b}+\frac{2}{3}\right) = \frac{2}{3} \text{ or } 2/3 \text{ sq units.}$$

3. $$A = \int_3^\infty \frac{1}{(x-2)^2}\,dx = \lim_{b\to\infty}\int_3^b (x-2)^{-2}\,dx = \lim_{b\to\infty} -\frac{1}{x-2}\Big|_3^b = \lim_{b\to\infty}\left(-\frac{1}{b-2}+1\right) = 1.$$

5. $$A = \int_1^\infty \frac{1}{x^{3/2}}\,dx = \lim_{b\to\infty}\int_1^b x^{-3/2}\,dx = \lim_{b\to\infty} -\frac{2}{\sqrt{x}}\Big|_1^b = \lim_{b\to\infty}\left(-\frac{2}{\sqrt{b}}+2\right) = 2.$$

7. $$A = \int_0^\infty \frac{1}{(x+1)^{5/2}}\,dx = \lim_{b\to\infty}\int_1^b (x+1)^{-5/2}\,dx = \lim_{b\to\infty} -\frac{2}{3}(x+1)^{-3/2}\Big|_0^b$$

    $$= \lim_{b\to\infty}\left[-\frac{2}{3(b+1)^{3/2}}+\frac{2}{3}\right] = \frac{2}{3}.$$

9. $$A = \int_{-\infty}^2 e^{2x}\,dx = \lim_{a\to-\infty}\int_a^2 e^{2x}\,dx = \lim_{a\to-\infty}\tfrac{1}{2}e^{2x}\Big|_a^2 = \lim_{a\to-\infty}\left(\tfrac{1}{2}e^4 - \tfrac{1}{2}e^{2a}\right) = \tfrac{1}{2}e^4.$$

11. Using symmetry, the required area is given by

    $$2\int_0^\infty \frac{x}{(1+x^2)^2}\,dx = 2\lim_{b\to\infty}\int_0^\infty \frac{x}{(1+x^2)^2}\,dx.$$

    To evaluate the indefinite integral $\int \dfrac{x}{(1+x^2)^2}\,dx$, put $u = 1+x^2$ so that

    $du = 2x\,dx$ or $x\,dx = \frac{1}{2}\,du$.

    Then $\displaystyle\int \frac{x}{(1+x^2)^2}\,dx = \frac{1}{2}\int \frac{du}{u^2} = -\frac{1}{2u}+C = -\frac{1}{2(1+x^2)}+C.$

    Therefore, $\displaystyle 2\lim_{b\to\infty}\int_0^b \frac{x}{(1+x^2)}\,dx = \lim_{b\to\infty} -\frac{1}{(1+x^2)^2}\Big|_0^b = \lim_{b\to\infty}\left[-\frac{1}{(1+b^2)}+1\right] = 1,$

    or 1 sq unit.

13. a. $I(b) = \int_0^b \sqrt{x}\, dx = \frac{2}{3} x^{3/2} \Big|_0^b = \frac{2}{3} b^{3/2}.$ b. $\lim_{b\to\infty} I(b) = \lim_{b\to\infty} \frac{2}{3} b^{3/2} = \infty.$

15. $\int_1^\infty \frac{3}{x^4}\, dx = \lim_{b\to\infty} \int_1^b 3x^{-4}\, dx = \lim_{b\to\infty} \left(-\frac{1}{x^3}\right)\Big|_1^b = \lim_{b\to\infty} \left(-\frac{1}{b^3} + 1\right) = 1.$

17. $A = \int_4^\infty \frac{2}{x^{3/2}}\, dx = \lim_{b\to\infty} \int_4^b 2x^{-3/2}\, dx = \lim_{b\to\infty} -4x^{-1/2}\Big|_4^b = \lim_{b\to\infty} \left(-\frac{4}{\sqrt{b}} + 2\right) = 2.$

19. $\int_1^\infty \frac{4}{x}\, dx = \lim_{b\to\infty} \int_1^b \frac{4}{x}\, dx = \lim_{b\to\infty} 4\ln x \Big|_1^b = \lim_{b\to\infty} (4\ln b) = \infty.$

21. $\int_{-\infty}^0 (x-2)^{-3}\, dx = \lim_{a\to-\infty} \int_a^0 (x-2)^{-3}\, dx = \lim_{a\to-\infty} -\frac{1}{2(x-2)^2}\Big|_a^0 = -\frac{1}{8}.$

23. $\int_1^\infty \frac{1}{(2x-1)^{3/2}}\, dx = \lim_{b\to\infty} \int_1^b (2x-1)^{-3/2}\, dx = \lim_{b\to\infty} -\frac{1}{(2x-1)^{1/2}}\Big|_1^b$

$$= \lim_{b\to\infty} \left(-\frac{1}{\sqrt{2b-1}} + 1\right) = 1.$$

25. $\int_0^\infty e^{-x}\, dx = \lim_{b\to\infty} \int_0^b e^{-x}\, dx = \lim_{b\to\infty} -e^{-x}\Big|_0^b = \lim_{b\to\infty}(-e^{-b} + 1) = 1.$

27. $\int_{-\infty}^0 e^{2x}\, dx = \lim_{a\to-\infty} \frac{1}{2} e^{2x}\Big|_a^0 = \lim_{a\to-\infty} \left(\frac{1}{2} - \frac{1}{2} e^{2a}\right) = \frac{1}{2}.$

29. $\int_1^\infty \frac{e^{\sqrt{x}}}{\sqrt{x}}\, dx = \lim_{b\to\infty} \int_1^b \frac{e^{\sqrt{x}}}{\sqrt{x}}\, dx = \lim_{b\to\infty} 2e^{\sqrt{x}}\Big|_1^b$ (Integrate by substitution: $u = \sqrt{x}$.)

$$= \lim_{b\to\infty}(2e^{\sqrt{b}} - 2e) = \infty, \text{ and so it diverges.}$$

31. $\int_{-\infty}^0 xe^x\, dx = \lim_{a\to-\infty} \int_a^0 xe^x\, dx = \lim_{a\to-\infty} (x-1)e^x\Big|_a^0 = \lim_{a\to-\infty}[-1 + (a-1)e^a] = -1.$
Note: We have used integration by parts to evaluate the integral.

33. $\displaystyle\int_{-\infty}^{\infty} x\,dx = \lim_{a\to-\infty} \tfrac{1}{2}x^2\Big|_a^0 + \lim_{b\to\infty}\tfrac{1}{2}x^2\Big|_0^b$ both of which diverge and so the integral

diverges.

35. $\displaystyle\int_{-\infty}^{\infty} x^3(1+x^4)^{-2}\,dx = \int_{-\infty}^{0} x^3(1+x^4)^{-2}\,dx + \int_{0}^{\infty} x^3(1+x^4)^{-2}\,dx$

$$= \lim_{a\to-\infty}\int_{a}^{0} x^3(1+x^4)^{-2}\,dx + \lim_{b\to\infty}\int_{0}^{b} x^3(1+x^4)^{-2}\,dx$$

$$= \lim_{a\to-\infty}\left[-\frac{1}{4}(1+x^4)^{-1}\Big|_a^0\right] + \lim_{b\to\infty}\left[-\frac{1}{4}(1+x^4)^{-1}\Big|_0^b\right]$$

$$= \lim_{a\to-\infty}\left[-\frac{1}{4}+\frac{1}{4(1+a^4)}\right] + \lim_{b\to\infty}\left[-\frac{1}{4(1+b^4)}+\frac{1}{4}\right]$$

$$= -\tfrac{1}{4}+\tfrac{1}{4}=0.$$

37. $\displaystyle\int_{-\infty}^{\infty} xe^{1-x^2}\,dx = \lim_{a\to-\infty}\int_{a}^{0} xe^{1-x^2}\,dx + \lim_{b\to\infty}\int_{0}^{b} xe^{1-x^2}\,dx$

$$= \lim_{a\to-\infty}-\tfrac{1}{2}e^{1-x^2}\Big|_a^0 + \lim_{b\to\infty}-\tfrac{1}{2}e^{1-x^2}\Big|_0^b$$

$$= \lim_{a\to-\infty}\left(-\tfrac{1}{2}e+\tfrac{1}{2}e^{1-a^2}\right) + \lim_{b\to\infty}\left(-\tfrac{1}{2}e^{1-b^2}+\tfrac{1}{2}e\right)=0.$$

39. $\displaystyle\int_{-\infty}^{\infty}\frac{e^{-x}}{1+e^{-x}}\,dx = \lim_{a\to-\infty}-\ln(1+e^{-x})\Big|_a^0 + \lim_{b\to\infty}-\ln(1+e^{-x})\Big|_0^b = \infty$, and it is divergent.

41. First, we find the indefinite integral $I = \displaystyle\int \frac{1}{x\ln^3 x}\,dx$. Let $u = \ln x$ so that

$du = \dfrac{1}{x}\,dx$. Therefore, $I = \displaystyle\int \frac{du}{u^3} = -\frac{1}{2u^2}+C = -\frac{1}{2\ln^2 x}+C.$ So

$$\int_{e}^{\infty}\frac{1}{x\ln^3 x}\,dx = \lim_{b\to\infty}\int_{e}^{b}\frac{1}{x\ln^3 x}\,dx$$

$$= \lim_{b\to\infty}\left[-\frac{1}{2\ln^2 x}\Big|_e^b\right] = \lim_{b\to\infty}\left(-\frac{1}{2(\ln b)^2}+\frac{1}{2}\right)=\frac{1}{2}$$

and so the given integral is convergent.

43. We want the present value PV of a perpetuity with $m = 1$, $P = 1500$, and $r = 0.08$.

We find $PV = \dfrac{(1)(1500)}{0.08} = 18,750$, or \$18,750.

45. $PV = \displaystyle\int_0^\infty (10,000 + 4000t)e^{-rt}\,dt = 10,000\int_0^\infty e^{-rt}\,dt + 4000\int_0^\infty te^{-rt}\,dt$

$= \displaystyle\lim_{b\to\infty}\left(-\frac{10,000}{r}e^{-rt}\Big|_0^b\right) + 4000\left(\frac{1}{r^2}\right)(-rt-1)e^{-rt}\Big|_0^b\right)$

(Integrating by parts.)

$= \dfrac{10,000}{r} + \dfrac{4000}{r^2} = \dfrac{10,000r + 4000}{r^2}$ dollars.

47. True. $\displaystyle\int_a^\infty f(x)\,dx = \int_a^b f(x)\,dx + \int_b^\infty f(x)\,dx$. So if $\displaystyle\int_a^\infty f(x)\,dx$ exists then

$\displaystyle\int_b^\infty f(x)\,dx = \int_a^\infty f(x)\,dx - \int_a^b f(x)\,dx$.

49. False. Let $f(x) = \begin{cases} e^{2x} & \text{if } -\infty < x \le 0 \\ e^{-x} & \text{if } 0 < x < \infty \end{cases}$. Then

$\displaystyle\int_{-\infty}^\infty f(x)\,dx = \int_{-\infty}^0 e^{2x}\,dx + \int_0^\infty e^{-x}\,dx = \frac{1}{2} + 1 = \frac{3}{2}$. But $2\displaystyle\int_0^\infty f(x)\,dx = 2\int_0^\infty e^{-x}\,dx = 2$.

51. $\displaystyle\int_0^\infty e^{-px}\,dx = \lim_{b\to\infty}\int_a^b e^{-px}\,dx = \lim_{b\to\infty}\left[-\frac{1}{p}e^{-px}\Big|_a^b\right] = \lim_{b\to\infty}\left(-\frac{1}{p}e^{-pb} + \frac{1}{p}e^{-pa}\right)$

$= \dfrac{1}{pe^{pa}}$ if $p > 0$ and is divergent if $p < 0$.

EXERCISES 15.4, page 998

1. $f(x) \ge 0$ on $[2,6]$. Next $\displaystyle\int_2^6 \frac{2}{32}x\,dx = \frac{1}{32}x^2\Big|_2^6 = \frac{1}{32}(36-4) = 1$,

and so f is a probability density function on $[2,6]$.

3. $f(x) = \frac{3}{8}x^2$ is nonnegative on [0,2]. Next, we compute

$$\int_0^2 \frac{3}{8}x^2\,dx = \frac{1}{8}x^3\Big|_0^2 = 1$$

and so f is a probability density function.

5. $\int_0^1 20(x^3 - x^4)dx = 20(\frac{1}{4}x^4 - \frac{1}{5}x^5)\Big|_0^1 = 20(\frac{1}{4} - \frac{1}{5}) = 20(\frac{1}{20}) = 1.$

Furthermore, $f(x) = 20(x^3 - x^4) = 20x^3(1 - x) \geq 0$ on [0,1].

Therefore, f is a density function on [0,1] as asserted.

7. Clearly $f(x) \geq 0$ on [1,4]. Next,

$$\int_1^4 f(x)\,dx = \frac{3}{14}\int_1^4 x^{1/2}\,dx = (\frac{3}{14})(\frac{2}{3})x^{3/2}\Big|_1^4 = \frac{1}{7}(8-1) = 1,$$

and so f is a probability density function on [1,3].

9. First, $f(x) \geq 0$ on $[0,\infty)$. Next, we compute

$$I = \int_0^\infty \frac{x}{(x^2+1)^{3/2}}\,dx = \lim_{b\to\infty}\int_0^b x(x^2+1)^{-3/2}\,dx.$$

Letting $u = x^2 + 1$, so that $du = 2x\,dx$, we find

$$I = \lim_{b\to\infty}\int_1^{b^2+1} u^{-3/2}\,du = \frac{1}{2}\lim_{b\to\infty} -2u^{-1/2}\Big|_1^{b^2+1} = \lim_{b\to\infty}\left(-\frac{1}{\sqrt{b^2+1}}+1\right) = 1.$$

So the given function is a probability density function on $[0,\infty)$.

11. a. $\int_0^4 k(4-x)\,dx = k\int_0^4 (4-x)\,dx = k(4x - \frac{1}{2}x^2)\Big|_0^4 = k(16-8) = 8k = 1$

implies that $k = 1/8$.

b. $P(1 \leq x \leq 3) = \frac{1}{8}\int_1^3 (4-x)\,dx = \frac{1}{8}(4x - \frac{1}{2}x^2)\Big|_1^3 = \frac{1}{8}[(12-\frac{9}{2})-(4-\frac{1}{2})] = \frac{1}{2}.$

13. a. We compute

$$\int_0^\infty kxe^{-2x^2}\,dx = k\lim_{b\to\infty}\int_0^b xe^{-2x^2}\,dx = k\lim_{b\to\infty}\left[-\frac{1}{4}e^{-2x^2}\Big|_0^b\right]dx$$

[Use the method of substitution with $u = -2x^2$.]

$$= k\lim_{b\to\infty}\left(-\frac{1}{4}e^{-2b^2}+\frac{1}{4}\right) = \frac{1}{4}k.$$

Since this value must be equal to 1, we see that $k = 4$.
b. The required probability is given by

$$P(x > 1) = \int_1^\infty f(x)\,dx = \lim_{b\to\infty} \int_1^b 4xe^{-2x^2}\,dx$$

$$= \lim_{b\to\infty}\left[-e^{-2x^2}\Big|_1^b \right] = \lim_{b\to\infty}(-e^{-2b^2} + e^{-2(1)}) = e^{-2} \approx 0.1353.$$

15. a. Here $k = \frac{1}{15}$ and so $f(x) = \frac{1}{15}e^{(-1/15)x}$.
    b. The probability is

$$\int_{10}^{12} \frac{1}{15}e^{(-1/15)x}\,dx = -e^{(-1/15)x}\Big|_{10}^{12} = -e^{-12/15} + e^{-10/15} \approx 0.06.$$

    c. The probability is

$$\int_{15}^\infty \frac{1}{15}e^{(-1/15)x}\,dx = \lim_{b\to\infty} \int_{15}^b \frac{1}{15}e^{(-1/15)x}\,dx$$

$$= \lim_{b\to\infty} -e^{(-1/15)x}\Big|_{15}^b = \lim_{b\to\infty} -e^{(-1/15)b} + e^{-1} \approx 0.37.$$

17. $\mu = \int_1^3 t\cdot\dfrac{9}{4t^3}\,dt = \dfrac{9}{4}\int_1^3 t^{-2}\,dt = -\dfrac{9}{4t}\Big|_1^3 = -\dfrac{9}{4}\left(-\dfrac{1}{3}+1\right) = \dfrac{3}{2}.$

So the expected reaction time is 1.5 seconds.

19. $\mu = \int_0^5 x\cdot\frac{6}{125}x(5-x)\,dx = \frac{6}{125}\int_0^5 (5x^2 - x^3)\,dx = \frac{6}{125}\left(\frac{5}{3}x^3 - \frac{1}{4}x^4\right)\Big|_0^5$

$= \frac{6}{125}\left(\frac{625}{3} - \frac{625}{4}\right) = 2.5.$

So the expected demand is 2500 lb.

21. The required probability is given by

$$P(0 \le x \le \tfrac{1}{2}) = \int_0^{1/2} 12x^2(1-x)\,dx = 12\int_0^{1/2}(x^2 - x^3)\,dx$$

$$= 12\left(\tfrac{1}{3}x^3 - \tfrac{1}{4}x^4\right)\Big|_0^{1/2} = 12x^3\left[\tfrac{1}{3} - \tfrac{1}{4}(x)\right]\Big|_0^{1/2}$$

$$= 12\left(\tfrac{1}{2}\right)^3\left[\tfrac{1}{3} - \tfrac{1}{4}\left(\tfrac{1}{2}\right)\right] - 0 = \tfrac{5}{16}.$$

23. $\mu = \int_0^\infty t\cdot 9(9+t^2)^{-3/2}\,dt = \lim_{b\to\infty} \int_0^b 9t(9+t^2)^{-3/2}\,dt$

$= \lim_{b\to\infty}\left[(9)(\tfrac{1}{2})(-2)(9+t^2)^{-1/2}\right]\Big|_0^b = \lim_{b\to\infty}\left[-\dfrac{9}{\sqrt{9+t^2}} + 3\right] = 3.$

15 Additional Topics in Integration

So the tubes are expected to last 3 years.

25. The probability function is $f(x) = \frac{1}{8}e^{-x/8}$. The required probability is

$$P(x \geq 8) = \frac{1}{8}\int_8^\infty e^{-x/8}\,dx = \lim_{b\to\infty} -e^{-x/8}\Big|_8^b = \lim_{b\to\infty}(-e^{-b/8}+e^{-1}) = e^{-1} \approx 0.37.$$

27. The probability function is $f(x) = 0.00001e^{-0.00001x}$. The required probability is

$$P(x \leq 20{,}000) = 0.00001\int_0^{20{,}000} e^{-0.00001x}\,dx = -e^{-0.00001x}\Big|_0^{20{,}000} = -e^{0.2}+1 \approx 0.18.$$

29. False. f must be nonnegative on $[a, b]$ as well.

31. False. Let $f(x) = 1$ for all x in the interval $[0, 1]$. Then f is a probability density function on $[0, 1]$, but f is not a probability density function on $\left[\frac{1}{2}, \frac{3}{4}\right]$ since $\int_{1/2}^{3/4} 1\,dx = \frac{1}{4} \neq 1$.

CHAPTER 15 REVIEW EXERCISES, page 1001

1. Let $u = 2x$ and $dv = e^{-x}\,dx$ so that $du = 2\,dx$ and $v = -e^{-x}$. Then

$$\int 2xe^{-x}\,dx = uv - \int v\,du = -2xe^{-x} + 2\int e^{-x}\,dx$$
$$= -2xe^{-x} - 2e^{-x} + C = -2(1+x)e^{-x} + C.$$

3. Let $u = \ln 5x$ and $dv = dx$, so that $du = \frac{1}{x}\,dx$ and $v = x$. Then

$$\int \ln 5x\,dx = x\ln 5x\,dx - \int dx = x\ln 5x - x + C = x(\ln 5x - 1) + C.$$

5. Let $u = x$ and $dv = e^{-2x}\,dx$ so that $du = dx$ and $v = -\frac{1}{2}e^{-2x}$. Then

$$\int_0^1 xe^{-2x}\,dx = -\frac{1}{2}xe^{-2x}\Big|_0^1 + \frac{1}{2}\int_0^1 e^{-2x}\,dx = -\frac{1}{2}e^{-2} - \frac{1}{4}e^{-2x}\Big|_0^1$$
$$= -\frac{1}{2}e^{-2} - \frac{1}{4}e^{-2} + \frac{1}{4} = \frac{1}{4}(1 - 3e^{-2}).$$

7. $f(x) = \int f'(x)\,dx = \int \dfrac{\ln x}{\sqrt{x}}\,dx.$ To evaluate the integral, we integrate by parts

with $u = \ln x$, $dv = x^{-1/2}\,dx$, $du = \frac{1}{x}\,dx$ and $v = 2x^{1/2}\,dx$. Then

$$\int \dfrac{\ln x}{x^{1/2}}\,dx = 2x^{1/2}\ln x - \int 2x^{-1/2}\,dx = 2x^{1/2}\ln x - 4x^{1/2} + C$$

$$= 2x^{1/2}(\ln x - 2) + C = 2\sqrt{x}(\ln x - 2) + C.$$

But $f(1) = -2$ and this gives $2\sqrt{1}(\ln 1 - 2) + C = -2$, or $C = 2$. Therefore,
$f(x) = 2\sqrt{x}(\ln x - 2) + 2.$

9. $\displaystyle\int_0^\infty e^{-2x}\,dx = \lim_{b\to\infty}\int_0^b e^{-2x}\,dx = \lim_{b\to\infty}\left(-\tfrac{1}{2}e^{-2x}\right)\Big|_0^b = \lim_{b\to\infty}\left(-\tfrac{1}{2}e^{-2b} + \tfrac{1}{2}\right) = \tfrac{1}{2}.$

11. $\displaystyle\int_3^\infty \dfrac{2}{x}\,dx = \lim_{b\to\infty}\int_3^b \dfrac{2}{x}\,dx = \lim_{b\to\infty} 2\,\ln x\Big|_3^b = \lim_{b\to\infty}(2\ln b - 2\ln 3) = \infty.$

13. $\displaystyle\int_2^\infty \dfrac{dx}{(1+2x)^2} = \lim_{b\to\infty}\int_2^b (1+2x)^{-2}\,dx = \lim_{b\to\infty}(\tfrac{1}{2})(-1)(1+2x)^{-1}\Big|_2^b$

$$= \lim_{b\to\infty}\left(-\dfrac{1}{2(1+2b)} + \dfrac{1}{2(5)}\right) = \dfrac{1}{10}.$$

15. $\Delta x = \dfrac{b-a}{n} = \dfrac{3-1}{4} = \dfrac{1}{2};\; x_0 = 1,\; x_1 = \dfrac{3}{2},\; x_2 = 2,\; x_3 = \dfrac{5}{2},\; x_4 = 3.$

Trapezoidal Rule:

$$\int_1^3 \dfrac{dx}{1+\sqrt{x}} \approx \dfrac{\frac{1}{2}}{2}\left[\dfrac{1}{2} + \dfrac{2}{1+\sqrt{1.5}} + \dfrac{2}{1+\sqrt{2}} + \dfrac{2}{1+\sqrt{2.5}} + \dfrac{1}{1+\sqrt{3}}\right] \approx 0.8421.$$

Simpson's Rule

$$\int_1^3 \dfrac{dx}{1+\sqrt{x}} \approx \dfrac{\frac{1}{2}}{3}\left[\dfrac{1}{2} + \dfrac{4}{1+\sqrt{1.5}} + \dfrac{2}{1+\sqrt{2}} + \dfrac{4}{1+\sqrt{2.5}} + \dfrac{1}{1+\sqrt{3}}\right] \approx 0.8404.$$

17. $\Delta x = \dfrac{1-(-1)}{4} = \dfrac{1}{2};\; x_0 = -1,\; x_1 = -\dfrac{1}{2},\; x_2 = 0,\; x_3 = \dfrac{1}{2},\; x_4 = 1.$

Trapezoidal Rule:

$$\int_{-1}^1 \sqrt{1+x^4}\,dx \approx \dfrac{0.5}{2}\left[\sqrt{2} + 2\sqrt{1+(-0.5)^4} + 2 + 2\sqrt{1+(0.5)^4} + \sqrt{2}\right]$$

$$\approx 2.2379.$$

Simpson's Rule:

$$\int_{-1}^{1} \sqrt{1+x^4}\, dx \approx \frac{0.5}{3}\left[\sqrt{2}+4\sqrt{1+(-0.5)^4}+2+4\sqrt{1+(0.5)^4}+\sqrt{2}\right]$$

$$\approx 2.1791.$$

19. $\frac{3}{128}\int_{0}^{4}(16-x^2)\, dx = \frac{3}{128}(16x-\frac{1}{3}x^3)\Big|_{0}^{4} = \frac{3}{128}(64-\frac{64}{3}) = \frac{3}{128}\left(\frac{192-64}{3}\right) = 1.$

Also, $f(x) \geq 0$ on $[0,4]$.

21. a. $\int_{0}^{2} kx\sqrt{4-x^2}\, dx = k\int_{0}^{2} x(4-x^2)^{1/2}\, dx = k(-\frac{1}{2})(\frac{2}{3})(4-x^2)^{3/2}\Big|_{0}^{2}$

$$= (-\frac{k}{3})(0-4^{3/2}) = \frac{k}{3}(8) = 1, \text{ or } k = 3/8.$$

  b. $\int_{1}^{2} \frac{3}{8}x\sqrt{4-x^2}\, dx = \frac{3}{8}(-\frac{1}{3})(4-x^2)^{3/2}\Big|_{1}^{2} = -\frac{1}{8}(0-3^{3/2}) = 0.6495.$

23. a. $\int_{0}^{3} kx^2(3-x)\, dx = k\int_{0}^{3}(3x^2-x^3)dx = k(x^3-\frac{1}{4}x^4)\Big|_{0}^{3}$

$$= k(27-\frac{81}{4}) = k(\frac{108-81}{4}) = k(\frac{27}{4}) = 1, \text{ or } k = 4/27.$$

  b. $\int_{1}^{2} \frac{4}{27}x^2(3-x)\, dx = \frac{4}{27}\int_{1}^{2}(3x^2-x^3)dx = \frac{4}{27}(x^3-\frac{1}{4}x^4)\Big|_{1}^{2}$

$$= \frac{4}{27}(8-4)-(1-\frac{1}{4}) = \frac{13}{27} \approx 0.4815.$$

25. Let $u = t$ and $dv = e^{-0.05t}$ so that $du = 1$ and $v = -20e^{-0.05t}$, and integrate by parts obtaining $S(t) = -20te^{-0.05t} + \int 20e^{-0.05t}\, dt = -20te^{-0.05t} - 400e^{-0.05t} + C$

$$= -20te^{-0.05t} - 400e^{-0.05t} + C = -20e^{-0.05t}(t+20) + C.$$

The initial condition implies $S(0) = 0$ giving -20(20) + C = 0, or C = 400.
Therefore, $S(t) = -20e^{-0.05t}(t+20) + 400$. By the end of the first year, the number of units sold is given by $S(12) = -20e^{-0.6}(32) + 400 = 48.761$, or 48,761 cartridges.

27. Trapezoidal Rule:
$A = \frac{100}{2}[0 + 480 + 520 + 600 + 680 + 680 + 800 + 680 + 600 + 440 + 0]$
$= 274{,}000$, or 274,000 sq ft.
Simpson's Rule:
$A = \frac{100}{3}[0 + 960 + 520 + 1200 + 680 + 1360 + 800 + 1360 + 600 + 880 + 0]$
$= 278{,}667$, or 278,667 sq ft.

CHAPTER 16

1. $f(0, 0) = 2(0) + 3(0) - 4 = -4.$ $f(1, 0) = 2(1) + 3(0) - 4 = -2.$
   $f(0, 1) = 2(0) + 3(1) - 4 = -1.$ $f(1, 2) = 2(1) + 3(2) - 4 = 4.$
   $f(2,-1) = 2(2) + 3(-1) - 4 = -3.$

3. $f(1, 2) = 1^2 + 2(1)(2) - 1 + 3 = 7;\ f(2, 1) = 2^2 + 2(2)(1) - 2 + 3 = 9$
   $f(-1, 2) = (-1)^2 + 2(-1)(2) - (-1) + 3 = 1;\ f(2, -1) = 2^2 + 2(2)(-1) - 2 + 3 = 1.$

5. $g(s,t) = 3s\sqrt{t} + t\sqrt{s} + 2;\ \ g(1,2) = 3(1)\sqrt{2} + 2\sqrt{1} + 2 = 4 + 3\sqrt{2}$
   $g(2, 1) = 3(2)\sqrt{1} + \sqrt{2} + 2 = 8 + \sqrt{2};$
   $g(0, 4) = 0 + 0 + 2 = 2,\ g(4,9) = 3(4)\sqrt{9} + 9\sqrt{4} + 2 = 56.$

7. $h(1,e) = \ln e - e \ln 1 = \ln e = 1;\ \ h(e,1) = e \ln 1 - \ln e = -1;$
   $h(e,e) = e \ln e - e \ln e = 0.$

9. $g(r,s,t) = re^{s/t};\ g(1,1,1) = e,\ g(1,0,1) = 1,\ g(-1,-1,-1) = -e^{-1/(-1)} = -e.$

11. The domain of f is the set of all ordered pairs (x, y) where x and y are real numbers.

13. All real values of u and v except those satisfying the equation $u = v$.

15. The domain of g is the set of all ordered pairs (r,s) satisfying $rs \geq 0$, that is the set of all ordered pairs where both $r \geq 0$ and $s \geq 0$, or in which both $r \leq 0$ and $s \leq 0$.

17. The domain of h is the set of all ordered pairs (x, y) such that $x + y > 5$.

19. The level curves of $z = f(x, y) = 2x + 3y$ for $z = -2, -1, 0, 1, 2$, follow.
    .

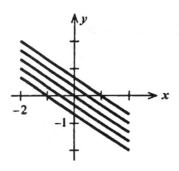

21. The level curves of $z = -4, -2, 2, 4$ are shown below.

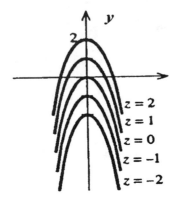

23. The level curves of $f(x,y) = \sqrt{16 - x^2 - y^2}$
     for $z = 0, 1, 2, 3, 4$

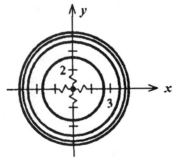

25. $V = f(1.5,4) = \pi(1.5)^2(4) = 9\pi$, or 9π cu ft

27.   a. $M = \dfrac{80}{(1.8)^2} = 24.69$.

     b. $\dfrac{w}{(1.8)^2} < 25;$ that is, $w < 2.5(1.8)^2 = 81$, that is less than 81 kg.

29.   a. $R(x,y) = xp + yq = x(200 - \frac{1}{5}x - \frac{1}{10}y) + y(160 - \frac{1}{10}x - \frac{1}{4}y)$

            $= -\frac{1}{5}x^2 - \frac{1}{4}y^2 - \frac{1}{5}xy + 200x + 160y.$

     b. The domain of R is the set of all points (x,y) satisfying

         $200 - \frac{1}{5}x - \frac{1}{10}y \geq 0,\ 160 - \frac{1}{10}x - \frac{1}{4}y \geq 0$

31.   a. $R(x,y) = xp + yq = 20x - 0.005x^2 - 0.001xy + 15y - 0.001xy - 0.003y^2$

            $= -0.005x^2 - 0.003y^2 - 0.002xy + 20x + 15y.$

     b. Since p and q must both be nonnegative, the domain of R is the set of all ordered pairs (x, y) for which

                 $20 - 0.005x - 0.001y \geq 0$

     and $15 - 0.001x - 0.003y \geq 0.$

33.   a. The domain of V is the set of all ordered pairs (P,T) where P and T are positive real numbers.

     b. $V = \dfrac{30.9(273)}{760} = 11.10$ liters.

35.   The number of suspicious fires is

$$N(100,20) = \frac{100[1000 + 0.03(100^2)(20)]^{1/2}}{[5 + 0.2(20)]^2} = 103.29,\ \text{or approximately } 103.$$

37.   a. $P = f(100{,}000,\ 0.08,\ 30) = \dfrac{100{,}000(0.08)}{12\left[1 - \left(1 + \dfrac{0.08}{12}\right)^{-360}\right]} \approx 733.76,$ or $\$733.76.$

       $P = f(100{,}000,\ 0.1,\ 30) = \dfrac{100{,}000(0.1)}{12\left[1 - \left(1 + \dfrac{0.1}{12}\right)^{-360}\right]} \approx 877.57,$ or $\$877.57.$

        16 Calculus of Several Variables

b. $P = f(100,000, 0.08, 20) = \dfrac{100,000(0,08)}{12\left[1-\left(1+\dfrac{0.08}{12}\right)^{-240}\right]} \approx 836.44,$ or \$836.44.

39. $f(M, 600, 10) = \dfrac{\pi^2(360,000)M(10)}{900} \approx 39,478.42\,M$

or $\dfrac{39,478.42}{980} \approx 40.28$ times gravity.

41. False. Let $h(x, y) = xy$. Then there are no functions f and g such that $h(x, y) = f(x) + g(y)$.

43. False. Since $x^2 - y^2 = (x + y)(x - y)$, we see that $x^2 - y^2 = 0$ if $y = \pm x$. Therefore, the domain of f is $\{(x, y) \mid y \neq \pm x\}$.

EXERCISES 16.2, page 1024

1. $f_x = 2,\ f_y = 3$
3. $g_x = 4x,\ g_y = 4$

5. $f_x = -\dfrac{4y}{x^3};\ f_y = \dfrac{2}{x^2}.$

7. $g(u, v) = \dfrac{u - v}{u + v};\ \dfrac{\partial g}{\partial u} = \dfrac{(u + v)(1) - (u - v)(1)}{(u + v)^2} = \dfrac{2v}{(u + v)^2}.$

   $\dfrac{\partial g}{\partial v} = \dfrac{(u + v)(-1) - (u - v)(1)}{(u + v)^2} = -\dfrac{2u}{(u + v)^2}.$

9. $f(s, t) = (s^2 - st + t^2)^3;\ f_s = 3(s^2 - st + t^2)^2(2s - t)$ and $f_t = 3(s^2 - st + t^2)^2(2t - s)$

11. $f(x, y) = (x^2 + y^2)^{2/3};\ f_x = \frac{2}{3}(x^2 + y^2)^{-1/3}(2x) = \frac{4}{3}x(x^2 + y^2)^{-1/3}.$ Similarly, $f_y = \frac{4}{3}y(x^2 + y^2)^{-1/3}.$

13. $f(x, y) = e^{xy+1};\ f_x = ye^{xy+1},\ f_y = xe^{xy+1}.$

15. $f(x,y) = x \ln y + y \ln x$; $f_x = \ln y + \dfrac{y}{x}$, $f_y = \dfrac{x}{y} + \ln x$.

17. $g(u,v) = e^u \ln v$. $g_u = e^u \ln v$, $g_v = \dfrac{e^u}{v}$.

19. $f(x,y,z) = xyz + xy^2 + yz^2 + zx^2$; $f_x = yz + y^2 + 2xz$,
    $f_y = xz + 2xy + z^2$, $f_z = xy + 2yz + x^2$.

21. $h(r,s,t) = e^{rst}$; $h_r = ste^{rst}$, $h_s = rte^{rst}$, $h_t = rse^{rst}$.

23. $f(x,y) = x^2 y + xy^2$; $f_x(1,2) = 2xy + y^2 \big|_{(1,2)} = 8$; $f_y(1,2) = x^2 + 2xy \big|_{(1,2)} = 5$.

25. $f(x,y) = x\sqrt{y} + y^2 = xy^{1/2} + y^2$; $f_x(2,1) = \sqrt{y}\ \big|_{(2,1)} = 1$,

    $f_y(2,1) = \dfrac{x}{2\sqrt{y}} + 2y\ \big|_{(2,1)} = 3$.

27. $f(x,y) = \dfrac{x}{y}$; $f_x(1,2) = \dfrac{1}{y}\bigg|_{(1,2)} = \dfrac{1}{2}$, $f_y(1,2) = -\dfrac{x}{y^2}\bigg|_{(1,2)} = -\dfrac{1}{4}$.

29. $f(x,y) = e^{xy}$. $f_x(1,1) = ye^{xy}\big|_{(1,1)} = e$, $f_y(1,1) = xe^{xy}\big|_{(1,1)} = e$.

31. $f(x,y,z) = x^2 yz^3$; $f_x(1,0,2) = 2xyz^3\big|_{(1,0,2)} = 0$; $f_y(1,0,2) = x^2 z^3\big|_{(1,0,2)} = 8$.
    $f_z(1,0,2) = 3x^2 yz^2\big|_{(1,0,2)} = 0$.

33. $f(x,y) = x^2 y + xy^3$; $f_x = 2xy + y^3$, $f_y = x^2 + 3xy^2$.
    Therefore, $f_{xx} = 2y$, $f_{xy} = 2x + 3y^2 = f_{yx}$, $f_{yy} = 6xy$.

35. $f(x,y) = x^2 - 2xy + 2y^2 + x - 2y$; $f_x = 2x - 2y + 1$, $f_y = -2x + 4y - 2$; $f_{xx} = 2$,
    $f_{xy} = -2$, $f_{yx} = -2$, $f_{yy} = 4$.

37. $f(x,y) = (x^2 + y^2)^{1/2}$; $f_x = \tfrac{1}{2}(x^2 + y^2)^{-1/2}(2x) = x(x^2 + y^2)^{-1/2}$;

16 Calculus of Several Variables

$$f_y = y(x^2 + y^2)^{-1/2}.$$

$$f_{xx} = (x^2 + y^2)^{-1/2} + x(-\tfrac{1}{2})(x^2 + y^2)^{-3/2}(2x) = (x^2 + y^2)^{-1/2} - x^2(x^2 + y^2)^{-3/2}$$

$$= (x^2 + y^2)^{-3/2}(x^2 + y^2 - x^2) = \frac{y^2}{(x^2 + y^2)^{3/2}}.$$

$$f_{xy} = x(-\tfrac{1}{2})(x^2 + y^2)^{-3/2}(2y) = -\frac{xy}{(x^2 + y^2)^{3/2}} = f_{yx}.$$

$$f_{yy} = (x^2 + y^2)^{-1/2} + y(-\tfrac{1}{2})(x^2 + y^2)^{-3/2}(2y) = (x^2 + y^2)^{-1/2} - y^2(x^2 + y^2)^{-3/2}$$

$$= (x^2 + y^2)^{-3/2}(x^2 + y^2 - y^2) = \frac{x^2}{(x^2 + y^2)^{3/2}}.$$

39. $f(x,y) = e^{-x/y}$; $f_x = -\dfrac{1}{y}e^{-x/y}$; $f_y = \dfrac{x}{y^2}e^{-x/y}$; $f_{xx} = \dfrac{1}{y^2}e^{-x/y}$;

$$f_{xy} = -\frac{x}{y^3}e^{-x/y} + \frac{1}{y^2}e^{-x/y} = \left(\frac{-x+y}{y^3}\right)e^{-x/y} = f_{yx}.$$

$$f_{yy} = -\frac{2x}{y^3}e^{-x/y} + \frac{x^2}{y^4}e^{-x/y} = \frac{x}{y^3}\left(\frac{x}{y} - 2\right)e^{-x/y}.$$

41. a. $f(x,y) = 20x^{3/4}y^{1/4}$. $f_x(256,16) = 15\left(\dfrac{y}{x}\right)^{1/4}\Bigg|_{(256,16)}$

$$= 15\left(\frac{16}{256}\right)^{1/4} = 15\left(\frac{2}{4}\right) = 7.5.$$

$$f_y(256,16) = 5\left(\frac{x}{y}\right)^{3/4}\Bigg|_{(256,16)} = 5\left(\frac{256}{16}\right)^{3/4} = 5(80) = 40.$$

  b. Yes.

43. $p(x,y) = 200 - 10(x - \tfrac{1}{2})^2 - 15(y-1)^2$. $\dfrac{\partial p}{\partial x}(0,1) = -20(x - \tfrac{1}{2})\big|_{(0,1)} = 10$;

At the location $(0,1)$ in the figure, the price of land is changing at the rate of \$10 per sq ft per mile change to the right.

$$\frac{\partial p}{\partial y}(0,1) = -30(y-1)\big|_{(0,1)} = 0;$$

At the location $(0,1)$ in the figure, the price of land is constant per mile change upwards.

45. $f(p,q) = 10,000 - 10p - e^{0.5q}$; $g(p,q) = 50,000 - 4000q - 10p$.

$\dfrac{\partial f}{\partial q} = -0.5e^{0.5q} < 0$ and $\dfrac{\partial g}{\partial p} = -10 < 0$

and so the two commodities are complementary commodities.

47. $R(x,y) = -0.2x^2 - 0.25y^2 - 0.2xy + 200x + 160y$.

$\dfrac{\partial R}{\partial x}(300,250) = -0.4x - 0.2y + 200|_{(300,250)}$

$$= -0.4(300) - 0.2(250) + 200 = 30$$

and this says that at a sales level of 300 finished and 250 unfinished units the revenue is increasing at the rate of \$30 per week per unit increase in the finished units.

$\dfrac{\partial R}{\partial y}(300,250) = -0.5y - 0.2x + 160|_{(300,250)}$

$$= -0.5(250) - 0.2(300) + 160 = -25$$

and this says that at a level of 300 finished and 250 unfinished units the revenue is decreasing at the rate of \$25 per week per increase in the unfinished units.

49. a. $T = f(32,20) = 35.74 + 0.6215(32) - 35.75(20^{0.16}) + 0.4275(32)(20^{0.16})$
≈ 19.99, or approximately $20°F$.

  b. $\dfrac{\partial T}{\partial s} = -35.75(0.16S^{-0.84}) + 0.4275t(0.16S^{-0.84})$

$$= 0.16(-35.75 + 0.4275t)s^{-0.84}$$

$\dfrac{\partial T}{\partial s}\Big|_{(32,20)} = 0.16[-35.75 + 0.4275(32)]20^{-0.84} \approx -0.285$;

  that is, the wind chill will drop by 0.3 degrees for each 1 mph increase in wind speed.

51. $V = \dfrac{30.9T}{P}$. $\dfrac{\partial V}{\partial T} = \dfrac{30.9}{P}$ and $\dfrac{\partial V}{\partial P} = -\dfrac{30.9T}{P^2}$.

  Therefore, $\dfrac{\partial V}{\partial T}\Big|_{T=300, P=800} = \dfrac{30.9}{800} = 0.039$, or 0.039 liters/degree.

$$\frac{\partial V}{\partial P}\bigg|_{T=300, P=800} = -\frac{(30.9)(300)}{800^2} = -0.015$$

or -0.015 liters/mm of mercury.

53. $V = \dfrac{kT}{P}$ and $\dfrac{\partial V}{\partial T} = \dfrac{k}{P}$; $T = \dfrac{VP}{k}$ and $\dfrac{\partial T}{\partial P} = \dfrac{V}{k} = \dfrac{T}{P}$; and

$P = \dfrac{kT}{V}$ and $\dfrac{\partial P}{\partial V} = -\dfrac{kT}{V^2} = -kT \cdot \dfrac{P^2}{(kT)^2} = -\dfrac{P^2}{kT}$

Therefore $\dfrac{\partial V}{\partial T} \cdot \dfrac{\partial T}{\partial P} \cdot \dfrac{\partial P}{\partial V} = \dfrac{k}{P} \cdot \dfrac{T}{P} \cdot -\dfrac{P^2}{kT} = -1$.

55. False. Let $f(x,y) = xy^{1/2}$. Then $f_x = y^{1/2}$ is defined at $(0, 0)$. But

$f_y = \dfrac{1}{2}xy^{-1/2} = \dfrac{x}{2y^{1/2}}$ is not defined at $(0, 0)$.

57. True. See Section 16.2.

USING TECHNOLOGY EXERCISES 16.2, page 1026

1. 1.3124; 0.4038 3. -1.8889; 0.7778 5. -0.3863; -0.8497

EXERCISES 16.3, page 1037

1. $f(x,y) = 1 - 2x^2 - 3y^2$. To find the critical point(s) of f, we solve the system
$$\begin{cases} f_x = -4x = 0 \\ f_y = -6y = 0 \end{cases}$$
obtaining $(0,0)$ as the only critical point of f. Next,
$$f_{xx} = -4, f_{xy} = 0, \text{ and } f_{yy} = -6.$$
In particular, $f_{xx}(0,0) = -4$, $f_{xy}(0,0) = 0$, and $f_{yy}(0,0) = -6$, giving
$$D(0,0) = (-4)(-6) - 0^2 = 24 > 0.$$
Since $f_{xx}(0,0) < 0$, the Second Derivative Test implies that $(0,0)$ gives rise to a relative maximum of f. Finally, the relative maximum of f is $f(0,0) = 1$.

3. To find the critical points of f, we solve the system

$$\begin{cases} f_x = 2x - 2 = 0 \\ f_y = -2y + 4 = 0 \end{cases}$$

obtaining $x = 1$ and $y = 2$ so that $(1,2)$ is the only critical point.
$$f_{xx} = 2, f_{xy} = 0, \text{ and } f_{yy} = -2.$$
So $D(x,y) = f_{xx}f_{yy} - f_{xy}^2 = -4$. In particular, $D(1,2) = -4 < 0$ and so $(1,2)$ affords a saddle point of f and $f(1,2) = 4$.

5. $f(x,y) = x^2 + 2xy + 2y^2 - 4x + 8y - 1$. To find the critical point(s) of f, we solve

the system
$$\begin{cases} f_x = 2x + 2y - 4 = 0 \\ f_y = 2x + 4y + 8 = 0 \end{cases}$$

obtaining $(8,-6)$ as the critical point of f. Next, $f_{xx} = 2, f_{xy} = 2, f_{yy} = 4$. In particular, $f_{xx}(8,-6) = 2, f_{xy}(8,-6) = 2, f_{yy}(8,-6) = 4$, giving $D = 2(4) - 4 = 4 > 0$. Since $f_{xx}(8,-6) > 0$, $(8,-6)$ gives rise to a relative minimum of f. Finally, the relative minimum value of f is $f(8,-6) = -41$.

7. $f(x,y) = 2x^3 + y^2 - 9x^2 - 4y + 12x - 2..$ To find the critical points of f, we solve the system
$$\begin{cases} f_x = 6x^2 - 18x + 12 = 0 \\ f_y = 2y - 4 = 0 \end{cases}$$

The first equation is equivalent to $x^2 - 3x + 2 = 0$, or $(x - 2)(x - 1) = 0$ which gives $x = 1$ or 2. The second equation of the system gives $y = 2$. Therefore, there are two critical points, $(1,2)$ and $(2,2)$. Next, we compute
$$f_{xx} = 12x - 18 = 6(2x - 3), f_{xy} = 0, f_{yy} = 2.$$
At the point $(1,2)$:
$$f_{xx}(1,2) = 6(2 - 3) = -6, f_{xy}(1,2) = 0, \text{ and } f_{yy}(1,2) = 2.$$
Therefore, $D = (-6)(2) - 0 = -12 < 0$ and we conclude that $(1,2)$ gives rise to a saddle point of f. At the point $(2,2)$:
$$f_{xx}(2,2) = 6(4 - 3) = 6, f_{xy}(2,2) = 0, \text{ and } f_{yy}(2,2) = 2.$$
Therefore, $D = (6)(2) - 0 = 12 > 0$. Since $f_{xx}(2,2) > 0$, we see that $(2,2)$ gives rise to a relative minimum with value $f(2,2) = -2$.

9. To find the critical points of f, we solve the system

$$\begin{cases} f_x = 3x^2 - 2y + 7 = 0 \\ f_y = 2y - 2x - 8 = 0 \end{cases}$$

Adding the two equations gives $3x^2 - 2x - 1 = 0$, or $(3x + 1)(x - 1) = 0$. Therefore, $x = -1/3$ or 1. Substituting each of these values of x into the second equation gives $y = 8/3$ and $y = 5$, respectively. Therefore, $(-\frac{1}{3}, \frac{11}{3})$ and $(1,5)$ are critical points of f.

Next, $f_{xx} = 6x$, $f_{xy} = -2$, and $f_{yy} = 2$. So $D(x,y) = 12x - 4 = 4(3x - 1)$. Then
$$D(-\tfrac{1}{3}, \tfrac{11}{3}) = 4(-1-1) = -8 < 0$$
and so $(-\frac{1}{3}, \frac{11}{3})$ gives a saddle point. Next, $D(1,5) = 4(3 - 1) = 8 > 0$ and since $f_{xx}(1,5) = 6 > 0$, we see that $(1,5)$ gives rise to a relative minimum.

11. To find the critical points of f, we solve the system
$$\begin{cases} f_x = 3x^2 - 3y = 0 \\ f_y = -3x + 3y^2 = 0 \end{cases}$$
The first equation gives $y = x^2$ which when substituted into the second equation gives $-3x + 3x^4 = 3x(x^3 - 1) = 0$. Therefore, $x = 0$ or 1. Substituting these values of x into the first equation gives $y = 0$ and $y = 1$, respectively. Therefore, $(0,0)$ and $(1,1)$ are critical points of f. Next, we find $f_{xx} = 6x$, $f_{xy} = -3$, and $f_{yy} = 6y$. So $D = f_{xx}f_{yy} - f_{xy}^2 = 36xy - 9$. Since $D(0,0) = -9 < 0$, we see that $(0,0)$ gives a saddle point of f. Next, $D(1,1) = 36 - 9 = 27 > 0$ and since $f_{xx}(1,1) = 6 > 0$, we see that $f(1,1) = -3$ is a relative minimum value of f.

13. Solving the system of equations
$$\begin{cases} f_x = y - \frac{4}{x^2} = 0 \\ f_y = x - \frac{2}{y^2} = 0 \end{cases}$$

we obtain $y = \frac{4}{x^2}$. Therefore, $x - 2\left(\frac{x^4}{16}\right) = 0$ and $8x - x^4 = x(8 - x^3) = 0$, and $x = 0$, or $x = 2$. Since $x = 0$ is not in the domain of f, $(2,1)$ is the only critical point of f. Next, $f_{xx} = \frac{8}{y^3}$, $f_{xy} = 1$, and $f_{yy} = \frac{4}{y^3}$. Therefore,

$$D(2,1) = \frac{32}{x^3 y^3} - 1 \Big|_{(2,1)} = 4 - 1 = 3 > 0 \text{ and } f_{xx}(2,1) = 1 > 0. \text{ Therefore, the relative}$$

minimum value of f is $f(2,1) = 2 + 4/2 + 2/1 = 6$.

15. Solving the system of equations $f_x = 2x = 0$ and $f_y = -2ye^{y^2} = 0$, we obtain $x = 0$ and $y = 0$. Therefore, $(0,0)$ is the only critical point of f. Next,
$$f_{xx} = 2, f_{xy} = 0, f_{yy} = -2e^{y^2} - 4y^2e^{y^2}.$$
Therefore, $D(0,0) = -4e^{y^2}(1+2y^2)\big|_{(0,0)} = -4(1) < 0$, and we conclude that $(0,0)$ is a saddle point.

17. $f(x,y) = e^{x^2+y^2}$
Solving the system
$$\begin{cases} f_x = 2xe^{x^2+y^2} = 0 \\ f_y = 2ye^{x^2+y^2} = 0 \end{cases}$$
we see that $x = 0$ and $y = 0$ (recall that $e^{x^2+y^2} \neq 0$). Therefore, $(0,0)$ is the only critical point of f. Next, we compute
$$f_{xx} = 2e^{x^2+y^2} + 2x(2x)e^{x^2+y^2} = 2(1+2x^2)e^{x^2+y^2}$$
$$f_{xy} = 2x(2y)e^{x^2+y^2} = 4xye^{x^2+y^2}$$
$$f_{yy} = 2(1+2y^2)e^{x^2+y^2}.$$
In particular, at the point $(0,0)$, $f_{xx}(0,0) = 2, f_{xy}(0,0) = 0$, and $f_{yy}(0,0) = 2$. Therefore, $D = (2)(2) - 0 = 4 > 0$. Furthermore, since $f_{xx}(0,0) > 0$, we conclude that $(0,0)$ gives rise to a relative minimum of f. The relative minimum value of f is $f(0,0) = 1$.

19. $f(x,y) = \ln(1+x^2+y^2)$. We solve the system of equations
$$f_x = \frac{2x}{1+x^2+y^2} = 0 \text{ and } f_y = \frac{2y}{1+x^2+y^2} = 0,$$
obtaining $x = 0$ and $y = 0$. Therefore, $(0,0)$ is the only critical point of f. Next,
$$f_{xx} = \frac{(1+x^2+y^2)2 - (2x)(2x)}{(1+x^2+y^2)^2} = \frac{2+2y^2-2x^2}{(1+x^2+y^2)^2}$$
$$f_{yy} = \frac{(1+x^2+y^2)2 - (2y)(2y)}{(1+x^2+y^2)^2} = \frac{2+2x^2-2y^2}{(1+x^2+y^2)^2}$$
$$f_{xy} = -2x(1+x^2+y^2)^{-2}(2y) = -\frac{4xy}{(1+x^2+y^2)^2}.$$
Therefore, $D(x,y) = \frac{(2+2y^2-2x^2)(2+2x^2-2y^2)}{(1+x^2+y^2)^4} - \frac{16x^2y^2}{(1+x^2+y^2)^4}.$
Since $D(0,0) = \frac{4}{1} > 0$ and $f_{xx}(0,0) = 2 > 0$, $f(0,0) = 0$ is a relative minimum

value.

21. $P(x) = -0.2x^2 - 0.25y^2 - 0.2xy + 200x + 160y - 100x - 70y - 4000$
$= -0.2x^2 - 0.25y^2 - 0.2xy + 100x + 90y - 4000.$

Then $\begin{cases} P_x = -0.4x - 0.2y + 100 = 0 \\ P_y = -0.5y - 0.2x + 90 = 0 \end{cases}$

implies that $\begin{cases} 4x + 2y = 1000 \\ 2x + 5y = \ \ 900 \end{cases}$. Solving, we find $x = 200$ and $y = 100$.

Next, $P_{xx} = -0.4, P_{yy} = -0.5, P_{xy} = -0.2$, and

$D(200,100) = (-0.4)(-0.5) - (-0.2)^2 > 0$. Since $P_{xx}(200, 100) < 0$, we conclude
that $(200,100)$ is a relative maximum of P. Thus, the company should
manufacture 200 finished and 100 unfinished units per week. The maximum
profit is
$P(200,100) = -0.2(200)^2 - 0.25(100)^2 - 0.2(100)(200) + 100(200) + 90(100) - 4000$
$= 10,500,$ or $\$10,500$ dollars.

23. $p(x,y) = 200 - 10(x - \frac{1}{2})^2 - 15(y-1)^2$. Solving the system of equations
$\begin{cases} p_x = -20(x - \frac{1}{2}) = 0 \\ p_y = -30(y - 1) = 0 \end{cases}$

we obtain $x = 1/2, y = 1$. We conclude that the only critical point of f is $(\frac{1}{2},1)$.
Next, $p_{xx} = -20, p_{xy} = 0, p_{yy} = -30$, so $D(\frac{1}{2},1) = (-20)(-30) = 600 > 0$.
Since $p_{xx} = -20 < 0$, we conclude that $f(\frac{1}{2},1)$ gives a relative maximum. So we
conclude that the price of land is highest at $(\frac{1}{2},1)$.

25. We want to minimize
$f(x,y) = D^2 = (x - 5)^2 + (y - 2)^2 + (x + 4)^2 + (y - 4)^2 + (x + 1)^2 + (y + 3)^2.$

Next, $\begin{cases} f_x = 2(x-5) + 2(x+4) + 2(x+1) = 6x = 0, \\ f_y = 2(y-2) + 2(y-4) + 2(y+3) = 6y - 6 = 0 \end{cases}$

and we conclude that $x = 0$ and $y = 1$. Also,
$f_{xx} = 6, f_{xy} = 0, f_{yy} = 6$ and $D(x,y) = (6)(6) = 36 > 0$.
Since $f_{xx} > 0$, we conclude that the function is minimized at $(0,1)$ and so $(0,1)$
gives the desired location.

27. Refer to the figure in the text.

$$xy + 2xz + 2yz = 300; \quad z(2x+2y) = 300 - xy; \quad \text{and} \quad z = \frac{300 - xy}{2(x+y)}.$$

Then the volume is given by

$$V = xyz = xy\left[\frac{300 - xy}{2(x+y)}\right] = \frac{300xy - x^2y^2}{2(x+y)}.$$

We find

$$\frac{\partial V}{\partial x} = \frac{1}{2}\frac{(x+y)(300y - 2xy^2) - (300xy - x^2y^2)}{(x+y)^2}$$

$$= \frac{300xy - 2x^2y^2 + 300y^2 - 2xy^3 - 300xy + x^2y^2}{2(x+y)^2}$$

$$= \frac{300y^2 - 2xy^3 - x^2y^2}{2(x+y)^2} = \frac{y^2(300 - 2xy - x^2)}{2(x+y)^2}.$$

Similarly, $\dfrac{\partial V}{\partial y} = \dfrac{x^2(300 - 2xy - y^2)}{2(x+y)^2}$. Setting both $\partial V / \partial x$ and $\partial V / \partial y$ equal to
zero and observing that both $x > 0$ and $y > 0$, we have the system

$$\begin{cases} 2yx + x^2 = 300 \\ 2yx + y^2 = 300 \end{cases}. \quad \text{Subtracting, we find} \quad y^2 - x^2 = 0; \quad \text{that is } (y-x)(y+x) = 0. \text{ So}$$

$y = x$ or $y = -x$. The latter is not possible since $x, y > 0$. Therefore, $y = x$.
Substituting this value into the first equation in the system gives

$$2x^2 + x^2 = 300; \quad x^2 = 100; \quad \text{and} \quad x = 10.$$

Therefore, $y = 10$. Substituting this value into the expression for z

gives $z = \dfrac{300 - 10^2}{2(10 + 10)} = 5.$ So the dimensions are $10" \times 10" \times 5"$. The volume is
500 cu in.

29. The heating cost is $C = 2xy + 8xz + 6yz$. But $xyz = 12{,}000$ or $z = 12{,}000/xy$.
Therefore,

$$C = f(x, y) = 2xy + 8x\left(\frac{12{,}000}{xy}\right) + 6y\left(\frac{12{,}000}{xy}\right) = 2xy + \frac{96{,}000}{y} + \frac{72{,}000}{x}.$$

To find the minimum of f, we find the critical point of f by solving the system

$$\begin{cases} f_x = 2y - \dfrac{72,000}{x^2} = 0 \\ f_y = 2x - \dfrac{96,000}{y^2} = 0 \end{cases}.$$

The first equation gives $y = 36000/x^2$, which when substituted into the second equation yields

$$2x - 96,000\left(\dfrac{x^2}{36,000}\right)^2 = 0, \quad (36,000)^2 x - 48,000x^4 = 0$$

$$x(27,000 - x^3) = 0.$$

Solving this equation, we have $x = 0$, or $x = 30$. We reject the first root because $x \neq 0$. With $x = 30$, we find $y = 40$ and

$$f_{xx} = \dfrac{144,000}{x^3}, \ f_{xy} = 2, \text{ and } f_{yy} = \dfrac{192,000}{y^3}.$$

In particular, $f_{xx}(30,40) = 5.33$, $f_{xy}(30,40) = 2$, and $f_{yy}(30,40) = 3$. So
$$D(30,40) = (5.33)(3) - 4 = 11.99 > 0$$

and since $f_{xx}(30,40) > 0$, we see that $(30,40)$ gives a relative minimum. Physical considerations tell us that this is an absolute minimum. The minimal annual heating cost is
$$f(30,40) = 2(30)(40) + \dfrac{96,000}{40} + \dfrac{72,000}{30} = 7200, \text{ or } \$7,200.$$

31. False. Let $f(x, y) = xy$. Then $f_x(0,0) = 0$ and $f_y(0,0) = 0$. But $(0,0)$ does not afford a relative extremum of $(0,0)$. In fact, $f_{xx} = 0$, $f_{yy} = 0$, and $f_{xy} = 1$. Therefore, $D(x, y) = f_{xx}f_{yy} - f_{xy}^2 = -1$ and so $D(0,0) = -1$ which shows that $(0, 0, 0)$ is a saddle point.

EXERCISES 16.4, page 1049

1. We form the Lagrangian function $F(x,y,\lambda) = x^2 + 3y^2 + \lambda(x + y - 1)$. We solve the

system
$$\begin{cases} F_x = 2x + \lambda = 0 \\ F_y = 6y + \lambda = 0 \\ F_\lambda = x + y - 1 = 0. \end{cases}$$

Solving the first and the second equations for x and y in terms of λ we obtain $x = -\dfrac{\lambda}{2}$ and $y = -\dfrac{\lambda}{6}$ which, upon substitution into the third equation, yields

$-\frac{\lambda}{2}-\frac{\lambda}{6}-1=0$ or $\lambda=-\frac{3}{2}$. Therefore, $x=\frac{3}{4}$ and $y=\frac{1}{4}$ which gives the point $(\frac{3}{4},\frac{1}{4})$ as the sole critical point of F. Therefore, $(\frac{3}{4},\frac{1}{4})=\frac{3}{4}$ is a minimum of F.

3. We form the Lagrangian function $F(x,y,\lambda)=2x+3y-x^2-y^2+\lambda(x+2y-9)$. We then solve the system

$$\begin{cases} F_x = 2-2x+\ \lambda=0 \\ F_y = 3-2y+2\lambda=0. \\ F_\lambda = x+2y\ -9=0 \end{cases}$$

Solving the first equation λ, we obtain $\lambda=2x-2$. Substituting into the second equation, we have $3-2y+4x-4=0$, or $4x-2y-1=0$.   Adding this equation to the third equation in the system, we have $5x-10=0$, or $x=2$. Therefore, $y=7/2$ and $f(2,\frac{7}{2})=-\frac{7}{4}$ is the maximum value of f.

5. Form the Lagrangian function $F(x,y,\lambda)=x^2+4y^2+\lambda(xy-1)$. We then solve the

system $\begin{cases} F_x = 2x+\lambda y=0 \\ F_y = 8y+\lambda x=0. \\ F_\lambda = xy-1=0 \end{cases}$

Multiplying the first and second equations by x and y, respectively, and subtracting the resulting equations, we obtain $2x^2-8y^2=0$, or $x=\pm2y$. Substituting this into the third equation gives $2y^2-1=0$ or $y=\pm\frac{\sqrt{2}}{2}$. We conclude that $f(-\sqrt{2},-\frac{\sqrt{2}}{2})=f(\sqrt{2},\frac{\sqrt{2}}{2})=4$ is the minimum value of f.

7. We form the Lagrangian function
$$F(x,y,\lambda)=x+5y-2xy-x^2-2y^2+\lambda(2x+y-4).$$
Next, we solve the system
$$\begin{cases} F_x = 1-2y-2x+2\lambda=0 \\ F_y = 5-2x-4y+\lambda=0 \\ F_\lambda = 2x+y-4=0 \end{cases}$$
Solving the last two equations for x and y in terms of λ, we obtain
$$y=\tfrac{1}{3}(1+\lambda) \text{ and } x=\tfrac{1}{6}(11-\lambda)$$
which, upon substitution into the first equation, yields
$$1-\tfrac{2}{3}(1+\lambda)-\tfrac{1}{3}(11-\lambda)+2\lambda=0$$
or $1-\tfrac{2}{3}-\tfrac{2}{3}\lambda-\tfrac{11}{3}+\tfrac{4}{3}+2\lambda=0$

or $\lambda = 2$. Therefore, $x = 3/2$ and $y = 1$. The maximum of f is
$$f(\tfrac{3}{2},1) = \tfrac{3}{2} + 5 - 2(\tfrac{3}{2}) - (\tfrac{3}{2})^2 - 2 = -\tfrac{3}{4}.$$

9. Form the Lagrangian $F(x,y,\lambda) = xy^2 + \lambda(9x^2 + y^2 - 9)$. We then solve
$$\begin{cases} F_x = y^2 + 18\lambda x = 0 \\ F_y = 2xy + 2\lambda y = 0 \\ F_\lambda = 9x^2 + y^2 - 9 = 0. \end{cases}$$

The first equation gives $\lambda = -\dfrac{y^2}{18x}$. Substituting into the second gives

$$2xy + 2y\left(-\frac{y^2}{18x}\right) = 0, \text{ or } 18x^2y - y^3 = y(18x^2 - y^2) = 0,$$

giving $y = 0$ or $y = \pm 3\sqrt{2}x$. If $y = 0$, then the third equation gives $9x^2 - 9 = 0$ or $x = \pm 1$. If $y = \pm 3\sqrt{3}/3$. Therefore, the points $(-1,0),(-\sqrt{3}/3,-\sqrt{6})$, $(-\sqrt{3}/3,\sqrt{6}),(\sqrt{3}/3,-\sqrt{6})$ and $(\sqrt{3}/3,\sqrt{6})$ give rise to extreme values of f subject to the given constraint. Evaluating $f(x,y)$ at each of these points, we see that $f(\sqrt{3}/3,-\sqrt{6}) = (\sqrt{3}/3,\sqrt{6}) = 2\sqrt{3}$ is the maximum value of f.

11. We form the Lagrangian function $F(x,y,\lambda) = xy + \lambda(x^2 + y^2 - 16)$. To find the critical points of F, we solve the system
$$\begin{cases} F_x = y + 2\lambda x = 0 \\ F_y = x + 2\lambda y = 0 \\ F_\lambda = x^2 + y^2 - 16 = 0 \end{cases}$$

Solving the first equation for λ and substituting this value into the second equation yields $x - 2\left(\dfrac{y}{2x}\right)y = 0$, or $x^2 = y^2$. Substituting the last equation into the third equation in the system, yields $x^2 + x^2 - 16 = 0$, or $x^2 = 8$, that is, $x = \pm 2\sqrt{2}$. The corresponding values of y are $y = \pm 2\sqrt{2}$. Therefore the critical points of F are $(-2\sqrt{2},-2\sqrt{2}),(-2\sqrt{2},2\sqrt{2}),(2\sqrt{2},-2\sqrt{2})(2\sqrt{2},2\sqrt{2})$. Evaluating f at each of these values, we find that $f(-2\sqrt{2},2\sqrt{2}) = -8$ and $f(2\sqrt{2},-2\sqrt{2}) = -8$ are relative minimum values and $f(-2\sqrt{2},-2\sqrt{2}) = 8$ and $f(2\sqrt{2},2\sqrt{2}) = 8$, are relative maximum values.

13. We form the Lagrangian function $F(x,y,\lambda) = xy^2 + \lambda(x^2 + y^2 - 1)$. Next, we solve the system

$$\begin{cases} F_x = y^2 + 2x\lambda = 0 \\ F_y = 2xy + 2y\lambda = 0 \\ F_\lambda = x^2 + y^2 - 1 = 0 \end{cases}.$$

We find that $x = \pm\sqrt{3}/3$ and $y = \pm\sqrt{6}/3$ and $x = \pm1$, $y = 0$. Evaluating f at each of the critical points $(-\frac{\sqrt{3}}{3}, -\frac{\sqrt{6}}{3}), (-\frac{\sqrt{3}}{3}, \frac{\sqrt{6}}{3})(\frac{\sqrt{3}}{3}, -\frac{\sqrt{6}}{3})(\frac{\sqrt{3}}{3}, \frac{\sqrt{6}}{3}), (-1,0)$, and $(1,0)$, we find that $f(-\frac{\sqrt{3}}{3}, -\frac{\sqrt{6}}{3}) = -\frac{2\sqrt{3}}{9}$ and $f(-\frac{\sqrt{3}}{3}, \frac{\sqrt{6}}{3}) = -\frac{2\sqrt{3}}{9}$ are relative minimum values and $f(\frac{\sqrt{3}}{3}, -\frac{\sqrt{6}}{3}) = \frac{2\sqrt{3}}{9}$ and $f(\frac{\sqrt{3}}{3}, \frac{\sqrt{6}}{3}) = \frac{2\sqrt{3}}{9}$ are relative maximum values.

15. Form the Lagrangian function $F(x,y,z,\lambda) = x^2 + y^2 + z^2 + \lambda(3x + 2y + z - 6)$. We solve the system

$$\begin{cases} F_x = 2x + 3\lambda = 0 \\ F_y = 2y + 2\lambda = 0 \\ F_z = 2x + \lambda = 0 \\ F_\lambda = 3x + 2y + z - 6 = 0. \end{cases}$$

The third equation give $\lambda = -2z$. Substituting into the first two equations gives

$$\begin{cases} 2x - 6z = 0 \\ 2y - 4z = 0. \end{cases}$$

So $x = 3z$ and $y = 2z$. Substituting into the third equation yields $9z + 4z + z - 6 = 0$, or $z = 3/7$. Therefore, $x = 9/7$ and $y = 6/7$. Therefore, $f(\frac{9}{7}, \frac{6}{7}, \frac{3}{7}) = \frac{18}{7}$ is the minimum value of F.

17. We want to maximize P subject to the constraint $x + y = 200$. The Lagrangian function is

$$F(x, y, \lambda) = -0.2x^2 - 0.25y^2 - 0.2xy + 100x + 90y - 4000 + \lambda(x + y - 200).$$

Next, we solve

$$\begin{cases} F_x = -0.4x - 0.2y + 100 + \lambda = 0 \\ F_y = -0.5y - 0.2x + 90 + \lambda = 0 \\ F_\lambda = x + y - 200 = 0. \end{cases}$$

Subtracting the first equation from the second yields
$0.2x - 0.3y - 10 = 0$, or $2x - 3y - 100 = 0$.

Multiplying the third equation in the system by 2 and subtracting the resulting equation from the last equation, we find $-5y + 300 = 0$ or $y = 60$. So $x = 140$ and the company should make 140 finished and 60 unfinished units.

19. Suppose each of the sides made of pine board is x feet long and those of steel are y feet long. Then $xy = 800$. The cost is $C = 12x + 3y$ and is to be minimized subject to the condition $xy = 800$. We form the Lagrangian function
$$F(x,y,\lambda) = 12x + 3y + \lambda(xy - 800).$$
We solve the system
$$\begin{cases} F_x = 12 + \lambda y = 0 \\ F_y = 3 + \lambda x = 0 \\ F_\lambda = xy - 800 = 0. \end{cases}$$
Multiplying the first equation by x and the second equation by y and subtracting the resulting equations, we obtain $12x - 3y = 0$, or $y = 4x$. Substituting this into the third equation of the system, we obtain
$$4x^2 - 800 = 0, \text{ or } x = \pm\sqrt{200} = \pm 10\sqrt{2}.$$
Since x must be positive, we take $x = 10\sqrt{2}$. So $y = 40\sqrt{2}$. So the dimensions are approximately 14.14ft by 56.56 ft.

21. We want to minimize the function $C(r,h)$ subject to the constraint $\pi r^2 h - 64 = 0$. We form the Lagrangian function $F(r,h,\lambda) = 8\pi rh + 6\pi r^2 - \lambda(\pi r^2 h - 64)$. Then we solve the system
$$\begin{cases} F_r = 8\pi h + 12\pi r - 2\lambda \pi rh = 0 \\ F_h = 8\pi r - \lambda r^2 = 0 \\ F_\lambda = \pi r^2 h - 64 = 0 \end{cases}$$
Solving the second equation for λ yields $\lambda = 8/r$, which when substituted into the first equation yields
$$8\pi h + 12\pi r - 2\pi rh(\tfrac{8}{r}) = 0$$
$$12\pi r = 8\pi h$$
$$h = \tfrac{3r}{2}.$$
Substituting this value of h into the third equation of the system, we find
$$3r^2(\tfrac{3r}{2}) = 64, \quad r^3 = \frac{128}{3\pi}, \text{ or } r = \frac{4}{3}\sqrt[3]{\frac{18}{\pi}} \text{ and } h = 2\sqrt[3]{\frac{18}{\pi}}.$$

23. Let the box have dimensions x' by y' by z'. Then $xyz = 4$. We want to minimize

$$C = 2xz + 2yz + \tfrac{3}{2}(2xy) = 2xz + 2yz + 3xy.$$

Form the Lagrangian function
$$F(x,y,z,\lambda) = 2xz + 2yz + 3xy + \lambda(xyz - 4).$$
Next, we solve the system
$$\begin{cases} F_x = 2z + 3y + \lambda yz = 0 \\ F_y = 2z + 3x + \lambda xz = 0 \\ F_z = 2x + 2y + \lambda xy = 0 \\ F_\lambda = xyz - 4 = 0. \end{cases}$$
Multiplying the first, second, and third equations by x, y, and z, respectively, we have
$$\begin{cases} 2xz + 3xy + \lambda xyz = 0 \\ 2yz + 3xy + \lambda xyz = 0 \\ 2xz + 2yz + \lambda xyz = 0. \end{cases}$$
The first two equations imply that $2z(x - y) = 0$. Since $z \neq 0$, we see that $x = y$. The second and third equations imply that $x(3y - 2z) = 0$ or $x = (3/2)y$. Substituting these values into the fourth equation in the system, we find
$$y^2\left(\tfrac{3}{2}y\right) = 4 \quad \text{or} \quad y^3 = \tfrac{8}{3}.$$
Therefore, $y = \dfrac{2}{3^{1/3}} = \dfrac{2}{3}\sqrt[3]{9}$ and $x = \dfrac{2}{3}\sqrt[3]{9}$, and $z = \sqrt[3]{9}.$

So the dimensions are $\dfrac{2}{3}\sqrt[3]{9} \times \dfrac{2}{3}\sqrt[3]{9} \times \sqrt[3]{9}$.

25. We want to maximize $f(x,y) = 100x^{3/4}y^{1/4}$ subject to $100x + 200y = 200{,}000$.
   Form the Lagrangian function
$$F(x,y,\lambda) = 100x^{3/4}y^{1/4} + \lambda(100x + 200y - 200{,}000).$$
We solve the system
$$\begin{cases} F_x = 75x^{-1/4}y^{1/4} + 100\lambda = 0 \\ F_y = 25x^{3/4}y^{-3/4} + 200\lambda = 0 \\ F_\lambda = 100x + 200y - 200{,}000 = 0. \end{cases}$$
The first two equations imply that $150x^{-1/4}y^{1/4} - 25x^{3/4}y^{-3/4} = 0$ or upon multiplying by $x^{1/4}y^{3/4}$, $150y - 25x = 0$, which implies that $x = 6y$. Substituting this value of x into the third equation of the system, we have
$$600y + 200y - 200{,}000 = 0$$
giving $y = 250$ and therefore $x = 1500$. So to maximize production, he should spend 1500 units on labor and 250 units of capital.

27. False. See Example 1, Section 16.5.

EXERCISES 16.5, page 1061

1. $\int_1^2 \int_0^1 (y+2x)\,dy\,dx = \int_1^2 \frac{1}{2}y^2 + 2xy \big|_{y=0}^{y=1}\,dx = \int_1^2 (\frac{1}{2}+2x)\,dx = \frac{1}{2}x + x^2 \big|_1^2 = 5 - \frac{3}{2} = \frac{7}{2}$.

3. $\int_{-1}^1 \int_0^1 xy^2\,dy\,dx = \int_{-1}^1 \frac{1}{3}xy^3 \big|_0^1\,dx = \int_{-1}^1 \frac{1}{3}x\,dx = \frac{x^2}{6}\big|_{-1}^1 = \frac{1}{6} - (\frac{1}{6}) = 0$.

5. $\int_{-1}^2 \int_1^{e^3} \frac{x}{y}\,dy\,dx = \int_{-1}^2 x\ln y\big|_1^{e^3}\,dx = \int_{-1}^2 x\ln e^3\,dx = \int_{-1}^2 3x\,dx = \frac{3}{2}x^2\big|_{-1}^2 = \frac{3}{2}(4) - \frac{3}{2}(1) = \frac{9}{2}$.

7. $\int_{-2}^0 \int_0^1 4xe^{2x^2+y}\,dx\,dy = \int_{-2}^0 e^{2x^2+y}\big|_{x=0}^{x=1}\,dy = \int_{-2}^0 (e^{2+y} - e^y)\,dy = (e^{2+y} - e^y)\big|_{-2}^0$

$= [(e^2 - 1) - (e^0 - e^{-2}) = e^2 - 2 + e^{-2} = (e^2 - 1)(1 - e^{-2})$.

9. $\int_0^1 \int_1^e \ln y\,dy\,dx = \int_0^1 y\ln y - y\big|_{y=1}^{y=e}\,dx = \int_0^1 dx = 1$.

11. $\int_0^1 \int_0^x (x+2y)\,dy\,dx = \int_0^1 (xy + y^2)\big|_{y=0}^{y=x}\,dx = \int_0^1 2x^2\,dx = \frac{2}{3}x^3\big|_0^1 = \frac{2}{3}$.

13. $\int_1^3 \int_0^{x+1} (2x+4y)\,dy\,dx = \int_1^3 2xy + 2y^2\big|_{y=0}^{y=x+1}\,dx = \int_1^3 [2x(x+1) + 2(x+1)^2]\,dx$

$= \int_1^3 (4x^2 + 6x + 2)\,dx = (\frac{4}{3}x^3 + 3x^2 + 2x)\big|_1^3$

$= (36 + 27 + 6) - (\frac{4}{3} + 3 + 2) = \frac{188}{3}$.

15. $\int_0^4 \int_0^{\sqrt{y}} (x+y)\,dx\,dy = \int_0^4 \frac{1}{2}x^2 + xy\big|_{x=0}^{x=\sqrt{y}}\,dy = \int_0^4 (\frac{1}{2}y + y^{3/2})\,dy$

$= (\frac{1}{4}y^2 + \frac{2}{5}y^{5/2})\big|_0^4 = 4 + \frac{64}{5} = \frac{84}{5}$.

17. $\int_0^2 \int_0^{\sqrt{4-y^2}} y\,dx\,dy = \int_0^2 xy\big|_0^{\sqrt{4-y^2}}\,dy = \int_0^2 y\sqrt{4-y^2}\,dy = -\frac{1}{2}(\frac{2}{3})(4-y^2)^{3/2}\big|_0^2$

$= \frac{1}{3}(4^{3/2}) = \frac{8}{3}$.

19. $\int_0^1 \int_0^x 2xe^y \, dy \, dx = \int_0^1 2xe^y \Big|_{y=0}^{y=x} dx = \int_0^1 (2xe^x - 2x) \, dx = 2(x-1)e^x - x^2 \Big|_0^1 = (-1) + 2 = 1.$

21. $\int_0^1 \int_x^{\sqrt{x}} ye^x \, dy \, dx = -\int_0^1 \int_x^{\sqrt{x}} ye^x \, dy \, dx = \int_0^1 -\tfrac{1}{2}y^2 e^x \Big|_{y=\sqrt{x}}^{y=x} dx = -\tfrac{1}{2}\int_0^1 (x^2 e^x - xe^x) \, dx$

$$= -\tfrac{1}{2}[x^2 e^x \big|_0^1 - 2\int_0^1 xe^x \, dx - \int_0^1 xe^x \, dx] = -\tfrac{1}{2}[x^2 e^x \big|_0^1 - 3\int_0^1 xe^x \, dx]$$

$$= -\tfrac{1}{2}[x^2 e^x - 3xe^x + 3e^x]\big|_0^1 = -\tfrac{1}{2}[e - 3e + 3e - 3] = \tfrac{1}{2}(3-e).$$

23. $\int_0^1 \int_{2x}^2 e^{y^2} \, dy \, dx = \int_0^2 \int_0^{y/2} e^{y^2} \, dx \, dy = \int_0^2 xe^{y^2} \Big|_{x=0}^{x=y/2} dy = \int_0^2 \tfrac{1}{2} ye^{y^2} \, dy$

$$= \tfrac{1}{4} e^{y^2} \Big|_0^2 = \tfrac{1}{4}(e^4 - 1).$$

25. $\int_0^2 \int_{y/2}^1 ye^{x^3} \, dx \, dy = \int_0^1 \int_0^{2x} ye^{x^3} \, dy \, dx = \int_0^1 \tfrac{1}{2} y^2 e^{x^3} \Big|_{y=0}^{y=2x} dx = \int_0^1 2x^2 e^{x^3} \, dx$

$$= \tfrac{2}{3} e^{x^3} \Big|_0^1 = \tfrac{2}{3}(e-1).$$

27. $V = \int_0^4 \int_0^3 (6-x) \, dy \, dx = \int_0^4 (6-x)y \Big|_{y=0}^{y=3} dx = 3\int_0^4 (6-x) \, dx$

$$= 3(6x - \tfrac{1}{2}x^2)\Big|_0^4 = 3(24-8) = 48 \text{, or 48 cu units.}$$

29. $V = \int_0^2 \int_0^{3-(3/2)z} (6-2y-3z) \, dy \, dz = \int_0^2 6y - y^2 - 3yz \Big|_{y=0}^{y=3-(3/2)z} dz$

$$= \int_0^2 [(6(3-\tfrac{3}{2}z) - (3-\tfrac{3}{2}z)^2 - 3(3-\tfrac{3}{2}z)z] \, dz$$

$$= -2(3-\tfrac{3}{2}z)^2 - \tfrac{2}{9}(3-\tfrac{3}{2}z)^3 - \tfrac{9}{2}z^2 + \tfrac{3}{2}z^3 \Big|_0^2$$

$$= (-18+12) - (-18+6) = 6 \text{ , or 6 cu units.}$$

31. $V = \int_0^1 \int_0^{-2x+2} (4-x^2-y^2) \, dy \, dx = \int_0^1 (4y - x^2 y - \tfrac{1}{3}y^3)\Big|_{y=0}^{y=2(1-x)} dx$

$$= \int_0^1 [8(1-x) - 2x^2 + 2x^3 - \tfrac{8}{3}(1-x)^3] \, dx$$

$$= [(8x - 4x^2 - \tfrac{2}{3}x^3 + \tfrac{1}{2}x^4) + \tfrac{2}{3}(1-x)^4]\Big|_0^1$$

$$= (8 - 4 - \tfrac{2}{3} + \tfrac{1}{2}) - \tfrac{2}{3} = \tfrac{19}{6}, \text{ or } \tfrac{19}{6} \text{ cu units.}$$

33. $V = \displaystyle\int_0^2 \int_0^2 5e^{-x-y} dx\, dy = \int_0^2 -5e^{-x-y}\Big|_{x=0}^{x=2} dy = \int_0^2 -5(e^{-2-y} - e^{-y})dy$

$$= -5(-e^{-2-y} + e^{-y})\Big|_0^2 = -5(-e^{-4} + e^{-2}) + 5(-e^{-2} + 1) = 5(1 - 2e^{-2} + e^{-4}) \text{ cu units.}$$

35. $V = \displaystyle\int_0^2 \int_0^{2x} (2x + y)dy\, dx = \int_0^2 2xy + \tfrac{1}{2}y^2\Big|_0^{2x} dx = \int_0^2 (4x^2 + 2x^2)\, dx$

$$= \int_0^2 6x^2\, dx = 2x^3\Big|_0^2 = 16, \text{ or } 16 \text{ cu units.}$$

37. $V = \displaystyle\int_0^1 \int_0^{-x+1} e^{x+2y}\, dy\, dx = \int_0^1 \tfrac{1}{2}e^{x+2y}\Big|_{y=0}^{y=-x+1} dx = \tfrac{1}{2}\int_0^1 (e^{-x+2} - e^x)\, dx$

$$= \tfrac{1}{2}(-e^{-x+2} - e^x)\Big|_0^1 = \tfrac{1}{2}(-e - e + e^2 + 1) = \tfrac{1}{2}(e^2 - 2e + 1) = \tfrac{1}{2}(e-1)^2 \text{ cu units.}$$

39. $V = \displaystyle\int_0^4 \int_0^{\sqrt{x}} \frac{2y}{1+x^2}\, dy\, dx = \int_0^4 \frac{y^2}{1+x^2}\Big|_0^{\sqrt{x}} dx = \int_0^4 \frac{x}{1+x^2}\, dx$

$$= \tfrac{1}{2}\ln(1+x^2)\Big|_0^4 = \tfrac{1}{2}(\ln 17 - \ln 1) = \tfrac{1}{2}\ln 17 \text{ cu units.}$$

41. $V = \displaystyle\int_0^4 \int_0^{\sqrt{16-x^2}} x\, dy\, dx = \int_0^4 xy\Big|_{y=0}^{y=\sqrt{16-x^2}} dx = \int_0^4 x(16 - x^2)^{1/2}\, dx$

$$= (-\tfrac{1}{2})(\tfrac{2}{3})(16 - x^2)^{3/2}\Big|_0^4 = \tfrac{1}{3}(16)^{3/2} = \tfrac{64}{3}.$$

43. $A = \displaystyle\frac{1}{\tfrac{1}{2}}\int_0^1 \int_0^x (x + 2y)dy\, dx = 2\int_0^1 xy + y^2\Big|_0^x dx = 2\int_0^1 (x^2 + x^2)dx = 4\int_0^1 x^2\, dx$

$$= \frac{4x^3}{3}\Big|_0^1 = \frac{4}{3}.$$

45. The area of R is $1/2$. The average value of f is

$$\frac{1}{1/2}\int_0^1 \int_0^x e^{-x^2}\, dy\, dx = 2\int_0^1 e^{-x^2} y\Big|_{y=0}^{y=x} dx = 2\int_0^1 xe^{-x^2}\, dx = -e^{-x^2}\Big|_0^1 = -e^{-1} + 1 = 1 - \frac{1}{e}.$$

47. The area of the region is, by elementary geometry, $[4 + \frac{1}{2}(2)(4)]$, or 8 sq units. Therefore, the required average value is

$$A = \frac{1}{8}\int_1^3 \int_0^{2x} \ln x \, dy \, dx = \frac{1}{8}\int_1^3 (\ln x)y \Big|_0^{2x} \, dx = \frac{1}{4}\int_1^3 x \ln x \, dx$$

$$= \frac{1}{4}(\frac{x^2}{4})(2 \ln x - 1)\Big|_1^3 \quad \text{(Integrating by parts)}$$

$$= \frac{9}{16}(2 \ln 3 - 1) - \frac{1}{16}(-1) = \frac{1}{8}(9 \ln 3 - 4).$$

49. The average population density inside R is $\dfrac{43,329}{20} \approx 2166$ people/sq mile.

51. The average weekly profit is

$$\frac{1}{(20)(20)}\int_{100}^{120}\int_{180}^{200}(-0.2x^2 - 0.25y^2 - 0.2xy + 100x + 90y - 4000)dx\,dy$$

$$= \frac{1}{400}\int_{100}^{120} -\frac{1}{15}x^3 - 0.25y^2x - 0.1x^2y + 50x^2 + 90xy - 4000x \Big|_{x=180}^{x=200} \, dy$$

$$= \frac{1}{400}\int_{100}^{120}(-144{,}533.33 - 5y^2 - 760y + 380{,}000 + 1800y - 80{,}000)dy$$

$$= \frac{1}{400}\int_{100}^{120}(155{,}466.67 - 5y^2 + 1040y)dy$$

$$= \frac{1}{400}(155{,}466.67y - \frac{5}{3}y^3 + 520y^2)\Big|_{100}^{120}$$

$$= \frac{1}{400}(3{,}109{,}333.40 - 1{,}213{,}333.30 + 2{,}288{,}000) \approx 10{,}460 \text{, or } \$10{,}460/\text{wk.}$$

53. True. This result follows from the definition.

55. True. $\iint\limits_R g(x,y)\,dA$ gives the volume of the solid bounded above by the surface $z = g(x,y)$. $\iint\limits_R f(x,y)\,dA$ gives the volume of the solid bounded above by the surface $z = f(x,y)$. Therefore,

$$\iint\limits_R g(x,y)\,dA - \iint\limits_R f(x,y)\,dA = \iint\limits_R [g(x,y) - f(x,y)]\,dA$$

gives the volume of the solid bounded above by $z = g(x,y)$ and below by $z = f(x,y)$.

CHAPTER 16 REVIEW EXERCISES, page 1066

1. $f(0,1) = 0;$ $f(1,0) = 0;$ $f(1,1) = \dfrac{1}{1+1} = \dfrac{1}{2}.$

    $f(0,0)$ does not exist because the point $(0,0)$ does not lie in the domain of f.

3. $h(1,1,0) = 1 + 1 = 2;$ $h(-1,1,1) = -e - 1 = -(e + 1);$
    $h(1,-1,1) = -e - 1 = -(e + 1).$

5. $D = \{(x, y) | y \neq -x\}$

7. The domain of f is the set of all ordered triplets (x,y,z) of real numbers such that
    $z \geq 0$ and $x \neq 1,\ y \neq 1,$ and $z \neq 1.$

9. $z = y - x^2$

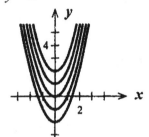

11. $z = e^{xy}$

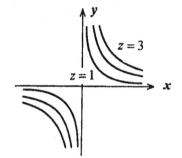

13. $f(x,y) = x\sqrt{y} + y\sqrt{x};\ f_x = \sqrt{y} + \dfrac{y}{2\sqrt{x}};\ f_y = \dfrac{x}{2\sqrt{y}} + \sqrt{x}$

15. $f(x,y) = \dfrac{x-y}{y+2x}.\ f_x = \dfrac{(y+2x)-(x-y)(2)}{(y+2x)^2} = \dfrac{3y}{(y+2x)^2}.$

    $f_y = \dfrac{(y+2x)(-1)-(x-y)}{(y+2x)^2} = \dfrac{-3x}{(y+2x)^2}.$

17. $h(x,y) = (2xy + 3y^2)^5;\ h_x = 10y(2xy + 3y^2)^4;\ h_y = 10(x + 3y)(2xy + 3y^2)^4.$

19. $f(x,y) = (x^2 + y^2)e^{x^2+y^2};$

    $f_x = 2xe^{x^2+y^2} + (x^2 + y^2)(2x)e^{x^2+y^2} = 2x(x^2 + y^2 + 1)e^{x^2+y^2}.$

$$f_y = 2ye^{x^2+y^2} + (x^2+y^2)(2y)e^{x^2+y^2} = 2y(x^2+y^2+1)e^{x^2+y^2}.$$

21. $f(x,y) = \ln\left(1+\dfrac{x^2}{y^2}\right)$. $f_x = \dfrac{\frac{2x}{y^2}}{1+\frac{x^2}{y^2}} = \dfrac{2x}{x^2+y^2}$; $f_y = \dfrac{-\frac{2x^2}{y^3}}{1+\frac{x^2}{y^2}} = -\dfrac{2x^2}{y(x^2+y^2)}$.

23. $f(x,y) = x^4 + 2x^2y^2 - y^4$; $f_x = 4x^3 + 4xy^2$; $f_y = 4x^2y - 4y^3$;
$f_{xx} = 12x^2 + 4y^2$, $f_{xy} = 8xy = f_{yx}$, $f_{yy} = 4x^2 - 12y^2$.

25. $g(x,y) = \dfrac{x}{x+y^2}$; $g_x = \dfrac{(x+y^2)-x}{(x+y^2)^2} = \dfrac{y^2}{(x+y^2)^2}$, $g_y = \dfrac{-2xy}{(x+y^2)^2}$.

Therefore, $g_{xx} = -2y^2(x+y^2)^{-3} = -\dfrac{2y^2}{(x+y^2)^3}$,

$$g_{yy} = \dfrac{(x+y^2)^2(-2x) + 2xy(2)(x+y^2)2y}{(x+y^2)^4} = \dfrac{2x(x^2+y^2)[-x-y^2+4y^2]}{(x+y^2)^4}$$

$$= \dfrac{2x(3y^2-x)}{(x+y^2)^3}.$$

and $g_{xy} = \dfrac{(x+y^2)2y - y^2(2)(x+y^2)2y}{(x+y^2)^4} = \dfrac{2(x+y^2)[xy+y^3-2y^3]}{(x+y^2)^4}$

$$= \dfrac{2y(x-y^2)}{(x+y^2)^3} = g_{yx}.$$

27. $h(s,t) = \ln\left(\dfrac{s}{t}\right)$. Write $h(s,t) = \ln s - \ln t$. Then $h_s = \dfrac{1}{s}$, $h_t = -\dfrac{1}{t}$.

Therefore, $h_{ss} = -\dfrac{1}{s^2}$, $h_{st} = h_{ts} = 0$, $h_{tt} = \dfrac{1}{t^2}$.

29. $f(x,y) = 2x^2 + y^2 - 8x - 6y + 4$; To find the critical points of f, we solve the
system $\begin{cases} f_x = 4x-8=0 \\ f_y = 2y-6=0 \end{cases}$ obtaining $x=2$ and $y=3$. Therefore, the sole critical
point of f is $(2,3)$. Next, $f_{xx} = 4, f_{xy} = 0, f_{yy} = 2$. Therefore,
$$D = f_{xx}(2,3)f_{yy}(2,3) - f_{xy}(2,3)^2 = 8 > 0.$$
Since $f_{xx}(2,3) > 0$, we see that $f(2,3) = -13$ is a relative minimum.

31. $f(x,y) = x^3 - 3xy + y^2$. We solve the system of equations $\begin{cases} f_x = 3x^2 - 3y = 0 \\ f_y = -3x + 2y = 0 \end{cases}$

obtaining $x^2 - y = 0$, or $y = x^2$. Then $-3x + 2x^2 = 0$, and $x(2x - 3) = 0$, and $x = 0$, or $x = 3/2$ and $y = 0$, or $y = 9/4$. Therefore, the critical points are $(0,0)$ and $(\frac{3}{2}, \frac{9}{4})$. Next, $f_{xx} = 6x, f_{xy} = -3$, and $f_{yy} = 2$ and $D(x,y) = 12x - 9 = 3(4x - 3)$. Therefore, $D(0,0) = -9$ so $(0,0)$ is a saddle point. $D(\frac{3}{2}, \frac{9}{4}) = 3(6-3) = 9 > 0$, and $f_{xx}(\frac{3}{2}, \frac{9}{4}) > 0$ and therefore, $f(\frac{3}{2}, \frac{9}{4}) = \frac{27}{8} - \frac{81}{8} + \frac{81}{16} = -\frac{27}{16}$ is the relative minimum value.

33. $f(x,y) = f(x,y) = e^{2x^2 + y^2}$. To find the critical points of f, we solve the system
$$\begin{cases} f_x = 4xe^{2x^2 + y^2} = 0 \\ f_y = 2ye^{2x^2 + y^2} = 0 \end{cases}$$
giving $(0,0)$ as the only critical point of f. Next,
$$f_{xx} = 4(e^{2x^2 + y^2} + 4x^2 e^{2x^2 + y^2}) = 4(1 + 4x^2)e^{2x^2 + y^2}$$
$$f_{xy} = 8xye^{2x^2 + y^2}$$
$$f_{yy} = 2(1 + 2y^2)e^{2x^2 + y^2}.$$
Therefore, $D = f_{xx}(0,0)f_{yy}(0,0) - f_{xy}^2(0,0) = (4)(2) - 0 = 8 > 0$
and so $(0,0)$ gives a relative minimum of f since $f_{xx}(0,0) > 0$. The minimum value of f is $f(0,0) = e^0 = 1$.

35. We form the Lagrangian function $F(x,y,\lambda) = -3x^2 - y^2 + 2xy + \lambda(2x + y - 4)$. Next, we solve the system
$$\begin{cases} F_x = 6x + 2y + 2\lambda = 0 \\ F_y = -2y + 2x + \lambda = 0. \\ F_\lambda = 2x + y - 4 = 0 \end{cases}$$
Multiplying the second equation by 2 and subtracting the resultant equation from the first equation yields $6y - 10x = 0$ so $y = 5x/3$. Substituting this value of y into the third equation of the system gives $2x + \frac{5}{3}x - 4 = 0$.

So $x = \frac{12}{11}$ and consequently $y = \frac{20}{11}$. So $(\frac{12}{11}, \frac{20}{11})$ gives the maximum value for f subject to the given constraint.

37. The Lagrangian function is $F(x,y,\lambda) = 2x - 3y + 1 + \lambda(2x^2 + 3y^2 - 125)$. Next, we solve the system of equations

$$\begin{cases} F_x = 2 + 4\lambda x = 0 \\ F_y = -3 + 6\lambda y = 0 \\ F_\lambda = 2x^2 + 3y^2 - 125 = 0. \end{cases}$$

Solving the first equation for x gives $x = -1/2\lambda$. The second equation gives $y = 1/2\lambda$. Substituting these values of x and y into the third equation gives

$$2\left(-\frac{1}{2\lambda}\right)^2 + 3\left(\frac{1}{2\lambda}\right)^2 - 125 = 0$$

$$\frac{1}{2\lambda^2} + \frac{3}{4\lambda^2} - 125 = 0$$

$$2 + 3 - 500\lambda^2 = 0, \text{ or } \lambda = \pm\frac{1}{10}.$$

Therefore, $x = \pm 5$ and $y = \pm 5$ and so the critical points of f are $(-5,5)$ and $(5,-5)$. Next, we compute

$$f(-5,5) = 2(-5) - 3(5) + 1 = -24.$$
$$f(5,-5) = 2(5) - 3(-5) + 1 = 26.$$

So f has a maximum value of 26 at $(5,-5)$ and a minimum value of -24 at $(-5,5)$.

39. $\displaystyle\int_{-1}^{2}\int_{2}^{4}(3x - 2y)\,dx\,dy = \int_{-1}^{2}\frac{3}{2}x^2 - 2xy\Big|_{x=2}^{x=4}\,dy = \int_{-1}^{2}[(24 - 8y) - (6 - 4y)]\,dy$

$$= \int_{-1}^{2}(18 - 4y)\,dy = (18y - 2y^2)\Big|_{-1}^{2} = (36 - 8) - (-18 - 2) = 48.$$

41. $\displaystyle\int_{0}^{1}\int_{x^3}^{x^2}2x^2y\,dy\,dx = \int_{0}^{1}x^2y^2\Big|_{y=x^3}^{y=x^2}\,dx = \int_{0}^{1}x^2(x^4 - x^6)\,dx$

$$= \int_{0}^{1}(x^6 - x^8)\,dx = \frac{x^7}{7} - \frac{x^9}{9}\Big|_{0}^{1} = \frac{1}{7} - \frac{1}{9} = \frac{2}{63}.$$

43. $\displaystyle\int_{0}^{2}\int_{0}^{1}(4x^2 + y^2)\,dy\,dx = \int_{0}^{2}4x^2y + \frac{1}{3}y^3\Big|_{y=0}^{y=1}\,dx = \int_{0}^{2}(4x^2 + \frac{1}{3})\,dx$

$$= (\frac{4}{3}x^3 + \frac{1}{3}x)\Big|_{0}^{2} = \frac{32}{3} + \frac{2}{3} = \frac{34}{3}.$$

45. The area of R is

$$\int_0^2 \int_{x^2}^{2x} dy\, dx = \int_0^2 y\Big|_{y=x^2}^{y=2x}\, dx = \int_0^2 (2x - x^2)\, dx = (x^2 - \tfrac{1}{3}x^3)\Big|_0^2 = \tfrac{4}{3}.$$

Then

$$AV = \frac{1}{4/3}\int_0^2 \int_{x^2}^{2x}(xy+1)dy\, dx = \frac{3}{4}\int_0^2 \frac{xy^2}{2} + y\Big|_{x^2}^{2x}\, dx$$

$$= \tfrac{3}{4}\int_0^2 (-\tfrac{1}{2}x^5 + 2x^3 - x^2 + 2x)\, dx = \tfrac{3}{4}(-\tfrac{1}{12}x^6 + \tfrac{1}{2}x^4 - \tfrac{1}{3}x^3 + x^2)\Big|_0^2$$

$$= \tfrac{3}{4}(-\tfrac{16}{3} + 8 - \tfrac{8}{3} + 4) = 3.$$

47. $f(p,q) = 900 - 9p - e^{0.4q}$; $g(p,q) = 20{,}000 - 3000q - 4p$.

We compute $\dfrac{\partial f}{\partial q} = -0.4e^{0.4q}$ and $\dfrac{\partial g}{\partial p} = -4$. Since $\dfrac{\partial f}{\partial q} < 0$ and $\dfrac{\partial g}{\partial p} < 0$

for all $p > 0$ and $q > 0$, we conclude that compact disc players and audio discs are complementary commodities.

49. Refer to the following diagram.

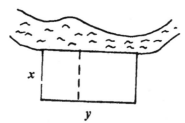

We want to minimize $C(x,y) = 3(2x) + 2(x) + 3y = 8x + 3y$ subject to $xy = 303{,}750$. The Lagrangian function is
$$F(x,y,\lambda) = 8x + 3y + \lambda(xy - 303{,}750).$$
Next, we solve the system
$$\begin{cases} F_x = 8 + \lambda y = 0 \\ F_y = 3 + \lambda x = 0 \\ F_\lambda = xy - 303{,}750 = 0 \end{cases}.$$
Solving the first equation for y gives $y = -8/\lambda$. The second equation gives $x = -3/\lambda$. Substituting this value into the third equation gives
$$\left(-\frac{3}{\lambda}\right)\left(-\frac{8}{\lambda}\right) = 303{,}750 \quad \text{or} \quad \lambda^2 = \frac{24}{303{,}750} = \frac{4}{50{,}625},$$
or $\lambda = \pm\frac{2}{225}$. Therefore, $x = 337.5$ and $y = 900$ and so the required dimensions of the pasture are 337.5 yd by 900 yd.